Lyophilization

Introduction and Basic Principles

Thomas A. Jennings

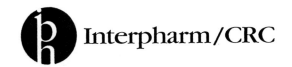

Interpharm/CRC

Boca Raton   London   New York   Washington, D.C.

Library of Congress Cataloging-in-Publication Data

Jennings, Thomas.
    Lyophilization : Introduction and basic principles / Thomas Jennings.
        p. cm.
    Includes bibliographical references and index.
    ISBN 1-57491-081-7
    1. Freeze-drying. I. Title.
RS199.F74J46   1999
6159.19—dc21   99-22039

Visit the CRC Press Web site at www.crcpress.com

© 1999, 2002 by CRC Press LLC
Interpharm is an imprint of CRC Press LLC

No claim to original U.S. Government works
International Standard Book Number 1-57491-081-7
Library of Congress Card Number 99-22039
Printed in the United States of America   5 6 7 8 9 0
Printed on acid-free paper

Contents

To
Marie

Preface

In 1978, I prepared my first set of detailed notes for a lyophilization seminar. In preparing those notes, it became apparent just how widely scientific information regarding lyophilization was disseminated throughout the scientific literature. In spite of the passage of time and an increase in the number of individuals engaged in the field of lyophilization, information regarding this technology continues to remain scattered throughout the literature in the form of papers, symposium proceedings, and chapters in books on related technologies.

With an increasing demand for knowledge of the lyophilization process, the novice and even the experienced practitioner are faced with the complex task of searching the literature for pertinent information. Such information may provide a basic insight into the principles governing this technology or provide guidelines for finding a solution to a particular problem area. However, the literature often involves case studies of the lyophilization of a particular formulation, and the results are often mistakenly taken as a general principle.

It is the intent of this treatise to provide the reader with a basic introduction to lyophilization and an understanding of the basic principles of this technology. While it is mainly directed to lyophilizing solutions, the principles of lyophilization will also be applicable to the lyophilization of more complex systems containing rigid cellular structures, such as those associated with tissue or plants.

Because diversity exists, I have made every effort to provide the reader with a fair and unbiased description of the analytical methods, the measurement of process parameters, and equipment. There will be instances throughout this text when I will cite a reference, not to support a statement but to serve as an illustration of the many pitfalls that await the unwary user of this technology. Because we tend to have such clear hindsight vision, the reader should regard these reference citations as stepping stones toward our understanding of a highly complex subject, not as a means of embarrassing an investigator(s). I will be careful to inform the reader of any commentary statements regarding a particular topic. I am well aware that the preparation of this text would not have been possible without the extensive contributions of others.

Thomas A. Jennings
Bala Cynwyd, Pennsylvania
October 1998

1

Introduction

HISTORICAL REVIEW

As the 19th century progressed, there was an increased study of microorganisms by biologists. Their efforts were often stymied by the heat-liable nature of the specimens, which prevented studies of a specimen over a period of time. Attempts to store biological specimens by air drying often proved unsuccessful. For example, the drying of specimens generally led to a considerable loss in the virulence or activity of the specimen [1].

This period also witnessed the development of the liquefaction of atmospheric gases and offered a means for preservation by freezing specimens to low temperatures. Initially, it was thought that extremely low temperatures were responsible for the loss in virulence or activity of the organism, but it was later shown that the thawing process also contributed to the destruction of the organism [2]. Although maintaining a substance at low temperatures offered then and now a means of storing biological specimens, a means for the preservation of heat-liable materials at near ambient temperatures was needed.

In 1890, Altman reported that he was able to obtain dry tissue, at subatmospheric pressures, at temperatures of about -20°C [3]. Neither how Altman conducted the drying process nor the nature of his drying apparatus is known; however, his landmark publication issued in an era of preserving biological specimens by drying under vacuum conditions and at temperature lower than 0°C.

The successful drying of tissue by Altman apparently did not immediately inspire others to preserve tissue under similar drying conditions, because it was not until 1905 that Benedict and Manning reported the drying of animal tissue at pressures less than 1 atm by means of a chemical pump [4]. The chemical pump achieved the low pressures in the drying vessel by first displacing the air in the system with boil-off vapors of ethyl ether. When all of the ethyl ether had evaporated, the system was sealed, and the residual ethyl ether was absorbed in a separate vessel containing concentrated sulfuric acid. The reduced pressure in the system was determined by means of a U tube. Although it is not known for certain, I presume that mercury was used in the construction of the U tube gauge.

As the pressure in the chamber was reduced, water vapor from the specimen was also absorbed by the concentrated sulfuric acid, thereby reducing the moisture content in the specimen. Although quite ingenious in design, the chemical pump was not very efficient in operation. Benedict and Manning reported that it required about 2 weeks to reduce the moisture content in a gelatin sample to 20% by weight.

A few years later Shackell [4] published a paper titled "An Improved Method of Desiccation . . . ." The apparatus used by Shackell was similar in design to that used by Benedict and Manning; however, the need for ethyl ether to displace the air in the chamber was eliminated by the use of a mechanical vacuum pump. With the aid of the vacuum pump, Shackell was able to achieve system pressures less than 1 mm Hg in about 2 min. Shackell's apparatus still used concentrated sulfuric acid in a separate vessel to prevent water vapor from entering the vacuum pumping system. It is of interest that the drying apparatus described by Shackell has the very same basic components as the *freeze-dryers* currently being used to produce lyophilized products: a drying chamber, a condenser chamber, and a vacuum system.

Shackell demonstrated the versatility of his drying method and apparatus by preparing a host of dried food products, such as honey, milk, and butter. He accomplished this by absorbing known amounts of the these materials on previously tared samples of cotton. After drying, the loss in weight was associated with the quantity of water vapor removed from the specimen. Beef specimens were also dried, but these specimens exhibited considerable shrinkage as a result of the drying process. Shackell found that the shrinkage could be reduced if the beef was mixed with dry sand prior to being placed in the drying apparatus. While Shackell offers no explanation for the role that sand played in preventing shrinkage in the beef samples, it is possible that the presence of the sand may have slowed the drying process and allowed the beef to reach a completely frozen state during the drying process.

But Shackell's major contribution was that he was able to achieve stable biological specimens and prevent chemical reactions from occurring for a significant period of time. He showed that the biological activity or chemical reactivity of such specimens could be restored by the addition of water. These results were achieved only if the specimen was frozen prior to being placed in the drying system and if the pressure in the drying system was reduced to less than 1 mm Hg.

The results achieved by Shackell's experiments were, in view of the nature of his apparatus, truly outstanding. Brain tissue from a rabbit that had been infected with rabies was dried while in a frozen state. The dried brain tissue was then converted into an aqueous emulsion solution and injected into a rabbit. The rabbit developed the usual symptoms of rabies and expired. In yet another experiment, Shackell froze dog's blood before coagulation could take place. The frozen blood was then dried. Clotting was observed only after the dried blood was reconstituted with water.

Shackell provided the scientific community with a new technique for stabilizing heat-liable biological substances. Although his experimental results were remarkable, they did not shed any insight regarding the mechanisms that were involved in the drying process. Shackell unwittingly set into motion a precedent for others to follow: freeze-drying or lyophilization is treated as an art rather than a science. Some practitioners of this stabilizing process find it is possible, even in today's highly technical environment, to dry materials successfully without understanding the basic principles involved in the drying.

EVOLUTION OF PROCESS AND EQUIPMENT

The purpose of this section is not to serve as a chronicle for the evolution of the lyophilization process; it is to provide the reader with an appreciation for the successful and, sometimes, unsuccessful efforts of the many individuals who have contributed to the development of the lyophilization technology. It should become apparent that, unlike other fields of technology, progress in the field of lyophilization can and continues to develop at a relatively slow pace. The reader should understand that the slow pace for gaining an understanding of this process does not stem from a lack of effort or the incompetency of the early investigators; it is from the sheer deceptive complexity and broad scope of diverse disciplines that encompass this technology.

It was not long after Shackell published his results that others were using his technique of drying while in the frozen state to stabilize heat-liable materials. For example, Hammer [5] showed that *Escherichia coli* (*E. coli*) could be stabilized for up to 54 days when dried in the frozen state, but was stable only a few days when air dried at ambient temperature. While the process, as developed by Shackell, was not named until sometime later, lyophilization was established, circa 1920, as a stabilizing process for heat-liable materials.

From the early 1920s to the present day, the apparatus and process first described by Shackell continue to undergo, when compared to such technologies as electronics, a relatively slow evolutionary process. Couriel [6], in his historical review of freeze-drying, provides some interesting insight to the evolutionary development of the process and the equipment. He points out that the first U.S. patent (U.S. Patent No. 1,630,485) was issued to Tival and made reference to the drying of frozen materials under vacuum conditions. While the patent makes reference to the drying process, it does not provide any detailed information regarding the actual method. A later U.S. patent issued to Elser in 1934 (U.S. Patent No. 1,979,956) describes drying equipment that replaces Shackell's sulfuric acid desiccating system with a cold trap chilled with dry ice. At about the same time, Flosdorf and Mudd used a mixture of methyl cellosolve and dry ice not only as a media for chilling the cold trap but also to freeze the material.

Couriel points out that it was not until 1939, some 30 years after Shackell's initial introduction of the lyophilization process, that Greaves published a paper describing the use of mechanical refrigeration in the drying equipment. He also provided the first scientific insight into the drying process by identifying the key operating parameters.

In any historical review, one must not overlook the efforts of Flosdorf and Mudd. While the pumping systems studied by these investigators did not find wide acceptance, their studies served as a stimulus for others. After numerous unsuccessful attempts to develop an effective pumping system for large quantities of water vapor, Struma and McGraw developed a drying apparatus that employed a mechanically refrigerated system that trapped the water vapor before it could enter into the vacuum pumping system.

The science associated with lyophilization, instituted by studies conducted by Rey and Merymann and others, advanced dramatically during the 1950s and 1960s. However, the pace I feel slowed in recent years with regards to further development of lyophilization as a science rather than an art. The slower pace may have been foreseen by Greaves [7]: *Although the pendulum swings in favor of freeze-drying, a certain mysticism has sprung up around the technique . . . .* Such mysticism stems not only from the deceptive complexity of this technology but also from a lack of understanding of the principles that serve as the foundation for the science of lyophilization.

DEFINITION OF LYOPHILIZATION

The coining of the term *lyophilization* is generally attributed to Rey [8] because of the porous nature of the dried product and its "lyophil" characteristic to rapidly reabsorb the solvent and restore the substance to its original state. Although Rey equated the term *lyophilization* with freeze-drying, the latter term has become more common because it is applicable to both aqueous and nonaqueous systems. It is interesting that lyophilization processes are often conducted in freeze-drying equipment, although the descriptive term *lyophilizer* is becoming more prevalent. In this book, the process will be defined as lyophilization, and the equipment will still be referred to as a freeze-dryer.

Although the steps in the lyophilization process outlined by Rey [8] have been generally accepted, only recently has a definition been given to the term *lyophilization* [9]. In its simplest form, *lyophilization* is defined as a stabilizing process in which the substance is first frozen and then the quantity of the solvent is reduced first by sublimation (*primary drying*) and then by desorption (*secondary drying*) to values that will no longer support biological growth or chemical reactions. Although this definition will be further refined later and the various processes described in greater detail, the key term in the above definition is that lyophilization is a *stabilizing* process.

A *stabilizing process* is one in which the natural kinetic clock of a substance has been greatly altered. Consider a vaccine that will retain 90% of its effectiveness for up to 2 days when stored at 4°C. If by reducing the moisture in the vaccine to levels where the kinetic clock is slowed down to where 1 sec is extended to 1 hr, then the vaccine will have a stability of nearly 20 years rather than just 2 days.

GENERAL DESCRIPTION OF THE PROCESS

The following is a general overview of the lyophilization process and freeze-drying equipment. The intent of this section is merely to familiarize the reader with the basic steps in the process and provide a general description of the equipment. This overview will also provide the reader with an understanding of my selection of topics and the order in which they will be presented in the remaining chapters.

Formulation

A *formulation* is defined as any system containing a solvent that, upon its removal, will enhance the stability of the substance. While the substance to be stabilized can include floral and food products, the majority of lyophilized formulations will consist of biological, biotechnology, diagnostic (in vivo and in vitro), pharmaceutical, and veterinary products. These latter formulations will generally consist of an active constituent and possibly other constituents that are added for stabilizing the formulation in its liquid state or for therapeutic reasons. In general, a glass container is filled with a specified quantity of the formulation and a special closure designed for lyophilization, as shown in Figure 1.1a, is placed into position.

Freezing

The principal function of the freezing process is to separate the solvent from the solutes. For an aqueous system, the water will form ice crystals, and solutes will be confined to the interstitial region between the ice crystals. The temperature necessary to achieve complete freezing of the formulation will be dependent on the nature of the solvent and other constituents that comprise the formulation. Freezing may be performed in an external freezing unit or on the shelves of the freeze-dryer. An example of the frozen ice–product matrix is given in Figure 1.1b. The freezing

---

Figure 1.1. Lyophilization in Glass Containers

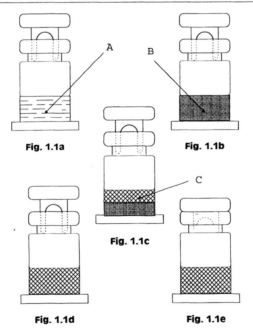

Fig. 1.1a

Fig. 1.1b

Fig. 1.1c

Fig. 1.1d

Fig. 1.1e

Figure 1.1a shows a fill volume of a liquid formulation, denoted by region defined by "A," in a glass container with a lyophilization closure positioned for the drying process; in Figure 1.1b, the frozen ice–product matrix of the formulation is signified by the region "B"; Figure 1.1c illustrates the primary drying process, and the interstitial cake portion is denoted by region "C"; the completion of secondary drying is shown by Figure 1.1d, and the final product with the closure in its stoppered position is illustrated by Figure 1.1e.

process is an important step in the lyophilization process, and Chapter 7 will present a more detailed discussion of the role that this stage plays in the lyophilization process.

Primary Drying

Once the formulation has reached a completely frozen state, the pressure in the freeze-dryer is reduced, and heat is applied to the formulation to initiate sublimation of the ice crystals. The sublimating solvent vapors pass through the opening in the closure. As the sublimation of the ice crystals proceeds, the ice-gas interface recedes through the cake (Figure 1.1c). Completion of the primary drying process occurs when all of the ice crystals have been removed from the formulation, and the volume occupied by the resulting cake is equivalent to that of the frozen matrix. Chapter 8 will examine the principles of the primary drying process in more detail.

Secondary Drying

At the completion of primary drying (Figure 1.1d), there will still be some water adsorbed onto the surface of the cake. This moisture may constitute, depending on the temperature and the nature of the constituents comprising the cake, 5–10% w/w of the dried product. In many cases, such moisture values may be too high, and the final product may not have the desired stability. The desired stability is obtained by reducing the moisture content in the product by desorbing the moisture from the cake without reducing the volume of the interstitial cake. The final desorption of the remaining water is usually accomplished by increasing the temperature of the product and reducing the partial pressure of water vapor in the container. The completion of secondary drying, which will be addressed in Chapter 9, is generally the final step in the lyophilization process.

Container-Closure System

After lyophilization, a formulation must be protected from the environment. In most cases, the formulation in the container is sealed by a stoppering mechanism contained in the freeze-dryer, which depresses the closure into the container (Figure 1.1e). The stoppering of the closure into the container temporarily protects the final product from the environment. Upon completion of the stoppering of the containers, the product can be safely removed from the freeze-dryer, and the stopper crimped sealed with a metal or colored plastic cap to provide a permanent seal for the product. A more detailed discussion of the container-closure system will be covered in Chapter 12.

Freeze-Drying Equipment

The following is a brief, general description of the essential components and their functions in a freeze-dryer. The general layout of a freeze-dryer is given in Figure 1.2.

Figure 1.2. A General Layout of a Freeze-Dryer

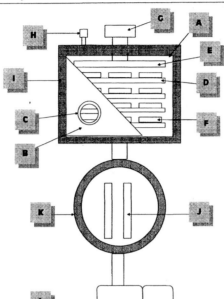

Key: A is the drying chamber; B is the door to the drying chamber, and C is a viewing port for a metal door; D represents the usable shelves of the dryer, while E is an unusable shelf; G represents the hydraulic ram system; H is the pressure gauge, and I is the thermal insulation for the chamber; J is the condenser surfaces housed in an insulated condenser chamber K; and L represents the vacuum pumping system.

The Freeze-Drying Chamber

The freeze-drying chamber serves two main functions: (1) provide a safe environment for the product during the entire lyophilization process and (2) provide the necessary temperatures and pressures to conduct each step of the lyophilization process.

The freeze-drying chamber (A), as illustrated by Figure 1.2, is a metal vessel, generally constructed from stainless steel, that can be accessed by a hinged door (B). The chamber door is either fabricated from metal such as stainless steel or from clear plastic. Figure 1.2 shows that access to the chamber is by means of a metal door containing a glass viewing port (C). The door is fitted with an elastomer gasket (not shown) to form a vacuum seal with the drying chamber.

The drying chamber contains usable shelves (D) and an unusable shelf (E). The trays (F) containing the product, either bulk or in glass containers, are loaded onto the usable shelves. The unusable shelf (E) serves as a radiation shield for the top shelf. All of the shelves are of a hollow construction that permits the serpentine flow of heat-transfer fluid. The heat-transfer fluid can be chilled to freeze the product or heated to provide the necessary energy for the primary and secondary drying processes. A hydraulic system (G) can move the shelves vertically to provide the necessary force to stopper the closures prior to opening the chamber door (B).

The drying chamber also contains a pressure gauge (H) and is equipped with an insulation covering (I) over the entire chamber surface to prevent heat transfer to the shelves (D) and the trays (F) during the drying process.

The Condenser Chamber

The main function of the condenser chamber is to house the condenser surfaces for the removal of water vapor from the gases that pass from the drying chamber. For the condenser plates to be effective, their operating temperatures must be a minimum of 20°C lower than the product temperature during the primary drying process. Unlike the shelves of the dryer which are chilled by a heat-transfer fluid, the condenser surfaces are generally refrigerated by direct expansion of a refrigerant.

Figure 1.2 illustrates an external condenser system in which the condenser surfaces (J) are housed in a separate insulated vacuum chamber (K). In some dryers, the condenser surfaces are housed in the drying chamber and are referred to as internal condensers.

The Vacuum Pumping System

The vacuum pumping system, in conjunction with the condenser system, provides the necessary pressures for conducting the primary and secondary drying processes. Typical mechanical vacuum pumps used in freeze-dryers are oil lubricated; however, oil-free mechanical pumping systems are available. The vacuum pump (L) shown in Figure 1.2 compresses the noncondensable gases that pass through the condenser chamber (K) and discharges these gases directly into the atmosphere.

PROPERTIES OF LYOPHILIZED MATERIALS

The following is a general description of the key properties of a lyophilized formulation. A more detailed discussion of these properties is given in Chapter 10. This section is designed merely to familiarize the reader with results of the lyophilization of a formulation.

Stability

The principal purpose for conducting a lyophilization process is to enhance the stability of a formulation (i.e., slow down the kinetic clock for the degradation or loss in potency of the active constituent). The dried formulation is considered stable as long as its activity or its potency remains within a given range of values. For example, the potency of an active constituent may range from 110% to 90%. The expiration date of a product will be determined from the length of time that all of the lyophilized formulation remains within the specified potency limits. There are two basic methods for determination of the stability of a lyophilized product: accelerated and long-term or real-time stability studies.

Long-Term or Real-Time Stability Testing

Long-term or *real-time stability studies* are conducted under the temperature and humidity conditions specified for storage of the product. At various time intervals, samples are removed and tested to determine the potency or activity of the dried product. Long-term or real-time stability testing extends over a period of years, and the results are used to determine a safe stability period and to assign the expiration date for a given batch.

Accelerated Stability Testing

The rate of the kinetic clock can be *accelerated* by maintaining the dried product at one or more elevated temperatures in a controlled-humidity environment. A typical accelerated test would be performed in an environment that is maintained at an elevated temperature of 40°C and a relative humidity of 50%. Samples are periodically removed from the accelerated study environment, and the potency or activity is determined. The dried formulation is considered stable if there is no significant change in the distribution of potency values over the duration of the test. This type of stability testing is helpful in assessing the presence of any thermal instability in a batch of dried product, but it cannot be used to determine the overall stability of the product and to assign an expiration date. The principal purpose of this form of accelerated stability testing is to correlate the results with the stability of the product as ascertained from long-term studies. A more detailed discussion of using the Arrhenius' equation to determine the stability of a lyophilized formulation is included in Chapter 10.

Cosmetic Properties

The *cosmetic properties* or appearance of a lyophilized product is dependent, in varying degrees, on each step of the lyophilization process (i.e., freezing and the primary drying processes). Examples of the impact that these steps can have on the cosmetic appearance of a *cake* are illustrated in Figure 1.3. In Figure 1.3a, the resulting final cake is uniform in nature. The uniform cake structure is a result of the ice structure that formed during the freezing process. The cake structure shown by Figure 1.3a is representative of the ideal structure of a lyophilized cake where the cake has a spongelike structure, and the cake volume is equivalent to the volume of the frozen matrix illustrated by "B" in Figure 1.1b. The means for forming such an ideal cake structure during the freezing process will be examined in greater detail in Chapters 3 and 7. The formation of the heterogenous structure illustrated by Figure 1.3b will be shown to be dependent on the freezing process.

The impact that the primary drying process can have on the cake structure is illustrated by Figures 1.3c and 1.3d. The conditions under which primary drying is conducted can result in partial collapse of cake structure, as illustrated by Figure 1.3c. The presence of such collapse is generally associated with the primary drying process when the ice-product matrix (see Figure 1.1b) is not in a completely frozen

Figure 1.3. Effect of Freezing and Primary Drying on the Cosmetic Properties of the Cake

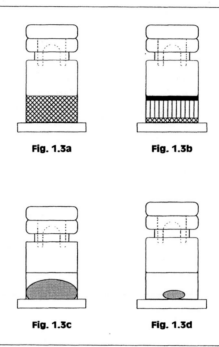

Figure 1.3a illustrates a cake matrix as a result of the formation of a uniform matrix during the freezing process. Figure 1.3b shows the formation of a nonuniform matrix as a result of heterogeneous ice formation and the presence of a crust or glaze on the upper cake surface. Figure 1.3c shows collapse of the cake structure as a result of the primary drying process. Figure 1.3d is an illustration of meltback of the product during the primary drying process.

state. An extreme case of cake *collapse* or *meltback* of cake is illustrated by Figure 1.3d. Meltback is a result of the presence of liquid states in the ice-product matrix. The conditions that result in collapse and meltback during the primary drying process will be examined in greater detail in Chapter 8.

Moisture

The main function of the lyophilization has been defined as enhancing the stability of a formulation primarily by the reduction of the solvent system, primarily water, to quantities that will no longer support biological growth or chemical reactions. The quantity of water remaining in the product, in order to achieve the desired stability, is product dependent. The reason for such product dependency is that there are basically two types of water that can be present in the lyophilized formulation.

Free Water

Free water is the water that is associated with formation of the ice-product matrix. It may be water that crystallized during the freezing process or water that formed a solid solution with the other constituents in the formulation. This water must be reduced by desorption in order to achieve the desired product stability. In general, the residual free water is reduced to less than 1% wt/wt of the final dried product; however, there are examples, such as in the lyophilization of vaccines, when the residual water content must be maintained within a range of values in order to achieve the desired stability [10].

Bound Water

Bound water is water that did not take part in the formation of the ice matrix or the solid solution containing the solutes. Bound water is the water necessary for the stability of the active constituent. For example, it may be the water that is involved in maintaining the folding configuration of a protein molecule. Removal of such water would alter not only the configuration of the protein but also its chemical functionality. It is, therefore, vital that the lyophilization process reduce the free water content of the product but not affect the bound water that is associated with the nature of the active constituent.

Reconstitution

Reconstitution is the process that is used to restore the lyophilized product to its original formulation. The process involves the addition of a known quantity of diluent to the dried cake. For the ideal cake, the addition of the diluent, namely water, results in rapid (less than 1 min) and complete dissolution of the dried cake. The properties (e.g., potency of the active constituent and *p*H) of the resulting solution should be within the prescribed limits of the original formulation. Excessively long reconstitution times, loss in potency, or the formation of a turbid solution are indications of an improper lyophilization process or a malfunction of the freeze-drying equipment.

APPLICATIONS

The following is a general discussion of the various applications of the lyophilization process in various industries. The intent of the discussion is not to provide the reader with a comprehensive review but with an overview of the scope of the lyophilization process. The reader should keep in mind that although the applications may vary, the basic principles involved in the lyophilization process remain the same.

Healthcare Industry

The most extensive use of the lyophilization process is involved in the healthcare industry. This includes the lyophilization of pharmaceuticals such as chemical compounds, parenteral formulations, vaccines, and also in vivo and in vitro diagnostic products. Included in this category are biotechnology products that consist mainly of protein-based products. A listing and discussion of these lyophilized healthcare products is beyond the scope of this text. Lyophilized products in the healthcare industry comprise about 8–10% of total costs, or in excess of $8 billion.

Veterinary

Perhaps the second most common use of lyophilization involves veterinary products. These products range from vaccines for individual household pets to large-scale applications, such as the inoculation of herds of cattle and sheep or large flocks of poultry. These lyophilized products serve not only to protect the animals and the investment of the producer but also to improve the quality of the product. These products also protect the consumer from diseases that may be transmitted by the consumption of animal products.

Food

The most widely known freeze-dried food product is coffee. However, freeze-dried coffee is not, in the true sense, a lyophilized product. Its dark brown color stems not from the coffee solution but from the manner in which the product is prepared. The coffee is first extracted from the coffee bean with hot water, and the water content is reduced to form a concentrated viscous extract. The extract is then carefully processed to form frozen granules. These coffee granules are then dried under conditions that generate partial collapse or *controlled meltback* [11], which provide the granules with the dark brown color. If the coffee were truly lyophilized, the color would appear golden brown because of the reflection of light from the surfaces of the frozen coffee solids. In addition, lyophilized coffee would require a cup of lyophilized powder for each cup of coffee instead of the single teaspoon that is currently used. As a result of the application of the term *freeze-dried* to a form of instant coffee, freeze-drying is no longer synonymous with lyophilization.

Other freeze-dried food products are manufactured. These products may be used as additives, such as berries in breakfast cereals, or are complete products that eliminate the need for refrigeration, such as ice cream that can be eaten without reconstitution. Freeze-drying of foods is also applicable for products where weight can be a factor, such as in backpacking. These freeze-dried meals, like freeze-dried beef, require reconstitution by the addition of water and enhancement of flavor by heating the product (12). Except in special cases like freeze-dried coffee, freeze-dried foods cannot compete economically with frozen or canned food products.

Other Applications

Other applications of lyophilization include the freeze-drying of floral product and taxidermy. There has only been limited success in the freeze-drying of floral products; only the flower portion of the plant appears to respond to the drying process, and there is often a slight change in color. For example, red roses tend to take on a purple tint. The stems of the flowers tend to assume a brown tint, and the leaves become brittle and curled. In the drying of plants, the nature of the plant plays an important role. For example, cacti, in order to conserve water, are equipped with an outer layer that is nearly impervious to the transport of water vapor. A plant such as a cacti would prove difficult or nearly impossible to dry without first disturbing the outer layer to permit the transport of water vapor. While the preservation of flowers by freeze-drying techniques could represent a major market, the growth of the market awaits the development of a better process. Pretreatment of the flower prior to freeze-drying when freshly cut or even during its growth may be necessary in order to achieve a more natural product with a relatively long shelf life.

Freeze-drying has been applied to the taxidermy of small animals. Many small mammals, such as squirrels and raccoons, have been successfully freeze-dried to provide a specimen having a lifelike appearance. In spite of the success of this technique, it is currently used on a very limited basis.

Freeze-drying has been instrumental in the rescue of precious manuscripts, books, and documents that incurred water damage as a result of a fire or a flood. By quickly freezing all of the documents and then freeze-drying the documents in batches, many documents and valuable manuscripts have been restored to nearly original condition. While such services are not often required, the fact that this technology has applications that can prove beneficial to later generations is indeed a rewarding legacy provided by this technology.

REFERENCES

1.	D. L. Harris and L. F. Shackell, Jr., *Am. Pub. Health Assn.*, 52: (1911).

2.	L. R. Rey in *Recent Research in Freezing and Drying* (A. S. Parker and A. V. Smith, eds.), Backwell Scientific Publications, Oxford, England, 1960, p. 40.

3.	R. Altman, *die Elementer Organism and ihre Beziehungen zu den Zellen*, Vert and Co. (1890).

4.	L. F. Shackell, *Am J. Physiol.*, 24: 325 (1909).

5.	B. W. Hammer, *J. Med Research*, 24: 527 (1911).

6.	B. Couriel, *J. Parenter. Drug Assoc.*, 34: 352 (1980).

7.	R. I. N. Greaves in *Biological Applications of Freezing and Drying* (R. J. C. Harris, ed.), Academic Press, New York, 1954, p. 87.

8.	L. R. Rey in International Symposium on freeze-drying of biological products, Washington, D.C. 1976. *Develop. Biol. Standard*, 36: S. Karger, Basel, 1977, p. 19.

9.	T. A. Jennings, *PDA J. Parenter. Sci. & Tech.*, 49: 272 (1995).

10.	L. Valette, et al., International Symposium on freeze-drying of biological products, Washington, D.C. 1976. *Develop. Biol. Standard*, 36: S. Karger, Basel, 1977.

11.	J. Flink, in *Freeze Drying and Advanced Food Technology* (S. A. Goldblith, L. Rey, and W. W. Rothmayr, eds.), Academic Press, New York, 1975.

12.	N. E. Bengtsson, ibid.

2

Product Formulation

INTRODUCTION

It is perhaps fitting that the second chapter is by far the most important chapter in the entire book. The importance of this chapter lies not in its content, but in the topic itself. For the nature of the formulation is the pivot upon which the rest of the lyophilization and its associated equipment and instrumentation revolves. The formulation may be equated to the sun of our solar system but, from my experience, it appears to be more analogous to a *black hole* (i.e., the composition does not shed light on the anticipated lyophilization process). Since each formulation is a unique system, I can only guide the reader in recognizing those areas of formulation preparation that need to be addressed to produce a stable product.

The previous chapter showed that the lyophilization process is applicable to a wide range of materials; however, the most widely used application involves products for the healthcare industry. These products would generally include pharmaceuticals, in vivo and in vitro diagnostics, biotechnology, and veterinary products. The common thread for each of these products is that they are prepared in an aqueous media and, unless stored at low temperature (–70°C), their shelf life is limited to a matter of days or weeks, even when stored at 4°C. The lyophilization process provides a means by which the shelf life of products can be greatly extended and ensures that on dissolution of the dried product with a suitable diluent, the functionality of the product is within specified limits.

In each lyophilized formulation, there will be at least one key constituent that is of a therapeutic or diagnostic nature. Such key constituent(s) will be denoted as the active constituent(s). In view of the vast number of active constituents, this chapter will categorize the active constituents according to their therapeutic functions in order to provide a general overview of the broad scope of active constituents found in lyophilized products.

The active constituent may require, for various reasons, the addition of other constituents. This chapter will consider the various roles that these other constituents play in constructing the formulation. These constituents may serve merely to provide bulk to retain the active constituent in the container during the drying process or interact with the active constituent to ensure stability. A principal objective of this chapter is to identify both the various constituents that may be found in a formulation and to identify the specific roles the various constituents play in formulations. For example, *p*H is often an important consideration in preparing the media in which the active constituent is stabilized. There are numerous occurrences in the literature where the active constituent is maintained in a *p*H buffered solution. However, in reality, the *p*H of the formulation was obtained by the addition of a substance that modified the *p*H but did not have any significant buffering capacity. Because of the widespread misuse of the term *pH-buffered* and in hopes of encouraging others to be more rigorous in their descriptions of their formulations, I have devoted a considerable portion of this chapter to reviewing the basic principles of *p*H and a *p*H-buffer system.

A major constituent that is often disregarded is the solvent system. For most formulations, the solvent system will be water but other solvents, such as alcohols, may be used separately or as part of the aqueous solution. The solvent system, as will be seen in later chapters, plays an active rather than a passive role in the lyophilization process.

For the lyophilization process to be reproducible, the composition and properties of the formulation must also be consistent from batch to batch. For this reason, it is necessary to characterize not only the potency of the active constituent but to show that other properties of the formulation that are colligative, optical, or electrical in nature are within a given range of values. As the information regarding the characterization of the formulation increases, so will the confidence of reproducibility from batch to batch.

In developing a formulation, the main focus is generally directed at preparing a stable environment for the active constituent. What is often overlooked during the development of the formulation are its key thermal properties, which will serve as the foundation for the lyophilization process. This chapter will introduce these key thermal properties and how they impact the lyophilization of the formulation. Since the lyophilization process is often the last major step in the manufacture of a product, the design of the formulation can have a major impact on the cost of manufacturing and on the profitability (bottom line) of the company.

ACTIVE CONSTITUENT

With respect to the lyophilization process, the active constituent of a formulation will fall within one of two general categories: synthetic and natural. The rationale for selecting these two general categories will become apparent when considering the impact that the active constituent can have on the thermal properties of the formulation.

Synthetic

Synthetic active constituents are those ingredients that are synthesized from organic or inorganic reactions. The chief advantages to synthesizing the active constituent of a formulation are that the compound can be prepared to a high degree of purity and the possibility that the final lyophilized product is in a crystalline state. Substances in a crystalline state tend to have greater stability than the same material in an amorphous state. Consider the following two examples of active constituents prepared by inorganic and organic reactions.

Cisplatin

A typical synthetic active constituent is the compound cisplatin [United States Adopted Names nomenclature for *cis*dichorodiammine-platinum(II)] that has been found to be an anti cancer drug by its activity on solid tumors [1]. The chemical structure for cisplatin is given as

$$\begin{array}{ccc} NH_3 & & Cl \\ & \diagdown \diagup & \\ & Pt & \\ & \diagup \diagdown & \\ NH_3 & & Cl \end{array}$$

[1]

The cisplatin compound is prepared by an inorganic reaction between NH_4OH and the $PtCl_4^{-2}$ ion. The compound is referred to as cisplatin because it has a significant dipole moment. When HCl reacts with $Pt(NH_3)_4^{+2}$ the isomer compound that is formed does not exhibit a permanent dipole moment. This trans-compound has the same composition as the cisplatin but has the following structure [2].

$$\begin{array}{ccc} Cl & & NH_3 \\ & \diagdown \diagup & \\ & Pt & \\ & \diagup \diagdown & \\ NH_3 & & Cl \end{array}$$

[2]

Sodium Salt of [2,3-Dichloro-4-(Methylenebutyryl)Phenox]Acetic Acid

The antibiotic organic compound sodium salt of [2,3-dichloro-4-(2-methylenebutyryl)phenox]acetic acid is prepared by neutralizing [2,3-dichloro-4-(2-methylenebutyryl)phenox]acetic acid with a stoichiometric quantity of 2% wt/vol sodium hydroxide solution. The nature of the compound formed is dependent on the drying method. The amorphous form is obtained if the aqueous solution is cooled to –50°C and then dried under a vacuum. A crystalline form of the compound can be formed by dissolving the amorphous form of the compound in 95% ethanol and then evaporating the ethanol [3]. The crystalline form of the compound is found to be more stable than the amorphous form, as is true for most active constituents [4–7].

Nature Derived

Nature-derived active constituents are substances that are obtained by a natural process such as fermentation or growth in a media. In some instances the active constituent contains constituents of the media that can vary in concentration and composition from batch to batch, while in others the active constituent is obtained in a highly purified state. The following are some examples of nature-derived active constituents.

Vaccines

The conventional means for preparing vaccines is to infect chick embryos with the virus. After a given inoculation period, the virus is harvested from the egg and pooled with viruses obtained from other fertilized eggs [8]. In some instances, the virus is purified and its infectivity is increased by centrifuging [9]. The pools, purified or unpurified, are then blended and mixed with a stabilizer solution to form the vaccine. Regardless of the treatment of the pooled virus, some constituents from the media will remain in the final virus pool.

Proteins

With the advent of biotechnology, there has been an explosive interest in various forms of proteins. The steps involved in producing these proteins is quite complex and can involve fermentation (e.g., *Escherchia coli* refractile paste), recovery by a solvent system and then purification by high performance liquid chromatography (HPLC) followed by precipitation, filtration, and further purification by ion exchange chromatography and gel exchange in order to produce the protein interleukin-2. Unlike the formation of vaccines, biotechnology exerts great efforts to produce high purity proteins [10].

This discussion shows that high purity active constituents can be formed by either inorganic or organic syntheses. Active constituents stemming from growth of a virus or organism in a medium can produce an active constituent with a given potency, but because of the complexity of the growth process, the composition of the active constituent can vary from batch to batch. Active constituents such as proteins produced from a fermentation process can be purified; however, the process is quite lengthy and complex.

Physical State

The nature of the physical state of the lyophilized active constituent is an important factor in determining its stability. In the *crystalline state*, the active constituent has a well-defined structure and other identifying physical characteristics such as a sharp melting temperature, density, and, under defined conditions, thermodynamic stability. However, in the *amorphous form*, there is short range order, and the material is thermodynamically unstable. Because lyophilized formulations tend to form *glassy states* in the interstitial region of the matrix, the physical state of active constituent tends to be amorphous rather than crystalline. This section will consider various means for transforming the active constituent from an amorphous form to a crystalline state to enhance stability.

Solvent System

Phillips et al. (3) approached the problem of producing a crystalline active constituent by using a selected solvent system. A group of active constituent compounds, identified as E-3816, were found difficult to lyophilize because they formed an amorphous structure when frozen in an aqueous solution. A crystalline form was obtained by dissolving the compound in a solvent system consisting of water and acetone. These compounds, however, were not soluble in acetone. The formation of ice during the freezing process increased the concentration of both the acetone and the compound in the interstitial region of the matrix. Since the compound was not soluble in acetone, the initial crystallization of the compound from the high concentration acetone solvent occurred at -62°C and was confirmed by an exotherm determined by differential thermal analysis (DTA). Crystallization of the compound occurred relatively slowly because of the viscosity of the residual solvent system. Once the crystallization of the compound is complete, the acetone, because its vapor pressure is higher than that of the ice, can be preferentially evaporated from the matrix. After the removal of acetone is complete, the ice can be sublimated from the matrix.

The presence of 1% to 5% isopropyl alcohol was shown to enhance the crystallization of cefazolin sodium [11]. With the addition of the isopropyl alcohol, crystalline cefazolin sodium can be obtained with and without thermal heat treatment as previously reported [12]. For a completely aqueous system, the thermal treatment at about -10°C required about 24 hr to increase the crystallization of the drug. However, the addition of the alcohol reduced the time necessary for thermal treatment to about 1 hr and an acceptable crystalline product was produced using a significantly shorter drying process [11].

Humidification

Humidification is a process in which the amorphous lyophilized formulation is crystallized by exposure to a humid atmosphere [4]. Kovalcik and Guillory [6] showed that humidification of amorphous lyophilized cyclophosphamide induced the formation of a crystalline monohydrate of the compound. The crystalline monohydrate form of cyclophosphamide was found to be more stable than the amorphous form.

Thermal Treatment

Crystallization of the active constituent can sometimes be induced by thermal treatment of the formulation during the freezing portion of the lyophilization process [11,12]. Gatlin and DeLuca [12] reported that drying of cefazolin sodium produces an amorphous cake that is unstable and becomes discolored on storage. However, these authors showed that after freezing the cefazolin sodium formulation at a rate of -10°C/min., there was, on warming, an exotherm with an onset temperature near -10°C. A crystalline form of cefazolin sodium could be obtained by first maintaining the previously frozen solution between -10°C and -4°C for a short period of time and then lowering the temperature and conducting a drying process [12].

OTHER CONSTITUENTS

The active constituent may require the presence of other constituents or excipients in order to obtain a lyophilized formulation. The addition of excipients may be required to provide (a), the correct range of pH values, (b) additional bulk to prevent physical loss of the active constituent, (c) a stabilizing environment for the active constituent in both its liquid and dried states, (d) protection of the active constituent during the freezing process, and (e) protection against oxidation [13,14]. The selection of the composition and concentration of the excipients requires a comprehensive understanding of the nature of the active constituent. Since the formulation has already been identified as a pivotal element and has such a major impact on the lyophilization process[15], formulating is without a doubt an important and often difficult step in the lyophilization process.

Buffers

This section will consider the role that pH buffers play in stabilizing the liquid phase of the formulation. It is hoped that the reader will bear with me in reviewing the *Law of Mass Action* and the definition of pH, both necessary to understanding the functionality of a pH buffer solution. An additional reason for having an understanding of the chemical nature of buffers will become evident in Chapter 7, "The Freezing Process."

Law of Mass Action

The *Law of Mass Action* states that for a general reaction such as

$$aA + bB \; \rightleftharpoons \; cC + dD \tag{1}$$

where the rate in the forward direction (k_f) is proportional to $[A]$ (concentration of a compound A is expressed in molarity (M) and denoted by $[A]$) and $[B]$ and the reverse rate (k_r) is dependent $[C]$ and $[D]$ [13, 14]. When the system is at equilibrium, $k_f = k_r$ or

$$k_f[A]^a[B]^b = k_r[C]^c[D]^d \tag{2}$$

The equilibrium constant K for reaction (2) is defined a

$$K = \frac{k_f}{k_r} = \frac{[C]^c[D]^d}{[A]^a[B]^b} \tag{3}$$

For a system that is in equilibrium, one should bear in mind Le Chatelier–van't Hoff's theorem: *Any alteration in the factors which determine chemical equilibrium causes the equilibrium to be displaced in such a way to oppose as far as possible the effect of the alternation* [16]. It will be shown that this theorem is particularity applicable to buffer systems.

Definition of pH

The term pH was originated by S. P. Sorensen in 1909 as a convenient means for expressing the concentration of *hydronium ions* $[H_3O^+]$ in a solution [16]. The hydronium ion is formed by the reaction

$$H_2O + H_2O \; \rightleftharpoons \; H_3O^+ + OH^- \tag{4}$$

According to the Law of Mass Action, the equilibrium constant K_w at 25°C is given as

$$K_w = \frac{[H_3O^+][OH^-]}{[H_2O]^2} = 1 \times 10^{-14} \qquad (5)$$

From equation 5, it is apparent that $[H_2O]^2$ is much greater than the product of $[H_3O^+][OH^-]$.

pH was defined by S. P. Sorensen as the negative log of $[H_3O^+]$. For example, the addition of a quantity of acid to water would result in an increase in the $[H_3O^+]$. For a $[H_3O^+]$ concentration of 6×10^{-4}, the pH is obtained as follows:

$pH = -(\log 6 - 4)$

$pH = -\log 6 + 4$

$pH = -.78 + 4$

$pH = 3.22$

and is limited to two decimal places.

Buffer Solution

A *buffer* is a solution that consists of a *weak acid* or *base* and its salt. The distinction between a strong acid or base and a weak acid or base is that the former completely dissociates in aqueous solution, but a weak acid or base only partially dissociates. The main characteristics of a buffer solution are its pH value and *capacity*. In order to understand the significance of these characteristics, consider the following examples.

Strong Acid and Strong Base. Strong acids are typically inorganic acids, such as hydrochloric, hydrobromic and nitric; strong bases include sodium and potassium hydroxides.

The reaction between 1 L of a 1 M solution of a strong acid and 1 L of a 1 M solution strong base can be expressed as follows:

$$H_3O^+ + OH^- \rightleftharpoons 2H_2O + 13,900 \text{ cal} \qquad (6)$$

The release of energy indicates that the reaction is *exothermic* in nature. From Le Chatelier–van't Hoff's theorem, it is clear that an increase in temperature of the water will increase the dissociation of the water, whereas a reduction in temperature will reduce the dissociation.

Weak Acid and Weak Base. For inorganic compounds, boric and carbonic acids are examples of weak acids and ammonium hydroxide is an example of a weak base. All organic acids, with the exception of sulfonic acids, are weak, and most organic bases are weak [17].

Weak Acid. An aqueous solution of a weak acid, such as acetic acid (HAc) can be expressed as

$$HAc + H_2O \rightleftharpoons H_3O^+ + Ac^-\qquad(7)$$

Assuming that $[H_2O]$ is constant, the Law of Mass Action defines the dissociation constant K_a (sometimes referred to as the ionization constant) as

$$K_a = \frac{[H_3O^+][Ac^-]}{[HAc]}\qquad(8)$$

K_a has a value of 1.754×10^{-5} at 25°C [17]. The reason for K_a having such a small value is that at equilibrium $[HAc] \gg [H_3O^+][Ac^-]$. If c is the initial concentration of acetic acid, $[HAc]$ will have the value of $c - [H_3O^+]$. Since $[H_3O^+] = [Ac^-]$, equation 8 can be written as

$$K_a = \frac{[H_3O^+]^2}{c - [H_3O^+]}\qquad(9)$$

Equation 9 can be arranged in the form of a quadratic equation

$$[H_3O^+]^2 + K_a[H_3O^+] - K_a c = 0\qquad(10)$$

and $[H_3O^+]$ determined from

$$[H_3O^+] = -\frac{K_a}{2}\sqrt{\frac{K_a^2}{2} - 2K_a c}\qquad(11)$$

In the case where $[H_3O^+] \ll [HAc]$, then equation 11 can be simplified as

$$[H_3O^+] = K_a c\qquad(12)$$

For a 0.1 M HAc solution when $K_a = 1.8 \times 10^{-5}$, the $[H_3O^+]$ will be 1.8×10^{-4} and by definition, the *p*H of the solution will be 3.74. When the molar concentration c is 1, the $[H_3O^+] = K_a$ and the *p*H is referred to as the *p*K. In the case of HAc, the *p*K = 4.74.

A weak acid (e.g., phosphoric acid $[H_3PO_4]$) may have more than one dissociation constant. These dissociation constants are based on the following Law of Mass Action

$$K_1 = \frac{[H_3O^+][H_2PO_4^-]}{[H_3PO_4]} = 7 \times 10^{-3}$$

$$K_2 = \frac{[H_3O^+][HPO_4^=]}{[H_2PO_4^-]} = 7 \times 10^{-8}$$

$$K_3 = \frac{[H_3O^+][PO_4^{-3}]}{[HPO_4^=]} = 4 \times 10^{-13}$$

Because the reactions for K_2 and K_3 will be suppressed by K_1 and they have relatively low dissociation constants, the $K_a = K_1$.

Weak Acid. The dissociation of a weak base solution can be expressed as

$$NH_4OH + H_2O \rightleftharpoons NH_4^+ + OH^- \tag{13}$$

and from the Law of Mass Action, the dissociation constant K_b can be obtained from the expression

$$K_b = \frac{[NH_4^+][OH^-]}{[NH_4OH]} \tag{14}$$

which can be simplified to

$$[OH^-] = K_b c \tag{15}$$

where c represents the initial concentration of NH_4OH.

By substituting the value of $[OH^-]$ from equation 5 into equation 15, one obtains the dissociation constant K_b

$$K_b = \frac{K_w}{[NH_4OH][H_3O^+]^2} \tag{16}$$

or in terms of the $[H_3O^+]$

$$[H_3O^+] = \sqrt{\frac{K_w}{K_b c}} \tag{17}$$

where c is the initial concentration of the base NH_4OH. From equation 17, it can be seen that $[H_3O^+]$ will decrease for a given K_b as the concentration of c increases.

The $[H_3O^+]$ of a 0.1 M solution of NH_4OH, where $K_b = 1.75 \times 10^{-5}$, is determined from equation 17 to be 7.55×10^{-5}. The pH of the NH_4OH solution is 9.12.

It should be pointed out that the simplified equations 12 and 17 are applicable only when $[H_3O^+]$ for the weak acid and $[OH^-]$ for the weak base are < 5% of the concentration c for the weak acid or weak base. Otherwise, the quadratic expression obtained from the Law of Mass Action must be used. Generally this will be necessary only when the value of c is < 0.001 M.

Salt Solutions. This section will consider the effect that the hydrolysis of salts formed from a strong acid and strong base, a weak acid and strong base, and a strong acid and a weak base, will have on the pH of a solution.

Salts of a Strong Acid and Strong Base. When the salt of a strong acid and a strong base (e.g., sodium chloride [NaCl]), is dissolved to form an aqueous solution, there is complete dissociation with the formation of cations [Na$^+$] and anions [Cl$^-$]. Because of the complete dissociation, there is no change in $[H_3O^+]$ and, therefore, no change in the pH of the solution.

Salts of a Weak Acid and Strong Base. The *hydrolysis* of a salt formed from the reaction of a weak acid with a strong base will cause a change in the *pH* of a solution. The hydrolysis of sodium acetate (NaAc) can be represented as

$$NaAc \rightarrow Na^+ + Ac^- \tag{18}$$

$$Ac^- + H_2O \rightleftharpoons HAc + OH^- \tag{19}$$

Applying the Law of Mass Action to reaction 19, one obtains an expression for (K_{hydr}) the *hydrolysis constant* [18], i.e.,

$$K_{hydr} = \frac{[HAc][OH^-]}{[Ac^-]} \tag{20}$$

and the $[H_2O]$ remains relatively unchanged by reaction 19. By multiplying the numerator and the denominator of equation 20 by $[H_3O^+]$, one obtains

$$K_{hydr} = \frac{[HAc][OH^-][H_3O^+]}{[Ac^-][H_3O^+]} \tag{21}$$

The terms in equation 21, $[OH^-][H_3O^+] = K_w$ while

$$\frac{1}{K_a} = \frac{[HAc]}{[Ac^-][H_3O^+]} \tag{22}$$

then K_{hydr} can be expressed as

$$K_{hydr} = \frac{K_w}{K_a} \tag{23}$$

Since $[HAc] = [OH^-]$ and $[Ac^-] = [Na^+] = c$, the concentration of the salt in the solution K_{hydr} can be written as

$$K_{hydr} = \frac{[OH^-]^2}{c} \tag{24}$$

or

$$K_{hydr} = \frac{K_w^2}{c[H_3O^+]^2} \tag{25}$$

By substituting for K_{hydr}, as defined in equation 23 and solving for $[H_3O^+]$, one obtains the following expression:

$$[H_3O^+] = \sqrt{\frac{K_a K_w}{c}} \tag{26}$$

For a 0.1 M NaAc solution, the dissociation constant *is* $K_a = 1.8 \times 10^{-5}$, the $[H_3O^+]$ from equation 26 is 1×10^{-9}, and the *p*H of the solution 9.0.

Salts of a Strong Acid and a Weak Base. The hydrolysis of a salt formed from the reaction of a strong acid with a weak base will cause a change in the *p*H of a solution. The hydrolysis of ammonium chloride (NH_4Cl) can be represented as

$$NH_4Cl \rightarrow NH_4^+ + Cl^- \tag{27}$$

and

$$NH_4^+ + H_2O \rightleftharpoons NH_4OH + H_3O^+ \tag{28}$$

Applying the Law of Mass Action to equation 26 gives

$$K_{hydr} = \frac{[NH_4OH][H_3O^+]}{[NH_4^+]} \tag{29}$$

Multiplying the numerator and the denominator by $[OH^-]$ one obtains

$$K_{hydr} = \frac{[NH_4OH][H_3O^+][OH^-]}{[NH_4^+][OH^-]} \tag{30}$$

Since $[H_3O^+][OH^-] = K_w$ and

$$\frac{[NH_4OH]}{[NH_4^+][OH^-]} = \frac{1}{K_b},$$

the term K_{hydr} can be expressed as

$$K_{hydr} = \frac{K_w}{K_b} \tag{31}$$

Because $[NH_4OH] = [H_3O^+]$ and $[NH_4^+] = [Cl^-] = c$, the concentration of the salt in the solution K_{hydr} can be written as follows:

$$K_{hydr} = \frac{[H_3O^+]^2}{c} \tag{32}$$

and $[H_3O^+]$ is expressed as

$$[H_3O^+] = \sqrt{\frac{K_w}{cK_b}} \tag{33}$$

For a 0.1 M NH_4Cl solution, the dissociation constant is 1.75×10^{-5}, the $[H_3O^+]$ is 7.55×10^{-5} and the *p*H of the solution is 4.12.

Buffer Solutions. A buffer solution has been previously defined as a solution consisting of a weak acid and its salt formed by a strong base or a mixture of a weak base and its salt formed by a strong acid. The main function of a buffer solution is to resist changes in the *pH* of the solution [19]. This section will first consider a buffer consisting of weak acid and its salt formed by a strong base and then a weak base and its salt formed by a strong acid.

Weak Acid and its Salt Formed by a Strong Base. From equation 8, the $[H_3O^+]$ for HAc can be expressed as

$$[H_3O^+] = \frac{K_a[HAc]}{[Ac^-]} \tag{34}$$

For the salt NaAc, there will be complete dissociation, and the concentration of Ac^- in the solution will be equal to the concentration of the salt (c_s). Because of the presence of the Ac^- ions from the salt, the dissociation of HAc will be repressed and, for all practical purposes, all of HAc will remain undissociated. Under these conditions, the concentration of HAc is c_a. By substituting in equation 34 c_s for $[Ac^-]$ and c_a for [HAc], one obtains an approximate expression for the $[H_3O^+]$ as

$$[H_3O^+] = \frac{K_a c_a}{c_s} \tag{35}$$

If concentration of HAc is 0.1 M and the salt is 0.2 M with $K_a = 1.8 \times 10^{-5}$, then the $[H_3O^+]$ will equal 9×10^{-6} and the *pH* of the solution will be 5.04. Let 1 mL of 1 M sodium hydroxide (NaOH) be added to 1 L of the solution. The NaOH will react with the HAc to form additional NaAc salt. Equation 35 now becomes

$$[H_3O^+] = \frac{0.099 K_a}{0.201} \tag{36}$$

The $[H_3O^+]$ concentration will be 8.86×10^{-6} and the *pH* will equal 5.05. If 1 mL of NaOH is added to pure water with a *pH* of 7.0, the $[H_3O^+]$ will be 1×10^{-11} and the *pH* will equal 11.00.

Now assume that 1 mL of 1 M hydrochloric acid (HCl) is added to 1 L of the original buffer solution. The addition of the acid will increase the [HAc] to 0.101 M, while the concentration of the salt $[Ac^-]$ is decreased to 0.199. The $[H_3O^+]$ is increased to 9.14×10^{-6}, and the pH becomes equal to 5.04. The addition of the 1 mL of 1 M HCl to the buffer solution did not significantly decrease the *pH* of the buffer solution.

If 1 mL of 1 M HCl is added to 1 L of pure water with a pH of 7.0, the $[H_3O^+]$ will increase to 1×10^{-4} and the *pH* will equal 4.00.

From the above, it can be seen that a buffer solution formed by a weak acid and its salt resists changes in *pH* resulting from the addition of a strong base or a strong acid.

Weak Base and its Salt Formed by a Strong Acid. From equation 14, the K_b for NH_4OH can be expressed as

$$K_b = \frac{[NH_4^+][OH^-]}{[NH_4OH]}$$

while

$$[OH^-] = \frac{[NH_4OH]K_b}{[NH_4^+]} = \frac{c_b K_b}{c_s} \tag{37}$$

For the salt NH_4Cl, there will be complete dissociation, and the $[NH_4^+]$ in the solution will be equal to the concentration of the salt (c_s). Because of the presence of the NH_4^+ ions from the salt, the dissociation of the NH_4OH will be repressed and, for all practical purposes, all of the NH_4OH will remain un-dissociated. Under these conditions, $[NH_4OH]$ will equal c_b. From equation 5, $[H_3O^+]$ can be substituted for $[OH^-]$ and equation 35 becomes

$$[H_3O^+] = \frac{c_s K_w}{c_b K_b} \tag{38}$$

If $[NH_4OH]$ is 0.2 M and the salt is 0.3 M with K_b at 1.75×10^{-5}, then the $[H_3O^+]$ will equal 8.57×10^{-10}, and the pH of the solution will be 9.07. Let 1 mL of 1 M NaOH be added to the 1 liter of solution. The NaOH will increase the $[NH_4OH]$ to .201 and decrease the $[NH_4^+]$ or c_s to .299 M. Equation 38 now becomes

$$[H_3O^+] = \frac{0.299 K_w}{0.201 K_b} \tag{39}$$

The $[H_3O^+]$ is now 8.50×10^{-10}, and the pH of the solution is 9.07. Thus, adding NaOH to the buffer did not significantly alter the pH of the solution.

If 1 mL of 1 M HCl is added to 1 liter of the buffer solution, the HCl will decrease the $[NH_4OH]$ or c_b to .199 and increase the $[NH_4^+]$ or c_s to .301 M. As a result, the $[H_3O^+]$ will be 8.64×10^{-10}, and the pH of the solution will be 9.06. Adding 1 mL of 1 M HCl to the buffer solution resulted in a decrease in the pH of only 0.01.

Buffer Capacity. In the last section, it was shown that mixtures of weak acids and their salts or weak bases and their salts produce buffer solutions. These solutions resist changes in pH when a strong acid or base is added. The *capacity of a buffer* to resist a change in pH is dependent on the ratio of the concentration of the weak acid (c_a) to the concentration of its salt (c_s) as shown in equation 35 or on the ratio of the concentration of the salt (c_s) to the concentration of the weak base (c_b) as represented by equation 38. The capacity of a buffer C_b is defined as

$$c_b = \frac{d(AB)}{d(pH)} \tag{40}$$

where AB represents the molarity of the strong acid or base added to 1 L of the buffer solution and $d(AB)/d(pH)$ is the slope of a plot of AB as a function of pH [19,20].

With the NH_4OH–NH_4Cl buffer solution described previously, Figure 2.1 shows a plot of acid molarity for HCl as a function of the pH of the solution as determined from equation 39. The pH capacity of this buffer solution, as determined from the slope of the plot is 0.005. The magnitude of the slope indicated that the buffer solution does not change significantly with the increase in the acid molarity.

Figure 2.1. Molarity of HCl as a Function of *p*H

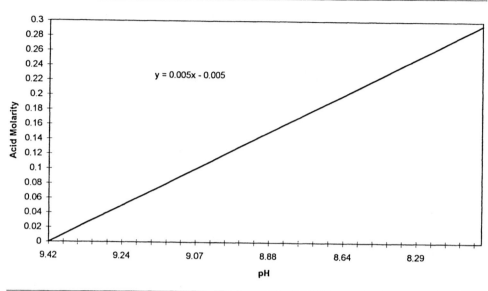

A plot of the molarity of HCl as a function of the pH of a buffer system consisting of 0.2 M NH_4OH and 0.3 M NH_4Cl. The slope of the plot is referred to as the buffer capacity.

The effect of the increase in the HCl in the buffer solution can be seen more graphically in Figure 2.2. This figure shows a plot of the *p*H of the buffer solution as a function of the acid molarity. In this plot, it can be seen that for acid molarity values < 0.1 M, there is relatively little change in the pH of the buffer solution. However, as the molality of the acid approaches the initial molarity of the base (0.2 M), the solution consists only of NH_4Cl and is no longer functioning as a buffer. An increase in the molarity of the HCl to greater than 0.2 M results in marked decrease in the *p*H of the solution.

Buffers in Formulations. Table 2.1 is a list of constituents that were used in formulations and identified as buffer systems [14,19,21–23]. An examination of Table 2.1 shows that of all reported buffer systems used in formulations, only two formulations represented true buffer systems, benzoic acid and sodium benzoate [14] and sodium phosphate dibasic and phosphoric acid [24]. For example, the adjustment of the *p*H of a solution containing sodium phosphate dibasic with NaOH represents a system consisting of the salt of a weak acid (phosphoric) and a strong base (NaOH).

Other solutions shown in Table 2.1 do not represent buffer solutions. The solution composed of citric acid and sodium phosphate certainly consists of a weak acid and the salt of a weak acid, but the latter is the salt of phosphoric acid rather than citric acid. While the addition of the above excipients will produce a solution having a given *p*H, the solution will not have the buffering capacity, as demonstrated in

Figure 2.2. A Plot of the *p*H of a Buffer System Consisting NH$_4$OH of 0.2 M NH$_4$OH and 0.3 M NH$_4$Cl as Function of the Molarity of HCl

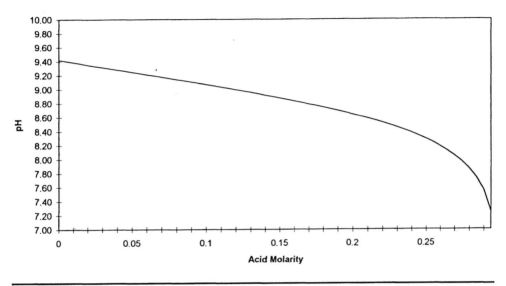

Figure 2.2. Except for the benzoic acid and sodium benzoate system, the other buffer solutions listed in Table 2.1 are best referred to as *pH-modified* solutions rather than as buffer solutions [13].

Bulking Compounds

Some active constituents are not very soluble in aqueous solutions. As a result, the percent solids in the formulation may be significantly < 2%. As the percent solids decreases, probability that the formulation will produce a poor self-supporting cake structure increases, which is illustrated in Figure 2.3. This figure shows that during the drying process, the poor structural properties of the cake resulted in some of the product being carried out of the container as a result of the gas flow during the drying process. Note cake material in the region of the closure. In order to prevent such a loss in product, the solid content of the formulation is increased by the addition of a *bulking agent* [14,23]. While numerous substances at times can serve as bulking agents, this text will consider only two of the most commonly used compounds. The intent of the following discussion of bulking agents is to show that other factors must be taken into consideration when selecting a bulking agent.

Mannitol

Perhaps the most common bulking agent used in formulations is mannitol [24–26]. An important advantage of mannitol is that if the formulation is frozen slowly, the

Table 2.1. Constituents Used in Formulations and Identified as Buffers

Buffers	References
Acetic acid	14,19
Adipic acid	14,19
Benzoic acid and sodium benzoate	14
Citric acid	14,19,23
Citric acid and sodium phosphate	23
Hydrochloric acid, sodium biphosphate, sodium phthalate and sodium chloride	21
Lactic acid	14,19
Maleic acid	14,19
Phthalate	19
Potassium phosphate	14,19
Potassium phosphate monobasic and sodium phosphate dibasic	23
Sodium hydrogenphosphate dihydrate and disodium hydrogenphosphate dodecahydrate	21
Sodium hydroxide, glycine and sodium chloride	23
Sodium phosphate monobasic	14
Sodium phosphate dibasic	14,23
Sodium phosphate dibasic and sodium hydroxide	24
Sodium phosphate dibasic and phosphoric acid	24
Sodium acetate	14
Sodium bicarbonate	14
Sodium citrate	14
Sodium tartrate	14,23
Tartaric acid	14,23
Tartaric acid and sodium tartrate	23

mannitol becomes crystalline [24–28]. The crystalline structure of the mannitol provides supporting structure for the active constituent and prevents loss of product from the container (Figure 2.3).

If a solution containing mannitol is frozen rapidly enough to prevent crystallization of the mannitol, then breakage of the vial can occur when the vial is warmed during the primary drying process. Differential scanning calorimetric (DSC) studies of noncrystalline mannitol solutions at rapid cooling rates (i.e., $-10°C/min$) showed

Figure 2.3. A Lyophilized Cake with Poor Self-Supporting Structure Resulting in the Loss of Product from the Container

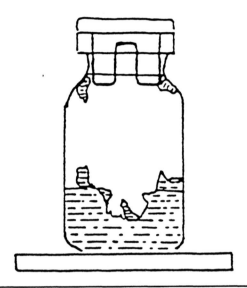

an exothermic peak at -28°C during warming of the frozen matrix. The result of this exotherm was to generate a pressure increase in the matrix that at times was responsible for the breakage of the glass containers [27, 29]. In some instances, only a few glass containers will be broken, while the breakage in others may exceed 50% of the total containers in the system.

Mannitol usually is nonreactive with the active constituent. Kovalcik and Guillory [6] studied the stability of cyclophosphamide and found that a bulking agent was necessary to achieve a cake *rather than a hard glass-like mass*. Exposing the cake to a small quantity of water was necessary to achieve stability. For the formulations prepared with equal quantities (3.2 mg/mL) of mannitol or lactose, the moisture treated cake containing mannitol maintained greater stability than a cake containing lactose as a bulking agent.

However, in formulations containing proteins such as gamma interferon, Inazu and Shina [30] found that the presence of mannitol in the formulation induced some unfavorable reactions. Such unfavorable reactions were determined by observing the reconstituted solutions, with and without other additives, at 25°C over a period of time. With only mannitol present, a reconstituted solution of gamma interferon exhibited a slightly hazy solution after 3 months and a turbid solution after 6 months of storage. If thiomalate of L-cysteine was used as an additive in the gamma interferon formulation, the presence of mannitol caused aggregation after 3 months, whereas the use of a maltose produced a clear solution even after 6 months of storage. While the above authors offer no explanations for the difference in clarity between the mannitol and the maltose solutions, their results demonstrate that the selection of an excipient depends on its interaction with the active constituent.

Polyvinyl Pyrrolidone

Bashir and Avis [31] showed that the addition of polyvinyl pyrrolidone (PVP) to a formulation greatly increased the strength of the cake when used in the presence of other excipients. These authors developed an apparatus, shown in Figure 2.4, for quantitatively measuring the force necessary to fracture a dried cake. The simple but elegant apparatus consisted of a standard single arm balance that supplied to a piston the force necessary to fracture a cake. The cake was prepared in a polystyrene cylinder with the base removed. Prior to filling the cylinder, the base was sealed with a polyethylene snap cap which provided a liquid tight seal. After drying was complete, the cap was removed and the vial was secured in place by a vial holder. With a 1 L beaker on the balance pan, the piston was then brought to rest, without applying any force, to the top of the cake. Force was applied by slowly adding water to the beaker until the cake fractured. The mass of water added to the beaker was equivalent to the force necessary to fracture the cake.

Figure 2.4. Plug Fracture Apparatus

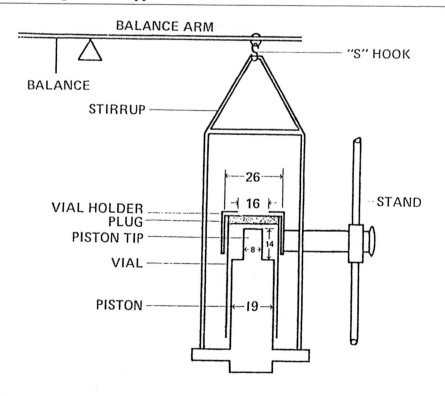

Plug fracture apparatus for measuring the fracturability of cakes of various excipients. (All dimensions given in mm). [Source: J. A. Bashir and K. E. Avis, *Bull. Parenter. Drug Assoc.*, 27 (2):71 (1973). Reprinted with permission.].

The results showed that the *fracturability* of the cake of a given excipient was directly related to its concentration. In the case of mannitol, the force necessary to fracture a cake from a 0.25 M solution was 175 g, although a force of 798 g was required to fracture the cake of a 1 M mannitol solution. PVP was also shown to produce strong cakes (e.g., the force required to fracture a cake containing 1.4% PVP required 142 g), but there was some evidence of cake shrinkage. The real advantage of PVP is when it is used in conjunction with another excipient. The addition of 0.8% PVP to a formulation containing mannitol or glycine increased the fracturability of the cake by a factor from 3 to 25 times and produced cakes with uniform structure and color and without shrinkage.

Other Bulking Agents

Table 2.2 lists additional bulking agents that have been used in lyophilized formulations [14,30,31]. While the principal role of the bulking agent is to provide supporting structure for the active constituent, interaction between the bulking agent and the active constituent [32] is possible. As a result, the bulking agent must be selected with consideration of the possible interactions between the bulking agent and the active constituent or other excipients in the formulation.

Stabilizing Agents

Excipients can also play a significant role in stabilizing the active constituent in an aqueous solution. When selecting excipients for a formulation, it is important to understand the nature of the active constituent and the role that the excipient will play in providing a stable environment for the active constituent. The following will be a description of the various modes in which the excipient affects the stability of the active constituent.

Law of Mass Action

The stabilization of an active constituent in an aqueous solution can be achieved by applying the Law of Mass Action. For example, consider the stability of cisplatin in an aqueous solution [1]. The *hydrolysis* of cisplatin in an aqueous solution has been represented by the following reaction:

Table 2.2. Additional Bulking Excipients

Bulking Agents	References	Bulking Agents	References
Lactose	14,31	Sodium chloride	14
Glycine	14,31	Sorbitol	14
Dextrose	14	Glucose	30
Maltose	30		

$$
\begin{array}{c}
NH_3 \quad Cl \\
\diagdown Pt \diagup \\
\diagup \quad \diagdown \\
NH_3 \quad Cl
\end{array}
+ H_2O \;\underset{k_{-1}}{\overset{k_1}{\rightleftarrows}}\;
\left[
\begin{array}{c}
NH_3 \quad Cl \\
\diagdown Pt \diagup \\
\diagup \quad \diagdown \\
NH_3 \quad OH_2
\end{array}
\right]^{+}
+ Cl^- \qquad (41)
$$

where k_1 is the hydrolysis reaction while k_{-1} is the formation of cisplatin. The cisplatin can undergo additional hydrolysis as represented by

$$
\begin{array}{c}
NH_3 \quad Cl \\
\diagdown Pt \diagup \\
\diagup \quad \diagdown \\
NH_3 \quad OH_2
\end{array}
+ H_2O \;\underset{k_{-2}}{\overset{k_2}{\rightleftarrows}}\;
\left[
\begin{array}{c}
NH_3 \quad Cl \\
\diagdown Pt \diagup \\
\diagup \quad \diagdown \\
NH_3 \quad OH_2
\end{array}
\right]^{+2}
+ 2Cl^- \qquad (42)
$$

From the Law of Mass Action, the overall dissociation constant K_{cis} can be expressed as

$$
K_{cis} = \frac{\left[
\begin{array}{c}
NH_3 \quad Cl \\
\diagdown Pt \diagup \\
\diagup \quad \diagdown \\
NH_3 \quad OH_2
\end{array}
\right]^{+2} [Cl^-]^2}{\left[
\begin{array}{c}
NH_3 \quad Cl \\
\diagdown Pt \diagup \\
\diagup \quad \diagdown \\
NH_3 \quad Cl
\end{array}
\right]} \qquad (43)
$$

By increasing the $[Cl^-]$ in the solution, the concentration of the cisplatin is increased or the stability of the cisplatin in the solution is enhanced. Atilla et al. [1] found that the addition of sodium chloride (NaCl) significantly increases the stability of cisplatin in an aqueous solution , even at concentrations as low as 0.1% or 1.7×10^{-2} M, but that the addition of excipients such as mannitol and dextrose did not enhance the stability of the cisplatin. The proper application of the Law of Mass Action can prevent or limit the hydrolysis and enhance the stability of an active constituent in a aqueous solution.

Excipient-Induced Crystallization of the Active Constituent

It has already been shown that a lyophilized active constituent has greater stability in a crystalline state than in an amorphous state. Several means were described (e.g., thermal treatment [11,12]) for obtaining a lyophilized crystalline active constituent. Here we shall consider the effect that excipients can have on the stability of lyophilized active constituents by inducing crystallization of the active constituent during the lyophilization process.

Korey and Schwartz [7] examined the effects of excipients on a number of active constituents, such as atropine sulfate, sodium cefoxitin, cefazolin sodium, procainamide HCl and others. These compounds could be obtained in the crystalline state, but they formed amorphous cakes when lyophilized. These authors found that excipients like glycine, alanine, serine, methionine, urea, and niacinamide were effective in inducing crystallization; however, crystallization was not induced when using excipients like mannitol and lactose. The degree of crystallization was found to increase as the mole fraction of the excipient was increased. For example, when the mole fraction of glycine in the solution was 0.41, the lyophilized cake of cefazolin sodium was amorphous. As the mole fraction of glycine was increased to 0.70, 0.75, and 0.80, the subsequent crystallization of the cefazolin sodium increased to 20%, 75%, and 100%, respectively.

The authors [7] offer three possible mechanisms for enhancing the crystallization of an active constituent through the use of an appropriate excipient.

1.   The excipient and the active constituent form a complex molecule that crystallizes during the freezing process.

2.   The excipient crystallizes during the freezing process and serves as a seed crystal for the active constituent.

3.   The active constituent and excipient formed a eutectic during the freezing process.

Examination of the cakes using DSC and X-ray diffraction studies should provide much needed information regarding the nature of the crystallization mechanism.

While crystallization of the active constituent tends to increase its stability, in the case of proteins, crystallization of the excipient may lead to a loss in stability, through the formation of oligomers [33]. For an excipient like mannitol, which can undergo crystallization during the lyophilization process, the protein molecules are concentrated on the surfaces of the mannitol where polymerization can occur rather than in a dispersed state as it would be if the mannitol were in an amorphous state. However, the selection of an excipient that does not undergo crystallization during the lyophilization process, such as sucrose, does not always provide the same stabilizing environment for proteins tending to form oligomers. The formulation of oligomers can be limited or prevented by the addition of a surfactant to the formulation [33].

Cryoprotectants

Cryoprotectants are excipients whose primary function is to protect the active constituent during the freezing process. In this, cryoprotectants differ from lyoprotectants, whose function is to prevent degradation of active constituents during the lyophilization process. The development of cryoprotectants evolved from an interest in preserving living cells. For example, preservation of the spermatozoa of a prized bull would permit the production of offspring long after the bull's demise.

An interesting account of the historical development of cryoprotectants has been summarized by Smith [34]. In this review, it was Lad in 1949 who first proposed that for living cells to survive a freezing process, the water in the cell must remain in a vitreous state. Smith [34] then describes how, almost by accident, the addition of glycerol to the solution containing the organism enabled an organism to survive the rigors of freezing to $-79°C$ and subsequent thawing to ambient temperature. This could be accomplished if the organism was cooled at a rate of $-1°C/min$

to -15°C and then to -4°C/min to -20°C or -79°C. Under these conditions frozen bull spermatozoa was found to remain active for up to 4 years at -79°C, however, after 6 years there was some decrease in mobility of the spermatozoa.

The freezing of larger organisms such as corneal epithelial cells showed that the cells disintegrated on thawing. However, if the concentration of glycerol in the solution is increased to 15%, the cells remain intact. It was Lovelock [34] who showed that the protection mechanism of the glycerol was its ability to penetrate the cell and prevent the formation of ice crystals. The formation of ice crystals in the cell resulted in an increase in the NaCl concentration, which caused chemical damage to the lipoprotein in the cell membrane.

While sucrose could not enter the cell to prevent the formation of ice crystals, dimethylsulfoxide (DMSO), because of its relatively small size, could penetrate red blood cells and prevent the formation of ice crystals during freezing. After thawing of the solution, the DMSO could be removed from the cell more rapidly than glycerol, thereby reducing the risk of hæmolysis (i.e., swelling and possible rupture of the cell membrane). Other excipients such as glucose were shown to enter red blood cells and prevent hæmolysis [35] on thawing. However, red blood cells containing glucose were found to be hypertonic with respect to normal plasma and would undergo immediate hæmolysis when infused into the bloodstream. If rapid freezing can be employed, then excipients such as PVP, which also does not enter the cell, can provide cell protection during the freezing and thawing process [35].

Because an excipient offers cryoprotection to the organism during freezing and thawing does not always ensure that the organism will survive the lyophilization process. MacKenzie [36] showed that the slow freezing of *Escherichia coli* in a 10% dextran-10 solution had a high percentage (> 84%) of surviving the freezing and drying process. However, for slow freezing of a 10% mannitol solution, 40% of the bacteria survived the freezing process but only 1% survived the drying process. The survival of *Escherichia coli* during rapid freezing showed that the survival was not dependent on the excipients (sucrose, dextran-10, glycerol and Ficol), i.e., the survival on thawing in just distilled water was 99%. However, on lyophilization 81% of the bacteria survived in a 10% sucrose solution while the survival of the bacteria in a 10% dextran-10 solution was only 1%. The selection of an excipient for lyophilization of an organism is dependent on not only the nature of the organism but also the freezing rate.

Lyoprotectants

In their review of the use of sugars to stabilize phospholipid biolayers and proteins, Crowe et al. [37] stressed that sugars play a major role in maintaining the stability of proteins in solution. Without the use of sugars as excipients, the lyophilization process was found to be more detrimental to the stability of proteins than was a freeze-thaw process. For example, the freeze-thawing of the enzyme myosin ATPase solution resulted in a 50% reduction in enzyme activity, whereas, the enzyme activity after lyophilization and reconstitution of a similar formulation showed a 70% reduction. However, when sugars were included in the protein formulation, lyophilized formulations had greater stability. When the enzyme myosin ATPase solution contained 100 mM of sucrose, the reconstituted solution retained 92% of its original activity. This is a dramatic increase in potency from a loss of 70% without the sugars present.

In another example, Crowe et al. [37] showed that the dehydrated-sensitive enzyme phosphofructokinase formulation without a sugar became completely and irreversibly inactivated when the formulation was subject to a lyophilization process. Lyophilizing a phosphofructokinase formulation containing disaccharides such as trehalose, maltose, and sucrose resulted in a marked increase in the enzyme's stability. The use of monosaccharides, such as glucose and galactose, did not provide a stabilizing medium for phosphofructokinase enzyme. The lack of stability of the enzyme was also observed when 0.9 mM of Zn^{+2} was used in the formulation; however, when the monosaccharide sugar was combined with the Zn^{+2}, there was a dramatic enhancement in the stability of the enzyme. The stability of the enzyme was found to be dependent on concentration and to decrease when the sugar concentration exceeds 150 mM.

Timasheff [37] proposed that the mechanism for the stability of proteins in the presence of disaccharides is a preferential exclusion of the sugar from the protein surface. Such a model may be applicable to solutions, but it appears to be inadequate in explaining the stability of the protein in the absence of a solvent, as is the case for a lyophilized product. Just what role the cation Zn^{+2} plays in the stability the lyophilized protein is also not apparent.

Townsend and DeLuca [38] introduced the term *lyoprotectant* meaning an excipient that enhances stability by preventing the degradation of proteins during the lyophilization process. They explain that the mechanism of the lyoprotectant is to form around the protein molecules an environment that provides a source of water molecules to prevent denaturation because of a change in the configuration of the protein resulting from the loss of *bridging* water in the molecule. For their studies, Townsend and DeLuca selected the protein bovine pancreatic ribonuclease (RNase) because the protein had been previously studied and its enzymatic activity can be quantitatively determined by spectrometric means. A change in assay was, therefore, associated with a change in the configuration of the protein molecule.

The lyophilized RNase formulations prepared from solution having *p*H values that between 3.0 and 10.0 by the addition or absence of sodium phosphate buffers were found to be more stable at 45°C in just distilled water or in the solution of sodium phosphate having a *p*H of 6.4. The effect of the addition of the excipients sucrose, Ficol 70, and PVP showed that stability of RNase at 45°C, at *p*H values ranging from 3.0 to 10.0, increased dramatically as the concentration of the given excipient exceeded 1% (w/v). Sucrose was found to be most effective at *p*H values > 7. Ficol 70 provides protection for both low and high *p*H values, but PVP is a more effective lyoprotectant at lower *p*H values.

SOLVENTS

Selecting a solvent system for a formulation requires the same degree of diligence as that used for the other constituents in the formulation. Because the preponderance of the solvent will be removed during the lyophilization process, the nature and the role that the solvent plays in the formulation is often neglected. Since water is the predominate solvent system used in most formulations, this section will examine the nature of water and its solubility and solvation properties. Other solvent systems, either absolute or containing water, will also be considered. Readers wishing to obtain additional information on the role of solvents are encouraged to consider an excellent review article on the principles of solubility in the development of drug formulation by Sweetana and Akers [39]. This review considers both aqueous and co-solvent systems and examples of formulations with *p*H values < 4 and > 8.

Nature of Water

In order to understand the role that water plays as a solvent, one must first have some understanding of the nature of the water molecule. The structure of the water molecule (Figure 2.5) consists of two hydrogen atoms chemically bonded to an oxygen. The strong chemical bond (220 kcal/mole) between the (1*s*) electrons of the hydrogen atoms and two unfilled (2*p*) orbitals of the oxygen atom results in the formation of a very stable molecule.

The angle between the 2*p* orbitals of oxygen is 90°. However, as a result of the formation of the O–H bonds, repulsive forces are generated between the two hydrogen atoms. These repulsive forces result in an angle of 105° between the hydrogen atoms and cause the water molecule to have an unsymmetrical distribution of charges, which are located along four arms of a tetrahedron like that shown in Figure 2.6. This unsymmetrical distribution of charges allows water molecules to form hydrogen bonding [40]. The formation of these hydrogen bonds between water molecules leads to a high surface tension at 20°C (72.75 dynes/cm), a high heat of fusion at 0°C (79.7 cal/g) and a high boiling point (100°C); for a compound like ethanol, which does not form hydrogen bonds, the surface tension at 20°C, heat of fusion, and boiling point are 22.89 dynes/cm, 26.05 cal/g at –114.5°C and 78.5°C, respectively [18].

In addition, the unsymmetrical distribution of charges of the water molecule produces a permanent dipole moment. Because of the dipole moment, the dielectric constant for water at 25°C is 78.54, which is relatively high when compared to other

---

Figure 2.5. The Structure of a Water Molecule

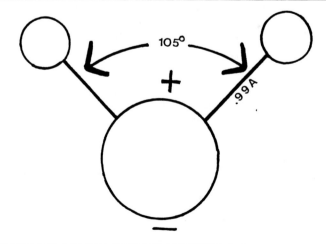

Figure 2.6. A Molecular Orbital for Water

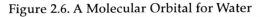

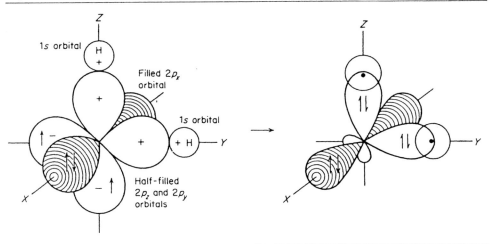

Formation of a molecular orbital for H_2O by overlap of $2p$ orbitals of oxygen and $1s$ orbitals of hydrogen. [Source: Moore, Walter J., *Physical Chemistry*, 3/e, 1962, p. 535. Reprinted by permission of Prentice Hall, Upper Saddle River, New Jersey.]

solvents like methanol (32.63) and ethanol (24.3)[18]. It is this high dielectric nature of water that permits it to interact with or dissolve hydrophilic molecules and molecules that contain polar groups like OH, CN, NO_2, and COOH, unless they are symmetrical in nature.

Solubility and Solvation Properties of Water

This section will consider the mechanism that makes water such an effective solvent. In an effort to maintain simplicity and brevity, the interaction of water with the dissolved species such as excipients, proteins, DNA, viruses, and cells will be described without resorting to rigorous thermodynamic derivations and discussions.

Solubility Mechanism

Salt crystals consist of an ordered structure that is composed of discrete ions. When water is present, the ions will dissolve in the water and form a solution as long as the energy required to break the ionic bonds of the crystal is less than the energy that is released when the ions become hydrated [2]. For NaCl, there will be an electrostatic interaction between the water molecules and the positive sodium ions (Na^+) and the negative chloride ion (Cl^-) like that illustrated in Figure 2.7. An examination of Figure 2.7 shows that electrostatic interaction with the Na^+ ion is associated with the negative region of the dipole moment of the water molecule and the electrostatic interaction with the Cl^- ion is associated with the positive region of the water molecule. As more ions are hydrated, the quantity of free water molecules decreases

Figure 2.7. Hydration

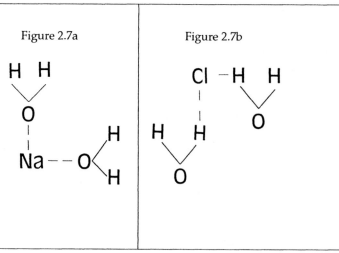

Figure 2.7a   Figure 2.7b

Figure 2.7a is an illustration of the hydration of Na^+ ion, while Figure 2.7b illustrates the hydration of the Cl^- ion.

until the energy released during the hydration of the ions approaches that required to break the ionic bonds of the crystal attains equilibrium or saturation. As water is removed from the isothermal system by evaporation, the loss of water results in a decrease in the heat of hydration and the formation of salt crystals; however, the concentration of NaCl in solution will remain constant.

Crystals consisting of molecules are bound by van der Waals type forces, such as dipole-dipole or dipole-polarization interaction between the molecules [17]. The energy associated with van der Waals type binding between molecules tends to be less than that found in crystals having ionic or covalent bonding. However, the solubility of these molecular crystals will be dependent on the van der Waals forces being of a lower energy than that of the hydrated molecule.

The energy associated with crystals formed by covalent bonding tends to be much greater than van der Waals or ionic bonds. Because of these higher binding energies, crystals formed with covalent bonds (e.g., diamond) have binding energy that far exceeds the energy for the hydrated carbon atom; consequently the diamond crystal is insoluble in water.

Solvation

Solvation is the process in which an ion species or polar molecule in an aqueous solution will form a layer of water molecules or a hydrated shell [40]. The orientation of this layer of water molecules will be dependent on the nature of the ionic charge or the unsymmetrical charge distribution of the molecule. As was illustrated in Figure 2.7a, the orientation of the hydrated shell is with the oxygen atom facing the ion. Because of the electrostatic field generated by the ionic charge, the hydrated shell may comprise several layers of water molecules. In the case of polar molecules, the orientation of the water molecules, as seen in Figure 2.6, will vary depending on the charge distribution of the molecule. Since the electrostatic fields generated by the dipole moment(s) of the molecule are considerably weaker than those created by the charge of an ion, the hydrated shell is limited to monolayer. The existence of these hydrated shells prevent the molecules from approaching one another or a similarly hydrated surface.

The use of a *surfactant* in a formulation serves to interact with protein molecules such that there is hydrogen bonding of the water molecules and, as a result, prevents the interaction between the protein molecules [40]. There may be occasions when it is desirable to remove the hydrated layer of a protein molecule by a process referred to as *salting out* [16]. In this case, the solubility of proteins can be reduced by the addition of an electrolyte. Because of the relatively large hydrated shell generated by the ions of the electrolyte, water is removed from the monolayer surrounding the protein, and the solubility of the protein is reduced.

The above discussion has shown that, because of its permanent dipole moment, water plays a major role in the formulation by providing an effective solvent. Perhaps the questions to ask when considering the use of an excipient in a formulation, are *What will be the nature of its solvation with water?* and *What effect will this hydrated molecule or ion have on the active constituent?*

Other Solvent Systems

Organic compounds, either with or without the presence of water, have been used as the solvent system for some active constituents. A list of organic solvents that have been used or examined in conjunction with a formulation is shown in Table 2.3 [3,41]. The following are reasons for selecting such solvent systems for a formulation.

Crystallization of the Active Constituent

When the sodium salt of [2,3-dichloro-4-(2-methylenebutyryl)phenox]acetic acid was dissolved in 95% ethanol and the solvent was removed by evaporation, a crystalline form of the compound was obtained [3]. It was previously shown that compounds in a crystalline state tend to be more stable. Organic compounds in which the compound is not solvent are added to an aqueous solution of the formulation.

Table 2.3. List of Organic Compounds Used as the Primary Solvent System or Added to an Aqueous Formulation [3,41]

Acetonitrile	Acetone	Dichloromethane
Dimethyl carbonate	Dioxane	Ethanol
Ethyl acetate	Isopropanol	Methanol
n-Butanol	n-Propanol	

By freezing the formulation, the water is removed from the solution in the form of ice, and the active constituent is crystallized out of the solution [3] as a result of the increase in the concentration of the organic compound. Sucrose normally has an amorphous form, but when dried from an aqueous solution containing ≥10% tertiary butyl alcohol is partial crystalline [41].

Stability of the Active Constituent

Nonaqueous solvents are sometimes found in lyophilized formulations. Their presence in the formulation may serve as a medium to reduce the rate of degradation of the active constituent in the presence of water [40,41]. If the active constituent were an insoluble crystalline organic free acid [41], NaOH would be added to increase solubility; however, the sodium salt may be unstable. Greater stability of the active constituent has been achieved by adding various concentrations of organic solvents such as 8% to 20% of the solvents listed in Table 2.3.

The solvent may, in the case of dichloromethane, represent residuals from the preparation of the active constituent [42]. In this case the solvent may play no role in stabilizing the active constituent but have an impact on the cosmetic properties of the final product. Solvents such as tertiary butyl alcohol have been added to the formulation in an effort to enhance the lyophilization process [43]. The use of multi solvent systems in the preparation of a formulation appears to be an area that can offer means for improving the quality of a product; however, the composition and concentration of the organic solvent will be dependent on the nature of the active constituent. It is my opinion that knowledge of the properties of the active constituent is the first step in determining the nature of the solvent system.

PHYSICAL CHARACTERISTICS

In the above discussions, the main emphasis was associated with the active constituent. A formulation is generally characterized by concentration and potency of the active constituent, but the lyophilization process is not just dependent on the nature of the active constituent but also on the combined effect of the entire formulation, which would include the excipients and other compounds (impurities) that may be present. In order for the lyophilization process to be reproducible, it is essential that the total formulation also be reproducible within a given set of limits. This section will consider various means by which the formulation can be characterized to ensure that its composition is within a given set of prescribed limits.

Colligative Properties

Colligative stems from the Latin term *colligatus*, which means *collected together* [16]. Therefore, the colligative properties of a formulation are dependent on the number of particles rather than on their composition. Knowledge of the colligative properties of a formulation would not only be a confirmation that the number of particles in the formulation fall within some given set of limits but would also provide confirmation that there has not been any change in the chemical composition as a result of the preparation of the formulation. For example, assume that a formulation contains 2% sucrose (2 g of sucrose/100 g of solution). Since sucrose has a molecular weight of 342.30 g, the number of sucrose molecules in 100 g of the formulation is 3.52×10^{21}. If during the preparation of the formulation, the pH of the solution were accidently momentarily decreased to a value of 4, acidic hydrolysis of the sucrose would form glucose and fructose and the number of particles would be increased to 7.04×10^{21}. In a worst-case scenario, the glucose may interact with the active constituent during the drying process and result in a loss in potency. However, by measuring a colligative property of the formulation, the change in the composition of the formulation could be determined prior to the lyophilization process. Assuming an aqueous solution, the colligative properties of solution that would be applicable to the formulation would be the depression of the freezing temperature of water and the osmotic pressure with respect to the solvent.

Freezing-Point Depression

The addition of a solute to a solvent will tend to lower the vapor pressure of the solvent by decreasing its chemical potential in the vapor phase (μ_S^{vapor}). Since at equilibrium $\mu_S^{vapor} = \mu_S^{soln}$, the chemical potential of the solvent is decreased. The decrease of μ_S^{soln} results in a depression of the freezing point (β) of the solution. The chemical potential determines the dependency of the free energy of the phase on its composition. A comprehensive discussion of the role of the chemical potential is considered to be outside the scope of this text. For a detailed discussion on the chemical potential the reader is referred to [16] or any other text on physical chemistry.

In the case of an aqueous 2% sucrose solution, the β will be 0.112°C. Acidic hydrolysis of the sucrose resulting in a 2% solution of fructose will give a β of 0.211°C, and a 2% glucose solution having a β of 0.214°C. The combined effect of the formation of fructose and glucose by the dissociation of sucrose is to decrease β for a 2% sucrose solution by an additional 0.313°C [18].

Osmolarity

Another colligative property of a formulation is the *osmolarity* of each constituent of the formulation. I would like to acknowledge that the following discussion of the subject of osmolarity of formulation is somewhat superficial; however, the main intent of its introduction is to acquaint the reader with its importance as a means for establishing the reproducibility of the formulation. Those wishing a more in-depth discussion of its principles and analytical methods should consult the ample review articles and texts on this subject that appear in the literature.

Osmolarity (O) is defined as the number of osmotically effective particles present in a solution. Osmolarity is generally expressed in terms of osmoles (Os) per liter or kg of water. In order to determine the osmolarity of a solution, one must first determine the *osmotic pressure* of the solution. The osmotic pressure is the force per unit area that is exerted on a semipermeable membrane resulting from $\beta\mu_S^{soln}$ generated by differences in concentrations across a semipermeable membrane. Since the osmotic pressure is dependent on particle concentration, pressures of the order of several hundred atmospheres have been reported at high particle concentrations [16]. The observed osmotic pressure is not only dependent on $\beta\mu_S^{soln}$ but also on the nature of the membrane.

Total Osmotic Pressure. For membranes that are permeable only to solvent molecules, the observed osmotic pressure (O) is a result of the total number of particles present. Because the depression of the freezing point and the osmolarity are both colligative properties that depend on the number of particles in the solution, the relationship between these properties is as follows:

$$O = \frac{\Delta}{K_f} \qquad (44)$$

where K_f is the molar freezing point depression constant [16]. The value of K_f for water is 1.86°C [18].

A comparison between the sensitivity of the depression of the freezing point and the osmotic pressure shows that the latter is far more sensitive to variations in particle concentration. Consider a change in the molarity of a sucrose solution from 0.1 to 0.2. The change in the depression of the freezing point is of the order of 0.220°C, while the change in the osmotic pressure at 20°C is 2.47 atm [16]. Thus, using high accuracy and precision pressure transducers, a measure of the osmotic pressure can determine relatively small changes in particle concentration in a formulation.

Colloid Osmotic Pressure. Colloid osmotic pressure (O_c) is obtained using a semipermeable membrane that is permeable to the solvent and low-molecular weight molecules. Such membranes have been prepared by treating cellophane with a 3% NaOH solution or cellulose nitrate with ammonium sulfate [20]. These membranes are particularity useful when used in determining the O_c in formulations consisting of a multi solvent system such as water and tertiary butyl alcohol, which is used to increase the solubility of proteins. Since the membrane will be permeable to the solvent system and any other lower-molecular weight species present, the observed O_c will be a result of large-molecular-weight species such as proteins or polysaccharides. In a dilute solution [16],

$$O_c \approx RTm' \qquad (45)$$

the O_c becomes directly proportional to the volume molarity m'. By plotting O_c as a function of m' and extrapolating to a very dilute solution, $m' \equiv c$, where c is the g-moles/L, the molecular weight of the a pure particle or the average molecular weight for a range of particle species can be determined.

Concentration Properties

Concentration properties of the formulation are dependent on the concentration of one or more of the constituents in the formulation, but their values are not a result of the colligative properties of the formulation. These properties include the pH of the solution; however, pH has already been discussed earlier in the chapter. The following are some relatively rudimentary optical measurements that can be used to further characterize the formulation.

Optical

The optical properties of a formulation are dependent on the composition and concentration of its constituents. However, such optical properties are not directly related to the colligative properties of the formulation. Although it was stated in the introduction of this chapter, it is worth repeating that the basic rationale for making these optical measurements is not to analyze the composition of the formulation but to establish that the observed optical properties fall within a range of values consistent with an acceptable formulation. By increasing the number of independent measurements of the characteristics of a formulation, one increases the confidence in the reproducibility of the formulation. While recognizing the value of spectrophotometry as an analytical method that can provide an abundance of information regarding the nature of the formulation, spectrophotometry will be excluded in favor of more simplified optical methods, such as the refractometry and polarimetry, that provide a single value rather than an interpretation of a spectrum.

Refractometry. When a beam of light passes from one medium to another there is a change in the speed of light. Because of the change in speed, there is also a change in the direction of the light beam, as illustrated by Figure 2.8. In this figure, the angle *i* between the incident light beam at the interface between the two mediums and the line *AB* which is drawn normal to the interface is referred to as the angle of incidence. The angle *r* between the line *AB* and the refracted beam is referred to as the angle of refraction. The *index of refraction* (*n*) is defined as

$$n = \frac{\sin i}{\sin r} \tag{46}$$

The immersion type of refractometer [20], as shown in Figure 2.9, is well suited for determining of the indices of refraction of aqueous solutions. The instrument consists mainly of a light source, mirror, prism, Amici compensator, objective, scale and ocular eyepiece. Using a white light source, the entire light beam that strikes the prism is refracted to give a light field. The Amici compensator will disperse all wavelengths of light except the sodium D line, consisting of two strong emission lines at 5.890 nm and 5.896 nm. The objective lens system focuses the D line light on a scale to form a sharp line of demarcation between light and dark fields. The index of refraction is determined directly from the scale. Since the D line light results in the formation of a line of demarcation between the light and dark fields, the refractive index is denoted as n_D. The index of refraction for an aqueous solution is generally reported relative to air for sodium yellow light [18]. The range of refractive indices for the immersion type refractometer is limited from 1.333 to 1.365. The more sophisticated Abbe refractometers can measure refractive indices over a wider range 1.3000 to 1.7000 with an accuracy of ± 0.0003.

Although the index of refraction is a useful measurement in verifying the consistency of a formulation, it must be used in conjunction with other measurements. Because the depression of the freezing point of water is dependent on the colligative properties of a solution, the acidic hydrolysis of a 2% sucrose solution resulted in an addition 0.313°C decrease in β. But n_D for 2% aqueous solutions of sucrose, glucose and fructose are 1.3359, 1.3358, and 1.3358, respectively. Since the index of refraction

Figure 2.8. Angle of Refraction

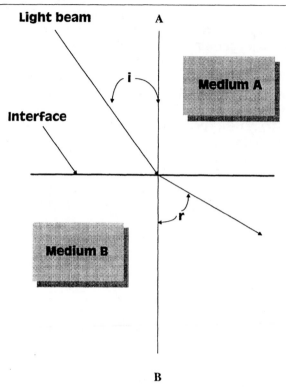

An illustration of the refraction of a light beam as it passes through the interface be-tween medium A and medium B, with respect to the line *AB* which is normal to the in-terface, where the angle *i* is the angle of incidence and the angle *r* is the angle of refraction. [Source: Daniels et al., *Experimental Physical Chemistry*, 5th ed. (1956), New York: McGraw-Hill Publishing Company . Reprinted with permission of the McGraw-Hill Companies.]

is dependent on concentration and group refractions and not on the colligative prop-erties of the solution, differentiating between a solution containing 2% sucrose and one comprising of 2% glucose and 2% fructose is difficult.

Polarimetry. *Polarimetry* is a very useful analytical technique when one or more of the constituents in the formulation can cause the rotation of a plane of polarized light that has passed through the solution. (An in depth explanation of the mecha-nism involved in the rotation of the polarized light as it passes through a solution containing an optically active constituent is beyond the scope of book.) For a given path length of the solution, solvent system and temperature, α, the sign and angle of the rotation of a substance is defined as [20]

$$\alpha = \frac{\alpha_\lambda^T lpd}{100}$$

(47)

Figure 2.9. Immersion Type of Refractometer

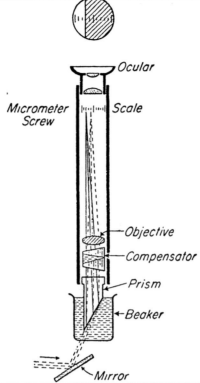

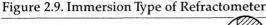

[F. Daniels, J. H. Mathews, J. W. Williams, P. Bender, and R. A. Alberty, *Experimental Physical Chemistry*, McGraw-Hill Book Company, Inc., New York, 1956, p. 17.]

where α_λ^T is the specific rotation of an optically active constituent at a given wavelength (λ) and temperature T, l is the length of the light path expressed in decimeters, p denotes the concentration of the solute in g of solute/100 g of solution, and d indicates the density of the solution, g/mL.

By measuring α, one can obtain the concentration of the optically active constituent in the formulation. The specific rotation α_λ^T for a substance is generally reported for sodium D line for a temperature range of 20°C to 25°C.

In the absence of any optically active substance, the polarimeter will produce a dark field when the instrument is set at 0°. In the presence of an optically active constituent, a light and dark region will appear in the viewing area. If the dark region appears on the right and the analyzer must be turned clockwise to restore the dark field, then the substance is said to be *dextrorotatory*; if the dark field is on the left and the instrument must be rotated counterclockwise, then the substance is said to be *levorotatory*.

Some constituents, like glucose, can be either dextrorotatory or levorotatory in the presence of a polarized sodium D line light source. For α-D-glucose, α_λ^T varies from 112 to 52.7 for concentrations varying from 1 g to 5 g/100 mL, however, one can also have α-L-glucose with α_λ^T values that vary from -95.5 to -51.4 for the same range of concentrations. In using polarimetry to confirm the consistency of a formulation, one would first prepare a plot of the sign and value of α for formulations containing a range of concentrations of the optically active constituent. From such a plot, one could then determine if the concentration of the optically active constituent is within the specified concentration limits of the formulation.

Particulate Matter

The infusion of particulate matter into human tissue can be injurious or cause toxicological conditions if the matter blocks the capillaries, which may be as small as 12 μm [44]. For example, it has been reported that individuals involved in drug abuse have suffered serious injuries or even death as a result of the injection of solids into their bodies [45].

In view of the above, the formulation should not only provide a stable environment for the active constituent but also ensure that the final formulation be free of any particulate matter, either from extrinsic or intrinsic sources [46] that could prove injurious to a patient. This section will consider visible particles and methods used for their detection; a discussion of subvisible particles, various means of their detection and potential sources of errors in the measurement; various sources from which visible and subvisible particles can arise; and means for eliminating particulate matter or preventing it from appearing in a formulation.

Visible Particles and Means for Their Detection

The 3rd supplement to the U. S. Pharmacopeia (USP), 19th edition, states that any injectable formulation that exhibits *visible foreign matter* be rejected. In other words, injectable formulations should be clear and free of any particle that can be seen by the unaided eye. Although visible particles are those particles that can be observed in a solution by the unaided eye, it is not uncommon that the examination be performed at a magnification of 2× and 3× [44]. In addition, visual inspection is performed by using a light source and examining the liquid against a contrasting white and black background.

It is obvious that the probability of detection of the presence of a particle in a solution will increase as the particle size increases and decrease as the particle size decreases; therefore, when considering the detection of a particle of a given size, one must express in terms of rejection probability (p). Rejection probability and can be divided into the following three ranges [45]:

$$0.0 \leq p < 0.3 \qquad \text{accept zone}$$

$$0.3 \leq p < 0.7 \qquad \text{gray zone}$$

$$0.7 \leq p \leq 1.0 \qquad \text{reject zone}$$

For a solution in the *accept zone*, the probability of rejection is quite low, and there is a high degree of confidence that the vial will not be rejected. In the *reject zone* there is a high probability that the vial will be rejected. The problem area is the *gray zone*. While there have been reports that particles of 50 μm have been visually detected [45], one must establish if such observations were in the gray zone or in the reject zone. It has been shown that with the aid of a magnifying lens of 3× and a diffused light source with white and black backgrounds, the rejection probability for a 65 μm particle was 0.7. When vials were examined without the aid of magnification, the rejection probability for vials having 100 μm particles was 0.59; and the probability increased to 0.82 when the particle diameter was 165 μm. False rejection of a vial can occur (e.g., mistaking air bubbles for particles) but such occurrences account for only 0.5% to 1.0% of rejected vials.

Visual inspection can also be performed by machines. When a machine is used, the containers are spun from 600 to 3500 rpm and then suddenly stopped. The solution continues to move, and the particles are detected from the shadows that they cast on a photo sensitive cell. Perhaps the chief advantage of machine inspection rather than a human visual inspection is that measuring the performance of the machine is easier. Yet in spite of the ability to measure the performance of the machine, the machine did not have a rejection probability of 1.0 when tested with vials having particles in the reject zone.

Subvisible Particles and Means for Their Detection

Subvisible particles are those particles in a formulation having dimensions that range from about 2 μm to > 25 μm or particulate matter whose dimension is not visible to the unaided eye. It is of interest to note that the USP has limits on the size and number per unit volume of subvisible particles for both large-volume parenterals (LVP) and small-volume parenterals (SVP), while the British and Japanese Pharmacopeias have limits on the size and number per unit volume of the subvisible particle for only LVPs [45].

There are currently five methods available for detecting subvisible particulate matter. The following will be a brief description and discussion of each of these methods.

Microscope. The microscope method is the official test method for U.S. and Japanese LVP solutions. The formulation is pumped through a test filter on which the particles are collected. The particles are counted with the aid of a binocular microscope having a 10× objective and 10× oculars with one of the oculars having a ruled reticle that allows the accurate measurement of 10 μm or 25 μm. For calibration, the microscope should be equipped with a calibrated micrometer that is traceable to National Institute of Standards and Technology (NIST) [44]. The size and number of particles is determined by first filtering the solution. The size of a given particle is defined by its longest linear dimensions [45]. The observed particles are then segmented into two groups (i.e., ≥ 10 μm and ≥ 25 μm).

Coulter Counter. The Coulter Counter® method is generally used in determining particulate matter in LVP. The measurement is made by pumping the solution, which must contain an electrolyte, through a region having an electric field defined by a set of electrodes. The presence of a particle will occupy a volume of the liquid and result in an increase in the resistance of the solution. Depending on the flow of the liquid, the magnitude of the change in resistance will be related to the size of the particle. The number of resistance changes of a given magnitude will be representative of the number of particles in a given segment.

Light Blockage. The light blockage method also requires the liquid formulation to pass a sensing device. In this method, the sensing device is a beam of light that is incident onto a photomultiplier tube. As a particle passes through the beam of light, there will be a reduction in the current output of the photomultiplier tube that will be proportional to the size of the particle. By determining the number and the magnitude of the decreases in current, the number of particles of a given size can be ascertained.

Light Scattering. In this method, the solution may be stationary or forced to pass through a beam of light. The beam of light is absorbed by a light trap. The presence of particles will cause light to be scattered at right angles to the solution. The intensity of the light that is scattered will be related to the size of the particle. From the number and intensity of pulses of light scattered, the number a particles of given range of particle sizes can be determined.

Hologram. In this technique for detecting subvisible particles in a solution, a laser beam is passed through a collimator/spatial filter assembly and then through a liquid cell. The cell contains the vial which is spun to cause movement of any particles in the solution. The beam that emerges from the cell is then projected onto a holographic plane. The hologram showing the particles is then projected onto a video screen [45].

Potential Sources of Measurement Errors. As with any analysis, there will be errors associated with the nature of the particles and errors associated with the measurement technique.

Configuration of the Particles. Ideally, all of the particles will have the same shape, preferably spherical in nature. For nearly spherical particles there will be relatively close agreement between the various measurement techniques. But particles can be present in a wide range of configurations from thin filaments to irregular flakes. With microscopic analysis, the particle is directly observed while the other techniques measure the particle's interaction with an electric field or a beam of light. Consequently, the configuration of a particle may result in an error in the measurement of the size of the particle. For example, a 25 μm long fiber observed by the microscope may be identified as a 9 μm particle by the light blockage method and as only a 6 μm particle when measured in an electric field [45].

Air Bubbles and Immiscible Liquid Droplets. The presence of air bubbles can result in false high particle counts with all measurement techniques other than microscopy. Immiscible droplets, while not true particles, still represent a foreign substance in the formulation. Such droplets can stem from either silicone or hydrocarbon-based oils and can be detected with optical and electrical measuring techniques. Unless the oil droplet does not wet the filter, such foreign substances cannot be observed with the optical microscope.

Handling and Storage. The handling of the solution can impact the observed number of particles. For example, shaking the solution can not only form bubbles, which some optical methods may erroneously include in the particle count; but it can also provide a false low count of the number of large particles with a concurrent false high count of the number of smaller particles. The false high particle count may be a result of the breakup of larger particles or of the dislodging of smaller particles that were adhered to the surface of the container.

The particulate count can also be affected by storage at a given temperature [45]. In some cases the count may decrease because of adherence of the particles to the wall of the container or show an initial increase that is followed by a decrease in the particle count. The constituents of the formulation, including the solvent system, are a potential source of extrinsic particular matter during storage. As discussed above, these constituents may be compounds that will serve a number of functions, such as bulking agents, *p*H modifiers, stabilizing agents, or electrolytes. Excipient-excipient interactions, excipient-active constituent interactions, or active constituent-active constituent interactions can produce particulate matter. The role of such interactions during the secondary drying process will be discussed further in Chapter 9. The key issue here is that the particle count should be performed under the same prescribed guidelines that will be used by those dispensing the formulation.

Various Sources of Extrinsic Particulate Matter in a Formulation

Extrinsic particulate matter in the formulation are foreign particles that enter the formulation from some external source during the formulation process. The following are a list of potential sources of extrinsic particulate matter.

Container. The glass container can be a source of particles in the formulation. The constituents in the formulation may interact with the container resulting in the formation of flakes. Formulations with high *p*H values may result in the formation of particulate matter by leaching of the glass. The glass may interact with one or more of the constituents to produce particulate matter. For example, vials treated with ammonium sulfate were found to produce a sulfate precipitate with GI147211, a water soluble analog of camptothecin [47]. Other particulate matter may stem from a lack of proper cleaning of the containers or from particulate materials in the environment during the filling operation. The presence of personnel in the filling area can represent a major source of particulate matter in the environment.

Closure. The closure can be a major source of contamination of lyophilized formulations. Pikal and Lang found that the source of haze in lyophilized formulations, paraffin wax and sulfur, was the elastomer closure [48]. Mass spectrum analysis of the outgassing from closures showed the presence of additional peaks having m/e values of 55, 57, 68, and 70 which were associated with the outgassing of closures at ambient temperatures. The outgassing was shown to produce a film on the surface of the container that prevented the adhesion of water to the glass surface or the formation of a meniscus [49]. The latter film was associated with the formation of a haze in a reconstituted lyophilized formulation. Particulate matter from the closure can also occur as a coring during the insertion of a needle. Lyophilized formulations sealed under a vacuum can cause the elastomer closures to be stressed and, thereby, enhance the likelihood of coring. Finally, a film of silicone oil is often applied to the closure to promote flow of the closures during the automatic filling operation and to assist in the seating of the closure during the stoppering operation. Silicone oil can creep down the sides of the container and possibly enter the formulation during the reconstitution process.

Formulation. The formulation itself can be a source of particulate matter, even after it has been passed through a 0.2 μm filter. A pH modifier such as sodium phosphate can form a precipitate $Na_2HPO_4 \cdot 12H_2O$ when the temperature of the formulation is lowered to 4°C [45]. Some active constituents require the presence of a surfactant to prevent aggregation of the active constituent.

Equipment. The freeze-drying equipment may be a source of particulate matter. Fine metal and carbide particles may be generated during the polishing of the dryer surfaces. Unless special precautions are taken to clean these surfaces, these fine particles may be trapped in such places as the stainless steel webbing that is used to protect the heat transfer lines to the shelves. These particles are difficult to remove and can become dislodged as a result of the turbulence generated during the evacuation of the dryer or during the back-filling of the chamber at the end of the drying process.

THERMAL PROPERTIES

The above discussion has been mainly focused on the considerations that go into the preparation of a combination of constituents that will provide an active constituent with a stable environment, in a liquid and in a dried state, for some finite length of time. The lyophilization process provides a means by which a formulation containing heat liable active constituents can be transferred from a stable liquid state to a dried state. The nature of the lyophilization process will, for the vast majority, be strongly dependent on the pseudo-colligative properties of the formulations. This section will briefly describe the function that the thermal properties play in the lyophilization process, the effect that the thermal properties can have on the cost of manufacturing and the impact that the formulation can have on the design, and construction of the freeze-dryer.

The Function of Thermal Properties

From the definition of the lyophilization (see Chapter 1) the first step is to freeze the formulation. During this freezing process, a matrix is formed as a result of the separation of the solvent and the solutes. For a typical aqueous solution, the water forms ice crystals and as the ice crystals grow, the concentration of the solutes is increased, and the interstitial solution becomes more viscous in nature until the mobility of the water (or transport of the water through the solution) approaches 0. As the mobility of the water approaches 0, the interstitial solution takes on glasslike properties.

After the freezing process is complete, the formulation matrix is slowly heated and the collapse temperature of the glassy interstitial region (i.e., that temperature where the mobility of the water in the interstitial region becomes significant) is determined. The heating process is continued to ascertain other thermal properties, such as the presence of any phase change in the interstitial region and the onset temperature of ice crystals of the matrix.

Although the thermal properties of the formulation will be examined in greater detail in Chapter 5, it is important to realize that the thermal properties will be dependent on the composition and concentration of the constituents that make up the formulation. In essence, the thermal properties of the formulation will not only govern the cosmetic properties of the final product but also the temperatures necessary to achieve a completely frozen state. It will be the collapse temperature that will determine the chamber pressure and shelf temperatures necessary to conduct the primary drying process.

Effect on the Cost of Manufacturing

Disregarding for the moment the cost of producing the active constituent, which in itself represents a major portion of the cost of manufacturing, selection of the other constituents that will make up the final formulation can also add considerably to the cost of manufacturing. The impact of the other constituents on the cost of manufacturing generally does not stem from their purchase price but from the impact that they have on the thermal properties of the formulation. For example, the low solubility of the active constituent may require a relatively large fill volume (≥ 25 mL). If such a large fill volume is combined with the formation of a glassy state in the interstitial region that has a low collapse temperature ($\leq -45°C$), the time required to complete the primary drying process could be quite excessive. By increasing the process time, one is also increasing the chances that some external event, such as a momentary loss in utilities (e.g., electric power, cooling water, or air) could impact the quality or potency of the final product. Productivity is related to process time, and the process time is related to the nature of the formulation.

Impact on the Freeze-Dryer

The formulation can certainly have a major impact on the demands related to the performance and cost of constructing and operating of the freeze-dryer. Lyophilization of a formulation with a large fill volume (≥ 25 mL) and a low collapse temperature ($\leq -45°C$) would require a freeze-dryer capable of providing shelf temperatures ($\leq -55°C$) in order to achieve a completely frozen matrix and maintaining shelf temperatures of $-40°C$ throughout the primary drying process. These features would not only add to the cost of the purchase of the dryer but, since the efficiency of the refrigeration system tends to decrease at lower temperatures, the cost of electrical power to operate the dryer under these conditions becomes significant.

The lower collapse temperature also demands lower chamber pressures during the primary drying process. For a collapse temperature of $-45°C$, the product temperature of the frozen matrix would have to be maintained at $\leq -50°C$ and require chamber pressures approaching 10 mTorr. Such a low pressure requirement not only demands a refrigeration system that has the capacity to maintain condenser temperatures $\leq -70°C$, but also requires special precautions to prevent the backstreaming [50,51] of hydrocarbon vapors from the vacuum pumping system from entering the dryer.

In order to increase the solubility of an active constituent, solvent systems containing solvents such as ethanol, tertiary butyl alcohol, and acetonitrile may be used in various concentrations. With significant quantities of ethanol present in the solvent system, shelf temperatures $\leq -120°C$ would be required to achieve a completely frozen state. Such requirements would place an increased demand on the refrigeration requirements for the shelves and the condenser.

The addition of tertiary butyl alcohol to increase the solubility of the active constituent has been reported not to lower significantly the collapse temperature [43] like that of ethanol. With substantially higher collapse temperatures, there will not be the additional demand on the refrigeration system. Typical condenser temperatures of $\leq -60°C$ would be sufficient to prevent water and tertiary butyl alcohol vapors from entering into the vacuum pumping system. However, environmental regulations may prevent disposal of the tertiary butyl alcohol solution into the public sewer system; so that disposing of the condensate from the condenser could be a waste disposal problem.

Acetonitrile can also be used to increase the solubility of some organic products; however, it has properties that make it especially difficult to use in a conventional freeze-dryer. First, acetonitrile is a highly toxic substance, and great care must be exercised in handling the condensate from the condenser. Second, special care must be taken to equip the dryer with seals that are impervious to the absorption of acetonitrile vapors.

The above has shown that one must look well beyond the first needs of a formulation (i.e., provide a stable environment for an active constituent) and consider the impact that it will have on the lyophilization process, which can have a direct impact on the cost of manufacturing and the design, construction, and performance of the freeze-dryer.

SYMBOLS

α angle of rotation (+ or –) of a polarized beam of light

α_λ^T specific rotation of an optically active compound at light wavelength (λ) and temperature T

C_b capacity of a buffer solution

c concentration expressed in molarity

c_a concentration of a weak acid

c_b concentration of a weak base

c_s concentration of a salt of a weak acid or weak base

d density of a solution (mass per unit volume)

β depression of the freezing point of a solution

i angle of incidence of a beam of light with a surface

k_f rate of a chemical reaction in the forward direction

k_r rate of a chemical reaction in the reverse direction

k_w equilibrium constant for water at a given temperature

K	equilibrium constant for a chemical reaction
K_a	dissociation constant for a weak acid in an aqueous solution at a given temperature
K_b	dissociation constant for a weak base in an aqueous solution at a given temperature
K_f	molar freezing point depression constant for a solvent
K_{hydr}	hydrolysis constant
l	length of a light path through a solution expressed in decimeters
m'	volume molarity
$\mu_S{}^{vapor}$	chemical potential of the vapor phase of a solvent
$\mu_S{}^{soln}$	chemical potential of the liquid phase of a solvent
M	molarity (i.e., moles of a compound per L)
n	refractive index of a system
n_D	refractive index using a D line light source
O	osmolarity of a solution expressed in osmoles (Os) per L or kg of water
O_c	colloid osmotic pressure
p	rejection probability
R	gas constant

REFERENCES

1. A. A. Hincal, D. F. Long, and A. S. Repta, *J. Parenter. Drug Assoc.*, 33: 107 (1979).

2. E. S. Gould, *Inorganic Reactions and Structure*, Henry Holt and Company, New York, 1955.

3. A. J. Phillips, R. J. Yardwood, and J. H. Collett, *Analytical Proceedings* 23: 394 (1986).

4. R. L. Alexander et al., U.S. Pat. 4,537,883.

5. J. B. Portnoff, M. W. Henley, and F. A. Resyaino, *J. Parenter. Sci. and Tech.*, 37: 180 (1983).

6. T. R. Kovalcik and J. K. Guillory, *J. Parenter. Sci. and Tech.*, 42: 165 (1988).

7. D. J. Korey and J. S. Schwartz, *J. Parenter. Sci. and Tech.*, 43: 80 (1989).

8. N. B. Finter, C. Burfoot, P. A. Young, and R. Ferris in International Symposium on freeze-drying of biological products, Washington, D.C. 1976. *Develop. Biol. Standard*, 36: S. Karger, Basel, 1977 p. 279.

9. M. Majoer, A. Herrmann, J. Hilfenhaus, E. Riechert, R. Mauler, and W. Hennessen, loc. cit., p. 285.

10. M. S. Hora, R. K. Rana, C. L. Wilcox, N. V. Katre, P. Hirtzer, S. N. Wolfe, and J. W. Thomson, in International Symposium on Freeze-Drying and Formulation, Bethesda, USA, 1990, *Develop. Biol. Standard*, 74: S. Karger, Basel, 1991, p. 295.

11. Y. Koyama, M. Kamat, R. J. De Anglelis, R. Srinivasan, and P. P. DeLuca, *J. Parenter. Sci. and Tech.*, 42: 47 (1988).

12. L. Gatlin and P. P. DeLuca, loc. cit., 34: 398 (1980).

13. K. P. Flora, A. Rahman, and J. C. Cradock, *J. Parenter. Drug Assoc.*, 34: 192 (1980).

14. Y. J. Wang and R. R. Kowal, loc. cit., p. 452.

15. T. A. Jennings, *IVD Technology*, Jan.–Feb. 1997, pp. 38–41.

16. W. J. Moore, *Physical Chemistry*, Prentice-Hall, Inc., Englewood Cliffs, N.J., 3rd ed., 1964.

17. I. M. Kolthoff and E. B. Sandell, *Textbook of Quantitative Inorganic Analysis*, the Macmillan Company, New York, 1945.

18. *Handbook of Chemistry and Physics* (R. C. Weast, ed.), 65th ed., CRC Press, Inc. Boca Raton, Florida , 1984.

19. G. L. Flynn, *J. Parenter. Drug Assoc.*, 34: 139 (1980).

20. F. Daniels, J. H. Mathews, J. W. Williams, P. Bender, and R. A. Alberty, *Experimental Physical Chemistry*, McGraw-Hill Book Company, Inc. New York, 1956.

21. C. Chiang, D. Fessles, K. Freeborn, R. Thirucote, and P. Tyle, *J. Pharm. Sci & Tech.*, 48: 24 (1994).

22. J. H. Beijnen, R. Van Gijn, and W. J. M. Underberg, *J. Parenter. Sci. & Tech.*, 44: 332 (1990).

23. I. Sugimoto, T. Ishihara, H. Habata, and H. Nakagawa, loc cit., 35: 88 (1981).

24. M. J. Pikal, K. Dellerman, and M. L. Roy, in International Symposium on Freeze-Drying and Formulation, Bethesda, USA, 1990, *Develop. Biol. Standard*, 74: S. Karger, Basel, 1991, p. 21.

25. L. Lachman, S. Weinstein, G. Hopkins, S. Slack, P. Eisman, and J. Cooper, *J. Pharm. Sci.*, 51: 224 (1962).

26. P. P. DeLuca, in International Symposium on freeze-drying of biological products, Washington, D.C. 1976. *Develop. Biol. Standard*, 36: S. Karger, Basel, 1977, p. 41.

27. N. A. Williams, Y. Lee, G. P. Poli, and T. A. Jennings, *J. Parenter. Sci. & Tech.*, 40: 135 (1986).

28. R. H. M. Hatley, in International Symposium on Freeze-Drying and Formulation, Bethesda, USA, 1990, *Develop. Biol. Standard*, 74: S. Karger, Basel, 1991, p. 105.

29. T. A. Jennings, *J. Parenter Drug Assoc.*, 34: 109 (1980).

30. K. Inazu and Z. K. Shiga, in International Symposium on Freeze-Drying and Formulation Bethesda, USA, 1990, *Develop. Biol. Standard*, 74: S. Karger, Basel, 1991, p. 307.

31. J. A. Bashir and K. E. Avis, *Bull. Parenter. Drug Assoc.*, 27: 68 (1973).

32. R. H. M. Hatley, in International Symposium on Freeze-Drying and Formulation, Bethesda, USA, 1990, *Develop. Biol. Standard*, 74: S. Karger, Basel, 1991, p. 6.

33. A. T. P. Skrabanja, A. L.J.de Meere, R A. de Ruiter, and P. J. M. van den Oetelaar, *J. Parenter. Sci. & Tech.*, 48: 311 (1994).

34. A. U. Smith, in *Aspects Théoriques et Indusriels de la Lyophilisation* (L. Rey, ed.), Herman, Paris, (1964), p. 257.

35. A. P. Rinfret, loc. cit., p. 297.

36. A. P. MacKenzie, in International Symposium on freeze-drying of biological products, Washington, D.C. 1976. *Develop. Biol. Standard*, 36: S. Karger, Basel, 1977, p. 263.

37. J. H. Crowe, L. M. Crowe, J. F. Carpenter, and C. A. Wistom, *Biochem J.* 242: 1 (1987).

38. M. W. Townsend and P. P. DeLuca, *J. Parenter. Sci. and Tech.*, 42: 190 (1988).

39. S. Sweetana and M. J. Akers, loc. cit., 50: 330 (1996).

40. J. Israelachvili and H. Wennerstrom, *Nature*, 379: 219 (1996).

41. H. Seager, C. B. Taskis, M. Syrop, and T. J. Lee, *J. Parenter. Sci. and Tech.*, 39: 161 (1985).

42. H. Seager, et al.; *Manufact. Chem. & Aersol News*, February 1979, p. 41.

43. P. P. DeLuca, M. S. Kamat, and Y. Koida, *Cong. Int. Technol. Pharm*, 5th, 1: 439 (1989).

44. J. M. Lanier, G. S. Oxborrow, and L. T. Kononen, *J. Parenter Drug Assoc.* 32: 145 (1978).

45. S. J. Borchert, A. Abe, D. S. Aldrich, L. E. Fox, J. E. Freeman, and R. D. White, loc. cit., 40: 212 (1986).

46. P. K. Gupta, E. Porembski, and N. A. Willaims, loc. cit., 48: 30 (1994).

47. W. Tong, J. Clark, M. L. Franklin, J. P. Jozjakowski, J. B. Lemmo, J. M. Sisco, and S. R. Whight, loc. cit., 50, 326 (1996).

48. M. J. Pikal and J. E. Lang, loc. cit., 32: 162 (1978).

49. K. S. Leebron and T. A. Jennings, loc. cit., 36: 100 (1982).

50. P. Larrat and D. Sierakowski, *Pharm Eng.*, Nov./Dec., 68 (1993).

51. H. Wycliffe, *J. Vac. Sci. Technol.* A5, 2608 (1987).

3

The Importance of Process Water

ROLE OF PROCESS WATER IN LYOPHILIZATION

Introduction

As illustrated in the previous chapter, water, as a universal solvent, plays a key role in the preparation of the formulation. In Chapter 2, the permanent dipole moment of the water molecule resulting from an unsymmetrical distribution of electrons and its tetrahedral configuration was described and shown to be a major factor in water's role as a solvent.

But the role of water in the lyophilization process extends well beyond just that of a solvent. From the present definition of the lyophilization process (see Chapter 1), the crystalline properties of water result in the separation of solutes in the interstitial region of the frozen matrix and eventually results in the formation of the dried cake. The phase diagram of water will also determine the chamber pressure during primary drying, and the heat of desorption will determine the residual moisture content and the stability of the dried product.

Purity

A survey of publications—not considered to be comprehensive—that describe the lyophilization of a formulation [1–27] revealed that 6 of the references [2,3,7,8,24,26] indicated distilled water was used as the solvent; water for injection (WFI)—prepared by five distillations—was cited as the solvent system in one reference [12]; USP grade WFI was used in the preparation of 3 formulations [6,21,25] and no specifications regarding the nature of the water used in the preparation were indicated in 14 lyophilized formulations [1,5,10,11,14–20,22,23,27]. Only one of the references [8] identified a property of the initial water source (i.e., the resistance of the water was 10 mV deionized water). Although not clearly stated by the authors [8], the unit mV is taken to represent Mv (1 3 10^6 v) and not mV (1 3 10^{-3} v), which would be an atypical electrical resistance for deionized water. In spite of the limited number of references examined, the results of the above survey clearly show that little attention is given to the nature of the water used in the formulation except to indicate if it was distilled, WFI or USP WFI. I intend to show in this and succeeding chapters that characterization of the properties of the water used in preparing the formulation is an important consideration in producing a consistent product.

When one thinks of purity, the concept of a substance without any foreign substances comes to mind; however, in terms of USP specifications for WFI, purity is determined as a matter of limits. Under the United States Pharmacopeia II, WFI must satisfy six basic criteria: total solids; chemical tests for calcium, sulfate, chloride, ammonia, and carbon dioxide; conductivity; *p*H; organic compounds; and the microbial and endotoxin levels.

Total Solids

The maximum *total dissolved solids* (TDS) permissible in WFI is 10 mg/L and is determined by the residual weight after the evaporation of a known quantity of water. This method does not distinguish between the presence of dissolved and undissolved substances. The presence of organic compounds with relatively high vapor pressures can also be determined by this test method; however, the method cannot distinguish between residual inorganic and organic substances.

Inorganic Compounds

USP XXII also requires that certain inorganic compounds not exceed defined limits (i.e., sulfate, 15 ppm; calcium 22 ppm; carbon dioxide 20 ppm; chloride 0.07 ppm; and total heavy metals, 0.06 ppm).

Conductivity

The *conductivity* of water is determined by a measure of electric current between two parallel plates whose voltage (DV) is maintained by an alternating power supply. The purity of the water is based on the premise that a reduction in the ion conductivity (i.e., lower ion concentration), will be an indication of the purity of the water. The conductivity (k) is defined as

$$\kappa = \frac{1}{\rho} = \frac{d}{R_o A} \qquad \frac{1}{\Omega\,\text{cm}} \tag{1}$$

where r is the volume resistivity of the water, A is the cross-sectional area of the plates, d is distance between the electrode plates, and R_o is the resistance of the cell. From Ohm's law, k can be expressed in terms of the applied voltage (DV) and current (I) with I expressed in amperes (A) as

$$\kappa = \frac{dI}{A\Delta V} \qquad \frac{1}{\Omega\,\text{cm}} \tag{2}$$

where the ratio d/A is referred to as the cell constant.

As purity of water increases, r will be $\geq 20 \times 10^6$ V cm (20 Mv) and k will approach 0 (i.e., $\ll 0.1$). The sensitivity of the measurement of k is increased by increasing the current I and the cell constant (d/A). The cell constant must be increased by maximizing d while decreasing the value of A, and the current I is increased by increasing DV. For a cell constant of 0.01 per cm [28] and $k = 5 \ 3 \ 10^{-8}$ per (V cm) or $r = 20$ MV, the value of DV, when the I is 10 mA, and the applied voltage will be 2.0 V.

The mobility of the ions in an aqueous solution and the ionization of water increases with temperature. Since USP XXII requires WFI to be stored and recirculated at temperatures higher than 80°C, there will not be a large differential in k between 20 MV and 4 MV water [28]. Although the conductivity measurements are made in-line while there is flow of the hot WFI, when the above measurements are compensated for a temperature of 25°C, there will be a significant differential (i.e., $k_{20 \ MV}$ = 0.05 per (V cm) while $k_{4 \ MV}$ = 0.2 per (V cm). As a result, the quality of the water must be evaluated at near ambient temperatures rather than a higher storage temperature.

pH Measurements

The importance of determining the *p*H is to ascertain the presence of absorbed carbon dioxide (CO_2). Ideally, pure water at a temperature of 25°C, will have a *p*H of 7, but the absorption of CO_2 will result in a decrease in *p*H. For example, 0.1 mg of CO_2 in 1 L of water will result in a decrease in *p*H to 6, and 1.0 mg of CO_2 in 1 L will lower the *p*H of the water to 5.5. One must resist the temptation to increase the *p*H by the addition of a base such as NaOH. The addition of the base will increase the *p*H but will also add a sodium carbonate ($NaHCO_3$ or Na_2CO_3) to the formulation.

Organic Compounds

Because of the volatility, the presence of organic compounds in the water cannot be determined from the TDS test. Also, the presence of organic compounds cannot be determined by measuring the conductivity of the water.

The presence of organic compounds in the water is determined by the *Pemanganate Oxidation Method* [28,29]. The method is qualitative and consists of adding 2.0 N sulfuric acid (H_2SO_4) to 100 mL of water and boiling the water of 10 min. Then 0.1 mL of 0.1 N potassium permanganate ($KMnO_4$) solution is added to the water to cause the water to take on a pink color. $KMnO_4$ is a relatively strong oxidizing agent and will oxidize organic compounds. As a result of the oxidation of the organic compounds, the permanganate is reduced to a manganous ion that is colorless. A change in the color of the water after the addition of $KMnO_4$ will indicate the presence of organic compounds but will not identify the nature or the concentration of the organic compounds.

Commercial organic analyzers have been developed to determine the total organic carbon (TOC) in the water. These instruments use various means to *oxidize* the organic compound into CO_2 and then measure the quantity of CO_2 present. While these methods give a more quantitative measurement of the total carbon present in the water, they cannot distinguish whether the CO_2 was produced from methane or from ethane dissolved in the water.

Crane et al. [29] did a comparison study of the sensitivity of the Permanganate Oxidation Method with that of TOC analyzers. These authors compared the two methods with various aqueous solutions containing various concentrations of potassium hydrogen phthalate (KHP). The reader should note that the unit mgC/L refers to the concentration of KHP, expressed in mg/L, not to the actual concentration of a given organic compound present in the water.

Crane et al. [29] showed that the Permanganate Oxidation Method did not indicate the presence of an organic compound like methanol until the KHP reached a concentration of 2,500 mgC/L, while the TOC analyzer detected 10 mgC/L of KHP when 12.5 mgC/L was added to WFI. The Permanganate Oxidation Method was also shown to fail to detect the presence of < 5000 mgC/L of acetonitrile. It is clear from these results that the qualitative Permanganate Oxidation Method lacks the sensitivity to detect the presence of substantial concentrations of some organic compounds in WFI.

Microbial and Endotoxin Levels

USP XXII also specifies certain *microbial* and *endotoxin* levels for WFI. Collentro [28] argues that since the WFI is stored and recirculated at temperatures higher than 80°C, it is not difficult for pharmaceutical manufacturers to meet these requirements. He cautions, however, that the reduction of organic compounds in the water could produce such a low-nutrient system that certain forms of bacteria could be stressed to transform into a spore state that has a dimension < 0.2 mm and, possibly, pass through a 0.2 mm filter. The incubation period for bacteria is typically 48 hr, but in the spore state it may be as long as 10 days [28].

Collentro [28] offers no lower limits for organic compounds in the WFI and appears not to take into account that there is a limit to how long the WFI can be stored. In addition, the water is not passed through the sterile filter until after the compounding of the formulation is complete. One would expect that the formulation would provide the necessary nutrient system to prevent the bacteria from reverting to a spore state. The presence of organic compounds in WFI will also present other problems that can impact the lyophilization process; these will be addressed in a later chapter.

HYDROGEN BONDING

In order to understand the physical properties of water, one must understand the nature of the *hydrogen bond*. In the previous chapter, hydrogen bonding was introduced as being a result of an unsymmetrical distribution of electron charge of the water molecule. But such a definition can hardly suffice when it comes to explaining the structure of ice or why the density of water goes through a maximum value at about 4°C. Because of the key role that water plays in the lyophilization process, this section will examine more closely the nature of the hydrogen bond and the effect that excipients or impurities can have on its properties.

Formation

In order to understand the formation of the hydrogen bond, one must return to the tetrahedron distribution of charges for the water molecule as illustrated by Figure

2.6. The two positive regions in Figure 2.6 result from two protons of the hydrogen atoms, and the two negative regions stem from unfilled $3p$ orbits of the oxygen atom. The high charge density of the hydrogen protons interacting with electronegative elements like N, O, and F leads to the formation of what is referred to as a hydrogen bond. The presence of a hydrogen bond in water is represented by the structure H-O...H where the H-O represents the covalent bond of the water molecule, which has a high bonding energy of about 120 kcal/mole, and the O...H represents the hydrogen bond and has a lower dissociation energy of about 5 kcal/mole [30]. Although the energy of the hydrogen bond is greater than Van der Waals intermolecular attractive forces, our comprehension of the complex nature of hydrogen bonding must employ sophisticated infrared and Raman spectroscopic techniques [31] and inelastic incoherent neutron scattering (IINS) measurements [32]. From such measurements, a greater understanding of the nature of the hydrogen bonds is emerging. For example, Li and Ross [32] have provided evidence, using the IINS measuring technique, that there are at least two different kinds of hydrogen bonding with different bond energies. Their neutron spectra suggests the presence of *strong* and *weak* hydrogen bonds in ice I and that the strong bonds outnumber the weaker bonds by a ratio of about 2:1. Since there is no reason to believe that strong and weak bonds do not exist in liquid water, Li and Ross [32] offer an explanation for the relatively high heat capacity of water. With an increase in the temperature of the water, the number of weak bonds increases while the number of strong bonds decreases. The conversion of a strong bond to a weak bond requires energy that would account for the high heat capacity of water.

Another anomaly of water in the liquid state is that as water is cooled, a maximum in density is attained at 4°C. As the temperature is further lowered, the water forms what has been referred to as *ice-like water clusters* [33] and the density of the water decreases. If one assumes the dual hydrogen bond model proposed by Li and Ross [32], at 4°C the number of weak bonds present reaches a maximum, and the formation of the ice-like water clusters is a result of a increase in the number of strong bonds. In order for a weak-strong bond conversion to take place, there must be release of energy. Such a release of energy with an onset temperature of about 4°C will be shown in Chapter 5.

Impact of Excipients or Impurities

The addition of excipients or the presence of impurities, such as gases, can affect the solvent structure or the formation of weak or strong hydrogen bonds. As the temperature of the water is increased, the presence of the foreign substances can have less effect on the solvent structure, although, even at 300°C, which is approaching the critical temperature of 374.1 at a pressure of 218.3 atm [34], the orientation between water molecules is not random [35]. By affecting the structure of water, excipients and impurities can impact the thermal properties of the water, such as the degree of supercooling and the degree of crystallization. The former will be considered in the following section, and the latter will be a topic in Chapter 5. The reader should be aware that water as a solvent interacts with materials, and this interaction is responsible for the changes in the properties of the water.

DEGREE OF SUPERCOOLING

The *degree of supercooling* can be defined as the temperature differential between the temperature at which nucleation ice crystals occurs [36]. From the phase diagram of water shown in Figure 3.1 [30], the triple point A, i.e., that temperature when all three phases (gas, liquid and solid [ice]) are present, is 0°C. The dotted line A–E represents the vapor pressure above liquid water at temperatures < 0°C. When nucleation occurs and there is a formation of hydrogen bonds, the temperature of the water-ice system is increased such that it approaches the triple point A. The following discusses the nucleation process, the effect that the degree of supercooling can have on the cake structure, and the impact that the degree of supercooling of water can have on the drying process.

Heterogeneous Nucleation

Heterogeneous nucleation is the nucleation of ice crystals on foreign substance. Such foreign surfaces may be the surface of the container, particulate matter present in the water, or even sites on large molecules such as proteins. The mechanism for heterogeneous nucleation is not completely understood. One proposed mechanism [37] is that there are adsorbed layers of water on the surface or on the particle. Because the

Figure 3.1. The Phase Diagram for Water (not drawn to scale) [30]

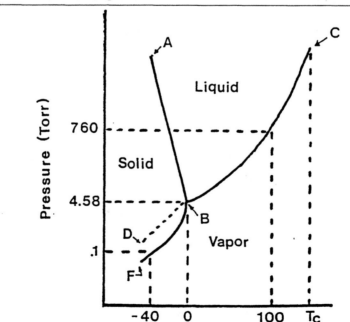

temperature of the water is below the triple point, these layers have structures with ice-like water clusters. These ice-like cluster layers may break away, perhaps as a result of some mechanical means, and, by retaining their structure, serve as nuclei for the formation of ice crystals.

Homogeneous Nucleation

Meryman [37] defines the *homogeneous nucleation* of ice as the development of a critical sized nucleus through the random aggregation of molecules. In order for nucleation to occur, the density (n) of aggregates or ice-like water clusters having a radius (r_m) must equal or exceed some critical density (n_j). It is reasonable to assume that there will be a frequency distribution of ice-like water clusters that will have a mean radius of $\bar{r}_m$. The equilibrium number of ice-like water clusters having a $\bar{r}_m$ at a temperature of T_i is proportional to [38]

$$\Delta F(T_i, \bar{r}_m)(kT)$$

where DF is the free energy of formation of the ice-like water cluster, k is Boltzmann's constant, and T is the temperature in K (Kelvin).

When $DF = 0$, then equilibrium will exist, and there will be no change in the distribution of ice-like water clusters. However, if DF is negative, then there will be a spontaneous increase in number of radii having a value of $\bar{r}_m$, and the result will be the nucleation of ice crystals.

Stephenson [38] defines $DF(T_i, r_m)$ as

$$\Delta F(T_i, r_m) = 4\pi r_m^2 \sigma - \frac{4}{3}\pi r_m^3 \Delta F' \tag{3}$$

where $DF9$ is defined as

$$\Delta F' = \frac{\Delta H_f S}{T_i} \tag{4}$$

where the molar DH_f is the heat fusion for ice and S is the degree of supercooling. The s term in equation 3 is surface free energy of the ice-like water cluster (units in energy cm) and will be dependent on the r_m. By differentiating DF with respect to r_m, one obtains the following expression:

$$\frac{d[\Delta F(T_i, r_m)]}{dr_m} = 8\pi r_m \sigma - 4\pi r_m^2 \Delta F' \tag{5}$$

for $DF = 0$, then r_m is defined as

$$r_m = \frac{2\sigma T_i}{\Delta H_f S} \tag{6}$$

Figure 3.2 illustrates a plot, determined from equation 6, of the ratio r_m/s as a function of the nucleation temperature (T_i). From this figure, one can obtain an insight about the mechanism for both heterogeneous and homogeneous nucleation. For temperatures higher than -15°C, r_m/s has relatively high values. These high values would be a result of water clusters having relatively large values of r_m and low surface energies (s). This plot would lead one to conclude that the formation of large

numbers of ice-like water clusters during steam distillation would be highly unlikely. Therefore, when nucleation occurs at temperatures > –15°C, the nucleation
sites must stem from hydrated particulate matter present in the water [33]. The presence of such hydrated particles would represent a site for the heterogenous nucleation of ice crystals.

   Homogeneous nucleation will occur at temperatures < –15°C, when the value
of the ratio r_m / s, as shown in Figure 3.2, is not strongly dependent on temperature.
At these lower temperatures, there will be relatively little change in the distribution
of the radii of the water clusters; therefore, the nucleation process will be governed
by the surface free energy (s) of the cluster. From the above model, the nucleation of
ice crystals at temperatures ≤ –45°C would require clusters to have smaller r_m and
higher s than those nucleating clusters at –21°C. The lower limit for nucleation of ice
crystals has been reported to be –46°C; when water is cooled very rapidly (i.e., about

Figure 3.2. A Plot of r_m / s as a Function of Temperature

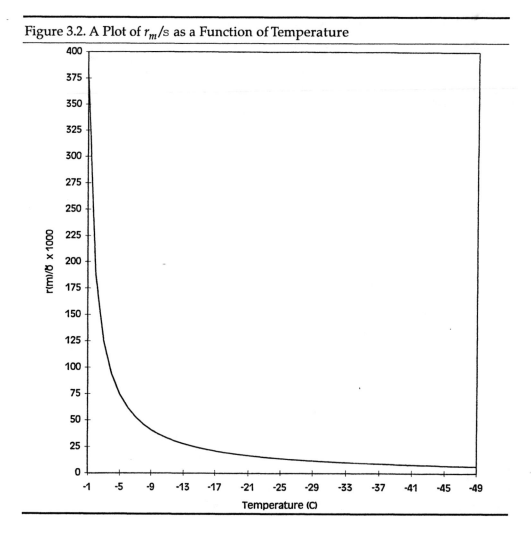

1 million °/sec [39]), nucleation did not occur even at –196°C or liquid nitrogen temperatures. The absence of nucleation is attributed not to the rapid cooling but to the formation of a second phase of liquid water. This example illustrates that our knowledge of the properties of water is still incomplete.

Rate of Ice Growth

The rate of ice crystal growth under supercooled conditions will be dependent on the magnitude or the degree of supercooling (i.e., the rate of ice growth increases with an increase in the degree of supercooling) [35,36,40]. Hallett [36] reported that the velocity crystallization of ice for supercooled water at –20°C was 23.7 cm/sec and the rate of ice crystal growth (U_a) as a function of the degree of supercooling can be expressed as

$$U_a = 0.08(T_o - T)^{1.9} \text{ cm/sec} \tag{7}$$

where T_o is the temperature at the triple point and T is the number of degrees below the triple point where nucleation occurs in the supercooled water.

Figure 3.3 is a plot of the data for the growth rate of ice as a function of nucleation temperature for supercooled water as reported in a log-log plot by Lindenmeyer et al. [40]. From this plot, the function, determined by the best fit for the rate of crystal growth as a function of the degree of supercooling $(T_o - T)$ was determined to be as follows:

$$U_a = 0.021(T_o - T)^{2.4} \text{ cm/sec} \tag{8}$$

and for supercooling of 20°C, the growth rate is determined to be 27.8 cm/sec. This value for ice growth rate at –20°C is within 11 percent of the rate of ice growth determined by Hallett [36] and calculated from equation 7. Although there is reasonable agreement between the calculated values of the rate of ice growth at –20°C and values obtained from equations 7 and 8, the differences are significant enough to indicate the experimental difficulties involved in making such measurements.

It has been my experience that most WFI in glass vials will have a degree of supercooling that is within the range of 10°C ± 3°C. From equation 8, the rate of ice crystal growth would be about 5.2 cm/sec or in relative terms 0.1 mph. Although such a rate would hardly be impressive when compared to our everyday movement, for a 1 mL fill volume, the ice structure would be distributed through the solution in 1/5 of a sec or the blink of an eye. For larger fill volumes, such as a 50 mL fill volume, there would be the problem of dissipating the latent heat released during the formation of the ice. Failure to dissipate the energy will cause the temperature of the solution to increase and further slow the rate of ice growth.

Effect on Cake Structure

The cake structure is directly related to ice formation. Consequently, the degree of supercooling will be a factor in determining the cosmetic nature of the product cake. This section will consider the effect that the degree of supercooling can have on the

Figure 3.3. Ice Growth Rate

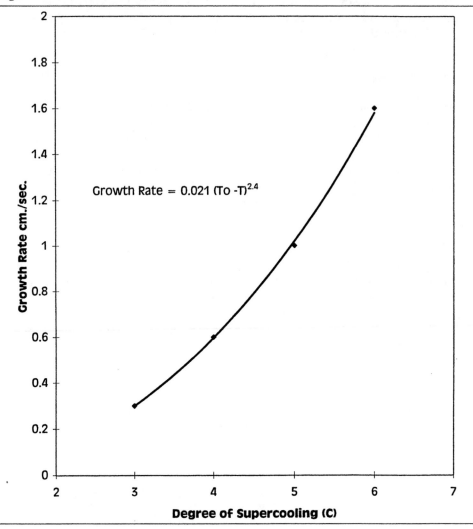

A plot of the ice growth rate as a function of the degree of supercooling for data obtained from Lindenmeyer et al. [40], indicated by the ¡ and solid line represented by the best fit shown by equation 8.

porosity of the cake and the impact that excipients of the formulation can have on the degree of supercooling.

Cake Structure Without Excipient Interaction

When ice forms in a supercooled solution and the excipients do not affect the properties of the water, the ice crystals take on a dendritic form, like that illustrated by

Figure 3.4. As can be seen from this figure, supercooling produces a tree-like ice structure in the water. The dendritic structure will continue to grow until the latent heat released during the formation of the ice causes the temperature of the system to approach the equilibrium freezing temperature or, in the case of pure water, the triple point, which is at 0°C. It has been observed that, as the degree of supercooling increased (i.e., nucleation occurs at temperatures significantly lower than the equilibrium freezing temperature), the crystal size and the angle between the branches (θ), as shown in Figure 3.4, will tend to decrease, and the rate of ice growth will increase. For example, at a degree of supercooling of only 0.2°C, θ has a value of 120° [36]. At higher degrees of supercooling, θ will decrease and the ice will assume the appearance of a fine powder.

For water having a low degree of supercooling, a coarse ice structure will occur, as shown by Figure 3.5. Even under slow freezing conditions that are controlled by a freeze-dryer (i.e., the shelf temperature is decreased at a rate of –1°C/min), only a small portion of the total volume of the water will undergo supercooling, while the remaining liquid will freeze near the triple point and form a solid ice structure. The result of such a freezing process is a heterogenous ice structure (Figure 3.5).

For water with a high degree of supercooling (i.e., ≥15°C), a relatively uniform ice structure will occur, as illustrated by Figure 3.6. The uniform ice structure occurs

---

Figure 3.4. Dendritic Ice Crystals

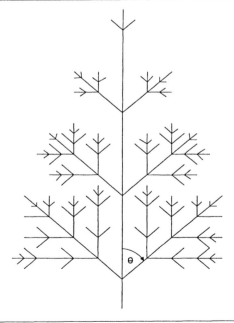

Dendritic ice crystals form as a result of nucleation in a supercooled solution with an angle θ between branches.

Figure 3.5. Ice Structure of Water with a Low Degree of Supercooling

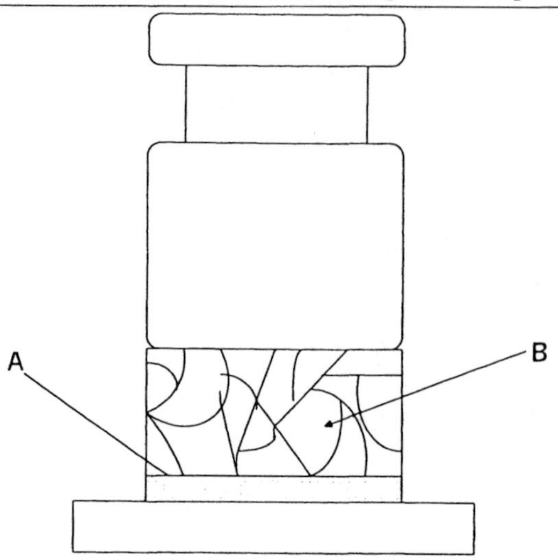

The freezing of water having a low degree of supercooling where region (A) represents the segment of the ice formed under supercooling condition and (B), that portion of the water that is frozen near the triple point.

because the entire volume of the water was below the triple point when ice nucleation occurred. The dense, fine structure of the ice is a result of the small angle θ between the branches, as illustrated by Figure 3.4.

Cake Structure with Excipient Interaction

The presence of excipients in a formulation can affect the degree of supercooling of the water component [36,37]. Since it has already been shown that a change in the degree of supercooling for a given volume of solution can affect the nature of the resulting cake, the change in the degree of supercooling of the water by the excipients can also impact the final cake structure. The effect of the excipients on the degree of supercooling appears to be a double-edged sword. Solutions having low concentrations of excipients can cause a slight increase in the degree of supercooling with respect to the degree of supercooling of pure water [36]. In the simultaneous measurement of the degree of supercooling of both pure water and the formulations performed in my laboratory (the actual number of thermal analyses of distinct formulations examined is not known for certain but is believed to be approaching 2,000), in a relatively large number of cases, there was only a small difference between the degree of supercooling of pure water and that observed in the formulation. When a small concentration of excipient was present in formulation, only in one formulation was there a significant increase in the degree of supercooling of at least 5°C with respect to that of the process water. The real lesson to be learned here is not so much that some excipients at given concentrations can increase the degree of supercooling but that this important thermal property of the water is best measured and not predicted at this time.

Hallett [36] indicates that for significantly larger concentrations of excipients, there will tend to be a decrease in the degree of supercooling of water in the formulation. For a molar solution, the degree of supercooling is decreased by a factor of 10. Yet the addition of higher concentrations of an excipient such as an alkali halide [36] will result in an increase in the degree of supercooling and an associated decrease in the angle θ between the dendrites or branches (see Figure 3.4). When supercooling occurs in viscous solutions from concentrated solutions of excipients, such as sucrose, the ice crystals will tend to grow in solid hexagonal prisms rather than the typical dendritic structure [36].

Hallett [36] has also found that the presence of ions like F^- and SO_4^{-2} can fit into the ice lattice. The result of the occlusion of such ions in the ice structure is a substantial increase in the degree of supercooling. The mechanism for increasing the degree of supercooling by the presence of these ions is not understood at this time. Because supercooling plays such an important role in the lyophilization process, there is a need for more basic research in the effects that excipients have on the supercooling of the water component of a formulation.

STRUCTURE OF ICE

Water is important to the lyophilization process, not only because of its universal solvent properties but also because of the unique structure of ice. These two properties make water an ideal choice for a component in a formulation. In this section, we

Figure 3.6. The Ice Structure Resulting from a Relatively High Degree of Supercooling (i.e., ≤ 15°C)

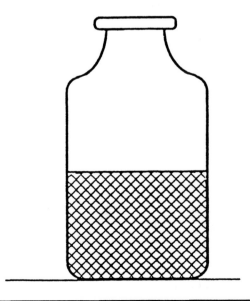

will first consider the *polymorphic forms* of ice. In addition, the structure of ice that occurs during the freezing process and the important role that the structure plays in the lyophilization process will be described. The section will conclude with a discussion of the effects that excipients and impurities can have on the physical properties of ice.

Polymorphic Forms of Ice

The physical properties of the 10 polymorphic forms of ice are listed in Table 3.1 [33]. This table shows that only two forms of ice (i.e., I_h and I_c) have densities that are < 1 . I_c can only exist at atmospheric pressures if the temperature is < -100°C; however, it will slowly transform to I_h at temperatures ranging from -80°C to -100°C. Because of the nature of the ice crystal structure, the remaining forms of ice (II to IX) have densities that are > 1; these ice crystals would not float if placed in a container of pure water. Transforming ice types I_h and I_c into a higher form (II to IX) requires a change in the crystal structure. To achieve such a change in crystal structure requires the use of high pressure. Such high pressures can be exerted naturally as a result of the weight exerted on ice that is located at the base of a glacier or artificially generated in a laboratory. Figure 3.7 illustrates the three-dimensional phase diagram for the various forms of ice [30]. One interesting feature shown in this figure is that at pressure < 5,000 atm, or a force of 73,500 psi, and temperature > 0°C, only water can exist. If pressure is increased to 20,000 atm, a system of ice VI and water exists for temperatures ranging from 20°C and 81.6°C. In addition, I_h and I_c can exist only at pressures < 5,000 atm. Since lyophilization is conducted at pressures < 1 atm and temperatures > -80°C, the crystal structure of ice will be limited to I_h.

Table 3.1. Properties of the Polymorphic Forms of Ice [33]

Type	Density (g/cm^3)	Crystal Structure
I_h	0.92	hexagonal
I_c	0.92	cubic
II	1.17	rhombohedral
III	1.14	tetragonal
IV	1.28	—
V	1.23	monoclinic
VI	1.31	tetragonal
VII	1.5	cubic
VIII	1.5	cubic
IX	1.14	tetragonal

Figure 3.7. A Water System at High Pressure

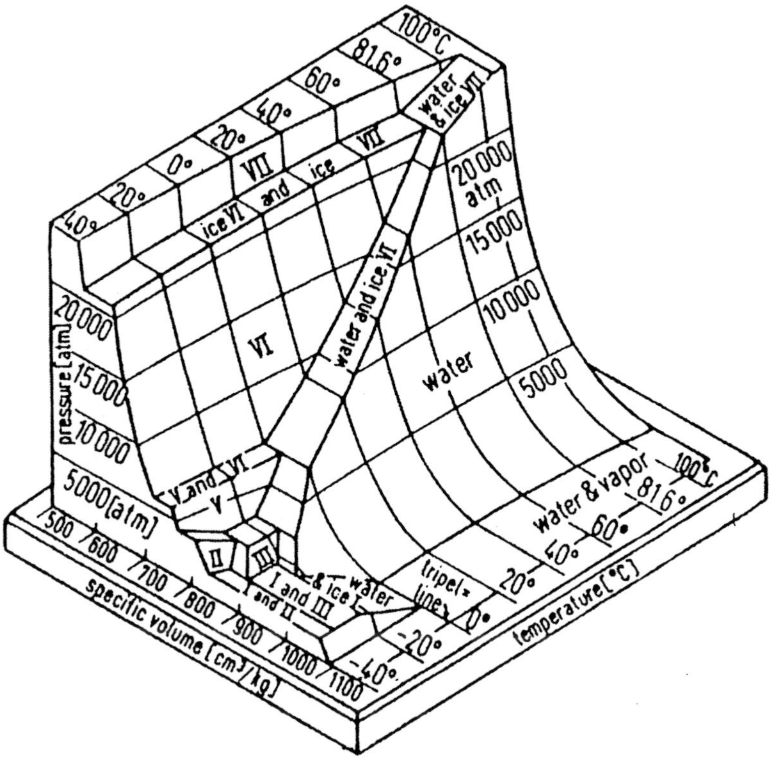

Water system at high pressure, showing the various forms of ice. Constructed from measurements of Bridgeman [Source: Semansky, *Heat and Thermodynamics, TE*, © 1997, Figure 4.9, page 112. Reprinted with permission of the McGraw-Hill Companies.]

Hexagonal Structure of I_h

The structure for I_h is illustrated by Figure 3.8, which shows that it has the same co-ordination as wortizite, the hexagonal form of zinc sulfide [30]. In ice I_h, each oxygen atom is tetrahedrally surrounded by four other oxygen atoms and the oxygen-oxygen distance is 2.67 Å (1 Å = 1×10^{-8} cm). The distance between the hydrogen-oxygen atoms that form the water molecule is 0.99 Å. As seen in this figure, the hydrogen bonds hold the oxygen atoms in place and account for the open structure of I_h and its density of 0.92 g/cm^3. The distance between the hydrogen-oxygen atoms that form the hydrogen bond is 1.77 Å.

It is this structure of ice (I_h) that plays such an important factor in the lyophilization process. When ice forms in a formulation, most solutes cannot fit into the I_h structure. Consequently, the concentration of the solute constituents of the formulation is increased in the interstitial region between the growing ice crystals. This separation of solvent and solutes is a key step in the formation of the frozen matrix.

Figure 3.8. Crystalline Structure of Ice

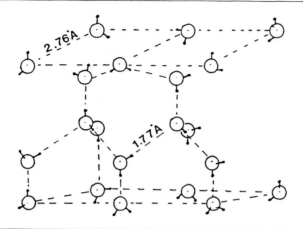

Crystalline structure of I_h where the ○ represent oxygen atoms, ● indicate hydrogen atoms, and the dashed lines signify hydrogen bonds.

If, during the freezing process, separation of the solvent from the solutes does not occur, a solid solution is formed. The vapor pressure of water in such a solid is greatly reduced and, for all practical purposes, the formulation cannot be lyophilized.

Affect of Solutes

While some solutes will have little or no effect on ice structure, other solutes can affect not only the ice structure but also its physical properties, such as melting temperature. This section will consider the formation and properties of clathrates and the impact that some solutes have on the inhibition of ice formation.

Clathrates or Gas Hydrates

Clathrates or gas hydrates are formed when the solute alters the structure of the water, such that the solute is surrounded by an envelope of water molecules [37]. The formation of the clathrate is associated with a release of a substantial quantity of energy, which is an indication that, once formed, the clathrate will have a relatively stable structure. Clathrates are also referred to as gas hydrates, because the solutes tend to be a result of dissolved gases such a methane, chlorine, and hydrogen sulfide. The structure of the clathrates formed from these gases tends to be cubic, with the unit cell containing 46 molecules [37].

Other compounds such as chloroform ($CHCl_3$) and ethyl chloride (C_2H_5Cl) can also form clathrates that have a cubic structure, but with a unit cell containing

136 molecules. Aggregates of clathrates can be formed and even crystallized out of water at temperatures higher than 0°C. About 80 known compounds have been identified as having formed clathrates when dissolved in water [43].

The clathrate or gas hydrate of chlorine was first observed by Davy in 1811. But the real impact of the unusual properties of these hydrates was not felt until the 1930s when natural gas (methane) pipes in the United States and Russia were found to be blocked with an ice like substance that existed at temperatures well above the melting temperature of water [43]. The problem was not solved until steps were taken to remove water vapor from the natural gas before it entered the pipeline. While the presence of most substances in water will tend to depress the freezing temperature of water, the formation of a clathrate will increase the melting temperature of the system.

Studies concerning clathrates have been confined to systems in which a high concentration of hydrate-forming compound exists. There are no known data showing the effects that low concentrations of clathrate-forming compounds can have on the various properties of water, especially during the freezing process, which is the first step in the lyophilization process. The following are some examples as to how the presence of clathrates could impact the lyophilization process.

- Clathrate-forming gas impurities, which would include such common gases as Ar, O_2, and N_2, could be absorbed when water is cooled during the preparation of a formulation. Sparging water with helium rather than nitrogen to remove unwanted oxygen would have less of an impact on the properties of the formulation.

- The presence of organic-forming clathrates could reduce the quantity of free water present. The reduction of the available free water, especially during the cooling process, causes a reduction in the solubility of a *p*H modifier and thereby alters the *p*H of the solution. A significant change in *p*H may result in a change in potency of the active constituent prior to completion of the freezing process.

- During freezing, the additional heat of formation for the clathrate could have an impact on the time required to achieve the temperatures necessary to complete the freezing of the formulation.

- The water associated with clathrates would have considerably lower vapor pressure and require substantially more energy to sublimate. By using calorimetric monitoring techniques during primary drying, which will be described in detail in Chapter 8, I was surprised on several occasions when the quantity of energy required to sublimate the ice during primary drying far exceeded the energy-based total amount of *free* water in the formulation. Such an increase in the heat of sublimation may have been associated with the formation of a clathrate with one of the constituents in the formulation

Clathrates or gas hydrates and their impact on the lyophilization process have been generally overlooked in the past; they are certainly a factor that needs to be considered during the preparation of the formulation.

Amorphous Frozen Water

Amorphous frozen water is water that is in a solid state but with short-range crystalline properties. A X-ray diffraction of such a system would produce a diffuse pattern showing no well-defined crystalline or polycrystalline structure. The production of frozen amorphous water from bulk water has not, as yet, been reported in the literature. Amorphous water can be formed, but it requires the slow deposition of water onto an extremely cold surface (i.e., at temperatures ≤–120°C). At these temperatures, the mobility of the water molecule is reduced to levels that prevent the formation of a *ground* state structure. Water molecules condensed under these conditions would be said to be in a *meltable* state [37]. Warming of the amorphous water film to temperatures > –100°C will provide the water molecules with sufficient energy to reach their ground state and form crystalline ice.

Excipients can also produce an amorphous frozen system at temperatures > –50°C. For example, I found that about 1/3 of the vials containing a 1% urea solution formed a completely amorphous frozen system (i.e., the system is completely solid and transparent at temperatures < –40°C). Attempts to lyophilize such frozen systems proved fruitless. After the vials containing ice crystals had completed primary drying, the vials containing the frozen solid solution produced a relatively thin film on the surface and gave no appearance of a cake. From these results, the presence of the urea may, at times, produce a solid frozen solution and also significantly lower the vapor pressure of the water to greatly inhibit the primary drying rate. When the solid solution was warmed to temperatures > –25°C, nucleation of the ice crystals occurred, and upon completion of the drying process, one could not distinguish between those vials that contained the frozen solid solution and those in which ice was formed during the initial freezing process.

Amorphous water is also formed in the interstitial regions between ice crystals during the freezing of most formulations, especially those that do not have a well-defined eutectic temperature. As a result of the freezing process, an aqueous solution consisting of the other constituents in the formulation are forced in the interstitial regions between the ice crystals. The solution goes from a liquid to a solid solution state or glassy state. The water in this glassy state is in an amorphous state, but, unlike the water content comprising the urea solid solution, the water contained in the interstitial region usually will not undergo any crystallization as a result of an increase in temperature.

ELECTRICAL CONDUCTIVITY OF ICE

An understanding of the *electrical conductivity* or *resistivity* of ice would, at first, appear to have little or no relationship with the lyophilization process, which consists mainly of a series of phase changes. But it will be shown in Chapter 6 that the electrical conductivity of ice will provide a means for determining some of the key thermal properties of a formulation, which are identified in Chapter 5. A knowledge of these thermal properties will define the temperatures necessary to produce a completely frozen matrix and the key operating parameters necessary to conduct the primary drying process. This section will first provide some historical background on the presence of a *liquid-like* layer on the surface of ice, followed by a review of various studies that have contributed to our present understanding of the conductivity in the disordered layer on an ice surface.

Disordered Surface Region

Jellinek [41] has prepared a rather comprehensive historical review regarding the liquid-like layer on an ice surface. The following brief summary of the Jellinek [44] paper provides the reader with a general understanding of the complexity of the electrical conductivity of ice.

Our discussion of the conductivity of ice starts with Michael Faraday's 1850 announcement that when two ice surfaces are brought in contact, freezing occurs at the interface. Because Faraday could not find similar behavior in other materials he attributed the unusual behavior of ice that he studied, to a water film on the ice surface. When the two ice surfaces, at 0°C, are brought together, freezing occurs. The freezing was explained as a result of *cohesive forces*. Tyndall accepted Faraday's mechanism and attempted to use it to explain that during the flow of glaciers, fractures in the glacier became fused. He termed the fusion or refreezing of these fractures as *regelation* or refreezing.

Others, such as J. Thompson, brother of Lord Kelvin, bitterly opposed Faraday's mechanism as being related to pressure melting, the same mechanism that generates the liquid layer at the base of the glacier. In spite of other experiments performed by Faraday, Thompson could not be convinced; however, he later stated that there was a defined explanation regarding the mechanism of regelation at that time. It was not until experiments were conducted at temperatures < 0°C that confirmed Faraday's concept of the presence of liquid film on the ice surface.

It was not until 1951 that Wey pointed out that, because of the permanent dipole moment of the water molecule, there should be a disturbance of the ice surface. Such a disturbance would lead to the surface of ice having different properties from that of the bulk ice. Wey postulated that such a *transition layer* on an ice surface could be hundreds of angstroms thick and have a liquid-like structure. Such a transition layer for ice would be present at temperatures < 0°C, and there would be equilibrium between the bulk ice structure of the ice and the liquid-like outer layer of the ice surface.

Nakaya and Matsumoto [41] were the first to test Wey's assumptions regarding the nature of the surface ice by performing the experiment illustrated in Figure 3.9. In Figure 3.9a, the two pure ice spheres, suspended by a thread, are positioned such that the surfaces are in direct contact, and the distance between the threads is d'. Upon increasing the distance between the threads by $d' + \Delta d'$, as illustrated in Figure 3.9b, an angle "a" is generated with respect to the original position of the supporting threads represented by the dashed line. The angle "a" confirms the existence of a cohesive force between the two ice spheres. Nakaya and Matsumoto also noted that upon the formation of the angle "a," the two spheres started to roll or slide as $\Delta d'$ was increased. Such rotation of the spheres was observed at temperatures as low as -7°C, indicating the existence of a liquid-like layer at temperatures well below the equilibrium melting temperature of ice.

In 1957, Hosler and Jensen performed similar adhesion experiments under saturated water vapor and dry atmospheres. Under saturated water vapor conditions, the adhesion between the two ice spheres was observed at temperatures as low as

Figure 3.9. The Interaction of Two Ice Surfaces

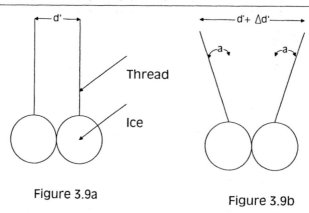

Figure 3.9a

Figure 3.9b

An illustration of the interaction of two ice surfaces as demonstrated by Nakaya and Matsumoto [41]. Figure 3.9a shows the two ice surfaces in direct contact with a distance d' between the support threads. Figure 3.9b shows that an angle "a," with respect to the original position of the supporting thread represented by the dashed line, is generated by the cohesive forces between the surfaces when the distance d' between the supporting threads is increased by a value $\Delta d'$.

–25°C; however, in a dry atmosphere there was no apparent adhesion between the ice spheres at temperatures < –3°C. Others have reported the presence of a liquid-like layer at even lower temperatures, e.g., –35°C [42]. These results support Wey's assumption of a transition layer [41]. In the dry atmosphere, the transition layer is removed by evaporation; in the saturated atmosphere, the transition layer is still prevalent at –25°C. Thus, there is no longer any doubt of a presence of a surface film on an ice surface as proposed by Michael Faraday in 1850. In recent years, the terms surface film and liquid-like layer has given way to the more descriptive term *disordered surface region* [43].

Conductivity of Ice

In order to better understand the nature of the electrical conductivity of ice, this section is divided into two parts. The first part examines the intrinsic conductivity of a single crystal (i.e., the electrical conductivity through a single crystal of ice). The second part is devoted to the electrical conductivity that occurs along the ice surface or in the disordered surface region.

Intrinsic Conductivity of Ice

Maidique et al. [44] studied the direct current conductivity of single crystals of ice of a known thickness. Electrical contact with the ice and a guard ring was made, depositing gold films on opposite sides of the ice crystal. When contact was made to the gold films and a voltage applied across the ice crystal, a steady current was attained after 2 sec. The applied voltages used in the study ranged from 5 mV to 2,000 V. A plot of the current as function of time showed the conductivity had ohmic characteristics (i.e., the current was directly proportional to the applied voltage).

A plot of the conductivity (σ') as a function of $1/T$ fitted the *activation energy* equation

$$\sigma' = \sigma_0' \exp^{-\epsilon/(kT)} \tag{9}$$

where ϵ is the activation energy. At temperatures near the melting temperature of ice, the activation energy was found to be approximately equal to 12 kcal/mole.

The results of the conductivity experiments by Maidique et al. [44] showed that the ideal ice crystal does not appear to have ionic charge carriers through its volume. These authors further state that *the mobility of the extrinsic carriers through the ice volume proved only slightly higher than that of protons in water and gave no evidence of any special proton-transfer mechanism across hydrogen bridges.* In conclusion, these authors have shown that single crystal ice has a relatively low intrinsic conductivity.

Surface Conductivity of Ice

During one of their measurements, Maidique et al. [44] observed that one of their samples cracked during the test and exposed a fresh, electrically unguarded surface. The result of the introduction of the new surface increased the conductivity of the ice, at a temperature near 0°C, by nearly an order of magnitude. Removing of guard ring resulted in the conductivity being mainly along the surface of the ice rather than through the ice crystal. The activation energy along the surface was found to be –33 kcal/mole, which is in agreement with surface conductivity measurements reported by others [45,46]. Maidique et al. [44] explain the observed –33 kcal/mole as not representing a true activation energy but as a change in conduction by a reduction in the thickness of the disordered surface region as a result of a decrease in the ice temperature.

From the above discussion, the electrical conductivity of polycrystalline ice measured between two electrodes embedded in the ice is dependent not only on the cell constant of the probes but also on the total surface area of the ice crystals. In the case of a formulation, the conductivity of ice crystals (σ_i') becomes significant only at temperatures where the mobility of the water in the interstitial region approaches zero and σ_i' is much greater than interstitial region (σ_r'). The significant role that the conductivity of the surface of the ice plays in the lyophilization process will become more evident in Chapters 5 and 6.

SUMMARY

This chapter has shown that, in spite of the lack of attention given to characterizing its properties, the nature of the water used in the preparation of the formulation plays a significant role in the lyophilization process. While WFI is used quite frequently in preparing a formulation and must meet certain criteria (i.e., total solids, conductivity, inorganic and organic specifications' and microbial, and endotoxin levels), it was shown that the presence of these impurities can impact the water's freezing characteristics (i.e, the degree of supercooling). Since the degree of supercooling can be affected by the excipients used in the formulation, we once again realize the importance of the composition of the formulation. A change in the nature of the ice structure will have a direct bearing on the cosmetic properties of the final product.

For most formulations, it was shown that some portion of the water will remain in an amorphous state upon completion of the freezing process. This water is part of the glassy interstitial region between the ice crystals that contains the other constituents in the formulation. It is the mobility of this interstitial water that will be shown to define the freezing and primary drying parameters.

The unusual presence of a disordered layer on the surface of the ice was shown to contribute to the relatively high electrical surface conductivity. The importance of such surface conductivity will become apparent during the discussion of the thermal properties of a formulation.

Perhaps the most revealing part of this chapter is not how much we know but how much more we need to learn about the properties of the most abundant compound on the surface of the earth.

SYMBOLS

"a"   an angle resulting from the interaction between two ice surfaces

A the cross-sectional area of the electrodes

d the distance between electrodes expressed in cm

d' the initial distance between threads of two ice spheres

$\Delta d'$ increase in distance between threads of two ice spheres

ϵ activation energy for the electrical conduction

ΔF free energy of formation

ΔH_f heat of fusion of ice

ΔV the applied voltage

I electrical current expressed in A

I_c polymorphic form of ice having a density < 1

I_h polymorphic form of ice having a density < 1

k Boltzmann's constant

κ the electrical conductivity expressed in Ω^{-1} cm^{-1}

n the density of ice-like water clusters of a given dimension

r_m radius of of ice-like water clusters

R_o the resistance of the cell

ρ the volume resistivity of a substance in Ω cm

S degree of supercooling

σ surface free energy of an ice-like water cluster

σ' electrical conductivity of ice

σ_i' conductivity of ice crystals in a matrix

σ_0' conductivity when ϵ/kT approaches 0

σ_r' conductivity of the interstitial region of a frozen matrix

T_i temperature of a system having a defined radius distribution of ice-like water clusters

T_0 temperature of water at the triple point

θ angle between the branches of ice crystals

U_α the rate of ice crystal growth expressed in cm/sec

REFERENCES

1.   S. Duggirala and P. P. DeLuca, *PDA J. Pharm, Sci. & Tech.*, 50: 297 (1996).

2.   N. A. Williams and D. L. Schwinke, loc. cit., 48: 135 (1994).

3.   T. R. Kovalcik and J. K. Guillory, loc. cit., 42: 165 (1988).

4.   L. Gatlin and P. P. DeLuca, loc. cit., 34: 398 (1980).

5.   I. Sugimoto, T. Ishihara, H. Habata, and H. Nakagawa, loc. cit., 35: 88 (1981).

6.   C. Chiang, D. Fessler, K. Freebern, R. Thirucrote, and P. Tyle, loc. cit., 48: 24 (1994).

7.   M. W. Townsend and P. P. DeLuca, loc. cit., 42: 190 (1988).

8.   P. P. DeLuca, M. S. Kamat, and Y. Koida, *Cong. Int. Technol. Pharm*, 5th, 1: 439 (1989).

9.   A. J. Phillips, R. J. Yardwood, and J. H. Collett, *Analytical Proceedings*, 23: 394 (1986).

10.   H. Seager, C. B. Taskis, M. Syrop, and T. J. Lee, *J. Parenter. Sci. and Tech.*, 39: 161 (1985).

11.   M. J. Pikal, S. Shah, D. Senor, and J. E. Lang, *J. Pharm Sci.*, 72: 635 (1983).

12.   K. A. Kagkadis, D. M. Rekkas, P. P. Dallas, and N. H. Choulis, *PDA J. Pharm, Sci. & Tech.*, 50: 317 (1996).

13.   M. J. Pikal, K. Dellerman, and M. L. Roy, in International Symposium on Freeze-Drying and Formulation, Bethesda, USA, 1990, *Develop. Biol. Standard*, 74: S. Karger, Basel, 1991, p. 21.

14.   M. J. Pikal and S. Shah, loc. cit., p. 165.

15.   J. F. Carpenter, T. Arakawa, and J. H. Crowe, loc. cit., p. 225.

16.   N. Hanafusa, loc. cit., p. 241.

17.   C. C. Hsu, C. A. Ward, R. Pearlman, H. M. Hguyen, D. A. Yeung, and J. G. Curley, loc. cit., p. 255.

18.   P. J. Dawson, loc. cit., p. 273.

19.   M. S. Hora, R. K. Rana, C. L. Wilcox, N. V. Katre, P. Hirtzer, S. N. Wolfe, and J. W. Thomson, loc. cit., p. 295.

20. K. Inazu and K. Shima, loc. cit., p. 307.

21. M. L. Roy, M. J. Pikal, E. C. Richard, and A. M. Maloney, loc. cit., p. 323.

22. S. Vemuri, loc. cit., p. 341.

23. J. Jang, S. Kitamura, and J. K. Guillory, *PDA J. Pharm, Sci. & Tech.*, 49: 166 (1995).

24. Y. Koyama, M. Kamat, R. J. De Anglelis, R. Srinivasan, and P. P. DeLuca, loc. cit., 42: 47 (1988).

25. K. Weijer, R. Augustine, C. Carleton, T. Haby, E. Kuspiel, R. Newman, D. Wadke, and S. Varia, loc. cit., 43: 286 (1989).

26. D. J. Korey and J. S. Schwartz, loc. cit., 43: 80 (1989).

27. S. Vemuri, C. Yu, and N. Roosdorp, loc. cit., 48: 241 (1994).

28. W. Collentro, *Ultrapure Water*®, April, p. 22 (1994).

29. G. A. Crane, M W. Mittelman, and M. Stephan, *J. Parenter. Sci. & Tech.*, 45: 20 (1991).

30. W. J. Moore, *Physical Chemistry*, Prentice-Hall, Inc. Englewood Cliffs, N.J., 3rd ed., 1964.

31. R. O. Watts in International Symposium on Freeze-Drying and Formulation, Bethesda, USA, 1990, *Develop. Biol. Standard*, 74: S. Karger, Basel, 1991, p. 41.

32. J. Li and D. K. Ross, *Nature*, 365: 327 (1993).

33. M. R. de Quervain, in *Freeze Drying and Advanced Food Technology* (S. A. Goldblith, L. Rey and W. W. Rothmayr, eds,) Academic Press, New York, 1975, p. 3.

34. *Handbook of Chemistry and Physics* (R. C. Weast, ed.), 65th ed., CRC Press, Inc. Boca Raton, Florida, 1984, p. F-64.

35. R. Crovetto, R. Fernández-Prini, and M. L. Japas, *J. Chem Phys.*, 76: 1077 (1982).

36. J. Hallett, in *Advances in Freeze-Drying—Lyophilisation* (L. Rey, ed.) Hermmann, Paris, 1966, p. 21.

37. H. T. Meryman, in *Cryobiology* (H. T. Meryman, ed.), Academic Press, New York, 1966, p. 1.

38. J. L. Stephenson, in *Recent Reaseches in Freezing and Drying* (A. S. Parks and A. V. Smith, eds.), Blackwell Scientific publication, Oxford, England, 1960, p. 121.

39. R. J. Speedy, *Nature*, 380: 289 (1996).

40. C. S. Lindenmeyer, G. T. Orrok, A. Jackson, and B. Chambers, *J. Chem Phys.*, 27: 822 (1957).

41. H. H. G. Jellinek, *J Colloid Interface Sci.*, 25: 192 (1967).

42. E. Mazzega, U. del Pennino, A. Loria, and S. Mantovani, *J. Chem Phys.*, 64: 1028 (1976).

43. I. Golecki and C. Jaccard, *J. Phys. C: Solid State Phys.* 11: 4229 (1978).

44. M. A. Maidique, A. von Hipple, and W. B. Westphal, *J. Chm. Phys.*, 54: 150 (1971).

45. N. H. Fletcher, *Phil. Mag.*, 18: 1287 (1987).

46. N. Maeno and H. Nishimura, *J. of Glaciology*, 21: 193 (1978).

4

Phase Changes

IMPORTANCE OF PHASE CHANGES

This chapter is one of the shortest chapters in this book; however, its brevity does not in any way detract from its importance in providing the concepts necessary for understanding some of the principles upon which the science of lyophilization is based. It is of fundamental importance to recognize that the entire lyophilization process is predicated on a series of phase changes:

liquid → solid (freezing),

solid → gas (primary drying), and

adsorbed state → gas (secondary drying).

An understanding of the lyophilization process requires a comprehension of these phase changes and how they relate to the nature of the formulation and interact with the operating parameters of the freeze-drying equipment. What makes lyophilization so deceptive is that it appears to be, as seen from the above, a relatively simplistic process. Its deception lies in the fact that it is often possible to obtain a dried product that meets the requirements for the potency of the active constituent and is within the required stability specifications without having a clear understanding of the mechanisms that are involved in the process. It would not be unusual if, in such cases, no rationale could be provided to substantiate that the product was actually lyophilized and not vacuum dried (i.e., during the primary drying process there was significant mobile water present in the interstitial region of the matrix). Failure to provide the rationale for the lyophilization process will, for the most part, stem directly from a lack of knowledge of the various phase changes that occur during the process.

This chapter is intended to provide the reader with the fundamentals involved in the phase changes during the lyophilization process. Based on an understanding of these fundamentals, the development of a lyophilization process will not be a result of a series of trial and error batches or sheer guesswork but will be based on a set scientific principles. One result of understanding the fundamental principles involved in the lyophilization process will be to make the reader more aware of its complexity.

Rather than take a purely academic approach to a discussion of phase changes, I have made every effort to illustrate the subject matter with examples that have some relationship with the lyophilization process.

HOMOGENEOUS AND HETEROGENEOUS SYSTEMS

Systems can be generally classified as being either *homogeneous* and *heterogeneous*. While the concept of a system being either homogeneous or heterogeneous represents no intellectual challenge, the issue is, *At what stage in the process does a given system represent a desirable state and under what condition does it present a problem?* The significance of understanding the relative importance of these systems will be more evident when the thermal properties of a formulation are discussed in Chapter 5.

Homogeneous System

A homogeneous system is one in which there is only one phase present. A typical homogeneous system would be the formulation prior to the freezing process. In this example, it would be most desirable if the formulation consisted only of solutes and solvent. For the most part, the presence of other phases in the formulation is undesirable, because they could represent visible and subvisible particles stemming from such sources as the water supply, the environment, interactions between excipients, or interactions involving the active constituent. However, it was shown in the previous chapter that 1% urea solutions are capable of complexing with the water to form a solid solution rather than an ice matrix during the freezing process. Since such a solid solution leads to lowering of the vapor pressure of water, the time required to complete the drying process for the solid solution would be greatly extended and the quality of the final product would be questionable. In this instance, the presence of a homogeneous system would be undesirable.

Heterogeneous System

A heterogeneous system is one in which more than one phase is present. For example, at the triple point of water (0°C) there are three phases present (i.e., gas, liquid, and solid). For a lyophilized product, the principal objective of the freezing process is to separate the solutes from the solvent. In order for drying to take place, the system must be heterogeneous in that during primary drying there will be a solid (ice) and vapor phase present, while secondary drying involves an adsorbed and vapor phase. However, after reconstitution of the lyophilized product, a heterogeneous solution could be an indication of turbidity and would be usually considered to be an undesirable result of the lyophilization process.

A fundamental understanding of when a system should be homogeneous or heterogeneous is paramount to fully grasping the lyophilization process. The importance of this understanding will become more evident in Chapter 5, especially when considering phase changes that occur in the interstitial region of a frozen matrix. For example, the frozen matrix will consist of ice crystals and a glassy system in the interstitial regions between the ice crystals. The frozen matrix is considered to be a heterogeneous system, while the glassy interstitial region could be homogeneous in nature. Because an active constituent in a crystalline state is more stable than one in an amorphous state, a heterogenous interstitial region may prove more desirable than a homogeneous one. For lyophilized biological products, however, the homogeneous interstitial region may prove to be more desirable. As a result, understanding the·nature of the active constituent is necessary to deciding whether the interstitial region should be homogeneous or heterogeneous.

CONSTITUENTS AND COMPONENTS

Constituent

A formulation must consist of at least two *constituents*, a solvent (water) and a solute (active constituent). Other formulations (e.g., those having a serum base) may consist of a large number of constituents, where each chemical species represents an individual constituent. A constituent is, therefore, defined as any solvent and solute that compose the formulation.

One must be particularly conscious of not changing the number of constituents during the formulation procedure. For example, an active constituent may require a given range of *p*H. The *p*H of the formulation can be decreased by adding an acid (e.g., HCl) or increased by adding a base such as NaOH. If the active constituent is the sodium salt of organic acid, then the addition of NaOH to increase the *p*H may have no effect on the number of constituents present in the formulation. But the addition of HCl to lower the *p*H will change the hydronium ion concentration and could introduce a new constituent to the formulation (i.e., NaCl). It has been my experience that a small change in the number and/or concentration of constituents in the formulation can have a major effect on a lyophilization process. We will see the effect that such changes have in subsequent chapters.

Component

In this section, we shall distinguish between a *component* and a constituent of a formulation. Every component of a formulation is a constituent, but not every constituent is a component. Components of a formulation are the minimum number of constituents required to define each phase that is present in the formulation during the lyophilization process. In most liquid formulations, ice is formed during the freezing process, and the remaining constituents (solutes) take the form of a glassy interstitial region between the ice crystals. Since only the water constituent underwent a phase change during the freezing process, the system is said to have but one component, i.e., water. If, as sometimes occurs with a dilute urea solution, no ice crystals are formed, then the system will have no components and, for all practical purposes, lyophilization of the system is not possible.

Consider a *saturated* solution of NaCl at temperatures ranging from 0.1°C to 100°C, which can be represented by the following reaction

$$\text{NaCl (saturated)} + 2H_2O \rightarrow \text{NaCl (anhydrous)} \qquad (1)$$

There are two constituents in the formulation, but the formation of the anhydrous NaCl is a result of only the NaCl constituent. In this case, NaCl is a component, while the water remains as a constituent.

However, at temperatures ranging from 0.1°C to −21.1°C, the following reaction occurs in a *saturated* NaCl solution:

$$\text{NaCl (saturated)} + 2H_2O \rightarrow \text{NaCl} \cdot 2H_2O \qquad (2)$$

where $\text{NaCl} \cdot 2H_2O$ crystallizes out of the saturated NaCl solution. The minimum number of constituents necessary to form the $\text{NaCl} \cdot 2H_2O$ is 2: NaCl and H_2O. In this example, all of the constituents in this solution are also considered components.

For a dilute NaCl solution (< 23 percent NaCl wt/vol) at temperatures between < 0°C and > −21.1°C, the presence of the salt will merely depress the freezing point of water. At some temperature < 0°C, only ice will form in the system. Under these conditions, there is only one component present; however, in this example, the component is water, not NaCl.

In a typical lyophilized formulation (e.g., a serum-based formulation), at temperatures < −40°C, the system will consist of ice crystals and a solid solution of serum constituents in the interstitial region of the matrix. Since the phase change in the system involved only water, there would be only one component in the system—water.

If in the above example aggregates were formed due to covalent bonding of a given serum constituent, then there would be two components present at lower < −40°C: water and the serum constituent forming the *aggregates*. A word of caution: The aggregates of the serum constituent would only be considered components if their presence is reversible. The term *component*, therefore, only applies to phase changes that are reversible; it does not include irreversible processes, such as the formation of a precipitate.

In general, for lyophilization to occur, there must be at least one component, namely water, present in the formulation during the freezing of the formulation to form a heterogeneous system. In the absence of a water component, the frozen system will be homogeneous and lyophilization becomes an arduous task. Although the liquid formulation is required to be homogeneous, a heterogeneous system is most desirable in the frozen state.

INTENSIVE STATE VARIABLES

Intensive state variables are those variables, namely pressure, temperature and concentration, that define the states or the phases present in the system.

Equilibrium Between Phases

In a heterogeneous system, equilibrium between the phases is independent of the mass associated with a given phase. For a macro-system at equilibrium, the existence of a phase is not dependent on mass but is related to the magnitude of the intensive state variables. The intensive state variables are defined by the thermo-dynamic conditions necessary to maintain the state of equilibrium. Therefore, a complete description of a system at equilibrium requires (a) knowledge of the nature of the phases present, (b) the identification of the components, and (c) the magnitude of the intensive state variables. For example, at the triple point of water, (see point B in Figure 3.1), there will be three phases (solid, liquid, and gas) and one component (water) present, and the intensive variables will be temperature (0°C) and vapor pressure of water (4.58 Torr). It is important to understand that, as long as the system is in equilibrium and all three phases are present, the temperature will be maintained at 0°C and the vapor pressure of water at 4.58 Torr. At temperatures < 0°C, the liquid phase will be absent but there will still exist an equilibrium between the solid phase (ice) and the gas phase above the ice that will be defined by the temperature of the ice and the vapor pressure of water vapor above the ice. Thus, setting the ice temperature will define the vapor pressure above the ice and, conversely, setting the vapor pressure above the ice will define the ice temperature.

Although the last statement is simple in concept, this relationship between the phases present and the intensive state variables is often completely ignored during consideration of the primary drying process. In a number of instances, investigators have reported data that violates the phase diagram of water or, as I often say, *it is not possible in this universe*. For example, the investigator may report that the product temperature and shelf temperatures are at -40°C, and the pressure in the system is 50 mTorr. If the pressure above the ice were indeed 50 mTorr, then the observed ice temperature would have to be < -40°C. One is then left to wonder which of the intensive variables (temperature or pressure) is in error.

Degrees of Freedom

Degrees of freedom (*f*) are defined as the number of intensive variables that can be varied without changing the number of phases present. Figure 3.1 illustrates (not to scale) the phase diagram for water. The intensive variables for the water system are pressure and temperature. The solid line B-C represents a state of equilibrium between the liquid and the gas phases of water. For line B-C, the pressure is defined as the vapor pressure water (i.e., that pressure of water vapor that is in equilibrium with the liquid phase for a given temperature). Based on line B-C, the water vapor pressure is defined by the temperature of the liquid. Since the vapor pressure of water cannot be changed without changing the temperature of the liquid water, the system along line B-C has only one degree of freedom.

Consider now the number of degrees of freedom in just the vapor region as shown by Figure 3.1. In this region, it is possible to change both temperature and water pressure without changing the number of phases present. Consequently, the system has two degrees of freedom.

Phase Rule

While professor of mathematics at Yale University in 1875 and 1876, J. Willard Gibbs developed what has become known as the *phase rule*. The phase rule is an elegantly simple expression for relating f to the number of components (C) and phases (P) present in a homogeneous or heterogeneous system at equilibrium:

$$f = C - P + 2 \qquad (3)$$

where the value of 2 is obtained from the derivation (1).

From an examination of the phase diagram of water (see Figure 3.1), C is 1 and P can vary from 1 to 3. For a homogeneous system, such as the previously used vapor phase, P will be 1 and f will have a value of 2. Along any of the solid or dotted lines that represent heterogeneous systems in which there is equilibrium between the phases, P will be 2 and f will be 1. At point B on the phase diagram for water—which represents the triple point (0°C)—all three phases will be present and f will be 0. With f at 0, there can be no change in pressure or temperature of the system.

The dotted line B–D in Figure 3.1 represents the vapor pressure–temperature relationship for supercooled water. The reason for supercooled water having a higher vapor pressure than the ice-temperature plot (B–F) for a given temperature is that the supercooled water has not lost the energy associated with the heat of fusion. Point C, which occurs at a pressure of 218 atm and a temperature of 374°C, is referred to as the critical point for the water system. At temperatures and pressures equal or greater than the critical point, the liquid phase can no longer be distinguished from that of the vapor phase.

The vapor pressure ice–temperature plot B–F is one of the most important relationships in the lyophilization process. It is a lack of understanding this relationship that often leads to reducing lyophilization to an art rather than a science. The importance of this relationship will become more evident in Chapters 8 and 13.

For a system where the pressure is constant and the intensive variables are temperature and concentration, Gibb's phase rule takes the form of

$$f = C - P + 1 \qquad (4)$$

The significance of this form of Gibb's phase rule will become evident in Chapter 5, when we discuss a two-component system, and Chapter 6, which reviews the various techniques for determining phase changes in a lyophilized formulation.

CLAPERYON–CLAUSIS EQUATION

Gibb's phase rule is very useful in understanding the relationship between the components and the phases present for both homogeneous and heterogeneous phase diagrams. However, it provides no information regarding the relationship between the intensive state variables. It will now be shown that such an interrelationship between intensive state variables can be derived from the *Claperyon-Clausis* equation. This function not only equates the intensive variables of a system but also provides a measure of the latent heat associated with the phase change. The general form of the Claperyon-Clausis equation is

$$\frac{dP'}{dT} = \frac{\Delta H}{T\Delta V'} \qquad (5)$$

where dP'/dT is the differential of the vapor pressure (P') with temperature (T) in K, $\Delta V'$ is the change in the molar volume, and ΔH is the latent heat associated with the phase change [1].

In the case where the phase change is from a solid to a gas (sublimation), equation 4 becomes

$$\frac{dP'}{dT} = \frac{\Delta H_s}{T(V'_g - V'_s)} \tag{6}$$

where ΔH_s represents the latent heat of sublimation. Since the molar volume of the gas (V_g') is much larger than the molar volume of the solid (V_s'), equation 6 simplifies to

$$\frac{dP'}{dT} = \frac{\Delta H_s}{T(V'_g)} \tag{7}$$

From the ideal gas law, equation 7 can be rearranged to obtain

$$\frac{dP'}{P'} = \frac{\Delta H_s}{RT^2} dT \tag{8}$$

where R is the gas constant. Integration of equation 8 gives

$$\ln P = \frac{\Delta H_s}{RT} \tag{9}$$

From a plot of $\ln P'$ as a function of $(1/T)$, the latent heat of sublimation is determined from the following equation:

$$\frac{d(\ln P')}{d(1/T)} = \frac{\Delta H_s}{R} \tag{10}$$

A plot of $\ln P'$ as a function of $1/T$ [2], as shown by Figure 4.1, will give a slope of $-\Delta H_s/R$, from which the heat of sublimation can be determined. Figure 4.1 shows that there is a marked increase in the slope of the function at temperatures $< -40°C$. The latent heat of sublimation for temperatures ranging from $-1°C$ to $-30°C$, represented by line A-B, is $-12,176.7$ cal/mole or -676.5 cal/g. For the temperature range from $-40°C$ to $-90°C$, represented by the line B-C, the latent heat of sublimation, determined from equation (9), is $-12,258.9$ cal/mole or -681.0 cal/g. The reason for the increase in the heat of sublimation may be attributed to the lack of a disordered layer at temperatures $< -35°C$ [3].

GLASSY STATES

A unique feature of most lyophilized formulations is the formation of *glassy states* as a result of the freezing process. It has been stressed previously that in order for lyophilization to occur, there must be a separation of the solutes from the solvent. Since water is generally the solvent of choice, the matrix formed (see Figure 1.3) will consist of ice crystals and the solutes constrained in the interstitial regions between the ice crystals. In this section, we shall briefly describe the glassy nature that can exist in the interstitial region and the latent heat of melting for such a glassy state.

Figure 4.1. A Plot of the ln P' as a Function of $1/T$ for Ice Temperatures Ranging from −1°C to 90°C

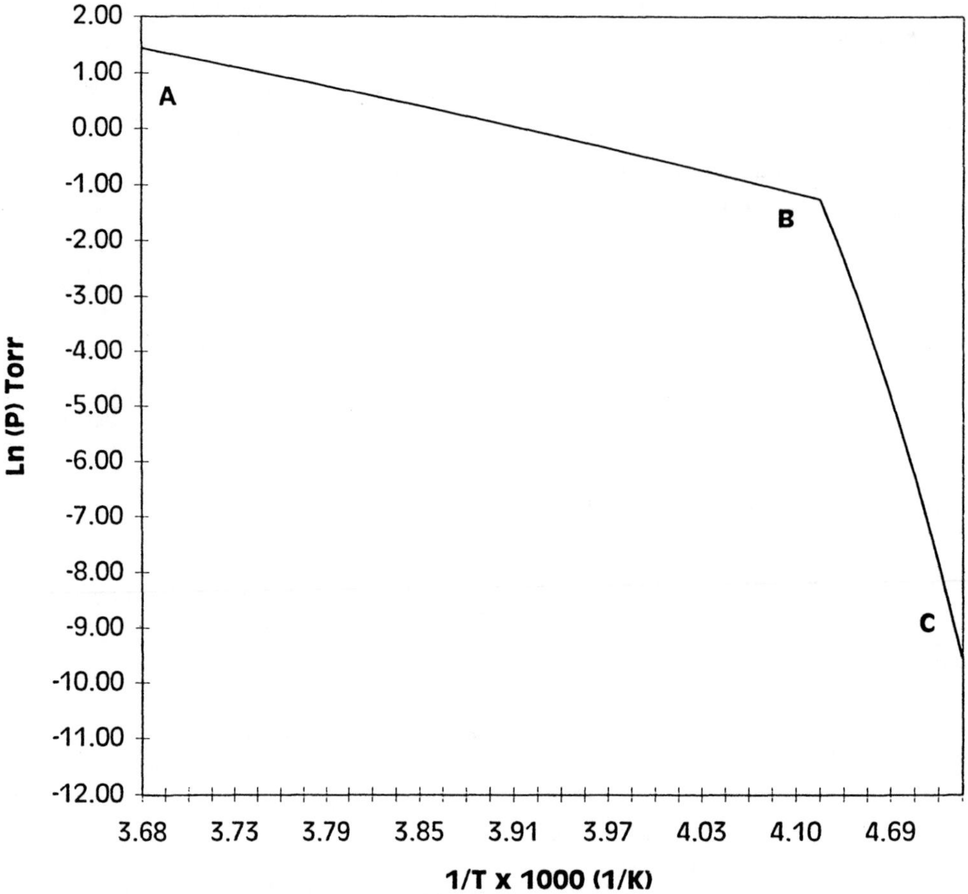

Formation

Homogeneous glassy states in the interstitial region of the matrix can be formed with excipients such as PVP [4,5], glucose, and sucrose [5]. Such states also contain a quantity of water that is noncrystalline and is often referred to as *unfreezable* water [5]. It is a combination of unfreezable water and excipients that comprise the interstitial glassy state.

Latent Heat of Melting

In the preceding discussion, the relationship between f and the intensive variables could be explained in terms of Gibb's phase rule. In addition, the latent heat of sublimation of ice could be determined from the Claperyon-Clausis equation. However, the phase rule is not applicable to describing the nature of interstitial glasses. Glasses have been described as a mysterious chaotic jumble of atoms (and molecules) that behave as if they were rigidly frozen [6]. In contrast to a change in phase, when interstitial glass is heated above the glass transition temperature (T_g'), a high viscosity fluid or *rubbery* system is formed [5] and f remains > 0. If these interstitial glasses represent frozen liquid systems, then there should be almost no latent heat associated with the transition temperature. However, the latent heat associated with T_g' associated with lyophilized formulations has been observed [5,7,8]. From the nature of these observations, the onset temperature for T_g' is not sharp and well defined, and the quantity energy associated with the transition temperature is relatively small. It has been the experience of this author, based on thermal analyses of about 2,000 formulations, that the energy associated with T_g' is seldom observed.

Glass Transition and Collapse Temperatures

When primary drying is conducted at temperatures $> T_g'$, there is sufficient mobile water present that the interstitial region can no longer maintain its original configuration, and a condition referred to as collapse occurs (see Figure 1.3c). Because of its association with a physical change in the physical nature of the final cake, T_g' will henceforth be referred to as the *collapse temperature* (*Tc*). The collapse temperature and its measurement will be considered further in Chapters 5 and 6.

SYMBOLS

C number of components in a system

ΔH_s heat of sublimation of ice

$\Delta V'$ the change in the molar volume of a system

f the number of intensive variables that can be changed without changing the number of phases present

P the number of phases in a system

P' the vapor pressure of water over ice

R the gas constant

Tc collapse temperature

T_g' glass transition temperature

V_g' molar volume of a substance in the gas phase

V_s' molar volume of a substance in the solid phase

REFERENCES

1. W. J. Moore, *Physical Chemistry*, Prentice-Hall, Inc. Englewood Cliffs, N.J., 3rd ed., 1964.

2. *Handbook of Chemistry and Physics*, (R.C. Weast, ed.), 65th Edition, CRC Press, Inc. Boca Raton, Florida, 1984, p. D-192.

3. E. Mazzega, et al., *J. Chem Phys.*, 64: 1028 (1976).

4. A. P. MacKenzie, in *Freeze Drying and Advanced Food Technology* (S. A. Goldblith, L. Rey, and W. W. Rothmayr, eds.), Academic Press, New York, 1975, p. 277.

5. H. Levione and L. Slade, *Cryo-Letters*, 9: 21 (1998).

6. P. G. Wolynes, *Nature*, 382: 495 (1996).

7. P. P. DeLuca, M. S. Kamat, and Y. Koida, *Cong. Int. Technol. Pharm*, 5th, 1: 439 (1989).

8. J. Jang, S. Kitamura, and J. K. Guillory, *PDA J. Parenter. Sci. and Tech.*, 49: 166 (1995).

5

The Thermal Properties
of Formulations

NEED FOR IDENTIFYING THERMAL PROPERTIES

Knowledge of the thermal characteristics serve two important functions: verifying the reproducibility of the formulation from batch to batch and providing the necessary rationale for obtaining a completely frozen matrix and the required operating parameters for the primary drying process.

Reproducibility of the Formulation

The first criteria of a formulation is to establish that it will be reproducible from batch to batch. In Chapter 2, reproducibility was expressed in terms of the concentrations of the active constituent and the excipients that were necessary to achieve a stable environment for the active constituent while in an aqueous solution. In the liquid phase, the formulation was characterized by properties such as the potency of the active constituent, pH, electrical conductivity, and osmolarity. The role of the process water in formulation was considered in Chapter 3. In this chapter, we will further expand our characterization of the formulation in terms of its thermal characteristics for temperatures ranging from 30°C to < -60°C. It is in this temperature range that the formulation will undergo the phase changes described in Chapter 4 and establish the batch-to-batch consistency of the formulation. This criterion is necessary for establishing a reproducible lyophilization process and providing the fundamental basis for process validation. Some thermal properties of the formulation may not have a direct bearing on the lyophilization process but provide a *fingerprint* that could prove valuable in characterizing the formulation. Should one find differences in a lyophilized product, the formulation should be beyond suspicion. Without batch-to-batch reproducibility of the formulation, the nature of the final lyophilized product could be left entirely to chance.

Rationale for the Freezing and Primary Drying Processes

The thermal properties of a lyophilized formulation determine the temperatures necessary to achieve a completely frozen state and the temperatures and pressures needed for the primary drying process. In the absence of such thermal characterization of the formulation, the establishment of a lyophilization process is left entirely to trial and error. Such drying processes are generally characterized as being excessively long and where the shelf temperature is increased in a stepwise fashion. All too often, the response to a request to explain the rationale for each of the steps is that the process works. The danger of using such a method to develop a lyophilization process is that if a problem arises in the final product (e.g., collapse of the cake), one would be at a loss to identify if the problem is associated with (a) the formulation, (b) the thermal characteristics of the formulation, (c) the lyophilization process, (d) the freeze-drying equipment or its associated instrumentation, or (e) operator error. Knowledge of the thermal properties of the formulation lifts lyophilization process development from an art and places it on a more rationale scientific basis.

The remainder of this chapter will describe the various thermal properties of a formulation. The objective of this chapter is to introduce the reader to these thermal properties. These thermal properties will be used in later chapters when describing the freezing process and the primary drying. The means for determining such thermal properties will be considered in the next chapter. I felt that describing the thermal properties and the various analytical techniques used to determine them in the same chapter would preclude an adequate treatment of both topics.

ICE–LIKE WATER CLUSTERS

As stated in Chapter 3, at temperatures near 4°C, the liquid state of water (at 1 atm) typically reaches its maximum density, 1 g/cm³. At temperatures < 4°C, hydrogen bonding occurs with the formation of ice-like water clusters, and there is a decrease in density. For reasons not completely understood at this time, not all process waters used in the preparation of a formulation undergo such a change in density. For those process waters that do exhibit such a thermal characteristic, it is of interest to note if the formulation also shows the formation of such ice-like water clusters and if there is agreement between the onset temperatures. The *onset temperature* is that temperature at which a phase change is first detected. It is not the formation of ice-like water clusters in the process water, but their absence in the formulation would indicate some complex interaction between the constituents of the formulation and the process water. When such interactions take place, there is little free water available to form the ice-like water clusters.

It has not been established if the presence or absence of the formation of ice-like water clusters has any effect on other key thermal properties, such as the degree of crystallization or collapse temperature. However, the importance of determining the formation of these clusters lies in establishing a baseline of thermal characteristics of the process water for future reference. For example, if there is a noticeable change in another thermal characteristic of the formulation, such as the degree of crystallization, and there is also an absence of ice-like water clusters in the process water and formulation, then one would be inclined to investigate what, if any, changes were made in generating, storing or piping the process water. The absence of ice-like water clusters and the change in the degree of crystallization may prove to be a coincidence but it is a logical starting point. Means for determining the formation of ice-like water clusters will be described in the next chapter.

DEGREE OF SUPERCOOLING

The degree of supercooling has been defined as the number of degrees below the equilibrium freezing temperature where nucleation occurs, i.e., primarily the formation of ice crystals. The nature of supercooling and the impact it can have on cake structure and the quality of the final product has already been described in detail in Chapter 3. The only reason for returning to the subject of supercooling is to emphasize that it is a key thermal property of the formulation. As a key thermal property, it will be important to establish that the range of temperatures at which supercooling occurs remains consistent from batch to batch. Major inconsistency in the degree of supercooling, either from batch to batch or vial to vial within a batch can lead to variations in the cosmetic appearance of the cake and quality of the final product.

EQUILIBRIUM FREEZING TEMPERATURE

The following discussion will show how observation of the equilibrium freezing temperature after supercooling a formulation can be used to determine the freezing-point depression. From the observed depression of the equilibrium freezing temperature, one can ascertain the total solute concentration in the system.

The Water System

The *equilibrium freezing temperature* for water was shown by the phase diagram for water (see Figure 3.1) as being the triple point at 0°C, where all three phases (gas, liquid, and solid) are present and, from Gibb's phase rule, the number of degrees of freedom (f) is 0. When supercooling occurs in pure water, the result is the formation of a system in which all three phases are present. The net result is that the temperature of the system will increase to 0°C. The system will remain at this temperature until one of the phases, namely water, is removed. The liquid phase is converted into the ice phase, and f for the system is increased to 1. When f is > 0, then the temperature of the system can be reduced to < 0°C. The time that the water system will remain at the triple point will depend on the mass of the water present and the rate that energy is removed from the system.

The measurement of the equilibrium freezing temperatures for some process waters indicates equilibrium freezing temperatures for water that are significantly > 0°C (e.g., 0.5°C). In view of the fact that the thermocouples were calibrated at 0°C and were found to have standard deviations ranging between 0.02°C to 0.05°C, the observed equilibrium freezing temperature would represent a deviation of 25σ to 10σ from the mean bath temperature. As a result, the likelihood that consistent measurements of a ice system having temperatures of the order of 0.5°C and still being within a frequency distribution of temperatures having a mean value of 0°C must be considered to be well beyond normal chance. The observation of equilibrium freezing temperatures > 0°C would suggest the presence of clathrates or gas hydrates in the water (see Chapter 3). Since low conductivity measurements would be an indication of the absence of ions in the water, one could eliminate clathrate-forming gases such as chlorine and hydrogen sulfide. Nonpolar gases, such as methane, could account for an equilibrium freezing temperature of water being > 0°C.

Formulation

For a formulation consisting of an active constituent and other excipients, the presence of these solutes results in a lowering of the vapor pressure of the water. This leads to depression of the freezing temperature. The reason for this depression can be seen in Figure 5.1, which shows that lowering the vapor pressure results in a reduction of the temperature along line A-B, which is associated with *depression of the freezing temperature* on the sublimation curve. The freezing point depression (ΔT_f) [1] is expressed as

$$\Delta T_f = \frac{1,000 K_f g_s}{W M_s} \tag{1}$$

Figure 5.1. Partial Phase Diagram for Water

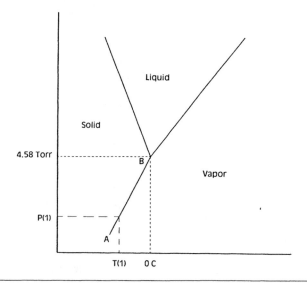

A segment of a phase diagram for water, not drawn to scale, where $P(1)$ represents the lowering of the vapor pressure of water as a result of solutes in the formulation and $T(1)$ is the depression of the freezing temperature on the sublimation curve A–B.

---

where K_f is the freezing-point constant for the solvent (water = 1.855), W is the weight of solvent, g_s is the number of grams of the solute, and M_s is the molecular weight of the solute. Because ΔT_f for a dilute formulation is dependent on the colligative properties of the solutes, ΔT_f can be expressed as

$$\Delta T_f = \frac{1,000 K_f}{W} \sum \frac{g_{s,i}}{M_{s,i}} \qquad (2)$$

where $g_{s,i}$ is the number of grams and $M_{s,i}$ is the molecular weight of the ith solute.

Although equation 1 is accurate only for ideal dilute solutions, it still provides a good estimate of the freezing-point depression of water for a formulation. This is illustrated by Figure 5.2, which shows the observed depression of the freezing point of water for sucrose, indicated by the solid line, while the dashed line represents those values calculated by equation 1. The solid line values represent an equilibrium temperature between the solid and ice phases for a given sucrose concentration. Figure 5.2 shows that the difference between the measured and calculated values for the freezing-point depression of water by sucrose becomes significant at a 3.5% sucrose concentration (wt/v). However, even at concentration of 9.8% (wt/v) the error in using equation 1 to determine the freezing-point depression is < 0.1°C.

When supercooling occurs in a formulation, the temperature will increase to the depression of the equilibrium freezing temperature as defined by either equation 1 or equation 2. A measure of the equilibrium freezing temperature indicates the total solute concentration in the system. In this regard, a measure of this thermal property represents a confirmation of the total solutes present in the formulation.

DEGREE OF CRYSTALLIZATION

The *degree of crystallization* (D) is defined as the ratio of the ice formed during the freezing process to the total freezable water in the formulation [3] or

$$D = \frac{\Delta H_{f,s}}{\Delta H_{f,i}} \qquad (3)$$

where $\Delta H_{f,s}$ is the heat of fusion identified with the formulation and $\Delta H_{f,i}$ is the heat of fusion associated with a water reference containing an equivalent quantity of *freezable water*. Freezable water is that water in the formulation that can crystallize during the freezing process. The other type of water that can be present is in a glassy state. The limits of D range from 0 to 1. By knowing D, one can ascertain the extent glass formation in the interstitial region of the matrix. Figure 5.3 illustrates frozen matrixes where D approaches 1 and where D is ≤ 0.5.

Figure 5.3a illustrates a frozen matrix where D approaches 1. As D approaches 1, the quantity of glassy states in the interstitial region approaches 0. When there is an absence of glassy states in the interstitial region, the solutes confined to the region are, in essence, already in a dried state as a result of the freezing process. The only water that would have to be removed would be that associated with adsorbed water on the surface of the cake during the primary drying process.

For $D \leq 0.5$, as depicted in Figure 5.3b, the quantity of the glassy states approaches or exceeds that of the ice crystals in the matrix. Under such conditions, the

Figure 5.2. Freezing-Point Depression of Water

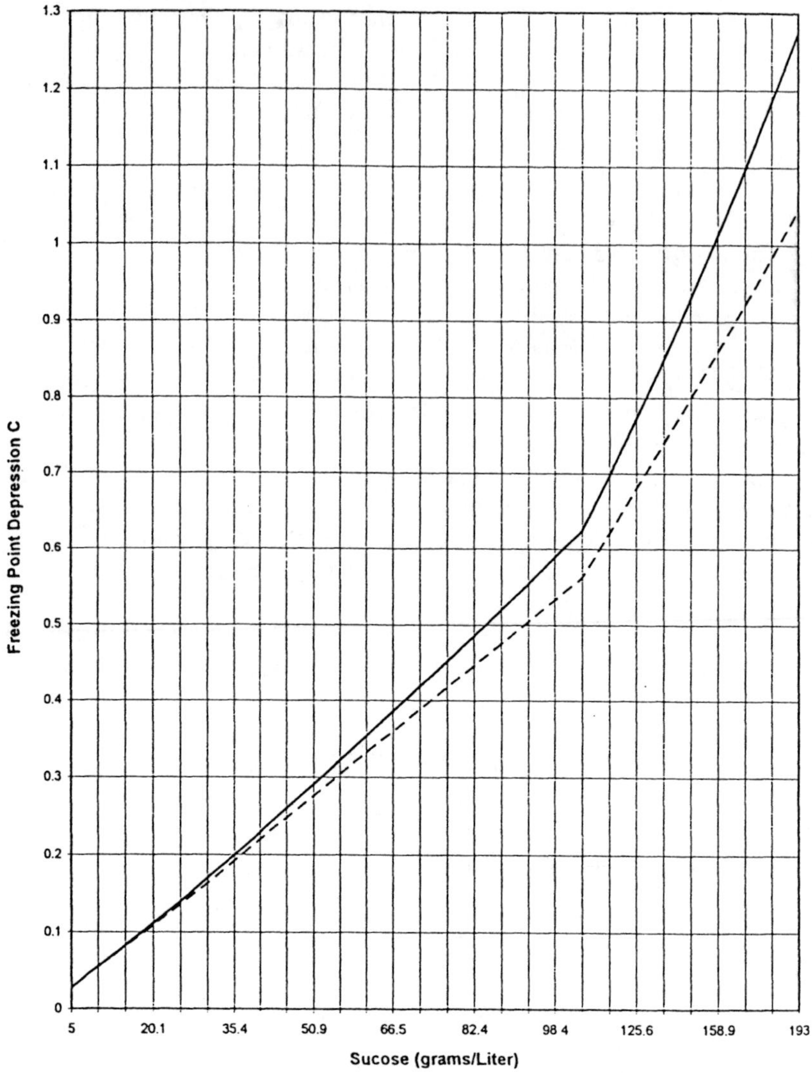

Depression of the freezing point of water as a function of the sucrose concentration expressed in g/L, where the solid line is experimental data obtained from reference [2] and the dashed line is calculated values obtained from equation 1.

ice crystals can be said to be confined in a *sea of glass*. From the phase diagram of water (see Figure 3.1), the dotted line represented the vapor pressure of the supercooled water, which was higher than the vapor pressure of water associated with the ice at the same temperature. However, the supercooled water associated with the interstitial glass also contains the solutes of the formulation. The presence of the solutes

Figure 5.3. Ice Matrices

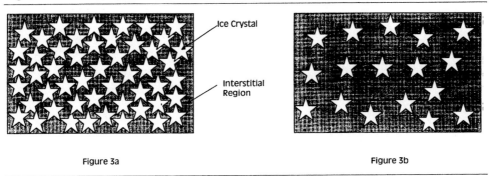

Figure 3a   Figure 3b

Figure 5.3a illustrates a matrix with a *D* approaching 1; Figure 5.3b shows a frozen matrix where *D* ≤ 0.5. The gray area indicates the interstitial region, and the ☆ indicates an ice crystal.

reduces the water vapor pressure of the glass. Under these conditions, the vapor pressure of the ice crystals, for a given temperature, will tend to be higher than that of the frozen glassy interstitial region.

The embedding of ice crystals in the *sea of glass* also greatly reduces the paths by which vapor can flow from the matrix during the primary drying process. Such a restriction on the vapor flow from the matrix will cause a reduction in drying rate. This reduction in the drying rate is a result of the lower vapor pressure of water associated with the glassy state and the impedance of the flow of water vapor from the matrix resulting from the sublimation of the ice crystals.

In the extreme case where *D* approaches 0, there is no formation of ice crystals, and the vapor pressure of the glassy system approaches 0. Under these conditions, primary drying occurs at an extremely low rate, and the formation of poorly defined amorphous layer in place of a porous cake is generally associated with a lyophilized product.

By knowing the *D* of a formulation, one can ascertain if the formulation is a suitable candidate for lyophilization without going through the frustrating trial and error process that is often used to establish a lyophilization process.

EUTECTIC TEMPERATURE

There is, in my opinion, no other term that has been more misused, misunderstood, or abused in the field of lyophilization than the *eutectic* temperature. This section will first define the term and then assess the frequency that a eutectic point is formed during the freezing of a lyophilized formulation.

Definitions

One of the most common definitions of an *eutectic point* is the lowest temperature at which the solution remains a liquid [4,5]. Although this definition is true in a sense, it is not rigorous enough for our purposes. There are two ways in which the eutectic point of a system can be more accurately defined.

Gibb's Phase Rule

The first definition makes use of Gibb's phase rule, which was introduced in the preceding chapter:

$$f = C - P + 2 \tag{4}$$

For a phase diagram of a two-component system ($C = 2$), as illustrated by Figure 5.4, the number of phases (P) present at the eutectic temperature [$T(e)$] is 4, i.e., vapor pressure, solution, A or B, and AB (the eutectic mixture). From equation 4, f at the eutectic point (E) is 0. When $f = 0$, there can be no change in temperature or composition of the solution, until there is a change in the number of phases present [6].

Figure 5.4. Phase Diagram for a Two-Component System

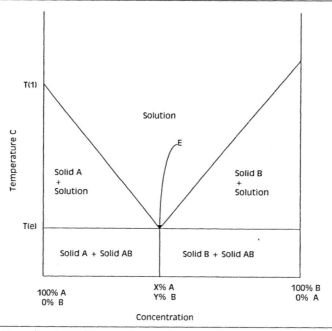

An illustration of a phase diagram for a two-component system (A and B) where E indicates the eutectic point, $T(1)$ is the freezing temperature of component A, $T(e)$ is the eutectic temperature, and AB is the composition of the eutectic.

Freezing-Point Depression

The second definition of the eutectic temperature is illustrated by Figure 5.5, the phase diagram for the NaCl system [7]. From Figure 5.5, the eutectic point can also be defined as the intersection between the freezing-point depression for water (line IE) and solubility curve for the dihydrate and the saturated brine solution denoted by line TE. This definition of the eutectic point was also proposed by DeLuca and Lachman [8].

Frequency of Eutectic Points in Formulations

It has been erroneously stated that most lyophilized drug products are organic electrolytes and exhibit eutectic points [9]. This viewpoint has been propagated by some equipment and instrument manufacturers who offer eutectic monitors for purchase. This general confusion has led some investigators from the FDA to inquire about the eutectic temperature of a formulation for which no phase change occurs in the interstitial region of the matrix.

Of the large number and wide range of formulations that I have examined, the number of formulations found to exhibit a true eutectic point was 1 to 2 per

---

Figure 5.5. Sodium Chloride–Water System

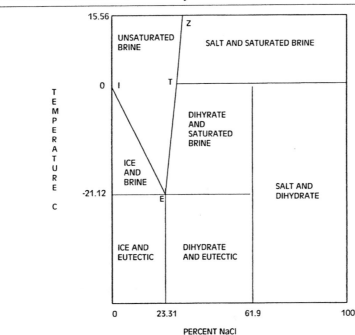

A sodium chloride–water system where E represents the eutectic point; line IE illustrates the depression of the freezing point; line TE is the solubility curve for the dihydrate, and line ZT is solubility curve for the salt [7].

1,000 analyses. Others have also observed an absence of the formation of an eutectic mixture as a result of the freezing process [10, 11]. In the case of sucrose, Franks [12] has proposed the possible existence of an eutectic point for sucrose at a 60% concentration. Such a eutectic point for sucrose has not been observed by this author nor by MacKenzie [11]. Perhaps the single most misleading information for falsely identifying the presence of an eutectic point is the observation of an endotherm peak, other than that associated with the melting of ice in the matrix, during the warming of a frozen formulation. The presence of such a peak may stem from the crystallization of a constituent from solution during the freezing of formulation (see line TE in Figure 5.5), prior to the constituent reaching the eutectic concentration. The onset temperature of the observed endotherm will represent a false high eutectic temperature and merely may be a result of the heat of solution for the constituent.

The key point to bear in mind is that the eutectic temperature can involve only the components in a formulation. If there are n number of constituents in a formulation and only one component (i.e., water), then the frozen system has neither a eutectic mixture nor an eutectic temperature. Even if there are two components in a system having n constituents, where $n > 2$, then an eutectic mixture could be a part of the interstitial region. The remaining constituents would also be in the interstitial region but in glassy state. Therefore, in order for a formulation that consists of n constituents to have a true eutectic temperature, then all n constituents must be components of the phase diagram for the eutectic mixture. As the number of constituents in a formulation is increased, the likelihood of forming such an eutectic system diminishes very rapidly.

COLLAPSE TEMPERATURE

Rey [13–15] showed that in an Earle's solution containing glucose and glycine, no real eutectic point is formed. But Rey refers to what he describes as a eutectic zone: starting at a temperature of -38°C, there was *slow progressive melting of the frozen interstitial material* . . . . The absence of an eutectic mixture in the interstitial region lead to our present understanding of the existence of glassy states that were described in the previous chapter.

Based on the knowledge of an amorphous system in the interstitial region of the matrix, MacKenzie [11], based on results obtained from a freeze-drying microscope, introduced the term *collapse,* which referred to the disappearance of an observed freezing pattern, for a given temperature, at the sublimating interface of the matrix. He found that such a collapse temperature was dependent on the constituents in the formulation. For example, the collapse temperature for dextran was found to be -9°C, whereas, a frozen matrix containing monosodium glutamate had a collapse temperature of -50°C. There also appears to be no simple relationship between the composition of the constituent and the collapse temperature. Sucrose, which consists of fructose and glucose was found to have a collapse temperature of -32°C, while the collapse temperatures for fructose and glucose were -48°C and -40°C, respectively.

I have used electrical measurements to determine the collapse temperature, which will be described in greater detail in the next chapter. For that reason, the following definition of the collapse temperature will differ from·that proposed by

MacKenzie [11]: The collapse temperature is that temperature at which the *mobility* of the water in the interstitial region of the matrix becomes significant. The basis for such a definition is that ion conduction in the interstitial region requires the presence of mobile water. If mobile water is present, then the interstitial region cannot be completely frozen.

In the absence of a well-defined eutectic temperature for the formulation during the freezing process, knowledge of the collapse temperature is essential in determining the temperature necessary to achieve a completely frozen state and the chamber pressure and operating temperatures necessary to conduct the primary drying process.

INTERSTITIAL MELTING TEMPERATURE

The *interstitial melting temperature* is defined by this author as that temperature at which there are known liquid states in the interstitial region of the matrix. The interstitial melting temperature is determined during the warming of the formulation and occurs when the electrical resistance of the formulation, generally at some temperature lower than the onset temperature for the melting of ice in the matrix, is equal to the electrical resistance of the process water used in the formulation at ambient temperature. The interstitial melting temperature is based on the rationale that if the resistance of the process water at ambient temperature is known, then an equivalent resistance for the formulation would also represent the presence of liquid states in the interstitial region of the matrix. The interstitial melting temperature is determined as a means of characterizing the thermal properties of the formulation and is the temperature at which severe collapse or meltback would occur during the primary drying process.

ICE MELTING TEMPERATURE

The ice melting temperature is that onset temperature for the melting of ice in the matrix. For most formulations, the onset melting temperature occurs between -1°C and -10°C. The onset temperature for the melting of ice differs from the equilibrium freezing temperature because the concentration of solutes is much greater than in the original formulation. This onset ice melting temperature is generally not sharp and well defined as that for a pure water sample. In addition, this onset temperature for the melting of ice is sometimes mistaken for the collapse temperature or an eutectic temperature. This thermal property plays no part in determining the parameters for the lyophilization process but, as in the case of other thermal properties, serves to characterize or *fingerprint* the formulation.

METASTABLE STATES

Metastable states are systems within a frozen matrix that are not in their normal ground state and that will revert to their ground state when provided with sufficient energy. The change of the system to the ground state is associated with a release of energy. Rey [13] and Luyet [16] were among the first to recognize the existence of metastable states in frozen aqueous solutions. Rey [13] observed that during warming of a frozen NaCl solution that an *exothermic phenomenon* was observed at temperatures lower than the eutectic temperature. He attributed the exotherm to *recystallization* of one of the components in the system.

Luyet [16] stressed *that crystallization is hindered by rapid cooling, in aqueous solutions, is well established.* The net effect of rapid cooling was to produce states that were semicrystalline or glassy and that could be an intermediate phase between the amorphous and crystalline phases. He also pointed out that the terms *vitreous, vitrification,* and *devitrification* were no longer appropriate and suggested that caution should be exercised in using such terms. For this reason, this author has adapted the physical chemistry approach by classifying such systems as metastable states.

The presence of metastable states has been reported by others [6,10,18,19]. Gatlin and DeLuca [18] froze formulations containing cefazolin sodium and nafcillin at a rate of –10°C/min to a final temperature of –70°C. They reported that upon warming, there were irreversible exotherms (metastable states) with onset temperatures at –11°C for the cefazolin sodium and –4°C for the nafcillin formulations. Thermal treatment of the formulation and removal of the metastable states resulted in lyophilized formulations that X-ray powder diffraction showed to be crystalline. Lyophilized formulations not thermally treated were found to be amorphous.

Formulations should be examined for the presence of metastable states and the impact that these states have on the quality of the lyophilized formulation. It has been the experience of this author that the presence of metastable states in formulations is not a common occurrence, especially when the freezing is performed at a relatively slow rate, < –10°C/min.

ACTIVATION ENERGY

Electrical resistance, in one form or another, has been used for an extended period of time to characterize the thermal properties of a lyophilized formulation (8,13–15,21–25); however, the literature is silent with regard to the conduction mechanism associated with the measurement. In Chapter 3, electrical conductivity was shown, for temperatures > –40°C to be dependent on the presence of a disordered layer on the ice surface. While not completely accepted [26], the *activation energy* along the surface was found to be –33 kcal/mole [27].

It has been proposed that when formulations are in a completely frozen state, the electrical resistivity of the interstitial region will far exceed that of the disordered surface region of the ice crystals [22]. When the formulation is warmed, the conductivity (the reciprocal of the electrical resistivity of a system) of the interstitial region will exceed that of the disordered layer of the ice surface. The temperature at which the conductivity of the interstitial region is significant, greater than that of the ice surface conductivity, is associated with the mobility of the water in the interstitial region and the collapse temperature.

The conductivity of the interstitial region (σ_i) at temperatures greater than the collapse temperature but where liquid states are not present (sometimes defined as the *rubbery* state [12]) can be defined as

$$\sigma_i = \sigma_0 \exp^{-\epsilon/RT} \quad \frac{1}{\Omega \text{ cm}} \tag{5}$$

where ϵ is the activation energy expressed in kcal/mole, R the gas constant, and T is expressed in K. ϵ is determined from the slope of the plot of ln (σ_i) as a function of $1/T$.

It has been observed by this author that, for a given formulation, there can be a relationship between the collapse temperature and the observed ϵ. For formulations having low collapse temperatures (< –40°C) and low values of ϵ (i.e., < 10 kcal/mole), the collapse temperature could be substantially increased (>10°C) by the addition of an excipient that would also increase ϵ to values approaching 30 kcal/mole. It has also been observed that the addition of excipients had no effect on the collapse temperature for those formulations having low collapse temperatures and high ϵ. While ϵ provides a useful thermal parameter to characterize a formulation, it provides no insight to the actual conduction mechanism that occurs with an interstitial region of the matrix in the *rubbery* state. It would appear, however, that with ϵ of the order of 30 kcal/mole, the conduction would be protonic, like that which occurs in the disordered layer of the ice crystal.

The relationship between the low collapse temperature and ϵ values < 10 kcal/mole is that one or more of the constituents in the formulation[1] increased conduction and thereby also depressed the temperature at which the mobility of the water in the interstitial region becomes significant (collapse temperature). The addition of an excipient that complexes with water to form a hydrate appears to enhance ϵ and increase the collapse temperature. Not all excipients have an effect on ϵ; the addition of mannitol has not, at least in the formulations studied, caused a significant increase in ϵ or the collapse temperature. Further discussion regarding the determination of ϵ and the effect of composition will be considered in Chapter 6.

Determination of ϵ in a formulation during a thermal analysis of the formulation can prove useful in selecting excipients and provide a means for increasing the collapse temperature. An increase in the collapse temperature will be shown in Chapter 8 to have a profound effect on reducing the time for the primary drying process.

SYMBOLS

D degree of crystallization during the freezing process

$\Delta H_{f,s}$ heat of fusion identified with the formulation

$\Delta H_{f,i}$ heat of fusion associated with a water reference

ΔT_f freezing-point depression of water

ϵ activation energy expressed in kcal/mole

---

1. The author apologizes to the reader for not providing more details, but confidentiality agreements prevent him from disclosing the nature of the formulations that were altered as a result of the study of activation energy.

f number of degrees of freedom

g_s number of grams of the solute

$g_{s,i}$ number of grams of the ith solute in an aqueous solution

K_f freezing-point constant for water (1.855)

M_s molecular weight of the solute

$M_{s,i}$ molecular weight of the ith solute is an aqueous solution

n number of constituents in a formulation

R gas constant

σ_i electrical conductivity in the interstitial region of the matrix

σ_o electrical conductivity when $RT \gg \epsilon$

W weight of solvent

REFERENCES

1. F. Daniels, et al., *Experimental Physical Chemistry*, McGraw-Hill Book Company, Inc., New York, 1956, p. 65.

2. *Handbook of Chemistry and Physics* (R.C. Weast, ed.), 65th Edition, CRC Press, Inc., Boca Raton, Florida, 1984, p. D-265.

3. T. A. Jennings, *IVD Technology*, March/April 1997, p. 2.

4. H. T. Meryman, *Cryobiology* (H. T. Meryman, ed.), Academic Press, New York, 1966, p. 2.

5. *Handbook of Chemistry and Physics* (R.C. Weast, ed.), 65th Edition, CRC Press, Inc., Boca Raton, Florida, 1984, p. F-80.

6. T. A. Jennings, *D&C*, November 1980, p. 43.

7. D. W. Kaufmann, *Sodium Chloride Production and Properties of Salt and Brine* (D. W. Kaufmann, ed.), American Chemical Society Monograph Series #145, Hafner Publishing Co., New York, 1968, p. 548.

8. P. DeLuca and L. Lachman, *J. Pharm. Sci.*, 54: 617 (1965).

9. P. P. DeLuca, International Symposium on freeze-drying of biological products, Washington, D.C. 1076. *Develop. Biol. Standard*, 36: S. Karger, Basel, 1977.

10. E. Maltini, in *Freeze Drying and Advanced Food Technology* (S. A. Goldblith, L. Rey, and W. W. Rothmayr, eds.), Academic Press, New York, 1975, p. 121.

11. A. P. MacKenzie, loc. cit., p. 277.

12. F. Franks, in International Symposium on Freeze-Drying and Formulation Bethesda, USA, 1990, *Develop. Biol. Standard*, 74: S. Karger, Basel, 1991, p. 9.

13. L. R. Rey, *Annals N.Y. Academy of Sci.*, 85: 510 (1960)

14. L. R. Rey, *Recent Research in Freezing and Drying* (A. S. Parkes and A. Smith, eds.), Blackwell Scientific Publications, Oxford, England (1960).

15. L. R. Rey, *Biodyamica*, 8: 241 (1961).

16. R. Luyet, *Recent Research in Freezing and Drying* (A. S. Parkes and A. Smith, eds.), Blackwell Scientific Publications, Oxford, England (1960).

17. L. R. Rey, International Symposium on freeze-drying of biological products, Washington, D.C. 1076. *Develop. Biol. Standard*, 36: S. Karger, Basel, 1977.

18. L. Gatlin and P. P. DeLuca, *J. Pharm, Drug Assoc.*, 34: 398 (1980).

19. R. H. M. Hatley, in International Symposium on Freeze-Drying and Formulation, Bethesda, USA, 1990, *Develop. Biol. Standard*, 74: S. Karger, Basel, 1991, p. 105.

20. P. DeLuca, L. Lachman, and R. Withnell, *J. Pharm. Sci.* 54: 1342 (1965).

21. P. P. DeLuca, in International Symposium on freeze-drying of biological products, Washington, D.C. 1076. *Develop. Biol. Standard*, 36: S. Karger, Basel, 1977, p. 41.

22. T. A. Jennings, *Parenter. Drug Assoc.*, 34: 109 (1980).

23. N. A. Williams and G. P. Polli, *J. Parenter. Sci. and Tech.*, 38: 48 (1984).

24. A. J. Phillips, R. J. Yardwood, and J. H. Collett, *Analytical Proceedings*, 23: 394 (1986).

25. P. P. DeLuca, M. S. Kamat, and Y. Koida, *Cong. Int. Technol. Pharm*, 5th, 1: 439 (1989).

26. M. A. Maidique, A. von Hipple, and W. B. Westphal, *J. Chem. Phys.*, 54: 150 (1971).

27. N. Maeno and H. Nishimura, *J. of Glaciology*, 21: 193 (1978)

6

Thermal Analytical Methods

INTRODUCTION

Shackell [1] was the first to determine that in order to stabilize various materials, one must freeze them prior to drying them at subatmospheric pressures (see Chapter 1). It is highly unlikely that he was aware of the role that the thermal properties of materials would play in determining the drying process. In addition, he did not have at his disposal any analytical instrumentation to determine the thermal properties that were described in the previous chapter.

The objective of this chapter is to review the various analytical methods that can be used to characterize the thermal properties of a formulation. Although this author has a preference with respect to a given analytical method, every effort will be made to present a fair and objective evaluation of other methods that have been or are currently being used to evaluate the thermal characteristics of a material. The evaluation of an analytical method will include (a) a general description of the apparatus and the analytical method, (b) those thermal properties that can be determined by the method, and (c) general comments regarding the accuracy and precision of the method in determining the various thermal characteristics.

THERMAL ANALYSIS

The most fundamental and simplest method for determining the thermal properties of a formulation is by a *thermal analysis* of the formulation [2]. A thermal analysis is conducted by plotting the temperature of a formulation as a function of temperature. In the case of a lyophilized formulation, the temperature span will typically be from ambient to < -50°C. A marked deviation in the plot is associated with an exothermic or endothermic phase change. This method was one of the first analytical methods used to characterize the thermal properties of a formulation, but it soon gave way to other methods [3]. The apparatus required to conduct a thermal analysis is relatively easy to construct and use, but, as will be seen, the number of thermal properties this method can determine is limited.

Description of the Apparatus

The configuration of the apparatus can vary, depending on the nature of the materials. For a lyophilized formulation, the thermal analysis apparatus can be constructed like that illustrated by Figure 6.1. This figure shows that the container used in the analysis is the same as the lyophilized formulation. The advantage of using the product container is that the results will include any impact that the container will have on the thermal properties of the formulation. The latter will be applicable provided that the container is not an exceptionally large (e.g., ≥ 1 L) bottle or is a tray. If large containers or trays will be used to lyophilize the formulation, then smaller containers of the same composition and finish should be used to determine the thermal characteristics.

Two type T (copper-constantan) thermocouples are used to determine the temperature of the formulation and that of the surface of the shelf on which the container rests. The thermocouple in the formulation, $T(p)$, is housed in an inert substance, such as Teflon®, so that it cannot chemically interact with the formulation during the course of the analysis. The junction point of thermocouple $T(p)$ (i.e., the point at which the two dissimilar metals are connected) is positioned in the middle of the container and about 1/8 in. from the bottom. The shelf thermocouple, $T(s)$, is attached directly to the shelf surface. The purpose of $T(s)$ is to record the shelf

Figure 6.1. A Thermal Analysis Apparatus for Determining the Thermal Properties of a Formulation

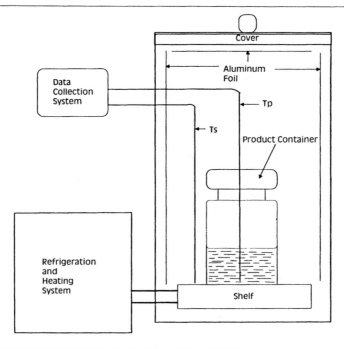

temperature during both the cooling and warming portions of the analysis. This plot will prove useful when comparing thermal properties of formulations (i.e., any difference in thermal properties is not attributed to the rate of cooling or warming). The output temperatures of the thermocouples are recorded by computerized data recording or a two-pen recorder.

The shelf of the apparatus is constructed so that it can be cooled and heated by the heat-transfer fluid from the refrigeration and heating system. The chamber and cover of the test chamber are lined with aluminum foil in order to minimize any radiant heat transfer to the sample during the course of the analysis. A small freeze-dryer can be used, in place of the apparatus shown in Figure 6.1, to conduct the thermal analysis.

Analytical Method

The following is a general description of the method used to conduct the thermal analysis of the formulation with the above apparatus.

Container

A container selected for the thermal analysis should be cleaned in the same manner that will be used during the manufacture of the lyophilized product.

Certification of the Thermocouples

The thermocouple output from the temperature recording instrumentation for $T(p)$ and $T(s)$ should be certified as being within the accuracy and precision limits specified by the instrumentation manufacturer at no less than two temperatures (e.g., an ambient temperature water bath using a NIST traceable thermometer and at 0°C using an ice-slush bath prepared from deionized or distilled water).

Fill Volume

A fill volume of the formulation is added to a clean container, preferably one cleaned in the same manner that will be used during the lyophilization of the formulation.

Setup of the Apparatus

The container containing the fill volume of the formulation is positioned in the middle of the shelf, and the Teflon® shielded thermocouple is positioned in the container as described above. $T(s)$ is attached to the shelf surface, and the lid to the apparatus is placed in position to reduce radiant heat transfer to the sample and prevent moisture from entering the test chamber.

Cooling

The refrigeration system is turned on, and the chilled heat-transfer fluid is used to lower the shelf temperature. The temperature recording instrumentation is turned on to record $T(p)$ and $T(s)$ as a function of time.

Warming

When the lowest formulation temperature is reached, the refrigeration system is turned off, and the heating system is used to warm the formulation to ambient temperature. It is recommended that the rate of warming be adjusted with respect to the fill volume so that there will not be a significant temperature difference between the bottom of the container and the top of the formulation.

Interpretation of the Data

A thermal analysis of an isotonic solution (0.9%) of NaCl was conducted using the above procedure and in an apparatus similar to that shown in Figure 6.1. A fill volume of 2 mL of the NaCl solution was added to a 5 mL tubing vial. The following are the results and interpretation of the cooling and warming data.

Cooling Data

The results of the cooling of the NaCl chloride solution are graphically[1] shown by Figure 6.2 and the tabular data are listed in Appendix A. Figure 6.2 and the tabular data reveal the following thermal properties during the cooling process.

Degree of Supercooling and Equilibrium Freezing Temperature. The degree of supercooling is shown in Figure 6.2 by the sharp increase in $T(p)$, indicated by the arrow A. On page 194, Appendix A, nucleation of ice crystals in the isotonic solution occurred at the 45.1 min mark when $T(p) = -10.5°C$ and the system attained an equilibrium freezing temperature of $-0.5°C$. The calculated equilibrium freezing temperature for an NaCl chloride solution (based on equation 1 in Chapter 5) is $-0.3°C$. From the definition of the degree of supercooling (see Chapter 5), the absolute difference in temperature between the onset of the nucleation of the ice crystals and the equilibrium freezing temperature is $10°C$.

Phase Change in the Interstitial Region. A further examination of Figure 6.2 shows that by arrow B there is a small change in rate of cooling of the isotonic solution. Page 196, Appendix A, shows that, for the time period between 79.4 min and 82.0 min, $T(p)$ remained between $-28.9°C$ and $-29°C$. This change in the rate of cooling of the solution is associated with supercooling and the formation of the eutectic mixture in the interstitial region of the matrix. Note that the latter supercooling refers to the *supercooling of the eutectic solution* that was formed in the interstitial region of the matrix as a result of the freezing process.

Warming Data

The results of warming the 0.9% NaCl solution are graphically shown by Figure 6.3 and the tabular data are listed in Appendix A. Figure 6.3 and the tabular data show the following thermal properties of the solution during the warming process.

---

1. Most graphs presented by the author in this text, unless cited elsewhere, were prepared using Microsoft® Excel, Version 5.0, produced by the Microsoft Corporation.

Figure 6.2. Temperature Versus Time for a Cooling Process

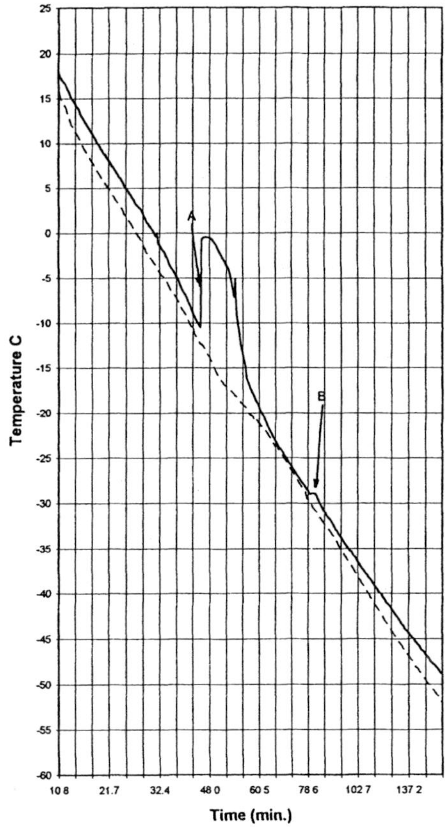

A plot of the temperature of an isotonic (0.9% wt/v) solution of NaCl, indicated by the solid line, and shelf-surface temperature, indicated by the dashed line, as a function of time during the cooling process.

Eutectic Point. The eutectic point for the isotonic NaCl solution is barely detectable during the warming process, as indicated by arrow A in Figure 6.3. An examination of the tabular data on page 199, Appendix A, shows that $T(s)$ increased from -19.9°C to -17.6°C, while the change in $T(p)$ was from -21.3°C to -20.1°C for the time span (70.3 min to 75.4 min). Because the temperature of the 0.9% NaCl solution changed more slowly than the temperature of the shelf surface is an indication that an eutectic point was present at a temperature approaching -21°C.

Ice Melting Temperature. Further examination of Figure 6.3 shows that the onset temperature for the melting of ice in the formulation occurs between -5°C and 0°C. However, the melting of ice in the system is not sharp and well defined. An examination of the tabular data (see pages 200–201, Appendix A) also does not provide any insight about the onset temperature for the melting of the ice crystals.

Figure 6.3. Temperature Versus Time for a Warming Process

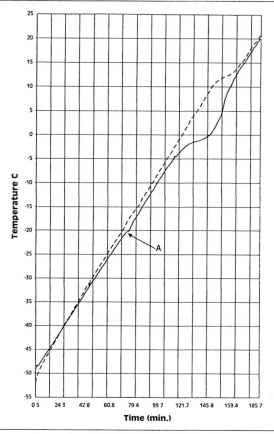

A plot of the temperature of an isotonic (0.9% wt/v) solution of NaCl, indicated by solid line, and shelf-surface temperature, indicated by the dashed line, as a function of time during the warming process.

Summary and Comments

The following is a summary of the thermal analysis apparatus, analytical method, and results for determining the thermal properties of an isotonic NaCl solution.

Apparatus

The apparatus required for conducting a thermal analysis of a formulation is relatively simple and easy to set up. An R&D–type freeze-dryer would serve as a suitable substitute for the apparatus illustrated in Figure 6.1. Using the actual product container with the appropriate fill volume will give a realistic assessment of the nature of the thermal properties that the formulation would experience during a lyophilization process.

Analytical Method

The analytical method is relatively simple and can be performed by personnel without any special training beyond normal laboratory practices.

Thermal Characteristics

Thermal analysis of the 0.9% NaCl solution did provide a measure of the following thermal characteristics:

- Degree of supercooling
- Equilibrium freezing temperature
- Phase change in the interstitial region during cooling
- Eutectic temperature

To determine the above thermal characteristics, graphic and tabular data are required. For example, accurately assessing the equilibrium freezing temperature and the eutectic temperature without the use of the tabular data would have been difficult.

The following thermal characteristics could not be ascertained from the thermal analysis data:

- Ice-like water clusters
- Degree of crystallization
- Collapse temperature
- Ice melting temperature
- Metastable states (Since the eutectic temperature was barely discernible from the data, it is unlikely that this method would be sensitive enough to determine the presence of metastable states, which may involve a lower quantity of latent energy.)
- Activation energy

While the apparatus and methodology for a thermal analysis of a formulation is relatively simple and easy to perform, the method would not be suitable for the vast majority of formulations that, during the cooling process, form a glassy interstitial region rather than an eutectic point.

DIFFERENTIAL THERMAL ANALYSIS

Differential thermal analysis (DTA) is an analytical technique that compares the difference in the thermal properties of a formulation with that of a reference substance. In the case of the thermal analysis of a lyophilized formulation at temperatures $< 0°C$, the reference substance would be one that does not undergo a phase change over the temperature span of the analysis; however, some investigators have selected water as a reference [4–6]. In one instance, the nature of the reference material was not specified [7]. Because the analysis will involve relatively low temperatures, reference materials such as sand or methanol would also be applicable. In conducting the analysis, the difference between the formulation temperature $T(p)$ and the reference temperature $T(r)$, or $T(p) - T(r)$ is plotted either as a function of time [3] or, as described below, with respect to the *absolute value of the temperature of the specimen, T(p)* [4,6,7]. Current practice is to plot the temperature difference as a function of $T(r)$ [8].

Description of the Apparatus

Although DTA has been used, and continues to be used (an expanded version of DTA analysis will be described in a later section in this chapter) as an analytical means for characterizing thermal characteristics, with the exception of Rey [3,6], most investigators did not describe their apparatus. The DTA sample holder used by Rey consisted of two parts fabricated from stainless steel. The sample and reference holders were blocks of stainless steel with four holes or chambers. A separate stainless steel block contained four thermocouple probes and was designed to fit the hole configuration of the lower block. The output of the thermocouples was recorded using a multipoint strip chart recorder. The following analytical method, used to perform DTA measurements, in order to characterize the thermal properties of a lyophilized formulation, will be described with respect to the apparatus and method described by Rey [3,6].

Analytical Method

The formulation under investigation was added to one chamber, and water reference was added to the second chamber. When the two sections of the apparatus were joined, one of the thermocouples measured the temperature of the formulation, while another thermocouple measured the temperature of the water reference. The other two thermocouples determined *the temperature of the central and peripheral parts of the block* [6].

The assembled sample holder was then placed in a constantly stirred bath of an unspecified liquid. In order to avoid contamination of the formulation or the water reference, the liquid of the bath contacted only a portion of the sample and reference holders. The temperature of the sample and reference holders was lowered by refrigeration of the bath and warmed by means of an electrical resistance heating coil. In some instances, the bath was replaced with liquid nitrogen. After 30 min in liquid nitrogen [6], the cell was removed and allowed to warm up in an isothermal atmosphere. Because supercooling occurred in both the formulation and the reference material, the thermal analysis was conducted only during the warming of the DTA cell [3].

Interpretation of the Data

The following are two examples of the use of DTA to characterize the thermal properties of a formulation using the above apparatus and analytical method. In all fairness to Rey, the above experimental apparatus was constructed in 1958, and since then there has been substantial improvement in the design and construction of DTA equipment [8]. It is to Professor Rey's credit that he published such a complete account of his apparatus, analytical method, and results while other investigators chose merely to report the results of their analyses. The objective in this section is to use his data merely as a basis for a discussion of the interpretation of DTA data for determining the thermal characteristics of formulations.

Sodium Chloride Solution

The DTA plot of a NaCl solution is shown in Figure 6.4. This DTA thermogram, taken as a function of time, shows that the onset temperature of an endotherm for the dilute NaCl solution (solid line) was –26°C and the temperature of the peak was found to be –21.6°C. For the more concentrated NaCl solution (broken line), the onset temperature was reported to be –27°C, while the peak temperature was –21.6°C. The accepted value for the eutectic temperature of the NaCl · H_2O system, at 1 atm,

Figure 6.4. Differential Thermal Analysis of the NaCl · H_2O System

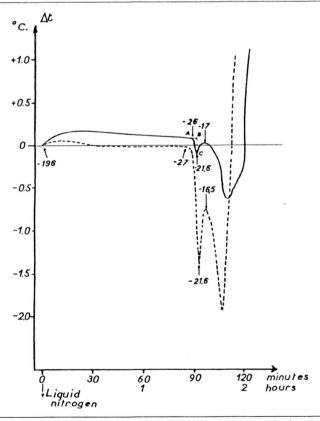

Solid line, 10 parts per 1,000 solution of NaCl in distilled water; *broken line*, 100 parts per 1,000 of NaCl in distilled water. [Source: New York Academy of Sciences, Vol. 85, 1960, Louis R. Rey, "Thermal Analysis of Eutectics in Freezing Solutions," page 518, figure 6.]

is –21.1°C [9]. The onset temperatures for the melting of ice crystals for the dilute and concentrated NaCl solutions were –17°C and –16.5°C, respectively.

Earl's Solution with Glucose and Glycine

The modified Earl's (Er.G.g) solution consists of the following constituents in g/L:

Sodium chloride	6.80
Potassium chloride	0.40
Calcium chloride	0.20
Magnesium sulfate	0.20
Monosodium sulfate	0.14
Sodium bicarbonate	2.20
Glucose	101.00
Glycine	100.00

The DTA thermogram for the solution is shown by Figure 6.5 [4]. This thermogram shows a small exothermic peak with an onset temperature of –65°C. Rey [4] reports that the exotherm was present in each DTA thermogram but varied in amplitude. He

---

Figure 6.5. Differential Thermal Analysis Diagram of the Er.G.g. Solution

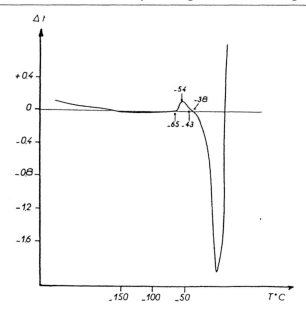

The reference solution was distilled water.

attributes the exotherm to *partial recrystallization of the interstitial material*; its presence was attributed to the nature of the freezing process and the presence of high concentrations of glucose and glycine in the solution. The onset temperature for the endothermic peak was –43°C but was found not to be sharp and well defined. Rey concluded that the Er.G.g solution did not form a eutectic point but did form what he referred to as an *eutectic zone*.

Summary and Comments

The following is a summary of the DTA technique for determining the thermal properties of a formulation and some comments regarding the above apparatus, method, and thermograms.

Apparatus

The sample and reference cells of the apparatus were constructed from blocks of stainless steel. Stainless steel is not a good thermal conductor, and warming the sample from liquid nitrogen temperatures to near 0°C required approximately 2 hr. Although the volume of the sample and the reference distilled water was not specified, the volume will not vary from analysis to analysis, so a reasonable comparison of the thermal properties of a formulation can be obtained. When the instrumentation available at the time this apparatus was constructed is taken into account, the design and construction of this DTA system is certainly more than adequate.

Analytical Method

There are three main concerns regarding the analytical method used to conduct the DTA analysis.

Distilled Water Reference. The first concern is the use of distilled water as a reference. Water may appear to be a logical choice for reference, however, supercooling will occur in both the formulation and the water reference. The use of water as a reference prevents any assessment of the thermal properties of the formulation during the freezing process.

DTA Thermogram. The temperature differential between the formulation and the water reference were plotted as a function of time or of the formulation temperature [3]. Since the degrees of freedom at the eutectic point are 0, the temperature of the formulation cannot change until one of the phases is removed by the warming process. The temperature of the water reference will continue to increase. If the system is heated at a relatively fast rate, the resulting thermogram will not consist of an endothermic peak but will have an endothermic spike. The amplitude of the spike will depend on the rate at which the water reference is heated and the quantity of the eutectic mixture present in the sample. In those DTA plots [3] that were made based on an ice temperature, an endothermic peak was obtained for the eutectic point.

Determination of the Eutectic Temperature. The amplitude of the endotherm peak for an eutectic point will vary with respect to the initial concentration of the NaCl in the sample. As the concentration of the NaCl is increased, the temperature at which the peak of the endotherm occurs will also increase. The determination of the eutectic temperature is defined by the onset temperature rather than the peak temperature. The onset temperatures for the two NaCl solution of –26°C and –27°C appear to be in error and may be a result of the nature of the apparatus.

Thermal Characteristics

The following is a discussion of the thermal properties that were determined by using the DTA method.

Eutectic Temperature. When the results of the DTA thermogram for the NaCl solution of 10 parts per 1,000, i.e., a 1% solution (see Figure 6.4), are compared to the thermal analysis warming plot for the 0.9% NaCl solution (see Figure 6.3), the eutectic endotherm for the DTA thermogram is much more apparent than that seen in the thermal analysis plot.

Metastable States. The presence of an exothermic peak associated with a metastable state, perhaps stemming from a recrystallization process, in the thermal properties of the Er.G.g solution is readily discernible in Figure 6.5. Since the grid spacing on y-axis was 0.4°C, it is highly unlikely that such an exotherm would have been observed on a warming thermal analysis plot as described in the previous section of this chapter.

Collapse Temperature. Although the concentrations of glucose and glycine were relatively high for a formulation, the onset temperature of –43°C for the observed broad endothermic peak could be representative of the collapse temperature. The amplitude of the endothermic peak is such that it would also have been apparent in a thermal analysis of the formulation.

Limitations

There are some limitations to the use of the DTA method in determining the thermal characteristics of a lyophilized formulation. Some of these limitations are a result of technique such as the use of water reference while other limitations are inherent in the nature of the analytical method.

Technique. As a result of using water reference and rapidly freezing the formulation by using liquid nitrogen to refrigerate the sample, the DTA method cannot determine the following thermal characteristics of a formulation:

- Formation of ice-like water clusters

- Degree of supercooling

- Equilibrium freezing temperature

- Degree of crystallization

It will be shown later in this chapter how DTA measurements can be used to ascertain these thermal characteristics of a formulation.

Inherent Limitations of DTA Measurements. There are some thermal characteristics that inherently cannot be ascertained by DTA measurements:

- Interstitial melting temperature

- Activation energy

- Collapse temperature (Although the results of Rey [3] indicate that it may be possible to determine the collapse temperature by DTA analysis, it has been my experience, based on analysis of about 2,000 formulations, that the collapse temperature could not be ascertained by this analytical technique.)

The measurement of these thermal characteristics will also be considered in a later section of this chapter.

DIFFERENTIAL SCANNING CALORIMETRY

Unlike the DTA method, *differential scanning calorimetry* (DSC) measures the difference in the heat flow (energy per unit time) between the sample and a reference rather than a difference in the sample and reference temperatures [10–13]. The heat flow is then plotted as a function of the reference temperature. This section will provide a general description of the apparatus and its function, the analytical method, interpretation of the data, and a summary and comments regarding the DSC method.

Description of the Apparatus

There are basically two types of DSC equipment available. The following is a brief description of the operating principle used in these systems and a discussion of the characteristics that are common to each system.

Heat Flux Design

A schematic drawing of a DSC cell that employs the *heat flux* design is illustrated by Figure 6.6 [14]. As seen from Figure 6.6, the sample and reference pans are mounted on raised platforms formed on a constantan metal disk. The constantan disk is connected to a silver heat-transfer block that controls the rate of cooling or heating of the sample pan by means of a special *feedback-control temperature controller* [14]. Under each of the pans is a chromel wafer connected to the raised portions of the constantan disk. Attached to each of the chromel wafers is a chromel-alumel thermocouple. As energy is transferred through the disk, the differential in heat flow between the sample and reference pans is determined.

The calorimetric sensitivity is determined from the electronic linearization of the cell calibration constant. Based on knowledge of the cell calibration constant, the temperature differential between the reference and sample cells can be equated to

Figure 6.6. A General Schematic Drawing of the DSC Sample Holder

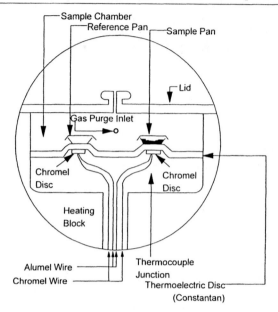

Based on the DuPont DSC 910 instrument. [Source: *The TA Hotline*, 1995, Vol. 3, Back page, Figure 1.]

the latent heat associated with a phase change or reaction occurring in the sample pan. Since the instrumentation is based on the sample pan temperature, the sample can be either maintained under isothermal conditions or heated or cooled at rates ranging from 0.01°C/min to 200°C/min.

Purge gas is passed through the cell. The gas enters the cell at the gas inlet, as shown in Figure 6.6, after being adjusted in temperature by prior circulation through the silver heat-transfer block. The function of the gas flow is to provide a uniform thermal environment for the cell throughout the analysis and to achieve heating rates of 50°C/min [14]. The pans can be sealed by crimping on a metal cover. A thermoelectric device is attached between the heat-transfer block and the sample pan. Based on the sign and magnitude of the temperature difference between the sample and reference pans, the control circuit of the DSC will determine the polarity and magnitude of the applied voltage to be supplied to the thermoelectric device. The power consumed by the thermoelectric device to either heat or cool the sample pan will be directly related to the energy flow to the sample pan necessary to compensate for the latent heat associated with sample. For example, if the sample passes through an eutectic point, the thermoelectric device will provide the energy to offset the endothermic energy of the melting of the eutectic mixture in order to maintain the temperature of the sample pan equal to that of the reference pan.

The temperature of the heat-transfer block can be controlled either by refrigeration or by electrical resistance heating. A lid is placed over the sample chamber to permit purging of the chamber with nitrogen or some inert gas prior to the start of the analysis. The temperatures of the pans are determined from thermocouples in contact with the bottom of the pans.

Null-Balance System

The control loops that permit the DSC instrument to use the *null balance* principle, the sampling technique, and the calibration of the system will be described in this section [15].

Control Loops. In this system, individual heaters are used to supply energy to the sample and reference containers. Using separate heaters allows for use of the *null balance* principle. This principle is illustrated by Figure 6.7, which shows that each pan holder has separate heaters and the system consists of two control loops. The

Figure 6.7. The Null-Balance Principle

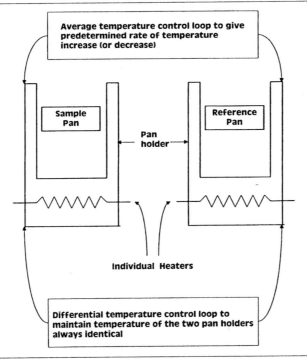

The null-balance principle for a DSC instrument showing the individual heaters for the pan holders and electronic control loops (average temperature control and differential temperature control).

function of the average temperature control loop is to ensure that the sample and reference temperatures are increased at some predetermined rate.

The second loop is referred to as a differential control loop. Its function is to remove any of the temperature difference between the sample (Tp) and the reference (Tr) that may result from an exothermic or endothermic reaction in the sample. It does this by adjusting the power input to the pans to remove any difference. Should an endotherm occur in the sample during the warming process, additional energy will be supplied to heat the sample to prevent any difference in temperatures between the pans. In the event of a metastable state during warming of the sample, additional energy will be provided to the reference pan in order to maintain a null balance.

Sampling Technique. In using the null-balance principle to ascertain the thermal properties of a formulation, there should be good thermal contact between the sample and the pan surface. Since the formulation will be in a liquid state, good thermal contact with the pan would require that the sample adhere to or wet the surface of the pan. Most formulations will wet the typical aluminum pan that is used to conduct the DSC analysis. However, for a formulation that exhibits poor wetting characteristics, pans constructed of a noble metal such as gold may be required. In order to obtain well-defined thermograms, the mass of the sample should range from 5 mg to 10 mg.

Calibration. As with any quantitative analytical instrument, calibration is an important step prior to performing a thermal analysis. The reference substance used should have a well-defined heat of fusion, ΔH_f (e.g., water is 79.7 cal/g). The calibration should be performed at the same heat rate [$dT(p)/dt$] and sensitivity that will be used for the sample. In calibration, the constant k' for the instrument is obtained from the following expression [15]:

$$k' = \frac{\Delta H_f m_c}{A_c} \quad \text{mcal/unit area} \tag{1}$$

where m_c is the mass of the reference substance and A_c is the area under the curve generated by the endotherm for the reference substance.

General Characteristics of DSC Systems

The sample and reference pans are fabricated from either aluminum or gold [16].

Analytical Method

The following is a general outline of the steps used to determine the thermal characteristics of a formulation.

Calibration and Certification

Prior to using the DSC instrument, it is essential to calibrate the instrument in accordance with the instructions provided by the instrument manufacturer. The accuracy and precision of the DSC instrument is checked by determining the melting temperatures of known reference substances: e.g., octane (melting point, -57°C) [17], n-decane (melting point -29.7°C), and water (melting point 0°C) [18].

Preparation of the Sample and Reference

The sample container is tared, and a mass of the formulation is added. The quantity of the formulation contained in the sample pan varies from 6 mg to 10 mg [11,17,18]. The sample pan is then crimped sealed to prevent a change in concentration of the formulation that may result from the evaporation of water. Increasing the sample size (e.g., > 20 mg) can result in a temperature gradient across the sample and an error in determining the onset temperature [11].

In most instances, the DSC measurements are conducted using an empty reference pan [17,18]; however, some investigators have suggested the use of an inert substance such as sand [11]. When using an inert substance in the reference pan, the mass of the reference should approach that of the sample.

Freezing of the Sample

The reported cooling rates to freeze the sample range from 1°C/min to 10°C/min [17,18]. Samples were, in general, frozen to temperatures lower than -70°C.

Warming Temperature Scanning Rate

The selection of a suitable temperature scan rate is generally based on the sample size. If the temperature scan rate is too fast, for a given sample size, the result can be a displacement of onset temperature to a higher temperature. Actual temperature scan rates have been reported as low as 0.625°C/min to a rate as high as 300°C/min [17,19].

Interpretation of the Data

The following will present DSC thermograms and the determination of the thermal properties of a formulation during cooling and warming of the sample.

Cooling Thermogram

None of the investigators using DSC to characterize the thermal properties of their lyophilized formulations showed thermograms taken during the cooling of the sample which does not eliminate the possibility of the existence of cooling thermograms. In addition, none of the investigators cited above offered any explanation regarding the absence of such thermograms.

Warming Thermogram

The various thermal properties that can be ascertained by DSC measurements are illustrated by the following thermograms.

Eutectic Temperature. The eutectic temperature for a 10% mannitol solution is seen in the DSC thermogram shown in Figure 6.8. The onset temperature for the small endotherm was not reported, but I estimated it to be approximately –24°C. The rather broad exothermic peak that occurs after the smaller endotherm is most likely a crystalline phenomenon associated with mannitol. The larger endothermic peak is quite symmetrical in nature and gives no indication of a dual peak composed of a reported eutectic point for mannitol at –5°C and the melting of the ice phase in the matrix. The exotherm related to the crystallization of the mannitol has been reported by others to have an onset temperature of –28°C. It is possible that the results shown by the thermogram in Figure 6.8 are a result of too rapid a warming temperature scanning rate for the sample size of 9.6 mg.

The thermogram shown in Figure 6.9 is an illustration of the determination of an eutectic point at temperatures near the melting temperature of ice. An examination of this thermogram shows the presence of two distinct, separate endothermic peaks for each of the concentrations. Support for the belief that the smaller endotherms represent a true eutectic point is that (a) the onset temperature of the peaks (–2.8°C) is independent of the concentration of the pentamidine isethionate and (b) the peak for the 15% pentamidine isethionate occurs at a higher temperature than that of the 5% concentration of the compound. Although the sample weight was 10 mg, similar to that used previously, the warming temperature scanning rate used

Figure 6.8. Power-Time Curve for a Frozen 10% Mannitol Solution

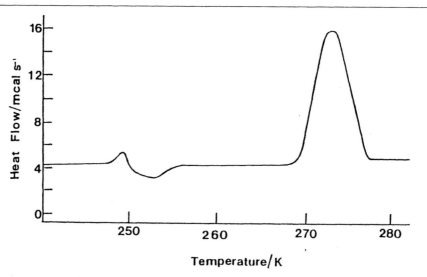

The power-time curve obtained on heating a frozen solution of 10% mannitol. This carbohydrate displays crystallization on warming. Again, heating and cooling rates of 5 K/min were used with a sample size of 9.6 mg. A crystallization can be seen following the glass temperature. The endotherm is a combined event: ice and eutectic melting. [Source: *Developments in Biological Standardization*, Vol. 74, Copyright 1992, Page 108, Figure 2. Courtesy of the International Association of Biological Standardization, Geneva, Swtizerland.]

Figure 6.9. Warming Thermograms for Pentamidine Isethionate

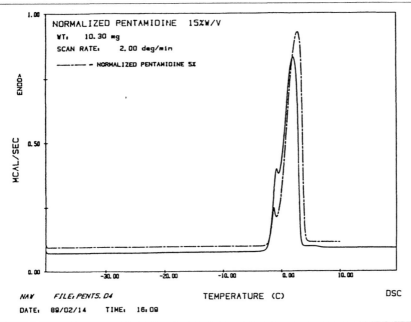

NORMALIZED PENTAMIDINE 15%W/V
WT: 10.30 mg
SCAN RATE: 2.00 deg/min
— — — = NORMALIZED PENTAMIDINE 5%

ENDO>

MCAL/SEC

NAW   FILE: PENTS. D4   TEMPERATURE (C)   DSC
DATE: 89/02/14   TIME: 16:09

Warming thermograms of 5% and 15% wt/v pentamidine isethionate after cooling at 2°C/min. The solid line represents the thermogram for the 15% wt/v and the broken line the 5% wt/v pentamidine isethionate solutions. [Source: *Journal of Parenteral Drug Association*, May–June 1994, Vol. 48, No. 3, Page 137, Figure 2.]

to obtain the thermogram in Figure 6.9 was 2°C/min and gives further credence to the possibility that the temperature scanning rates (5 K/min) used to obtain the thermogram shown in Figure 6.8 was excessive.

Metastable States. The presence of metastable states during warming can be determined from a DSC thermogram. An example of a thermogram indicating the presence of a metastable state in a frozen 29.7% wt/wt cefazolin sodium solution is shown in Figure 6.10 [17]. The metastable state is represented by the exothermic peak defined by points C, D, and E. The onset temperature for the exotherm at point C was -11°C. It was shown that if the frozen cefazolin sodium solution was warmed to a temperature just beyond the exothermic peak (e.g., -6°C) and then cooled to -25°C, the resulting DSC thermogram did not exhibit an exothermic peak during the warming process [17]. The absence of the exotherm in the second warming thermogram is an indication that the initial exothermic peak was a result of an irreversible process (i.e., the presence of a metastable state).

Ice Melting Temperature. The onset temperature for the ice melting temperature can be ascertained from the DSC thermogram. An example of such a thermogram is shown in Figure 6.10, where the onset temperature, indicated by point F, was determined to be -4°C [16].

Figure 6.10. DSC Thermogram for the Warming of a Frozen Cefazolin Sodium Solution

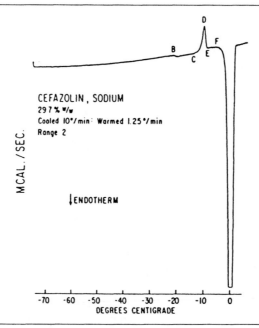

[Source: *Journal of Parenteral Drug Association*, Sept.–Oct. 1980, Vol. 34, No. 5, Page 402, Figure 2.]

Degree of Crystallization. The degree of crystallization from a DSC thermogram is obtained by dividing the area under the ice melting endotherm, as shown by Figure 6.10, by the heat of fusion for pure water [11]. From such a measurement, the quantity of unfrozen water in the formulation at the onset temperature for the melting of the ice can be determined. Errors in making such a determination have been pointed out by Hatley [11].

The first error, and perhaps the most serious of all errors, is that the onset temperature indicates the onset of the presence of water in the sample. As further ice is melted, a temperature gradient may occur across the sample. Should such a temperature gradient occur, which will be dependent on the sample size and warming temperature scanning rate, then the energy associated with the heat capacity of that portion of the water at temperatures > 0°C should be taken into account when determining the heat of fusion.

Hatley [11] states that as a result of depression of the *melting* temperature of ice by the solutes in the solution, there will be a corresponding depression of the latent heat of fusion for the pure ice crystals. For example, the heat of fusion for water at 0°C is 79.7 cal/g, while at –20°C the heat of fusion is reduced to 67.7 cal/g; a 1°C suppression of the melting temperature will result in a heat of fusion of 79.2 cal/g.

I find two difficulties with Hatley's [11] assertions. The first is that solutes will tend to depress the freezing point of water and not the melting temperature of frozen system. Second, if one accepts this model for temperature dependence of the heat of fusion, then at a temperature of $\leq -133°C$ the heat of fusion of water would be 0 and a liquid state would exist at 0 K which would contradict the Third Law of Thermodynamics. In reality, the heat of fusion of ice will increase rather than decrease at temperatures lower than 0°C. A close approximation of the heat of fusion of ice at -20°C, based on the specific heat for ice of 0.4668 cal/g °C [20], would be 89.1 cal/g.

Hatley [11] asserts that the heat of dilution does not contribute significantly to the area under the endothermic peak for the melting of the ice. But the presence of a crystalline excipient, such as mannitol or pentamidine isethionate [18] can increase the area under the curve and provide a false low value for the unfrozen water.

Hatley [11] abandons the DSC thermogram as a means of determining the unfrozen water and recommends the determination of the unfrozen water from *a* T_g *(the glass temperature of the dried product)/moisture profile using very concentrated mixtures.* It is not clear to me just what is meant by the latter statement.

Collapse Temperature. The determination of the collapse temperature, *Tc*, from DSC thermograms is illustrated in Figure 6.11. This figure shows that *Tc* is dependent on the protein concentration. The thermogram shows that the lower concentration of protein had a lower *Tc* than the freezing temperature that was routinely used to freeze the formulation. *Tc* for the higher protein concentration was -47°C; *Tc* for the lower concentration was not reported.

Summary and Comments

The above has described the DSC apparatus, the analytical method, and the thermal characteristics that can be obtained from a sample of a formulation. Like all analytical methods, it has its distinct advantages and limitations that should be considered when using the DSC.

Apparatus

The DSC instrument is a sophisticated instrument and is used not only in determining the characteristics of lyophilized formulations but also in many other fields, such as ceramics, petroleum products, and materials used in the electronics industry. As a result, the upper temperature limit for DSC measurements is in excess of 200°C [13]. The design of the instrument is such that energy associated with the latent heat associated with phase change or chemical reaction is displayed as a function of temperature. Since a computer is used not only to control the instrumentation but also to store and display the data, today's DSC instruments can be considered user-friendly.

Analytical Method

In spite of current design, the use of DSC to characterize the thermal properties of a lyophilized formulation does require some expertise on the part of the user. The following are some advantages and disadvantages of the analytical method.

Figure 6.11. DSC Heat Scans for Two Protein Concentrations

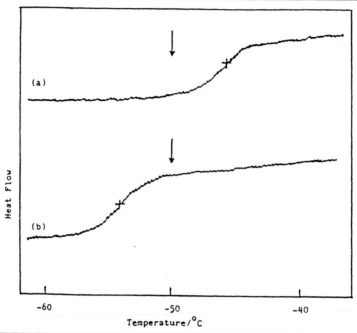

DSC heating scans for two different protein concentrations where thermogram (a) is for the higher protein concentration and thermogram (b) is the lower protein concentration. The collapse temperatures are denoted by the + symbols, and by the ↓ the routine freezing temperatures. [Source: *Developments in Biological Standardization*, Vol. 74, Copyright 1992, Page 117, Figure 11. Courtesy of the International Association of Biological Standardization, Geneva, Swtizerland.]

Advantages. A chief advantage of DSC analysis is that it requires a relatively small sample for analysis. The sample can be crimped sealed during the analysis to protect it from any extraneous effects from the atmosphere. Preparing the sample for analysis is relatively uncomplicated and using temperature scanning rates of 2°C/min to 10°C/min, the time required to obtain a DSC thermogram of a formulation is typically less than 1 hr. The analytical procedure can be quite flexible, such as in the determination of the presence of metastable states when, after completing passing through the exotherm, the sample is cooled and rewarmed to demonstrate that the exothermic peak was irreversible [17].

Another important advantage of the analytical method is that the instrument can be calibrated and certified from thermograms of pure substances having known melting temperatures.

Disadvantages. There are also some disadvantages inherent in the analytical method. For example, the typical sample size for a lyophilized formulation had a mass of the order of 10 mg. Since a drop of water can have a mass of 50 mg, the required sample is relatively very small; obtaining such a small sample, by adding a partial drop of the solution to the pan, requires some skill on the part of the investigator. Because of the low sample mass, the sample can be easily contaminated either from the dispensing device or from the sample holder. In addition, the fill volume of the final lyophilized formulation will be orders of magnitude greater than the DSC sample size, giving rise to the question of whether the actual fill volume might have an impact on thermal properties.

The selection of the proper temperature scanning rate is an important consideration. Selecting too fast a scan rate can result in a broadening of the thermogram and lack the definition needed to determine the eutectic point [11]. I feel that this disadvantage can be avoided through training and experience.

Since the pans used in DSC analysis are constructed of either aluminum or gold, and the lyophilized formulation is most often manufactured in a type I glass container, what impact, if any, will the composition of the pan have on the observed thermal properties? Since the degree of supercooling of water in a glass vial will be on the order of 10°C, determining the supercooling of water for a given cooling rate would provide some indication of whether the metal pan had any impact on the observed thermal properties. For example, an observed degree of supercooling of 20°C would mean that the matrix of the formulation in the DSC pan would differ significantly from that obtained during the freezing of the formulation in a vial.

Thermal Characteristics

The following is a summary of the thermal characteristics that can be determined from a DSC thermogram.

Eutectic Temperature. The presence of an eutectic point can be determined from a DSC thermogram. Even an eutectic temperature (-2.8°C) near the onset of the melting of the ice in the matrix can be clearly distinguished from a DSC thermogram [18] (see Figure 6.9).

Metastable States. The presence of metastable states in a formulation can be identified (see Figure 6.10) and shown to be irreversible by DSC thermograms [17].

Ice Melting Temperature. The onset temperature for the melting of the ice in the formulation can also be determined from a DSC thermogram.

Degree of Crystallization. Attempts to relate the area under the ice melting endotherm to the degree of crystallization have been shown to have a number of sources of error [11]. Most investigators using DSC thermograms to determine the thermal characteristics of their formulations do not report the degree of crystallization of the formulation.

Collapse temperature. The collapse temperature can be ascertained from a warming DSC thermogram (see Figure 6.11) [11]. However, this figure shows that the endotherm associated with the collapse temperature was not sharp and well defined for either protein concentration.

Other Thermal Characteristics During Warming. From DSC thermograms, it was not possible to determine the interstitial melting temperature or the activation energy for the conduction mechanism in the interstitial region when the formulation was in the rubbery state. This should not necessarily be viewed as a limitation of the DSC thermogram because the analytical method does not involve the electrical properties of the formulation.

Thermal Properties During Cooling. Regardless of the cooling rate used to freeze the formulation [17], DSC thermograms are not generally reported as part of the thermal characteristics of the formulation. As a consequence, the DSC method, as reported by the above investigators, does not provide information regarding the degree of supercooling, the equilibrium freezing temperature of the formulation, or any phase changes that may occur in the interstitial region of the matrix during the freezing process.

FREEZING MICROSCOPE

As lyophilization started its rather slow transformation from an art to a science, investigators started to seek various means for characterizing a formulation. This was prompted by findings that a lyophilization process would produce an acceptable product for one formulation but an unacceptable product for a different formulation [21]. It was shown earlier in this chapter that some investigators have employed thermal analysis, DTA or DSC to characterize their formulations. Other investigators chose to ascertain the difference in formulations by visual observation using a specially equipped microscope. The remainder of this section will describe (a) the apparatus used to examine the nature of the formulation, (b) the analytical method for determining the thermal characteristics of the formulation, (c) the interpretation of the results of such microscopic examination, and (d) a general summary and comments regarding this analytical method for characterizing the thermal properties of lyophilized formulations.

Description of the Apparatus and Analytical Method

Several apparatuses have been used to photomicrographically investigate the thermal characteristics of various formulations. For the sake of expediency, only one apparatus will be described in detail; a brief description should suffice for the others. For those readers wishing more detailed information on apparatuses not shown in this section, the respective references should be consulted.

Photomicrography and Cinemicrography

One of the first pioneers in *photomicrography* was Luyet [22] who assembled an apparatus that allowed observation of the freezing process at temperatures ranging from 0°C to -150°C. This apparatus consisted of a controlled temperature bath into which a specimen, a drop of the formulation contained between two cover glasses, was placed in a prefocused position for a microscope. The images from the microscope could be either photographed or cinematographed during the course of the freezing process.

Freeze-Drying Microscope

While the previous apparatus was designed solely to observe the freezing process for a formulation at a given bath temperature, a *freeze-drying microscope* was designed to observe the freezing process and the change in the nature of the frozen structure during the drying process [23,24]. The freeze-drying microscope described by MacKenzie [23] is shown in Figure 6.12, which shows that the sample S, consisting of the sample solution trapped between two glass surfaces [21], was in contact with the wall of the vacuum chamber F. The temperature of the vacuum chamber was controlled by a stirred liquid bath that was refrigerated by cooled, dry air. The images obtained from the system were by transmission micropsy (i.e., the light passes through the specimen rather than being reflected from its surface). Figure 6.12 does not indicate the location or the type of sensor that was used to determine the bath temperature.

A schematic of a similar freeze-drying microscope was described by Willemer [24]; however, the sample rested on a sample carrier that was not in contact with a bath. The sample carrier and, hence, the sample were refrigerated with liquid nitrogen and warmed by an electric heating element. Nitrogen gas was also used to

Figure 6.12. Schematic Views of the Freeze-Drying Microscope

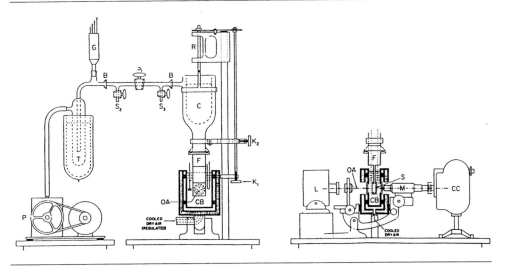

Left drawing: Front elevation of the apparatus showing the cooling bath in cross-section. Key: B,B: ball joints; C: condenser; CB: cooling bath; F: freeze-drying chamber; G: pirani gauge; K_1, K_2: stage control knobs; OA: optical axis of the system; P; two-stage mechanical pump; R: movable rods; S_1, S_2, S_3: stopcocks; T: refrigerated trap. Right Drawing: side elevation of part of the apparatus showing both the freezing chamber and its cooling in cross-section. Key: CB: cooling bath; CC: cine camera; F: freeze-drying chamber; L: lamp; M; microscope tube; OA: optical axis of the system; S: sample. [Source: *Bulletin of Parenteral Drug Association*, April 3, 1978, "Basic Principles of Freeze-Drying for Pharmaceuticals," A. P. Mackenzie, Page 122, Figure 7.]

prevent condensation or frost buildup on the microscope objective during examination of the formulation. The sample solution appears to be contained between a microscope slide and a cover glass. The temperature span for the instrument was from 100°C to -100°C. Maximum cooling and warming rates of the sample were given as 25°C/min. No information was provided regarding the type and location of the sensor used to determine the sample temperature. Like the previous freeze-drying microscope, the image obtained by the Willemer [24] freeze-drying microscope was by transmission microscopy.

Williams and Schwinke [18] provide a brief description of the freeze-drying microscope that was used in conjunction with their study of the thermal properties of pentamidine isethionate. The system consisted of an Nikon Optiphot-Pol® optical microscope. A specially constructed cooling chamber (4 in. in diameter and 0.5 in. in height) was positioned on the stage of the microscope. A glass-covered opening in the center of the cooling chamber provided a means for viewing the sample during the freezing and warming processes. From the description of the apparatus, it is not apparent during the drying process if the sample was observed using reflective or transmission microscopy. The sample consisted of a drop of the pentamidine isethionate solution that was placed on the slide and then covered with a glass cover. The sample was cooled by passing a refrigerated fluid through the cooling chamber. These authors experienced some difficulty in controlling the cooling rate of the system but were able to obtain a cooling rate of 1°C/min. The temperature of the sample was determined by attaching a thermocouple to the *slide support within the cooling stage*. The thermocouple was checked using an ice bath and was found to have an accuracy of within 1°C of the bath temperature. When the desired temperature was reached, the specimen chamber was evacuated to pressures ranging from 50 to 100 mTorr. The authors did not indicate the type of pressure transducer that was used or whether it was attached to the cooling block or was located on the foreline to the pumping system. There was no means provided for warming the specimen chamber, and the authors [18] had to rely on heat transfer between the microscope stage and the ambient room temperature. Under these conditions, a typical warming rate was 0.5°C/min and sublimation of the ice from the sample could be observed.

Interpretation of the Observations

An interpretation of the observations made during the freezing and drying processes of the sample will be considered here.

Freezing Process

The work of Luyet [22] provides us with a detailed, visual insight into the nature of the growth patterns of ice. Freezing was observed in solutions containing sucrose, glycerol, gelatin and albumin using a photomicrography and cinemicrography apparatus. These studies led to the following information regarding the combined effect that the composition and concentration of a solute and temperature can have on the nature of ice formation.

Effect of Solutes on Ice Formation at High Temperatures. The freezing of a 35% bovine albumin solution at temperatures ranging from -1.5°C to -3°C produced a number of different symmetrical forms of hexagonal ice. Luyet [22] observed that during ice formation, there was a change in the symmetry of the ice crystals. At temperatures approaching 0°C, the first ice appeared in hexagonal forms; concentrated solutions took a solid prism configuration. As the prisms enlarged, there was apparently a faster rate of ice growth along three of the axes, which led to the formation of six-sided stars and rosettes. After additional ice growth, secondary branches formed dendritic structures. A picture of the dendritic formation of ice is shown in Figure 6.13. Although the ice formation in Figure 6.13 was shown in the vapor phase, it does show the dendritic formation that also occurs during the freezing of solutions.

Effect of Solutes on Ice Formation at Lower Temperatures. When ice nucleation occurred in a 50 percent sucrose solution at temperatures ranging from -20°C to -25°C, faster rates of ice growth occurred, and the hexagonal symmetry of the ice crystals became more irregular in nature. At still lower temperatures, -55°C to -70°C, ice growth in a 6 M glycerol solution initially was in the form of a disk. When warmed the disks grew and came in contact with one another, and recrystallization occurred (i.e., a dense cloud of minute ice crystals formed).

Figure 6.13. The Formation of Dendritic Ice in the Gas Phase

[Reproduced with permission from Dr. Kenneth G. Libbrecht, Professor of Physics, California Institute of Physics, 1998.]

Freeze-Drying During Warming

The interpretation of thermal analysis data obtained from microscopic observations during warming of a formulation in a freeze-drying microscope will be considered here.

Collapse Temperature. The work of MacKenzie [21,23] revealed that each formulation had one temperature at which collapse occurred at the freeze-drying front. The collapse was accompanied by a loss of the freezing pattern, i.e., the material took on a random appearance. In that portion of the sample awaiting the passing of the front, the appearance of the sample retained the original structure of the frozen system. If the sample were cooled sufficiently, the passage of the drying front produced a material with very much the same structure as that of the original frozen matrix. By varying the temperature of the sample, MacKenzie was able to determine the observed Tc to a precision of ±1°C; for some formulations, a precision of ±0.5°C was obtained. For the Tc values of mainly sugar formulations, the reader is referred to Table I in reference [21].

Eutectic Temperature. The results obtained from a freeze-drying microscope allow differentiation between an interstitial region consisting of a glass state and one composed of an eutectic mixture. As described above, the loss of structure for a system composed of a glassy state occurs at the freeze-drying front, and the remainder of the ice matrix retains its original structure. When an eutectic is present in the interstitial region, the entire matrix undergoes a structural change when the sample temperature reaches the eutectic temperature [18,21]. The only eutectic temperature determined by means of a freeze-drying microscope was reported by Williams and Schwinke [18] for pentamidine isethionate. The eutectic temperature of the pentamidine isethionate determined from a DSC thermogram was -2.8°C. When examined with a freeze-drying microscope, the frozen matrix of the sample melted at -4°C.

Summary and Comments

The following is a summary and comments regarding the advantages and some key limitations of the freeze-drying microscope as a means of ascertaining the thermal characteristics of a formulation.

Advantages of the Freeze-Drying Microscope

The freeze-drying microscope offers some distinct advantages in the determination of the thermal properties of a formulation. One advantage is that the cost of a freeze-drying microscope is less than the cost of other analytical systems, such as DSC. A second advantage is that freeze-drying microscopes are relatively easy to use and require only a small quantity of the formulation. Consequently, the thermal properties of several formulations can be examined in a single day. The results provide the investigator with a visual image of the configuration of the frozen matrix. Thermal properties, such as Tc, can often be determined to a precision of ±0.5°C, and the measurements do not vary significantly between investigators [21].

Limitations of the Freeze-Drying Microscope

Instrument and Methodology. There are some concerns regarding the thermal characteristics obtained from an examination of a specimen using a freeze-drying microscope. The first is that the formulation is confined between two glass surfaces. What effect does the thickness of the sample film have on the observed thermal properties? And what impact, if any, do the glass surfaces have on the freezing of the formulation? It is well known, especially within the electronics industry, to which we owe the convenience of the digital computer, that the properties of materials change substantially when the material is in a thin film form. Are the thermal characteristics obtained from microscopic sample of a formulation representative of the thermal characteristics of that formulation when it is in a glass vial?

The *nucleation temperature* for the freezing of the water is a function of the thickness of the film of water (i.e., the nucleation temperature decreased as the film thickness was reduced [25]). Reducing the film thickness of the water to 0.01 mm reduced the nucleation temperature to –30°C. If the water film was reduced even further, the formation of ice crystals required temperatures below –100°C. The actual film thickness of the formulation between the glass plates in the freeze-drying microscope is not known with any certainty. However, based on the effect that the thickness of a water film can have on the nucleation temperature and the resulting ice structure, the thickness of the formulation film may be a factor that can effect the thermal properties obtained by a freeze-drying microscope.

Not only is the nucleation temperature dependent on the water film thickness, but the vapor pressure of water is also reduced as the film thickness is decreased [25]. It would also be logical to assume that if there was a reduction in the vapor pressure of the water for a given film thickness, there would also be a corresponding reduction in the vapor pressure of the ice. For a given freeze-drying microscope pressure, a significant reduction in the vapor pressure would require a higher temperature to initiate sublimation. As a result, frozen thin films of a formulation in which there has been a reduction in the vapor pressure will require higher temperatures to sublimate the ice crystals. Under these conditions, the observed collapse temperature or eutectic temperature will be false high. Thus, the observed collapse or eutectic temperature of a formulation could be dependent on the design and construction of the vacuum system associated with the instrument, as illustrated in Figure 6.12.

Another aspect that could impact the results obtained from a freeze-drying microscope is the preparation of the surfaces of the glass slide and/or cover slides. Glass microscope slides, for example, are sometimes packaged with a thin lubricant film on the surface. This lubricant film prevents adhesion between two flat glass surfaces. Since none of the above investigators mention any prior cleaning of the glass microscope slide or glass cover, there is a possibility that the surface film could contaminate the sample. Because of the small amount of formulation used in the analysis, the contamination resulting from a lubricant may have an impact on the observed thermal properties. I have found that removing the lubricating film from the surface of a glass slide required immersing the segregated slides in hot (near boiling) concentrated nitric acid for a period of no less than 0.5 hr. Such an operation is conducted under a fumehood, and safety precautions must be taken so that personnel are not exposed to the nitric acid vapors. The acid bath is allowed to cool to room temperature, and the slide is removed and rinsed by allowing steam to condense on the surface. The segregated slides are allowed to dry and are then stored in a dry dust-free container.

A final concern regarding the use of the freeze-drying microscope as a means of characterizing the thermal properties of a formulation is the measurement of the sample temperature. Williams and Schwinke [18] determined the sample temperature from a thermocouple attached to the sample cell support. In the infrared region of the spectrum, the emissivity of glass approaches that of a black body, so heat from the light source for the microscope may cause an increase in the sample temperature. This may explain why Williams and Schwinke [18] observed the eutectic temperature of pentamidine isethionate as –4°C with the freeze-drying microscope, while DSC gave a value of –2.8°C. This concern could be resolved if those using this technique to ascertain the thermal properties would verify the accuracy of their system by determining the eutectic temperatures of known salt solutions such as NaCl and KCl.

Thermal Properties. Although the freeze-drying microscope can provide information about the collapse and eutectic temperatures, the method does not provide information about the following thermal characteristics of a formulation:

- Formation of ice-like water clusters
- Equilibrium freezing temperature
- Degree of crystallization
- Interstitial melting temperature
- Metastable states
- Activation energy

The freeze-drying microscope, while providing the investigator with a visual representation of a formulation during freezing and warming, is rather limited in its characterization of the formulations complete thermal properties.

ELECTRICAL RESISTANCE

The next analytical method for determining the thermal properties of a lyophilized formulation to be considered in this chapter will be resistance measurements. A brief introduction describing the basic underlying chemistry and physics regarding such measurements may prove helpful in understanding the terminology that will be used in this section and the apparatus and the analytical method that will be described.

Introduction

The basic relation between current, applied voltage, and resistance of a circuit obeying Ohm's Law will be reviewed here. The nature of the resistance of a substance, the conduction mechanism, and cell constant will also be considered.

Ohm's Law

When a dc voltage (V) is applied across an electrically conductive material having a length (l) and a cross-sectional area (A), like that shown in Figure 6.14, the current (I) in amperes through the conductor is defined according to Ohm's Law

$$I = \frac{V}{R'} \tag{2}$$

where R' is the electrical resistance. From equation 2, I will be directly proportional to V and inversely proportional to R' of the material.

Relationship Between Resistance and Resistivity

R' of a substance in ohms is defined by the following expression:

$$R' = \rho \frac{l}{A} \tag{3}$$

where ρ is the resistivity of the material and has the units of Ω-cm, l is the length of the material, and A is its cross-sectional area. The resistivity of a homogeneous substance is a bulk property of the substance, very similar to density, and is independent of the configuration of the conductor. However, the resistance of a conductor is not only proportional to ρ but also dimensionally dependent. For example, in Figure 6.14, the resistance of the conductor defined by the area A (CD $\times$ CE) and the length (l) (CF) will be greater than the resistance having a cross-section B (CE $\times$ CF) and a length (CD). In defining the resistance of a system, the dimensions of the system should also be specified so that others can reproduce the results.

Figure 6.14. Electrical Current Through a Substance

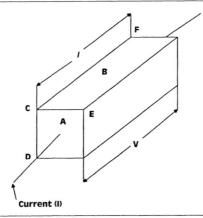

Current (I)

An illustration of the electrical current I through a substance having a resistivity ρ; a length (l) and cross-sectional area A, and obeying Ohm's Law when a voltage V is applied across the conductor.

Conduction Mechanism

Understanding the role of *lines of force* resulting from an electric field is necessary to understanding conduction in a liquid medium. Figure 6.15 illustrates two examples of the lines of force resulting in equal-potential lines associated with a conductor in an aqueous solution. When opposite potentials are applied to two cylindrical conductors, as illustrated by Figure 6.15a, there will be a series of regions surrounding each conductor where the potential is equal; hence the lines are described as *equal-potential surfaces*. Normal to these surfaces are lines of force. It is along these lines of force that charged particles move from one electrode to another. In the case of a

---

Figure 6.15. Lines of Force in an Aqueous Solution

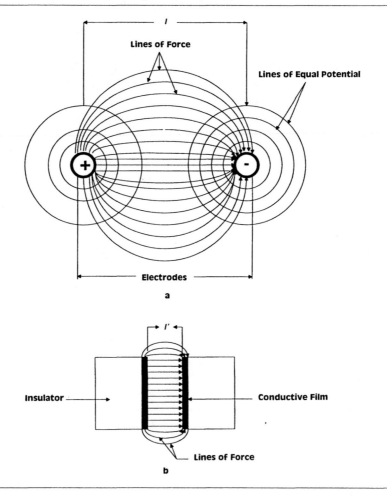

An illustration of surfaces of equal potential, denoted by the solid lines and electric lines of force indicated by the arrows, where 6.15a represents a system composed of cylindrical electrodes separated by a distance *l* and 6.15b a system where the width of the conductors exceeds the distance (*l'*) between the electrodes.

solution, negatively charged ions, such as OH^-, Cl^-, NO_3^- and SO_4^{-2}, will flow along the lines of force to the anode (the positively charged surface); the positive ions like H_3O^+, Na^+, K^+, and Ca^{2+} will flow to the cathode (negatively charged surface). As a consequence, to define the resistivity of a system, the cross-sectional area A in equation 3 is that area defined by the electric field where *l* represents the distance between the charged surfaces shown in Figure 6.15.

In Figure 6.15b, the conductors are metal films, such as gold, deposited onto an electrically insulated surface. Because the width of the conductor (not shown) is much greater than the distance *l'* between the metal films and because of the presence of electrical insulator layer, the equal-potential lines will tend to be parallel to the metal surface. As a result, most of the lines of force will be confined between the two electrodes, which define the cross-sectional area A of the system.

The conduction mechanism varies significantly during the course of a thermal analysis. At temperatures > 0°C, the conduction is governed by the principles of electrochemistry. As ice forms and a *rubbery state* exists between ice crystals, conduction becomes protonic along the surfaces of ice crystals and a combination of ionic and protonic in the interstitial region of the matrix. When the interstitial region reaches a completely frozen state, conduction is primarily along the surfaces of the ice crystals in the matrix.

Parker Effect

When determining the configuration and dimensions of a resistance cell, any error resulting from the *Parker Effect* must be insignificant. The Parker Effect occurs in high resistance solutions, such as those often found in lyophilized formulations. In such solutions, errors resulting from polarization of the electrodes will be insignificant, but as the resistance of the solution (R_0) increases, there appears to be a decrease in the cell constant (*K*) which is defined as [26]

$$K = \frac{l}{A'} \tag{4}$$

where *l* is distance between the electrodes and *A'* is the cross-sectional area of the solution through which the current will pass. It was first thought that such a change in *K* was a result of the adsorption of electrolyte on the surface of the electrodes. However, it was subsequently shown that the decrease in resistance ($-\Delta R$) was not a result of a change in the cell constant but was related to the following expression:

$$-\Delta R \propto R_o^3 \omega^2 C_p \tag{5}$$

where ω is the *frequency* of the alternating current and C_p is the *shunt capacitance* of the cell. The shunt capacitance results from the walls of the cell and the opposite electrode. This shunt capacitance can be eliminated by changing the cell constant (i.e., increasing the distance between the walls of the cell and the electrodes and decreasing the distance between the electrodes). Resistance cells with small cell constants will tend not to exhibit any Parker Effect and be relatively independent of frequency. As a result, a knowledge of the design and construction of the resistance cell is an important criterion in using resistance measurements to determine the thermal characteristics of a lyophilized formulation.

Description of the Apparatus

Multielectrode System

In 1954, Greaves [3] was one of the first investigators to use electrical resistance measurements to ascertain the thermal properties of materials at temperatures $< 0°C$. In a series of publications [3,4,6] in the early 1960s, Rey furthered the use of electrical measurements to characterize the thermal properties of frozen systems. The resistance cell described by Rey [4] is illustrated in Figure 6.16 and consists of five platinum electrodes housed in a specially designed cylindrical insulator that fits into a glass container. As seen by Figure 6.16, four of the electrodes were electrically connected in parallel and arranged around a second electrode that was located in the center of the cell. Although not specifically stated, the temperature of the sample during the course of the analysis was determined by a measurement of the temperature of the bath into which the resistance cell was immersed. Rey [6] also used a

---

Figure 6.16. Resistance Cell

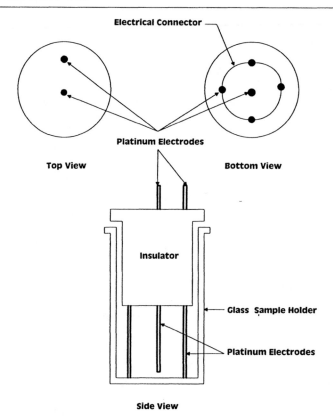

Illustration of the resistance cell used by Rey [4] to ascertain the thermal properties of a formulation showing the top, bottom, and side views.

two-platinum electrode cell configuration confined in a glass container. This electrode system was placed in a special stainless steel cell, and the temperature of the cell measured by means of a thermocouple. In order to prevent polarization of the electrodes, the measurements were conducted at a frequency of 1,000 Hz. The data were reported in terms resistivity [3,6] as a function of temperature, where resistivity measurements were reported on a log scale ranging from 2 to > 10. The resistivity units were reported in terms of $M\Omega/cm$. Other electrical measurements reported by Rey [4] are in terms of electrical resistance as a function of temperature. With no intent to criticize the pioneering work done by Rey in using resistance measurements to characterize the thermal properties of a formulation, I would like to point out how easy it is to confuse the terms *resistance* and *resistivity*.

Dual Electrode System

Lachman et al. [27] also used electrical resistance measurements to characterize the thermal properties of formulations. These authors used a dc bridge circuit that was capable of measuring resistance values as high as 10^{12} Ω. Since the actual currents in the sample were quite low because the applied voltage was less ≤ 3 V, the effects of polarization of the electrodes were considered to be negligible.

The conductivity cell consisted of two platinum electrodes and a grounded chromel-alumel thermocouple inserted through a Teflon® cap. The configuration and dimensions of the electrodes were not specified. The cell was placed in a specially constructed Plexiglas® container that was cooled by the expansion of liquid carbon dioxide. The Teflon® cap was also equipped with a vent in order to equalize the pressure in the chamber with that in the cell.

Vertical Resistivity Cell

While most investigators used cylindrical metal electrodes of platinum, Jennings and Powell [28] investigated the characteristics of a vertical resistivity cell like that shown in Figure 6.17. The electrode consisted of a glass tube connected to flat glass plate that was coated with gold. The glass plate had a length of 3 cm and a width of 0.467 cm. A metric scale on the edge of the electrode served to measure the electrode surface in contact with liquid. A copper rod on the top of the electrode served as an electrical connector and a platinum wire served as an electrical connection between the copper rod and the deposited gold film.

The electrodes were mounted in two retaining blocks like that shown in Figure 6.18. The side view of the resistivity cell is illustrated by Figure 6.18a. In this figure, two electrodes are supported by retaining blocks A and B. The retaining blocks also allowed the gold surfaces of the electrodes to be parallel with each other. Electrical insulation between the cells is provided by the glass tubing of the electrodes. The retaining blocks are mounted on a metal frame that is supported by a clear plastic cylinder. Three set screws threaded in the plastic cylinder serve to retain the bottle containing the liquid. As shown in Figure 6.18b, the position of the electrode in block A can be accurately moved with respect to the electrode supported by block B by means of a micrometer drive mechanism, which can vary the distance between the electrodes from 0.5 mm to 3.0 cm.

With the above apparatus, Jennings and Powell [28] investigated the effect that the distance between the electrodes (i.e., a change in the cell constant [see equation

Figure 6.17. Front and Side Views of a Vertical Electrode

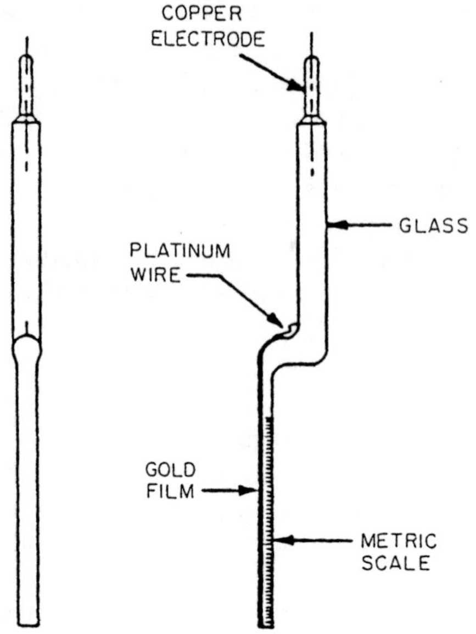

[Reprinted with permission from *Medical Device & Diagnostic Industry*, "Use of the Vertical Resistivity Cell in Lyophilization," March 1980. Copyright © 1980 Canon Communications LLC.]

4]), had on the observed resistivity of distilled water at 16°C. The measurements were made at a frequency of 120 Hz and a water depth of 3 cm in the glass container. The volume of water between the electrodes increased as the distance between the electrodes was decreased. Figure 6.19 shows the resistivity of water as a function of the distance between the two parallel electrodes. For electrode distances > 0.25 cm, the resistivity is independent of the distance between the electrodes. Because of the constriction of the electrodes, the lines of force between the electrodes, as illustrated by Figure 6.14b, will be confined between the electrodes so that the cell constant is well defined. For resistance cells using cylindrical metal electrodes, as illustrated by Figure 6.15a, the area defined by the lines of force is more difficult to define, and there is an uncertainty regarding the actual cell constant. Without knowledge of the cell constant, the resistivity of a solution cannot be determined.

There is, however, as shown by Figure 6.19, a marked increase in the measured resistivity when the distance between the electrodes is less than 0.1 cm. Since resistance cells having small cell constants will tend not to exhibit any Parker Effect, the increase in resistivity of the water appears to result from the decrease in the thickness of the water layer between the electrodes. Such a decrease in the water layer may result from the presence of a double layer formed on each of the electrode surfaces. As the distance between the electrodes is decreased, the thickness of the double layer will remain relatively constant but the free water layer not associated with the double layer will decrease. At some electrode distance, the thickness of the free water layer reaches a threshold value that results in an increase in the resistance of

Figure 6.18. Side View (a) and Top View (b) of a Vertical Resistivity Cell

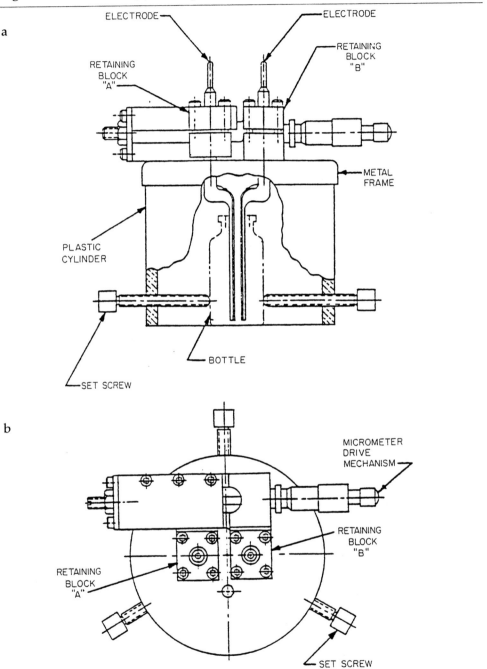

Figure 6.19. Resistivity Versus Distance

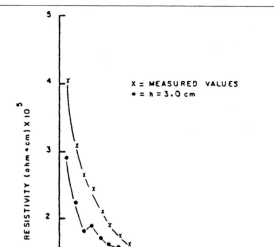

The effect of distance between distance between parallel, vertical electrodes on the resistivity of distilled water at 16°C where x indicates the resistivity determined from the total volume of water between the electrodes and the ● represents the resistivity values based on a depth of 3 cm. [Reprinted with permission from *Medical Device & Diagnostic Industry*, "Use of the Vertical Resistivity Cell in Lyophilization," March 1980. Copyright © 1980 Canon Communications LLC.]

the free water layer. The reader should realize that the presence of a double layer on the surface of an electrode is quite complex and is still not fully understood; further discussion of the phenomenon of the double layer is beyond the scope of this text.

To determine the thermal characteristics of a formulation using a vertical resistivity cell, the sample is placed in a test chamber that contains a hollow stainless steel plate designed for serpentine flow of a heat-transfer fluid to simulate cooling and warming of a freeze-dryer shelf. The temperature of the heat-transfer fluid to the shelf is pumped from a methanol reservoir. The temperature of the methanol reservoir is controlled by a separate refrigeration and heating system. Shelf temperatures can then be varied from 30°C to approaching -70°C.

Analytical Methods

Multielectrode System

The analytical method used by Rey [3,4] involved placing the solution in the glass container and then positioning the electrodes as shown in Figure 6.16. Using liquid nitrogen as a refrigerant, the solution was then frozen to temperatures approaching –80°C. The time required to freeze the solution ranged from 2 min to 20 min. Resistance measurements as a function of temperature were recorded as the freezing and warming progressed. The mechanism or means of warming the sample was not given.

Dual Electrode System

The resistance measurements in the dual electrode cell [27] were conducted by placing 3 ml of the solution in a 10 mL sample cell. The Teflon® cap was designed so that the platinum electrode and the temperature sensor were positioned just below the surface of the solution [29]. Sample temperatures ranging from –30°C to –50°C could be obtained in less than 5 minutes by introduction of solid carbon dioxide into the Plexiglas® container. Although warming required 2 hr to reach the freezing temperature of the solution [29,30], the means for warming of the sample was not described by the authors. Resistance measurements as a function of temperature were recorded during both the freezing and warming of the sample.

Vertical Resistivity Cell

To determine the thermal properties of a solution, a 1 mL to 50 mL sample of the solution was placed into glass tubing or a molded vial. The vial was then positioned in the vertical resistivity cell as shown in Figure 6.18 and the electrodes were adjusted so that they contacted the bottom of the glass container. The distance between the electrodes was adjusted to 0.2 cm, and, from the depth of the solution, the cell constant was determined. A copper-constantan thermocouple was positioned so that it was within 1/16 in. to 1/8 in. from the bottom of the container. The thermal properties of the solution were recorded during the cooling and warming of the sample. The resistance of the sample was plotted as a function of temperature.

Interpretation of the Data

The following will examine the thermal properties of various formulations using the above described resistance apparatus and analytical methods.

Multielectrode System

With the multielectrode system, Rey [3,4] examined salt solutions with known eutectic points and more complex systems that would typify a lyophilized formulation. Examples of these results and interpretation of the data for each of these systems will now be described.

Eutectic Systems. The resistance of a NaCl solution, consisting of 10 parts NaCl and 1,000 parts water as a function of temperature during cooling and warming is shown by Figure 6.20 [3]. Rey points out that the data show that there is a major difference in the log resistivity-temperature plots during the cooling and warming phases. The differences in the plots are attributed to the supercooling of the eutectic solution. For this reason, Rey suggests that, when using resistance measurements, the eutectic point should be determined from the warming of the frozen matrix rather than during the freezing of the solution.

Figure 6.20. Variations in Electrical Resistance for Distilled Water

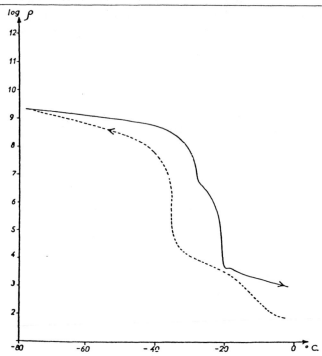

Variation of the electrical resistance of a 10 parts per 1,000 solution of NaCl in distilled water according to low temperature changes. *Solid line*, measurements taken in the course of rewarming; *broken line*, measurements taken during the cooling process. [Source: New York Academy of Sciences, Vol. 85, 1960, Louis R. Rey, "Thermal Analysis of Eutectics in Freezing Solutions," page 521, figure 9.]

Figure 6.20 shows that the sample was cooled from about 0°C to –80°C, which is well below the known eutectic temperature of –21.1°C [9]. As the system approached –80°C the resistivity of the system was > 1,000 MΩ-cm. The eutectic temperature is determined as being –21.6°C by a 10-fold decrease in the resistivity of the formulation (i.e., the resistivity of the system changed from about 100 kΩ-cm. to 10 kΩ-cm. Rey acknowledged that there was substantial decrease in the resistivity of the system that occurred at temperatures between –50°C and –40°C; he attributed this change in resistivity to recrystallization process occurring in the system.

Complex Formulations. Figure 6.21 shows the log of the resistivity of a sample of Earl's solution, whose composition was described in earlier in this chapter. This figure shows, as did the previous figure, that the freezing plot differs significantly from the plot obtained during the warming of the formulation. The warming plot shows that there is no sharp decrease in resistivity to mark the onset temperature for the mobility of water in the interstitial region of the matrix.

Figure 6.21. Variations in Electrical Resistance for Earl's Solution

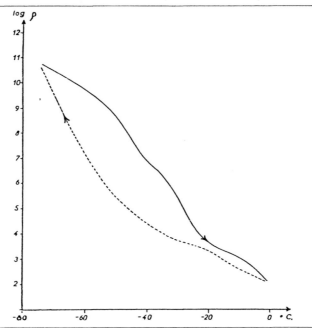

Variation of the electrical resistance of Earl's balanced salt solution according to low temperature changes. *Solid line,* measurements taken during rewarming period; *broken line,* measurements taken during the cooling period. [Source: *PDA Journal of Pharmaceutical Sciences,* Vol. 54, No. 4, April. 1965, "Lyophilization of Pharmaceuticals 1" by P. DeLuca and L. Lachman, Page 621, Figure 8.]

Dual Electrode System

A plot of log resistivity as a function of temperature for a dual electrode system for NaCl, potassium bromide (KBr), and potassium chloride (KCl) solutions [29] is shown in Figure 6.22. In this figure, the eutectic temperature is marked by the point where there is a sharp change in the slope or *knee* for the plot of log resistivity as a function of temperature. Measured by this method, eutectic temperatures for solutions of NaCl, KBr, and KCl are -21.6°C, -12.9°C, and -11.1°C, respectively, and are in agreement with other published values of these solutions.

   Figure 6.23 shows the cooling and warming plots of the resistance as a function of temperature for a 0.3 M solution of an organic salt, methylphenidate HCl. In this figure, DeLuca and Lachman [29] show the effect that the warming rate can have on the observed resistance properties of the formulation. In the case of rapid warming, the sample temperature increased from -35°C to -2°C in about 20 min; whereas, in the case of slow warming, the time required to increase the temperature ranged from 1.5 hr to 2 hr. In the case of fast warming, there is no apparent change in the slope of the resistance-temperature plot to signify the onset of the mobility of water in the interstitial region of the matrix.

Figure 6.22. Log Resistivity Versus Temperature for Three Salt Solutions

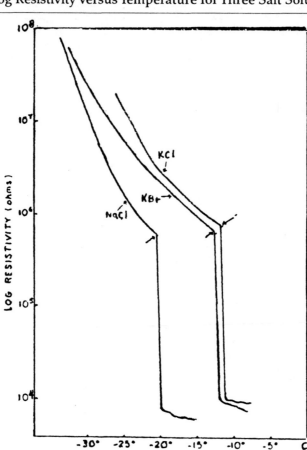

Determination of the eutectic temperature by plotting log resistivity as a function of temperature. Values at → indicate specific points on the curves area: NaCl, -21.6°C; KBr, -12.9°C; and KCl, -11.1°C. [Source: *PDA Journal of Pharmaceutical Sciences*, Vol. 54, No. 4, April. 1965, "Lyophilization of Pharmaceuticals 1" by P. DeLuca and L. Lachman, Page 622, Figure 10.]

For slow warming, the resistance plot was marked with a region where there was an increase in resistance similar to that associated with rapid warming. This was followed by a relative plateau of resistance values followed by a relatively sharp decrease in resistance. The resistance-temperature plot then underwent a sharp decrease in resistance, which had an onset temperature of -11.7°C, and was followed by gradual decrease in resistance as the temperature approached 0°C. Resistance plots similar to that shown during the slow warming of methylphenidate HCl were also observed by DeLuca and Lachman [29] for aqueous solutions having various concentrations of phentolamine methanesulfonate. The extent of the sharp decrease

Figure 6.23. Cooling and Warming Curves for a 0.3 M Methylphenidate HCl Solution (0–10^8 Ω decade)

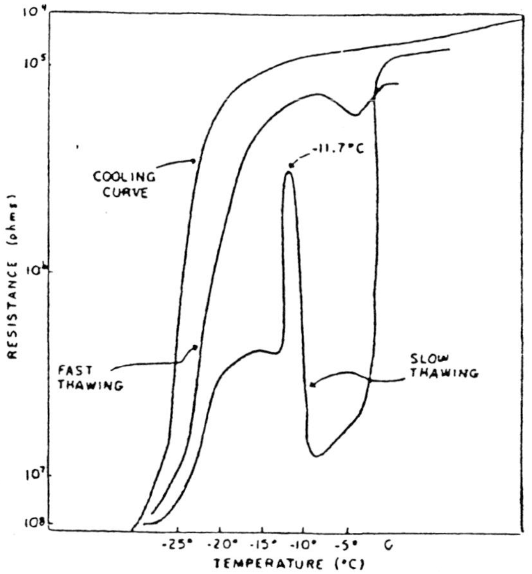

[Source: *Journal of Pharmaceutical Science*, P. DeLuca and L. Lachman, Vol. 54, Page 617, 1965. Reprinted with permission.]

in resistance was observed to depend on concentration, i.e., the decrease in resistance became more pronounced at lower concentrations.

Vertical Resistivity Cell

The results of the cooling and warming of a 1% NaCl solution, using the apparatus shown in Figure 6.18 is shown in Figure 6.24. The cell constant for these results was 23.6 cm. The freezing and warming curves are similar to those by Rey [3] in Figure 6.20. The results show that by taking the knee in the plot as the onset temperature of the eutectic point as illustrated by Figure 6.22, the eutectic temperature was found to range from -23°C to -24°C. However, using the sharp decrease in resistance as suggested by Rey [3], the eutectic point approaches -21°C. Additional results regarding the use of the vertical cell to ascertain the thermal characteristics of a formulation will be considered later in this chapter.

Summary and Comments

Electrical resistance measurements do provide a means for characterizing the thermal properties of a formulation. Investigators using electrical resistance measurements tend to use resistance and resistivity interchangeably. For example, the plots shown by Rey [3] indicate that the resistivity of the system is plotted as a function of

Figure 6.24. Resistivity Versus Temperature for an Ice Solution of Sodium Chloride

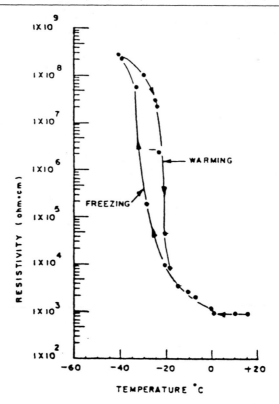

The resistivity of a solution and an ice matrix, formed from a 1% solution of NaCl, as a function of temperature. [Reprinted with permission from *Medical Device & Diagnostic Industry*, "Use of the Vertical Resistivity Cell in Lyophilization," March 1980. Copyright © 1980 Canon Communications LLC.]

temperature while the caption indicates resistance. DeLuca and Lachman [29] determine the eutectic temperatures of NaCl, KBr, and KCl from a plot of the resistivity as a function of temperature; however, the cell constant was not reported. When using cylindrical electrodes either in a multielectrode or dual electrode configuration, the electric field is quite complex, and defining the resistivity of the system is difficult. Only when using a vertical resistivity cell with the distance between the parallel electrodes between 2 mm and 3 mm and the width of the electrodes greater than the distance between the electrodes will the electric field be principally confined to the dimensions of the cell.

In each of the plots, there was a difference between the cooling and warming plots. This difference stems from a temperature differential existing across the sample during the freezing process. As a result, the electrical resistance properties of the sample can vary from an ice matrix at the bottom of the sample container to a completely liquid system at the top of the solution. DeLuca and Lachman [29] attempted to eliminate such an effect by inserting the electrodes and temperature sensor just below the surface of the solution. However, in spite of their precaution, a differential in the resistance-temperature plots during cooling and warming persisted.

The use of resistance plots to characterize a formulation appears to be most applicable when the formulation consists of single a constituent and forms an eutectic mixture. Rey [3] showed that for a more complex solution, such as an Earl's solution, the resistance measurements fail to indicate the temperature at which the mobility of the interstitial region is significant (i.e., the collapse temperature). It has been my experience that the thermal property of a formulation, on which the lyophilization process is based, will be that of the collapse temperature rather than an eutectic temperature.

While resistance measurements have been shown to determine the eutectic temperature of a formulation, this technique provides no information regarding other thermal properties, such as the following:

- formation of ice-like water clusters

- degree of supercooling

- equilibrium freezing temperature

- degree of crystallization

- collapse temperature

- interstitial melting temperature

- metastable states

Resistance measurements can prove useful in identifying formulations that form an eutectic mixture; however, as seen by Rey's [3] results of Earl's balanced salt solution, resistance measurements do not provide a definitive indication of the collapse temperature. In addition, resistance measurements do not provide information regarding a number of thermal properties that can prove useful in characterizing a lyophilized formulation.

DIELECTRIC ANALYSIS

Introduction

Another analytical method using the electrical properties of a formulation to determine its thermal properties was introduced by Morris et al. [31] and involves the measurement of the dielectric properties of the formulation. The basis for *dielectric analysis* (DEA) can be summarized as follows:

When an electric field (E) is applied to a material having a permanent dipole (e.g., water), the displacement force (D) is defined as

$$D = E + 4\pi P \tag{6}$$

where P is the polarization and is the sum of P_d and P_o, where P_o is the polarization induced by molecules having a permanent dipole, and P_d is the induced polarization of nonpolar molecules resulting from the applied electric field.

If the applied voltage has a frequency ω, the time required for alignment of the permanent or induced dipoles with the applied electric field, known as the relaxation time (τ), will be dependent on temperature. For a material like ice, the value of τ at –10°C has been reported to have a value 56 μsec [32]. The alignment of the dipoles with the alternating electric field can cause an energy loss to occur because of heat generated from the alignment process.

Morris et al. [31] define the *dielectric permittivity* (ε') as

$$\epsilon' = \epsilon_\infty + \frac{\epsilon_0 - \epsilon_\infty}{1 + \omega^2\tau^2} \tag{7}$$

and the loss ε" resulting from the heat generated by the alignment process by

$$\epsilon'' = (\epsilon_0 - \epsilon_\infty)\frac{\omega\tau}{1 + \omega^2\tau^2} \tag{8}$$

where ϵ_0 is the low frequency relative permittivity where τ >> ω and ϵ_∞ is the high frequency relative permittivity where ω >> τ.

Figure 6.25. Theoretical Debye Plots

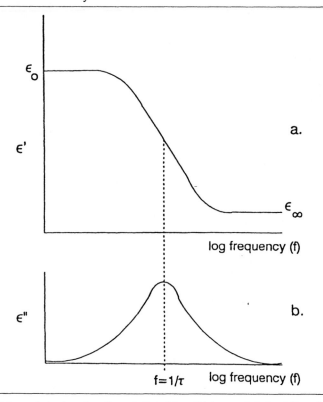

Theoretical Debye plots showing the relationship of (a) permittivity (ε') and (b) loss (ε"), both with respect to frequency. [Source: *PDA Journal of Pharmaceutical Science and Technology*, K. R. Morris, S. A. Evans, et al., Nov.–Dec. 1994, Vol. 48, No. 6, Page 320, Figure 2.]

A plot of ϵ' and ϵ_∞ as a function of frequency for the ideal dipole system is illustrated by Figure 6.25. This plot shows that at low frequencies, ϵ' has a high constant value, and ϵ'' is at a minimum because $\tau \gg \omega$. At high frequencies where $\omega \gg \tau$, ϵ' is at a minimum and ϵ'' is at a minimum because the dipoles do not have sufficient time to respond to the imposed electric field. The frequency at the maximum value of ϵ'' is equal to the reciprocal of τ.

The energy $\Delta H'$ for a given temperature is obtained from

$$\tau(T) = \tau_0 \exp \frac{\Delta H'}{RT} \tag{9}$$

by determining the slope of the plot of $\ln\tau$ as a function of $1/T$ where T is expressed in K and R is the gas constant. At about –10°C, the value of $\Delta H'$ is the 57 kJ/mole [32]. From a measure of ϵ' and ϵ'', Morris et al. [31] were able to predict the collapse temperature for water and a number of other solutions.

Description of the Apparatus

The DEA instrument consists of a single-surface, co-planar, interdigitated-comb sensor whose construction is illustrated by Figure 6.26. With the sensor assembled as shown in Figure 6.26, the sensor was calibrated by lowering a ram so that springs from the ram were brought into contact with the sensor pads. With a force transducer, the ram/sensor spacing was determined to ensure good sample-sensor contact. Johari and Whalley [32], in their study of the dielectric properties of ice, found it necessary to use a piston to exert a force of 1.5 kbar on the ice

Figure 6.26. DEA Instrument Schematics

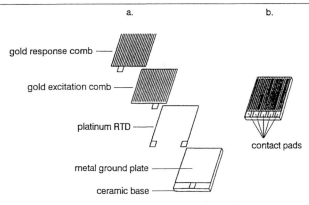

a.   b.

gold response comb

gold excitation comb

platinum RTD

contact pads

metal ground plate

ceramic base

Schematics showing (a) the individual components of a single surface, coplanar, inter-digitated-comb sensor, and (b) a complete sensor. [Source: *PDA Journal of Pharmaceutical Science and Technology*, K. R. Morris, S. A. Evans, et al., Nov.–Dec. 1994, Vol. 48, No. 6, Page 321, Figure 3.]

sample. The force was necessary to ensure good contact between the sample and the electrodes and prevent cracking of the ice sample. The capacitance of their empty cell was ~3 pF.

At the start of an analysis, Morris et al. [31] applied about 1 mL of the sample solution to the sensor. This volume of sample was found sufficient to fill all of the channels of the sensor. The channels of the sensor were 125 μm wide and 12.5 μm deep. The ram was then positioned 2.5 mm above the sensor. This distance ensured good ram/sample/sensor contact and prevented extrusion of the liquid from the sensor. After completing the permittivity check of the instrument and prior to the start of the analysis, the ram was raised so that no force.was applied to the sample during the analysis.

The measurements were performed over frequencies ranging from 0.03 Hz to 300,000 Hz. The ac voltage was applied to the gold excitation comb and the output current was detected by the gold respond comb. The current was then amplified and transformed to an output voltage. The temperature of the sample was determined by means of a resistance thermal device (RTD). Dielectric measurements were made under a nitrogen atmosphere. The instrument was purged with nitrogen at a flow of 500 mL/min. No details were supplied regarding the nature of the sample holder. Since this is a relatively new analytical technique, it is hoped that additional information regarding the configuration of the dielectric cell and the capacitance of the empty cell will be supplied in subsequent publications. Test data were stored so that ϵ' and ϵ'' could be determined for each temperature.

Analytical Method

Preparation of the Sample

The sample solutions were prepared on a weight/weight basis. All of the solutions were 10% wt/wt except for Azactum™, which was supplied by Bristol-Myers Squibb and was 11% wt/v.

Test Method

The samples were first cooled to temperatures that were low enough to freeze the sample. No details were provided by the authors regarding the method used to freeze the sample.

Once frozen, the sample was then either warmed at a rate of 1°C/min or warmed and maintained at a given temperature long enough to obtain a complete set of data for a range of frequencies. It is hoped that in subsequent publications, Morris et al. [31] will provide more details regarding the methods used in the freezing and warming of the sample.

Interpretation of the Data

Water

The results of the DEA analysis of the melting of 18 MΩ water is shown in Figure 6.27. Morris et al. [31] state that the data show that the onset is dependent on frequency, and they attribute this dependency to the fact that the measurement does

Figure 6.27. DEA Plots for Water

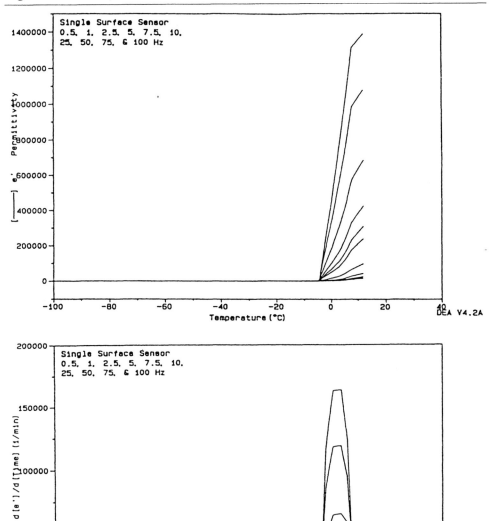

DEA plots of (a) permittivity (ϵ') and (b) derivative (with respect to temperature) of the permittivity, both versus temperature for 18 MΩ water showing the frequency independent of the melt. Each curve represents a different frequency, with the lower frequencies giving the greater responses. [Source: *PDA Journal of Pharmaceutical Science and Technology*, K. R. Morris, S. A. Evans, et al., Nov.–Dec. 1994, Vol. 48, No. 6, Page 322, Figure 4.]

not occur under equilibrium conditions. The nonequilibrium conditions are attributed to *a time lag from the heating rate* and the high conductivity of the liquid state. As a result, the DEA method was unable to ascertain the melting temperature of the ice. It would appear from these results that the DEA method is not applicable for determining the eutectic temperature or phase changes involving liquid states in the formulation.

Relaxation Time

Morris et al. [31] first attempted to determine the collapse temperature based on a plot of the $\ln \tau$ as a function of $1/T$ (K). At the collapse temperature, there should be a change in the slope. When such plots were made with data collected from several systems, two inflection points of the plot were observed, and neither agreed with collapse values determined by the freeze-drying microscope. In addition, the authors [31] report that the inflections were neither sharp nor well defined.

DCC and TOF Methods

It was assumed that when the frequency is low, there was, at a given temperature, an increase in the mobility of the ions and dipoles (i.e., at low frequencies, the ions and dipoles had time to respond to the electric field, as in the case of a dc field). The effect of low frequencies was found to be most prominent in plots of ϵ'' rather than plots of ϵ'. At the latter frequencies, a plot of ϵ'' as a function of ω has two characteristics: the y intercept, which is referred to as the DCC, and a point in the minimum that is referred to as the *take off frequency* (TOF). It was shown that DCC and/or the TOF could be used to ascertain the collapse temperature of the system.

DCC Method for Determining the Collapse Temperature. The DCC was expressed in terms similar to equation 9:

$$DCC(T) = DCC_o \exp\frac{-\Delta H'}{RT} \tag{10}$$

Plots of ln(DCC) as a function of $1/T$ (K) give a plot with two distinct, linear regions. The intersection of these regions gives collapse values that are similar to those obtained from the freeze-drying microscope.

TOF Method for Determining the Collapse Temperature. It was shown that the TOF could also be expressed in terms similar to equation 9:

$$TOF(T) = TOF_o \exp\frac{-\Delta H'}{RT} \cdot \tag{10}$$

The plot of ln(TOF) as a function of $1/T$ (K), as shown in Figure 6.28, shows that the plot has two linear segments. The intersection of these two segments gives temperature values that are in agreement with those determined a freeze-drying microscope. Morris et al. [31] felt that collapse temperatures determined from the TOF method were closer to those determined from the freeze-drying microscope than those were obtained from the DCC method. Table 6.1 shows collapse

Figure 6.28. Collapse Temperature Determination

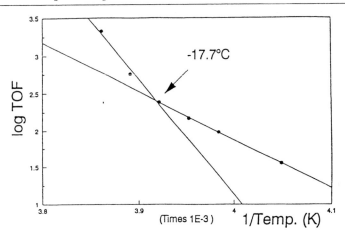

Collapse temperature of ln(TOF) versus 1/T (K) according to equation 10. The extrapolated intersection of the two linear portions identifies the collapse temperature of the system . [Source: *PDA Journal of Pharmaceutical Science and Technology*, K. R. Morris, S. A. Evans, et al., Nov.–Dec. 1994, Vol. 48, No. 6, Page 328, Figure 10.]

Table 6.1. Comparative Summary of Collapse Temperature Predictions [31]

System	TOF	DCC	Freeze-Drying Microscope
10% wt/wt Sucrose	–35.0°C	–35.6°C	–32°C
10% wt/wt Trehalose	–33.4°C	–28.0°C	–32°C
10% wt/wt Sorbitol	–47.2°C	–46.4°C	–45°C
11% wt/v Azactum™	–17.7°C	–20.3°C	–17°C

temperatures for several systems determined by DCC, TOF, and the freeze-drying microscope.

Summary and Comments

The following is a summary and some general comments regarding the DEA method for determining the thermal properties of a formulation.

Theoretical Basis for DEA Measurements

Morris et al. [31] provide an excellent introduction on defining ϵ' and ϵ'' and on the means for determining such values from equations 7 and 8. In addition, the energy associated with relaxation time can be ascertained from equation 9. It is, however, unfortunate that the introduction does not provide the reader with any sound reasons why such dielectric measurements would provide a basis for characterizing the thermal properties of a formulation.

Apparatus

The description of the sensor is certainly clear and very informative, however, a complete description of the cell that is used to house the sensor should also have been included. For example, during the analysis, the system is purged with a flow of nitrogen gas at 500 mL/min. It is hoped that in some future publication the authors will provide additional information as to the role of such a gas flow. Such information should include whether the temperature of the gas provided the basis for freezing and warming the sample.

Analytical Method

Except for the Azactum™ formulation, which was 11% wt/v, the concentration of all of the other solutions (sucrose, trehalose, and sorbitol) was 10% wt/wt. Many formulations have total concentrations that are significantly less than 10% wt/wt. It

Figure 6.29. Complex Plane Plots of the Permittivity of Ice

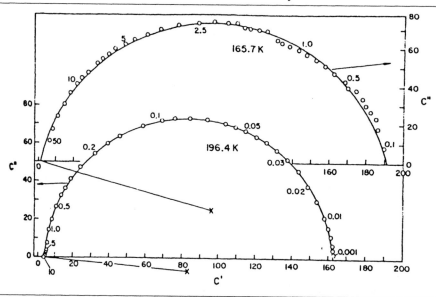

Complex plane plots of the permittivity of ice at the temperatures indicated on the curves. The numbers attached to the points on the upper curve are frequencies in Hz, and the numbers attached to the points on the lower curve are in kHz. [Reprinted with permission from *The Journal of Physical Chemistry*, Vol. 75, No. 3, 1981, Page 1335, Figure 1. Courtesy of the American Chemical Society, Washington, D.C.]

would have been helpful Morris et al. [31] had indicated their reasons for selecting such concentrations and whether the DEA method would be applicable to solutions at lower concentrations.

Interpretation of the Data

Ice. In view of the amount of literature on the dielectric properties of ice, Morris et al. [31] should have provided data on the permittivity of ice, like that shown in Figure 6.29, which is a plot of ϵ'' as a function of ϵ' for a given temperature. By this means, they could have demonstrated that the data obtained from the DEA were a measure of the dielectric properties of the formulation and not associated with an artifact of the sensor system.

Prior to providing the data shown in Figure 6.27, they should have shown that ϵ' and ϵ'' data as a function of time were similar to that shown by the theoretical plot in Figure 6.25. In Figure 6.27, they state that the onset is dependent on frequency. A frequency-dependent onset is not clearly apparent from the plots shown in Figures 6.27a or 6.27b. For Figure 6.27a, the frequency values for the permittivity of the ice at temperatures lower than 0°C had values that exceed 100,000. The units for permittivity are $coul^2/(newt\text{-}m^2)$. The reported value for the permittivity of ice at -5°C is only 26 $coul^2/(newt\text{-}m^2)$. They should have indicated the units for the permittivity used in their investigation. In addition, Figure 6.27a shows a plot of $d(\epsilon')/[\text{Time}]$ (1/min), while the caption for this plot indicates that the derivative of the permittivity is taken with respect to temperature.

The determination of the dielectric properties of water gave the authors an opportunity to show that their sensor system was compatible with those used by other investigators. It would have been useful had the authors [31] determined the dielectric properties of ice at -76.7°C to ascertain if the permittivity of ice determined in the sensor illustrated by Figure 26 gave similar results to those shown in Figure 6.29 [32].

DCC and TOF Methods. The authors [31] indicate that the plots of the ln(DCC) as a function of $1/T$ (K) were similar to those shown for the TOF in Figure 6.28. It is my opinion that a plot of ln(DCC) as a function of $1/T$ (K) should also have been presented to verify that the DCC method does indeed provide a measure of the collapse temperature.

The plot in Figure 6.28 shows six data points. While the data points for the low frequency measurements fall on a straight line, the two data points at the higher temperatures do not fall on the line; it is stated, however, that the TOF method provides a collapse temperature with a precision to within 0.1°C. From the limited number of data points, it would appear that such precision is not justified.

Dielectric measurement of the thermal properties of a formulation is certainly a new and novel analytical method. It is hoped that there will be additional publications concerning this method that will provide additional information regarding the nature of the test cell and the characterization of the sensor. From the low temperature portion of the DCC and the TOF plots, one should be able to determine the value of $\Delta H'$, which should be of the order of 57 kJ/mole for ice.

As in the case of the freeze-drying microscope, it appears that the DEA method is limited to determining only the collapse temperature of the formulation. Because of the low conductivity of the system, it would also appear that the above dielectric method would not, particularly in the case involving electrolytes, be applicable in determining an eutectic temperature.

D_2 AND DTA ANALYTICAL SYSTEM

Introduction

The D_2 and DTA analytical system uses two independent analytical methods to ascertain the thermal properties of a formulation. D_2 enhances the sensitivity of resistance measurements to determine thermal properties that involve low energy levels for a change in state, and an expanded version of DTA analysis is used to characterize thermal properties involving energy. This section will first consider the term D_2 and then discuss the expanded role of DTA analysis in determining the thermal properties of a formulation.

D_2 *Analysis*

The concept of D_2 measurements is based on the following model [33, 34]. Consider a small region in a frozen matrix that has a configuration like that shown in Figure 6.30. The frozen matrix consists of layers of ice in between layers of interstitial material composed of the constituents of the formulation and, for the majority of formulations, a small quantity of supercooled water.

The electrical resistance (R') of the matrix n layers between faces A and B is expressed as

$$\frac{1}{R'} = \Sigma \left(\frac{1}{R'_{ice}} \right)_i + \Sigma \left(\frac{1}{R'_s} \right)_j \quad \frac{1}{\Omega} \tag{12}$$

where R'_{ice} is the resistance of an ice layer in a system having i ice layers and R'_s is the resistance of j interstitial layers in a system having n layers where $n = i + j$.

If the resistivity of the ice and interstitial layers are ρ_{ice} and ρ_s, respectively, then using equations 3 and 12 one obtains

$$\frac{1}{R'} = \frac{1}{\rho_{ice}} \Sigma \left(\frac{A}{l} \right)_i + \frac{1}{\rho_s} \Sigma \left(\frac{A}{l} \right)_j \quad \frac{1}{ohms} \tag{13}$$

Figure 6.30. Matrix Model for D_2 Analysis

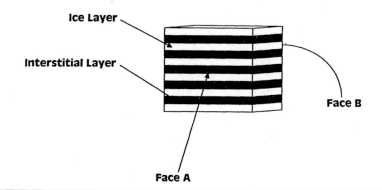

where ρ_{ice} and ρ_s are the resistivity of the ice and interstitial layers, respectively and the terms l and A as previously defined.

For i layers of ice, the parallel resistance can be expressed as

$$\frac{1}{R'_{ice}} = \frac{D_o}{\rho_{ice}} \quad \frac{1}{\Omega} \tag{14}$$

where

$$D_o = \frac{\sum\limits_{j=1}^{i}\left[\prod\limits_{n=1,n\neq j}^{i}\left(\frac{l}{A}\right)_n\right]}{\prod\limits_{n=1}^{i}\left(\frac{l}{A}\right)_n} \tag{15}$$

where Π denotes the sum of product of the terms in the bracket. As seen from equation 15, D_o is a highly complex term.

For k layers of the interstitial material, the parallel resistance can be expressed as

$$\frac{1}{R'_s} = \frac{F}{\rho_s} \quad \frac{1}{\Omega} \tag{16}$$

where F is a highly complex term defined by

$$F = \frac{\sum\limits_{j=1}^{k}\left[\prod\limits_{j=1,n\neq i}^{k}\left(\frac{l}{A}\right)_n\right]}{\prod\limits_{n=1}^{k}\left(\frac{l}{A}\right)_n} \tag{17}$$

From equations 14 and 16, the resistance of the model matrix shown by Figure 6.30 and represented by equation 12 can now be simplified as

$$R' = \frac{\dfrac{\rho_{ice}}{D_o} \times \dfrac{\rho_s}{F}}{\dfrac{\rho_{ice}}{D_o} + \dfrac{\rho_s}{F}} \tag{18}$$

As the temperature is decreased, the mobility of the water in the interstitial region of the matrix approaches 0; because of the conductivity of the ice surface, $\rho_s \gg \rho_{ice}$. Because the thickness (d_c) of the ice layer will tend to be much greater than that of the interstitial region (d_i), D_o tends to be much greater than F. As a consequence ρ_s/F will be much greater than ρ_{ice}/D_o, and equation 18 simplifies to

$$R' = \frac{\rho_{ice}}{D_o} \quad \Omega \tag{19}$$

where the term D_o can be broken down into a product of two terms,

$$D_o = D_1 \times D_2 \quad cm \tag{20}$$

D_1 can be defined as the cell constant for the system,

$$D_1 = \frac{A}{l} \quad \text{cm}$$

(21)

and from equations 19, 20 and 21, D_2 is expressed as

$$D_2 = \frac{\rho_{ice}}{R' \times D_1} = \frac{\rho_{ice}}{\rho_{meas}}$$

(22)

since $\rho_s = R' \times D_1 \ \Omega\text{-cm}$ and ρ_{meas} is the resistivity determined by the cell.

D_2 is a dimensionless term involving the ratio of fundamental properties of a material (i.e., a property that is independent of quantity and configuration). Because the conductivity of ice occurs primarily along the surface, a volume of ice consisting

Figure 6.31. Apparatus for D_2 and DTA Analysis

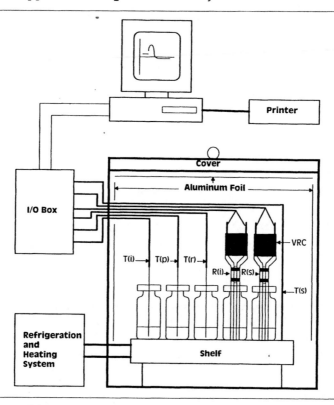

Illustration of the apparatus used in performing D_2 and DTA analysis of a formulation, where $T(i)$ is the water reference temperature; $T(p)$ is the formulation temperature, $T(r)$ is the methanol reference temperature, $T(s)$ is the shelf-surface temperature, $R(i)$ is the resistance of the water reference, $R(s)$ is the resistance of the formulation, and VCR is the vertical resistance cell.

of small ice crystals will have a lower resistivity than an equivalent volume of ice consisting of large ice crystals. If ρ_{ice} represents the resistivity of the ice formed by the process water used in the preparation of the formulation and ρ_{meas} is the resistivity of a formulation, then when the mobility of the water in the interstitial region of the matrix approaches zero, $\rho_s \gg \rho_{ice}$ and ρ_{meas} approaches ρ_{ice} and D_2 approaches 1 or a constant value, depending on the size of the ice crystals in the water reference and that formed in the formulation. A determination of the onset temperature, where D_2 increases with temperature, is a measure of the temperature where $\rho_s \leq \rho_{ice}$ is the temperature at which the mobility of the water in the interstitial region of the matrix becomes significant, which is defined as the collapse temperature.

Description of the Apparatus

The apparatus used in making D_2 and DTA measurements of a formulation is illustrated in Figure 6.31. The apparatus is similar to that shown in Figure 6.1; however, there are some significant differences.

Test Containers, Sensors, and Resistance Cells

Test Containers. The test containers are usually glass vials similar to those that are used in the lyophilization of the formulation. The vials are cleaned by rinsing in an inverted position with a stream of hot (80°C to 90°C) WFI or deionized water and then allowed to air dry in an inverted position. The reason for the hot water rinse is to passivate the fresh glass surface and remove any soluble silicates that could contaminate the liquid formulation. Prior to use, each container is labeled to identify the contents.

Sensors. Four copper-constantan thermocouples are used as temperature sensors. The shelf-temperature sensor [$T(s)$] is constructed from 20-gauge thermocouple wire, and the thermocouple junction is electrically insulated. The shelf-temperature sensor is taped to the surface of the shelf. The remaining sensors [$T(p)$, $T(r)$, and $T(i)$] are made from 0.010 in. diameter bare thermocouple in thin-walled Teflon® tubing. $T(i)$ is the temperature of the water reference. The tubing is bent so the small thermocouple junction is located at the crease in the bend. The use of bare thermocouple wire with such a small diameter will decrease the response time of the thermocouple to temperature changes. Teflon® is used to provide electrical insulation of the thermocouple and an inert environment when the sensor is placed in the formulation or reference material.

Resistance Cells. The resistance cells are fabricated from a pair of electrodes like that shown in Figure 6.17. In place of the electrode holder shown in Figures 6.18a and 6.18b, the cell is constructed so that electrodes are separated by an insulator having a thickness of 0.020 in. At this electrode spacing, the resistance determined by the cells is not subject to variations in the distance between the electrodes like that seen in Figure 6.19.

Data Collection and Computer System

Data Collection. The output of the thermocouples and the voltage (5 V low frequency) for the resistance cells are collected and supplied by the I/O box. The output of the analog signals from the temperature sensors and the resistance cells are digitized by an A/D converter contained in the I/O box.

Computer System. The digitized signals from the thermocouple sensors and resistance cells are sent to the computer system for processing and storage. The computer system provides real-time display of the data in both tabular and graphic form throughout the entire analysis. The operating program for the system is stored on a hard disk drive while the analytical data are stored on floppy disks (identified as "DATA") in units of 10 data points.

Analytical Method

The following is a brief description of the analytical method used in the D_2 and DTA analysis.

Run Identification

The first step in the analytical method is to identify and store on the DATA disk the following information concerning the thermal analysis:

- Identification of the formulation
- Test date
- Operator
- Fill volume of the formulation
- Identification of the reference material
- Nature of the process water
- Notebook number
- Notebook page

The Run Identification program also allows the operator to access the Parameter Table. This table sets the operating parameters that will be used during the analysis. The major parameters include the following:

- *Temperature Read Count:* The number of readings for each of the thermocouples that will be used to determine the average temperature of the sensor.

- *Resistance Read Count:* The number of readings for each of the resistance cells that will be used to determine the average resistance of the sensor.

- *Grid Spacing:* The temperature interval for the y-axis grid spacing for the DTA thermograms. Note that these settings can be changed when printing out a hard copy of the thermogram.

Calibration

The calibration procedure not only calibrates the thermocouple sensors at two points but also checks the resistance measuring system and the cell constants of the probes.

Calibration of the Thermocouple Sensors. The thermocouples are first checked in a water bath at ambient temperature against a NIST–traceable thermometer. The gain in the amplifier system contained in the I/O box is adjusted such that the formulation temperature $[T(\dot{p})]$ is in agreement with that of the thermometer.

The thermocouples are then placed into an ice slush bath (checked with an NIST–traceable thermometer) prepared from deionized water. After the output of the thermocouple sensors has reached a steady state, the zero adjustment on the thermocouple amplifier is adjusted so that the output of $T(p)$ approaches 0°C.

The calibration of each sensor is obtained from 25 readings of each thermocouple sensor. The average temperature and standard deviation for each sensor are calculated. If the standard deviation on any of the sensors is ≥ 0.1°C, then the steps in the calibration procedure must be repeated. The difference between the average sensor temperature and the temperature of the ice slush bath is determined, and the temperature of the sensor will be adjusted by the addition of the offset temperature. A printout of the individual data for each of the sensors is provided. The calibration data are stored on the DATA disk.

Resistance Check. The resistance check is used to ascertain that the programming and the circuitry for the resistance measurements are performing within specifications. The first test is a short-circuit test in which the connections to the resistance leads are connected. A resistance value greater than 10 Ω for either resistance lead will indicate an error in the system and must be corrected before the system will allow the operator to proceed to the next test. The values for the short test are stored on the DATA disk and provided with the printout of the data.

The second resistance test is an open-circuit test. In this test, the resistance leads are connected across the resistance cells that will be used for the resistance measurements. Typical open-circuit values are of the order of 10^{11} Ω; resistance values $\leq 1 \times 10^9$ Ω will indicate a malfunction in the system or a low parallel resistance in either of the resistance cells. The malfunction must be corrected before proceeding with the analysis. The values of the open-circuit test will be stored on DATA disk and provided with the printout of the data.

Cell Constant. The final resistance check is to ascertain if the cell constants of the resistance cells are compatible, i.e, $\leq 10\%$ difference, in resistance between the cells. The rationale for such a check is as follows:

It has been shown that D_2 is defined as

$$D_2 = \frac{\rho_{ice}}{R' \times D_1} = \frac{\rho_{ice}}{\rho_{meas}} \qquad (22)$$

where ρ_{ice} is the resistivity of the water reference and ρ_{meas} is the resistivity of the formulation. Since the units of ρ are ohms-cm, the height of the liquid in each of the resistance cells would have to be determined and entered into computer system. To avoid the necessity of such a measurement and the errors associated with such a measurement, the measurements are simplified by the following:

The resistivity of the water reference (ρ_w) can be expressed as

$$\rho_w = \frac{R_w}{K_w} \quad \Omega\text{-cm} \tag{23}$$

where K_w is the cell constant for the ice resistance cell as defined by equation 4, and R_w is the resistance of the process water.

Likewise, the resistivity of the formulation can be expressed as

$$\rho_{meas} = \frac{R_{meas}}{K_{meas}} \quad \Omega\text{-cm} \tag{24}$$

where K_{meas} is the cell constant for the formulation cell and R_{meas} is the measured resistance of the process water. From equations 23 and 24, D_2 can now be defined by

$$D_2 = \frac{\rho_w}{\rho_{meas}} = \frac{R_w}{K_w} \times \frac{K_{meas}}{R_{meas}} \tag{25}$$

If the resistance cells for the ice reference and the formulation are place into similar containers having solution with the same resistivity (e.g., 1×10^{-4} M KCl), and the observed resistance values are equal, the respective cell constants must also be equal. For equal cell constants, D_2 can now be represented by

$$D_2 = \frac{R_w}{R_{meas}} \tag{26}$$

R_w is the resistance of the cell used for the water reference and R_{meas} is the resistance of the cell for the sample. Care must be taken to rinse any residual KCl from the resistance cells prior to their use in determining the thermal properties of a formulation.

Equation 22 can now be rewritten as

$$D_2 = \frac{\rho_{ice}}{R' \times D_1} = \frac{\rho_{ice}}{\rho_{meas}} = \frac{R_{ice}}{R_{meas}} \tag{27}$$

Setup of the Analysis

Upon completion of instrument calibration, the samples are installed in the test chamber as shown in Figure 6.31. A given fill volume of the formulation is added to the sample vial, and sensor $T(p)$ is positioned in the center of the vial approximately 1/16 in. to 1/8 in. from the bottom. A vertical resistance probe is placed in the middle of a vial containing a fill volume of the formulation, such that it contacts the bottom for determining R_{meas}. A fill volume of the process water is placed in a vial that will contain the resistance cell for determining R_{ice}. A quantity of process water equal to the same amount of water contained in the fill volume of the formulation is added to the vial containing the sensor to measure $T(i)$, and the sensor is positioned like the one for determining $T(p)$. The vial for the reference contains a fill volume of methanol, and the temperature sensor $T(r)$ is positioned like those described above.

Cooling Analysis

The collection of data during the cooling portion of the analysis is determined by a temperature *window*. When the change in temperature of the formulation, $T(p)$, or the methanol reference, $T(r)$, is equal to or greater than the last recorded temperature by a temperature value specified by the *window*, then the computer will record, as an event, all of the current sensor temperatures and resistance values. The magnitude of the *window* is determined from the standard deviation obtained during the calibration of the thermocouples. Standard deviations for the thermocouple sensors range from 0.02°C to 0.05°C. For such standard deviations, a typical *window* value is 0.2°C. As a result, the $T(p)$ or $T(r)$ must have a temperature change that would represent 4σ to 10σ of the previous temperature. The odds that such a temperature change would occur without a change in the mean temperature of $T(p)$ or $T(r)$ is beyond normal chance. With the selection of the temperature *window*, the refrigeration system is turned on and the temperature of the formulation is lowered from ambient temperature to $\leq -50°C$. During the course of the cooling process, more than 300 events are stored.

Warming Analysis

The shelf-surface temperature is increased by heating the heat-transfer fluid that passes through the shelf (shown in Figure 6.31). Depending on the fill volume, the warming analysis requires 2 to 3 hr to complete (i.e., all of the temperature sensors are 20°C ± 2°C).

Printout of Data

The data are printed out in three segments.

General Information. The first page of the printout contains the general information regarding the identification of the product and the results of the setup of the instrumentation prior to the start of the analysis. This segment contains all of the information that was entered by the operator regarding Run Identification, the values of Parameter Table, results of the calibration of the thermocouples, resistance circuit checks, resistance values for the cell constants, and the value of the temperature window for the recording of the events.

Tabular Data. All data stored during the cooling and warming analyses are printed out by event number. The headings and an example of the data for the tabular printout are as follows:

EVENT	ELAPSED TIME	R(KOHM) ICE	R(KOHM) SAMPLE	D_2	T(C) ICE	T(C) SAMPLE	T(C) REF	T(C) SHELF	DT1	DT2	DT3
0	00:10:45	216.556	7.572	29.59	21.933	21.924	21.572	19.664	.352	−.008	.360

where the ELAPSE TIME is denoted in hours:minutes:seconds, R(KOHM) is the resistance expressed in $k\Omega$, DT1 is the temperature difference $[T(p) - T(r)]$, DT2 is the temperature difference $[T(p) - T(i)]$, and DT3 is the temperature difference $[T(i) - T(r)]$.

Data Plots. The printout also displays the data by a series of graphic plots.

- R_{ice}, R_{sample}, and D_2 as a function of $T(p)$
- DT1 as a function of $T(r)$
- DT2 as a function of $T(r)$
- DT3 as a function of $T(r)$
- R_{ice}, R_{sample}, and D_2 as a function of EVENTS
- DT1 as a function of EVENTS
- $T(p)$ and $T(s)$ as a function of ELAPSED TIME in hours

Using these plots in conjunction with the tabular data for determining the thermal properties of a formulation will be demonstrated and discussed in the next section.

Interpretation of the Data

The following sections will illustrate the use of the D_2 and DTA in determining the thermal properties of a formulation during the cooling and warming processes.

Cooling

The following are the thermal properties of the formulation that are obtained during the cooling process.

Formation of Ice-like Water Clusters. The formation of ice-like water clusters in deionized water is shown in Figure 6.32. This figure shows a small, relatively broad exotherm that is related to the formation of the ice-like water clusters. The onset temperature for the formation of these clusters occurred at a reference temperature of 6°C, in which $T(i)$, from the tabular data shown in Appendix B, had a value of about 7°C.

The DTA thermogram for a solution composed of 3% lactose (D-(+)-Lactose Monohydrate Powder "Baker Analyzed" Reagent supplied by J. T. Baker Inc.) and 0.9% NaCl (the NaCl used was certified A.C.S. and supplied by Fisher Scientific Co.) is shown in Figure 6.33 and the data obtained from Appendix C. This figure shows that there was some variation in the baseline but no definitive exothermic peak to indicate the formation of ice-like water clusters. The absence of such clusters indicates that a significant quantity of the water in the solution had complexed with the solutes so that there was insufficient *free water* to form ice-like water clusters.

While the formation of ice-like water clusters is not taken into consideration when defining the lyophilization process, nevertheless, the presence or absence of such clusters is a fingerprint of the nature of the process water and the formulation. The absence of the formation of these clusters in the process water at some later date would indicate a change in the nature of the thermal properties of the deionized water or WFI.

Degree of Supercooling. The degree of supercooling of a formulation, consisting of 3% (wt/v) lactose and 0.9% (wt/v), NaCl is illustrated by the discontinuity in the plot of $[T(p) - T(r)]$ versus $T(r)$ shown by Figure 6.34. The sharp increase in

Figure 6.32. DTA Thermogram Plot During Cooling and With Ice-like Cluster Formation

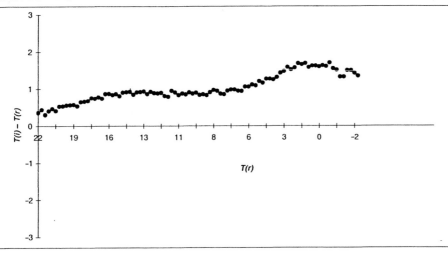

DTA thermogram plot of $T(i) - T(r)$ as a function of $T(r)$ for temperatures $> -10°C$ during cooling, showing the formation of ice-like water clusters in 1.92 mL of deionized water contained in a 5 mL tubing vial, where $T(i)$ is the water reference temperature and $T(r)$ is the methanol reference temperature. Data obtained from Appendix B.

Figure 6.33. DTA Thermogram Plot During Cooling Without Ice-like Cluster Formation

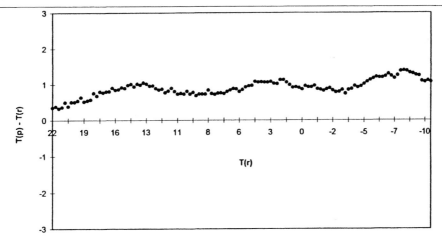

DTA thermogram plot of $T(p) - T(r)$ as a function of $T(r)$ for temperatures $> -10°C$ during cooling, showing no formation of ice-like water clusters in 2.00 mL a solution containing 3% (wt/v) lactose and 0.9% (wt/v) NaCl in a 5 mL tubing vial, where $T(p)$ is the temperature of the isotonic lactose solution and $T(r)$ is the methanol reference temperature. Data obtained from Appendix C.

$[T(p) - T(r)]$ is a result of an increase in $T(p)$ and resulted from the latent heat associated with the formation of ice in the formulation. The tabular data listed in Appendix C show that during cooling the nucleation of the ice crystals in the formulation occurred when $T(p)$ was $\leq -9.42°C$ and $\geq -9.62°C$. While Figure 6.34 clearly shows the presence of supercooling in the formulation during the cooling process, the tabular data provide a measure of the nucleation temperature. Based on the equilibrium freezing temperature of $-0.35°C$, as described below, the degree of supercooling for this formulation is $-9.1°C$. The large exothermic peak shown in this thermogram is associated with the formation of ice crystals in the formulation.

Equilibrium Freezing Temperature. Upon nucleation of the ice crystals, as seen in Figure 6.34, the temperature of the sample will momentarily increase to the equilibrium freezing temperature or the freezing point depression of water due to the colligative properties of lactose and NaCl. From the tabular data for $T(p)$ listed in Appendix C, the equilibrium freezing temperature, based on the average of $T(p)$ values from event 123 to event 129 was $-0.35°C$. The calculated equilibrium freezing temperature, based on the following equation (from Chapter 5):

$$\Delta T_f = \frac{1,000 K_f}{W} \sum \frac{g_{s,i}}{M_{s,i}}$$

was $-0.44°C$.

Figure 6.34. DTA Thermogram Plot During Cooling for Lactose and Sodium Chloride Solution

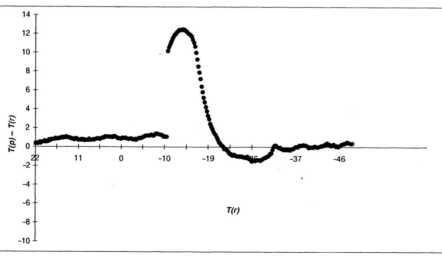

DTA thermogram plot of $T(p) - T(r)$ as a function of $T(r)$ during cooling of 2.00 mL of a solution containing 3% (wt/v) lactose and 0.9% (wt/v) NaCl in a 5 mL tubing vial where $T(p)$ is the temperature of the isotonic lactose solution and $T(r)$ is the methanol reference temperature. Data obtained from Appendix C.

Examination of the freezing temperature for the deionized water reference showed that the equilibrium freezing temperature (see events 116 to 126 in Appendix C) was > 0°C or an average value of 0.56°C. Since the standard deviation for the ice temperature measurements was 0.030°C, the odds against one that the actual temperature being 0°C and the above values being part of the normal distribution would be about 19σ and of galactic dimensions. Such an equilibrium freezing temperature for the water reference would signify the presence of organic compounds in the water that were not removed and formed hydrates with the water. Such hydrates have already been shown to increase the equilibrium freezing temperature of water.

Interstitial Phase Changes During Cooling. In spite of the presence of NaCl in the formulation, the thermogram shown in Figure 6.34 indicates that there is no apparent phase change in the interstitial region of the matrix during the cooling process at temperatures lower than –20°C. The absence of a phase change in the interstitial region would signify that in the presence of the lactose, the concentration of the NaCl never reached a concentration of 23% wt/wt.

The thermogram of $[T(p) - T(r)]$ as a function of $T(r)$ for a 0.9% (wt/v) NaCl solution during cooling is shown by Figure 6.35 and is based on the tabular data in Appendix B. Examination of this thermogram and the tabular data shows a small, broad exothermic peak with an onset temperature of 3.8°C (formation of ice-like water clusters), a degree of supercooling of 10.1°C, and an average equilibrium freezing temperature of –0.46°C [see values of $T(p)$ for events 112 to 117]. The calculated equilibrium freezing temperature, from equation 27, for an isotonic NaCl solution is

Figure 6.35. DTA Thermogram Plot During Cooling for a Sodium Chloride Solution

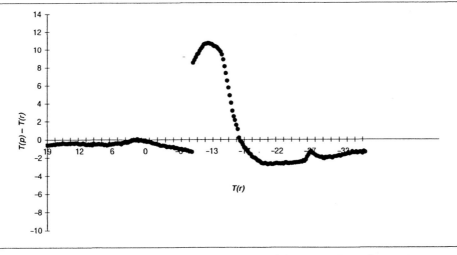

DTA thermogram plot of $T(p) - T(r)$ as a function of $T(r)$ during cooling of 2.00 mL a solution containing 0.9% (wt/v) NaCl in a 5 mL tubing vial, where $T(p)$ is the temperature of the isotonic solution and $T(r)$ is the methanol reference temperature. Data obtained from Appendix B.

–0.29°C. This thermogram shows a second small exothermic peak at temperatures lower than –20°C, with an onset temperature of –28.9°C [see values of $T(p)$ for events 196 to 200]. Because the large exothermic peak is associated with the formation of ice crystals, the latter peak indicates a phase change in the interstitial region of the matrix.

Warming

Collapse Temperature. The collapse temperature Tc is determined from a linear plot of D_2 as a function of $T(p)$ in the region, where the $\rho_{ice} > \rho_{meas}$ or $R(i)^2 > R(s)^3$. The equation for the function and Tc are determined by extrapolating to $D_2 = 1$ or a constant value resulting from differences in the size of the ice crystals in the formulation and the water reference. The extrapolated value for $T(p)$ indicates the temperature where the mobility of the water in the interstitial region of the matrix is significant. Using this means for determining the Tc of a formulation is demonstrated by determining the Tc for solutions containing 3% lactose, 3% lactose + 0.9% NaCl, and a 0.9% NaCl.

A 3% Lactose Solution. A plot of $R(s)$, $R(i)$, and D_2 as functions of $T(p)$ is shown in Figure 6.36. At temperatures lower than –20°C, $R(s) \geq R(i)$, which indicates that the

---

Figure 6.36. Warming Plot for a Lactose Solution

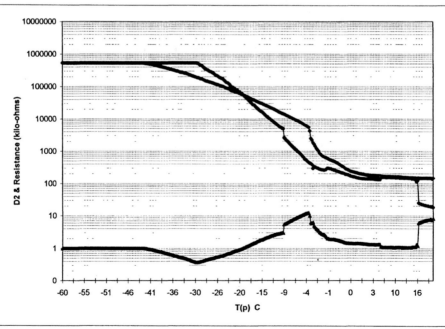

A warming plot of $R(s)$, $R(i)$, and D_2 as a function of $T(p)$ for a 3% lactose solution, where ● denotes $R(s)$, ◆ indicates $R(i)$, and ▲ the D_2 values. The log plot was used only to display the data.

---

2. $R(i) = R_{ice}$ 3. $R(s) = R_{meas}$

mobility of the water in the interstitial region of the matrix approached or was equal to 0. A plot of D_2 as a function of $T(p)$ for $R(i) > R(s)$ is shown in Figure 6.37. This plot shows a linear relationship between D_2 and $T(p)$. The equation for the line was found to be

$$D_2 = 0.2115x + 4.9594 \qquad (28)$$

where $x = T(p)$. Figure 6.36 and the tabular data shown in Appendix D indicate that $D_2 = 1$ when the matrix of the lactose solution was in a completely frozen state. Based on equation 28, the extrapolation of $D_2 = 1$ gave a Tc value of $-19°C$.

A 3% Lactose and 0.9% NaCl Solution. The warming plot for $R(s)$, $R(i)$, and D_2 as functions of $T(p)$ is shown in Figure 6.38. A comparison of this figure with Figure 6.36 shows that the addition of the NaCl resulted in a significant change in $R(s)$ at temperatures lower than $-20°C$. Figure 6.38 and the tabular data listed in Appendix C show that during warming there was a sharp increase in D_2 at a $T(p)$ of $-25°C$. Such

Figure 6.37. Plot of D_2 as a Function of $T(p)$ During Warming of a 3% Lactose Solution

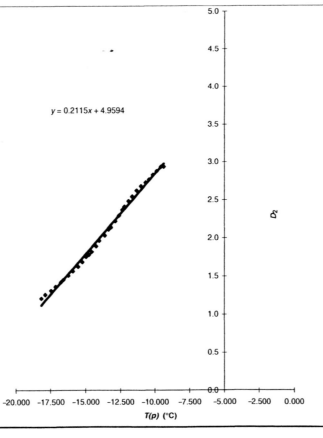

$y = 0.2115x + 4.9594$

Figure 6.38. Warming Plot for a Lactose and Sodium Chloride Solution

---

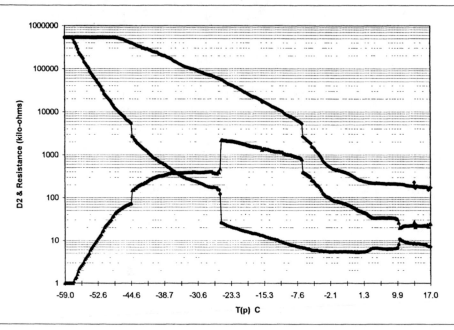

A warming plot of R(s), R(i), and D_2 as a function of T(p) for a 3% lactose and 0.9% NaCl solution, where ● denotes R(s), ◆ indicates R(i), and ▲ the D_2 values. The log plot was used only to display the data.

---

a change in D_2 is attributed to a marked decrease in the electrical resistance, which is associated with an increase in the mobility of the water in the interstitial region of the matrix. A further change in electrical resistance properties of the formulation is illustrated by Figure 6.39. This latter figure shows a linear plot of D_2 as a function of T(p). The equation for the line shown in Figure 6.39 is

$$D_2 = 9.8597x + 522.79 \qquad (29)$$

and, for $D_2 = 1$, the extrapolated value of T(p) or Tc is found to be -53°C. This result shows that even small quantities of electrolytes, such as NaCl, can have a major impact on the thermal characteristics of a formulation.

Isotonic (0.9%) Solution of NaCl. The results of the warming of a 0.9% NaCl solution is shown in Figure 6.40. This figure is a plot of R(s), R(i), and D_2 as functions of T(p). This figure shows that D_2 did not approach 1 or a constant value. When compared with the results in Figure 6.38, the initial slope of the D_2 versus T(p) was not as steep at T(p) values lower than -40°C. There was however, a sharper increase in D_2 at temperatures near -21°C, which signifies the presence of more mobile water in the interstitial region of the matrix and will be shown to be associated with the melting of the eutectic mixture of ice and sodium dihydrate. Figure 6.41 is a linear plot of D_2 as a function of T(p). This latter plot can be expressed by

Figure 6.39. Plot of D_2 as a Function of $T(p)$ During Warming of a 3% Lactose and 0.9% NaCl Solution

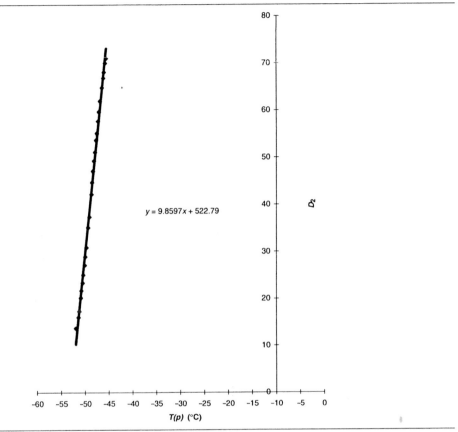

$$D_2 = 7.3405x + 350.61 \qquad (30)$$

from which a Tc of $-48°C$ can be obtained by extrapolation to $D_2 = 1$. The significance of these results is that during the freezing process, only a portion of the NaCl formed an eutectic mixture. A small quantity of electrolyte was present in the interstitial region in a glassy or rubbery state. The presence of an endothermic peak associated with the presence of an eutectic mixture is no guarantee that all of the electrolyte formed in dilute solutions formed an eutectic mixture.

Note the similarity between Figure 6.38 and the resistance data shown in Figure 6.22 by DeLuca and Lachman [29]. The presence of the resistance of the water reference is a clear indication that the decrease in resistance prior to the onset of the eutectic temperature is a result of the mobility of water in the interstitial region of matrix, not a result of a change in the resistance of the ice in the matrix.

Interstitial Melting Temperature. The interstitial melting temperature is that temperature at which known liquid states are present in the interstitial region of the

Figure 6.40. Warming Plot for a Sodium Chloride Solution

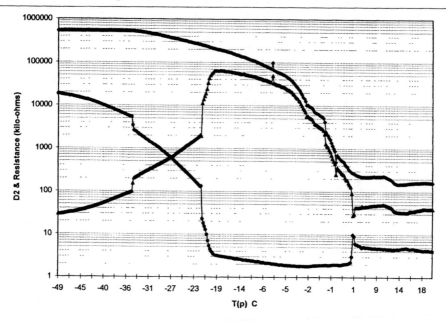

A warming plot of $R(s)$, $R(i)$, and D_2 as a function of $T(p)$ for a 0.9% NaCl solution, where ● denotes $R(s)$, ◆ indicates $R(i)$, and ▲ the D_2 values. The log plot was used only to display the data.

matrix. The interstitial melting temperature is of interest for those formulations that do not have an eutectic point and generally occurs at a temperature lower than the onset temperature for the melting of the ice crystals. The melting temperature is that temperature at which, during the warming process, the resistance of the formulation approaches the resistance of the process water at ambient temperature. I have arbitrarily selected 100 kΩ as the resistance value for the interstitial melting temperature. It should be stressed that I do not place a great deal of significance on the interstitial melting temperature other than it serves as a benchmark or fingerprint for the formulation. In the case of the 3% lactose and 0.9% NaCl solution, the onset interstitial melting temperature was found to occur during warming at event 140 where $T(p)$ = –26°C where $R(s)$ = 122.8 kΩ (see Appendix C).

Phase Change in the Interstitial Region. The determination of a phase change in the interstitial region by a DTA thermogram of the warming of a 0.9% NaCl solution is shown by Figure 6.42. From this figure, the onset temperature for eutectic temperature is denoted by the letter "A", and $T(e)$ is the eutectic temperature in terms of the methanol reference temperature. $T(r)$ was determined to be –19.17°C; from the tabular data listed in Appendix B (see event 128 on page 220), this is equivalent to a $T(p)$ of –21.1°C. This method provides an accurate means for determining the onset of an endothermic process; the method is also (although it is not shown here) capable of

Figure 6.41. Plot of D_2 as a Function of $T(p)$ During Warming of a 0.9% NaCl Solution

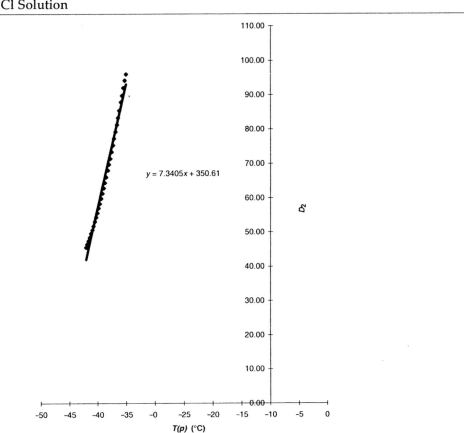

$$y = 7.3405x + 350.61$$

$T(p)$ (°C)

determining the onset temperature for exothermic processes such as the presence of metastable states.

Degree of Crystallization. The degree of crystallization is determined from the ratio of the areas under the endothermic peaks during warming for the DTA water reference and the formulation. The quantity of water in the DTA water reference is equivalent to that contained in the formulation. Figure 6.43 shows the DTA thermogram for a 3% lactose and 0.9% NaCl solution, and Figure 6.44 is the thermogram for the equivalent amount of deionized water. The area under the endotherm shown in Figure 6.43 was 52.97, while the area of the endotherm shown in Figure 6.44 was 67.53. From these areas, the degree of crystallization was found to be 0.78, which indicates that 22% of water in the 3% lactose and 0.9% NaCl solution did not crystallize during the freezing process. The reader should also note that in Figure 6.43 there was no additional endotherm to indicate the presence of an eutectic mixture. For this solution, the interstitial region of the frozen matrix consisted of a glassy state comprised of uncrystallized deionized water, lactose, and NaCl.

Figure 6.42. DTA Thermogram Plot for a Sodium Chloride Solution

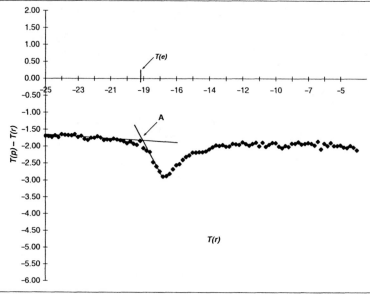

DTA thermogram plot of $T(p) - T(r)$ as a function of $T(r)$ during warming of 2.00 mL a solution containing 0.9% (wt/v) NaCl in a 5 mL tubing vial, where $T(p)$ is the temperature of the isotonic solution and $T(r)$ is the methanol reference temperature. The letter "A" denotes the onset temperature for the endotherm and $T(e)$ the eutectic temperature in terms of $T(r)$.

Figure 6.43. DTA Thermogram Plot of a Sodium Chloride Solution

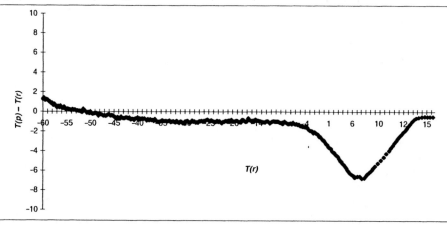

DTA thermogram plot of $T(p) - T(r)$ as a function of $T(r)$ during warming of 2.00 mL a solution containing 0.9% (wt/v) NaCl in a 5 mL tubing vial, where $T(p)$ is the temperature of the isotonic solution and $T(r)$ is the methanol reference temperature.

Figure 6.44. DTA Thermogram Plot of Deionized Water

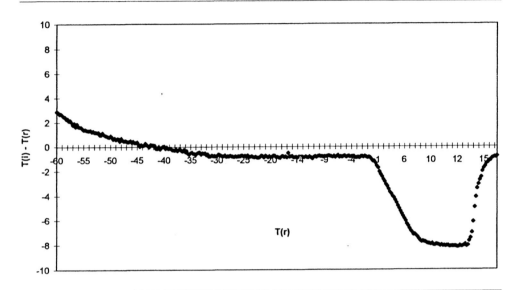

DTA thermogram plot of $T(i) - T(r)$ as a function of $T(r)$ during warming of 1.92 mL of the deionized water used to prepare the solution containing 0.9% (wt/v) NaCl in a 5 mL tubing vial, where $T(i)$ is the temperature of the deionized water and $T(r)$ is the methanol reference temperature.

Onset Temperature for Ice Melting. Another thermal property of a formulation that can serve as a benchmark or fingerprint is the onset temperature for the melting of the ice crystals in the formulation and the deionized water reference. Figure 6.45a is a plot of $[T(p) - T(r)]$ as a function of $T(r)$ in the region of the onset temperature for the melting of the ice crystals shown in Figure 6.43, while Figure 6.45b is a plot of $[T(i) - T(r)]$ as a function of $T(r)$ in the region of the onset temperature for the melting of the ice in Figure 6.44. The onset temperature $[T(m)]$ for the melting of the ice was obtained from the tabular data in Appendix C by determining the temperature of the methanol reference at the point of intersection (A). For the deionized water, the onset temperature was determined to be –0.4°C. Examination of the tabular data for $T(i)$ for events 244 to 250 showed that the equilibrium melting temperature approached 0°C.

The $T(m)$ for the melting of the ice for the isotonic 3% lactose (wt/v) solution was –2.7°C, which Appendix C shows to be equivalent to a $T(p)$ of –4.6°C. Further examination of $T(p)$ for events from 224 to 264 showed no indication of a well-defined equilibrium melting temperature.

The onset temperature of –0.4°C for the melting of the deionized water would suggest that there was some contamination present. Such contamination could be of an organic nature, which would pass through the deionized column and absorbed CO_2 from the atmosphere. The key point is that evaluation of the process water

Figure 6.45. DTA Thermogram Plot for Deionized Water

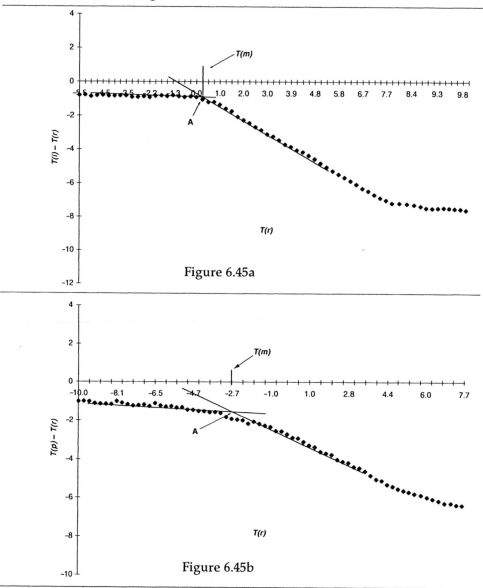

Figure 6.45a

Figure 6.45b

Figure 6.45a is a DTA thermogram plot of $T(i) - T(r)$ as a function of $T(r)$ during warming of 1.92 mL of deionized water. Figure 6.45b is a DTA thermogram plot of $T(p) - T(r)$ as a function of $T(r)$ during warming of 2.00 mL a solution containing 3% lactose (wt/v) and 0.9% (wt/v) NaCl. The DTA analysis of the deionized water and the isotonic lactose solution were conducted in 5 mL tubing vials, where $T(i)$ is the temperature of the deionized water, $T(p)$ is the isotonic solution, and $T(r)$ is the methanol reference temperature. The arrow "A" indicates the point of intersection and $T(m)$ the onset temperature for the melting in terms of $T(r)$.

based on its resistivity would not detect the presence of hydrocarbons; however, as a result of the freezing process, such hydrocarbons would be incorporated into the formulation and affect thermal properties.

The calculated and measured equilibrium freezing temperature for the isotonic 3% lactose solution were –0.44°C and –0.35°C, respectively. The onset temperature of –4.6°C would provide a clue as to the concentration of the lactose and the NaCl in the interstitial region of the matrix. Correcting for the onset temperature of the deionized reference, T(m) becomes –4.2°C. From the assumption that the ratio of the concentration of the lactose to the concentration of the sodium chloride is equal to that of the original formulation, the concentration of constituents in the interstitial region can be estimated from equation 27 (see Chapter 5)

$$\Delta T_f = \frac{1,000 K_f}{W} \sum \frac{g_{s,i}}{M_{s,i}}$$

where $K_f = 1.855$ and W is assumed to be 1,000 g. From this equation, the concentration of the lactose in the interstitial region is estimated to be 81.5% (wt/v) and NaCl 13.7% (wt/v). The highest reported concentration of lactose in a aqueous solution is only 18% (wt/wt) at 20°C [20]. At $T(p) \le -10°C$, however, the low values of $R(s)$ would indicate, as seen in Appendix C, the presence of a liquid state in the interstitial region of the matrix. In addition, the onset of the melting of the ice is not an associated with a significant decrease in $R(s)$ but with a slight increase. The electrical properties of the solution would indicate that high concentrations of lactose can exist in the interstitial region of the matrix.

Additional evidence for the existence of high concentrations of lactose stem from the estimated low concentration (13.7% wt/v) of NaCl. This concentration of NaCl is lower than the 23% required to form an eutectic mixture. The DTA thermogram shown in Figure 6.43 shows no indication of an eutectic mixture present in the interstitial region. If lactose was not part of the glassy and liquid state that occurs in the interstitial region, then the concentration of NaCl could have reached 23% and an endotherm, like that shown in Figure 6.42, would have been present in the warming DTA thermogram for the frozen isotonic 3% (wt/v) lactose solution.

Activation Energy. The measure of the activation energy (ΔE) at temperatures greater than Tc provides insight to the nature of conductivity in the interstitial region of the matrix. The conductivity (σ') of the system, as shown in Chapter 5, can be expressed as

$$\sigma' = \sigma'_o \exp^{-\Delta E / RT} \qquad \frac{1}{\Omega \, \text{cm}} \tag{31}$$

where ΔE is the activation energy and $\sigma' = \sigma'_o$ when $RT \gg \Delta E$. From equation 27, D_2 can be expressed as

$$D_2 = \frac{\rho_{\text{ice}}}{\rho_{\text{meas}}} = \frac{R_{\text{ice}}}{K_{\text{ice}}} \times \frac{K_{\text{meas}}}{R_{\text{meas}}} \tag{32}$$

or when the cell constants $K_{\text{meas}} \approx K_{\text{ice}}$, D_2 can be expressed as

$$D_2 = \frac{\rho_{ice}}{\rho_{meas}} = \frac{R_{ice}}{R_{meas}} \tag{33}$$

Since the σ' is the reciprocal of ρ, then, in terms of σ', D_2 is expressed as

$$\frac{1}{D_2} = \frac{\sigma_{meas}}{\sigma_{ice}} \tag{34}$$

Where σ_{meas} is the conductivity in the interstitial region of the matrix and σ_{ice} represents the conductivity of the ice. Substituting equation 31 for σ' in equation 34 gives

$$\frac{1}{D_2} = B_o \exp^{-\frac{\Delta E_{meas} - \Delta E_{ice}}{RT}} \tag{35}$$

Where ΔE_{meas} is the activation energy for the conductivity in the interstitial region of the matrix, ΔE_{ice} is the activation energy for the conductivity of the ice, and B_o is the ratio of $\sigma_{o,meas}$ to $\sigma_{o,ice}$.

Assuming that ΔE_{ice} has a value of 33,000 kcal/mole [35], ΔE_{meas} can be determined from an exponential plot of $1/D_2$ as a function of $1/T$ were T is expressed in K.

The following are two examples for determining the activation energy for two solutions.

A 3% Lactose and 0.9% NaCl Solution. A exponential plot of $1/D_2$ as a function of $1/T(p)$ is shown by Figure 6.46 (obtained from the tabular data shown in Appendix B). From the slope of the 18.745, ΔE_{meas} can be expressed as

$$-\Delta E_{meas} = 18.745 \times R \times 1,000 - \Delta E_{ice} \tag{36}$$

where R has a value of 1.987 cal/K-mole. From equation 36, the value of ΔE_{meas} is –4,246 kcal/mole.

A 3% Lactose Solution. Figure 6.47 shows the exponential plot for $1/D_2$ as a function of $1/T(p)$ for the 3% lactose solution (see tabular data in Appendix D). The slope for this plot was determined to be $0.9531x$.

$$-\Delta E_{meas} = 0.9531 \times R \times 1,000 - \Delta E_{ice} \tag{37}$$

Equation 37 gives a ΔE_{meas} value of –31,106 kcal/mole. The relative scatter of the points about the line is attributed to the low value of the slope.

The Tc values for the 3% lactose and the isotonic 3% lactose solution were found to be –19°C and –53°C, respectively. These results indicate that the ΔE of the matrix at temperatures greater than the Tc is directly related to the magnitude of Tc, i.e., a low ΔE is associated with a low Tc.

Shelf-Surface and Product Temperatures as a Function of Time. It is important in any thermal analysis to maintain a permanent record of the temperature-time functions for the shelf-surface and product temperatures during both the cooling and the warming portions of the analysis. This record provides a reference for comparing the thermal properties of a formulation and separates the thermal properties with that

Figure 6.46. An Exponential Plot of $1/D_2$ as a Function $1/T(p)$ During Warming of a Matrix Formed from a Solution of 3% Lactose and 0.9% NaCl

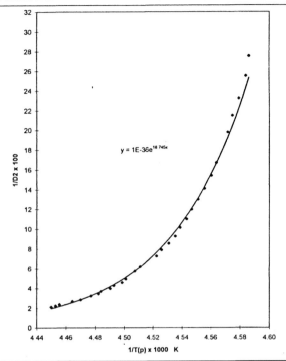

Figure 6.47. An Exponential Plot of $1/D_2$ as a Function $1/T(p)$ During Warming of a Matrix Formed from a Solution of 3% Lactose

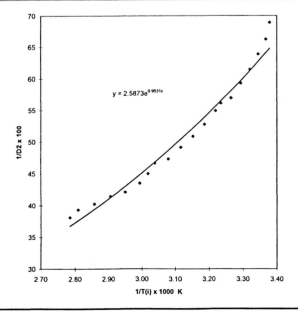

of the method used in performing the analysis. The $T(p)$ and $T(s)$ plots as a function of time for an 0.9% NaCl solution, during both cooling and warming, are shown in Figures 6.2 and 6.3.

Summary and Comments

There is a distinct advantage to using two independent analytical methods for determining the thermal properties of a formulation. The expanded DTA analysis (i.e, using more than one reference material) provides a means for characterizing the thermal properties of a formulation that involves phase changes during both the cooling and warming processes. The D_2 portion of the analysis provides key information when a change in state does not involve sufficient quantities of energy to be determined by the DTA analysis. From the electrical measurements, one is able to evaluate the impact that an excipient can have on a formulation's Tc. A third feature of this method for determining the thermal properties of the formulation is the tabular data. The use of tabular data has been shown to be invaluable in determining the thermal characteristics of a formulation.

Apparatus

One of the key aspects of this apparatus is that the thermal analysis of the formulation is conducted in the same containers that will be used to lyophilize the formulation. As a result, the thermal analysis will include the impact of the interior-surface and the heat-transfer properties of the container on the observed thermal properties. The interior-surface properties of the container may impact the degree of supercooling and the heat-transfer properties may affect product temperature during the cooling and warming portions of the analysis.

The software program requires that the operator calibrate the thermocouple sensors, check the electrical circuits for the resistance measurements, and confirm that there is agreement between the cell constants prior to conducting each analysis. These temperature calibrations and electrical measurements are recorded and included with the tabular data. The data are recorded according to the value of the event *window* (i.e., that change in the sample or methanol reference temperature that is considered to be well beyond the normal distribution for the mean value of the last recorded temperature). With this technique, the system is capable of detecting the sudden changes in temperature that are associated with processes, such as the degree of supercooling, and yet limits the amount of stored data to only that which is considered essential, i.e., less than 400 events per cooling or warming analysis.

Interpretation of the Data

The following is a list of the key thermal properties of formulation that are determined with this instrumentation and the significance of such properties with regard to the lyophilization process.

Ice-Like Water Clusters. The formation of ice-like water clusters in the formulation and the water reference sample is identified by a small exothermic peak during the cooling process, at temperatures $< 0°C$. The absence of the formation of ice-like water clusters in the formulation would signify the complexing of water with the constituents or, in the case of process water, the presence of some form of contamination. The presence or absence of ice-like water clusters is a benchmark or fingerprint for the formulation or the water reference.

Degree of Supercooling. The degree of supercooling in the formation and the water reference is typified by discontinuity in the exothermic peak associated with formation of ice crystals during the cooling process. The value for the degree of supercooling is best obtained from the tabular data. The degree of supercooling of a formulation will govern the cake structure. When supercooling occurs throughout the entire fill volume, the cake will be uniform. If only a small volume of the formulation supercools, then the cake may contain a crust or glaze.

Equilibrium Freezing Temperature. When supercooling occurs, the temperature of the formulation or the process water will increase to that temperature defined by the freezing point depression of water due to the colligative properties of the constituents in the formulation. Although this thermal property does not enter into the determination of the parameters for the freezing or primary drying process, it does serve as a fingerprint and will be a useful benchmark for characterizing the future batches of the same formulation.

Interstitial Phase Changes During Cooling or Warming. Exothermic DTA peaks during cooling or endothermic or exothermic peaks during warming are useful in identifying phase changes, like those associated with an eutectic point or the presence of metastable states. Metastable states may indicate the crystallization of one or more constituents in the formulation. Such crystallization, if it occurs with the active constituent, may serve to enhance the stability of the formulation.

Collapse Temperature. Tc is defined by this analytical method as that temperature at which the mobility of the water in the interstitial region of the matrix becomes significant. It is determined during warming from a plot of D_2 as a function of $T(p)$. It is also obtained from a linear regression analysis of the data as that temperature obtained from an extrapolation of D_2 to a value of 1 or to some constant value where the matrix of the formulation is known to be frozen. In the vast majority of formulations, Tc will govern the temperature necessary to achieve a completely frozen state and the product temperature during the primary drying process.

Interstitial Melting Temperature. The interstitial melting temperature is that value of $T(p)$ where the resistance of the sample during warming, generally 100 kΩ, approaches the resistance of the water reference at ambient temperature. I place minor significance on such a value in determining the parameters of a lyophilization process. The interstitial melting temperature does serve as a fingerprint or benchmark of a thermal property that can be used to compare the thermal properties of other batches of the same formation.

Degree of Crystallization. The use of two reference materials, methanol and process water, where the water reference contains the same quantity of water as that in the formulation, allows for a determination of the degree of crystallization. From the areas under the endothermic peaks for water and the formulation, one obtains a ratio of the quantity of ice formed in the formulation during freezing to the total freezable water. As the degree of crystallization becomes ≤ 0.50, ice crystals are contained in a sea consisting of the constituents and uncrystallized water. Such a matrix would probably impede the drying process. As the degree of crystallization approaches 0, there is no separation of solutes and solvent, and the drying rate can approach 0.

Onset Temperature for Ice Melting. The measure of the onset temperature for the melting of the ice in the formulation provides a means by which one can ascertain the concentration of the constituents in the interstitial region of the matrix. In a formulation consisting of 3% lactose and 0.9% NaCl, the onset melting temperature was determined to be $-4.6°C$. From this onset temperature, the concentration of lactose in the frozen interstitial region was estimated to be 81.5% (wt/v), and the NaCl attained a concentration of 13.7% (wt/v). The NaCl concentration was not sufficient to form an eutectic mixture. The absence of a phase change in the interstitial region of the matrix further confirms that the concentration of the NaCl did not reach the 23% necessary to form an eutectic mixture.

Activation Energy. Because of the definition of D_2 and because there is agreement between the cell constants, the reciprocal of D_2 can be used to define the σ' of the system. As a result, it is possible to determine ΔE for the conduction mechanism at temperatures greater than the Tc. It was shown that, based on the assumption that ice has a ΔE of 33,000 kcal/mole [35], the ΔE for a formulation is directly related to Tc, i.e., a higher Tc results in a higher ΔE.

SYMBOLS

A	cross-sectional area
A'	cross-sectional area of the solution through which the ion current will flow
B_o	ratio of $\sigma_{o,meas}$ to $\sigma_{o,ice}$
ac	alternating voltage (V [volt])
A_c	area under an endothermic curve
C_p	shunt capacitance (F [farad])
d_c	thickness of the ice layer in a frozen matrix
d_i	thickness of the interstitial region layer in a frozen matrix
$dT(p)/dt$	rate of change in product temperature in DSC measurements
dc	direct current (A [ampere])
D	displacement force
D_o	is a highly complex term involving the ice layers in a frozen matrix
DCC	the y intercept obtained from a plot of the dielectric loss (ϵ'') as a function of frequency (ω)

D_2 the ratio of the resistivity of ice reference to the resistivity of a formulation

ΔE conductivity activation energy (kcal/mole)

ΔE_{ice} conductivity activation energy for ice (33,000 kcal mole)

ΔE_{meas} conductivity activation energy for the formulation (kcal/mole)

ΔH_f heat of fusion of the reference material used for calibrating the DSC instrument (cal/g)

$\Delta H'$ Arrhenius energy of a dielectric medium at a given temperature (kcal/mole)

DT1   temperature difference $[T(p) - T(r)]$

DT2   temperature difference $[T(p) - T(i)]$

DT3   temperature difference $[T(i) - T(r)]$

E electric field (V/cm)

ϵ' dielectric permittivity

ϵ'' loss resulting from heat generated by the alignment of molecules in an electric field

ϵ_o low frequency permittivity

ϵ_∞ high frequency permittivity

F a highly complex term involving the interstitial layers in a frozen matrix

I electrical direct current (A)

k Boltzmann's constant

k' calibration constant for a DSC cell

K electrical cell constant

K_{ice} cell constant for the process water resistance probe

K_{meas} cell constant for the formulation resistance probe

l length of the material

m_c mass of the reference substance (g)

ω frequency (Hz)

P total polarization

P_d induced polarization of nonpolar molecules

P_o polarization of polar molecules

R gas constant

R' electrical resistance of a conductor (Ω)

$R(i)$ measured water reference resistance (kΩ)

R_{ice} resistance of volume of ice (Ω)

R_{meas} resistance of volume of ice formulation (Ω)

R_o electrical resistance of a solution (Ω)

$R(s)$ measure resistance of the sample (kΩ)

R_s resistance of a volume of the interstitial region of a matrix (Ω)

R_w resistance of the process water using a resistance probe having cell constant of K_{ice}

R'_w resistance of the process water using a resistance probe having cell constant of K_{meas}

ρ resistivity of a substance (Ω-cm)

ρ_{ice} resistivity of a volume of ice (Ω-cm)

ρ_{meas} resistivity of a formulation (Ω-cm)

ρ_s resistivity of a volume of interstitial material (Ω-cm)

ρ_w resistivity of the process water

σ' electrical conductivity

σ'_o electrical conductivity when $RT \gg \Delta E$

σ_{ice} he conductivity of ice

σ_{meas} conductivity in the interstitial region of the matrix

τ relaxation time

Tc collapse temperature (°C)

$T(i)$ water reference temperature (°C)

T_g glass transition temperature of a dried material

$T(m)$ onset temperature for the melting of the ice in the formulation

$T(p)$ product temperature in °C

Tp sample pan temperature for DSC analysis

$T(r)$ temperature of a reference material (°C)

T_r reference pan temperature for DSC analysis

$T(s)$ shelf-surface temperature (°C)

TOF   *take off frequency* [31] from a plot of the dielectric loss (ϵ'') as a function of frequency (ω)

wt/v   weight per unit volume

REFERENCES

1. L. F. Shackell, *Am J. Physiol.*, 24: 325 (1909).

2. F. Daniel, J. H. Mathews, J. W. Williams, P. Bender, and R. A. Alberty, *Experimental Physical Chemistry*, McGraw-Hill Book Co., Inc., New York, 1956, p. 108.

3. L. R. Rey, *Annals New York Academy of Sciences*, 85: 510 (1960).

4. L. R. Rey, *Biodynamics*, 8: 241 (1961).

5. A. J. Phillips, R.J. Yarwood, and J. H. Collett, *Analytical Proc.*, 23: 16 (1986).

6. L. R. Rey, *Recent Research in Freezing and Drying* (A. S. Parkes and A. Smith, eds.), Blackwell Scientific Publications, Oxford, England (1960).

7. E. Maltini, in *Freeze Drying and Advanced Food Technology* (S. A. Goldblith, L. Rey, and W. W. Rothmayr, eds.), Academic Press, New York, 1975.

8. M. S. Feder, et al., *American Laboratory*, January 1993, p. 22.

9. D. W. Kaufmann, *Sodium Chloride Production and Properties of Salt and Brine* (D. W. Kaufmann, ed.), American Chemical Society Monograph Series #145, Hafner Publishing Co., New York, 1968, p. 548.

10. N. A. Williams and G. P. Poli, *J. Parenter. Sci. and Tech.*, 38: 38 (1984).

11. R. H. M. Hatley, in International Symposium on Freeze-Drying and Formulation Bethesda, USA, 1990, *Develop. Biol. Standard*, 74: S. Karger, Basel, 1991, p. 105.

12. S. Dauerbrunn and P. Gill, *American Laboratory*, September, 1993, p. 54.

13. B. Cassel, B. Twombly, and F. Summers, loc. cit. January 1995, p. 42.

14. Courtesy of TA Instruments.

15. J. L. McNaughton and C. T. Mortimer, in *IRS; Physical Chemistry Series 2*, Butterworth, London, 10: 1976.

16. M. P. Di Vito, W. P. Brennan, and W.A. Kunze, loc. cit., September, 1985, p. 100.

17. L. Gatlin and P. P. DeLuca, *J. Parenter. Drug Assoc.*, 34: 398 (1980).

18. N. A. Williams and D. L. Schwinke, *J. Parenter. Sci. and Tech.*, 48: 136 (1994).

19. B. C. Burros, *American Laboratory*, January 1986, p. 20.

20. *Handbook of Chemistry and Physics*, (R.C. Weast, ed.), 65th ed., CRC Press, Inc. Boca Raton, Florida, 1984.

21. A. P. MacKenzie, in *Freeze Drying and Advanced Food Technology* (S. A. Goldblith, L. Rey, and W. W. Rothmayr, eds.), Academic Press, New York, 1975, p. 277.

22. R. Luyet, *Recent Research in Freezing and Drying* (A. S. Parkes and A. Smith, eds.), Blackwell Scientific Publications, Oxford, England, 1960.

23. A. P. MacKenzie, *Parenter. Drug Assoc. Inc.*, 20: 101 (1966).

24. H. Willemer, presented at *Joint Meeting I.I.R: Commissions C1 and C2*, Karlsruhe (Fed. Rep. of Germany), 1977.

25. H. T. Meryman, *Cryobiology* (H. T. Meryman, ed.), Academic Press, New York, 1966.

26. S. Glasston, *Introduction to Electrochemistry*, Norstrand, New York, 1960, pp. 37–41.

27. L. Lachman, P. P. DeLuca, and R. Withnell, *J. Pharm. Sci.*, 54: 1342 (1965).

28. T. Jennings and H. Powell, *Medical Device and Diagnostic Industry*, 2: 35 (1980).

29. P. DeLuca and L. Lachman, *J. Pharm. Sci.*, 54: 617 (1965).

30.   P. DeLuca and L. Lachman, *J. Pharm. Sci.*, 54: 1411 (1965).

31.   K. R. Morris, S. A. Evans, A. P. MacKenzie, D. Scheule, and N. G. Lord, *PDA J. Pharm. Sci. & Tech.*, 48: 318 (1994).

32.   G. P. Johari and E. Whalley, *J. Chem. Phys.* 75: 133 (1981).

33.   T. A. Jennings, *J. Parenter. Drug Assoc.*, 34: 109 (1980).

34.   A. Sheu and J. Willson, *J. Parenter. Sci. & Tech.*, 47: 180 (1993).

35.   N. Maeno and H. Nishimura, *J. of Glaciology*, 21: 193 (1978)

APPENDIX A

Data for Thermal Properties of a 0.9% (wt/v) NaCl Solution During Cooling and Warming

Cooling Tabular Data for a 0.9% NaCl Solution

Time (min)	$T(p)$ (°C)	$T(s)$ (°C)	Time (min)	$T(p)$ (°C)	$T(s)$ (°C)
10.8	17.94	15.80	18.6	10.7	7.40
10.5	17.62	15.40	18.8	10.4	7.20
11.3	17.3	14.90	19.1	10.14	6.90
11.6	17.1	14.60	19.3	9.9	6.70
11.8	16.89	14.30	19.6	9.7	6.50
12.0	16.7	14.20	19.8	9.4	6.20
12.5	16.3	13.70	20.1	9.2	6.00
13.0	15.95	13.20	20.3	8.9	5.70
13.2	15.76	12.90	20.6	8.8	5.50
13.7	15.34	12.40	20.8	8.5	5.30
14.0	15.06	12.00	21.0	8.3	5.10
14.2	14.85	11.80	21.7	8.1	4.90
14.5	14.65	11.60	21.5	7.9	4.70
14.7	14.41	11.30	21.7	7.6	4.40
15.0	14.15	11.10	22.0	7.4	4.20
15.2	13.94	10.80	22.3	7.2	4.00
15.5	13.68	10.50	22.5	6.9	3.70
15.9	13.22	10.10	22.7	6.7	3.50
16.2	13	9.80	23.0	6.4	3.30
16.4	12.79	9.60	23.2	6.2	3.10
16.7	12.53	9.30	23.5	6.1	2.90
16.9	12.3	9.10	23.7	5.8	2.70
17.2	12.08	8.90	24.0	5.6	2.40
17.4	11.84	8.60	24.5	5.2	2.00
17.6	11.59	8.40	24.7	5	1.80
17.9	11.35	8.10	24.9	4.8	1.60
18.1	11.13	7.90	25.2	4.6	1.40
18.4	10.9	7.70	25.4	4.4	1.20

Time (min)	T(p) (°C)	T(s) (°C)	Time (min)	T(p) (°C)	T(s) (°C)
25.7	4.2	1.00	36.1	–4.3	–6.8
25.9	3.9	0.70	36.3	–4.5	–7
26.1	3.8	0.6	36.6	–4.8	–7.2
26.6	3.4	0.2	37.1	–5.1	–7.5
26.9	3.2	0	37.5	–5.4	–7.8
27.1	3	–0.3	38.0	–5.8	–8
27.4	2.8	–0.4	38.5	–6.1	–8.4
27.6	2.7	–0.7	40.0	–6.4	–8.6
27.8	2.5	–0.9	40.5	–7.5	–9.6
28.3	2.2	–1.3	40.9	–7.8	–9.9
28.8	1.8	–1.6	41.4	–8.1	–10.2
29.0	1.6	–1.8	41.9	–8.5	–10.5
29.5	1.2	–2.2	42.4	–8.8	–10.2
29.8	1.1	–2.4	42.9	–9.1	–11.1
30.0	0.8	–2.6	43.3	–9.4	–11.4
30.3	0.6	–2.8	43.8	–9.7	–11.6
30.5	0.4	–3	44.3	–10	–11.9
30.8	0.2	–3.2	44.8	–10.3	–12.2
31.0	–0.1	–3.3	45.1	–10.5	–12.3
31.5	–0.4	–3.7	45.3	–0.7	–12.4
32.0	–.9	–4	45.5	–0.5	–12.5
32.2	–1.1	–4.2	46.0	–0.5	–12.8
32.4	–1.3	–4.4	46.5	–0.4	–13
32.7	–1.5	–4.6	47.0	–0.5	–13.3
32.9	–1.8	–4.7	47.5	–0.5	–13.6
33.2	–2	–4.9	48.0	–0.5	–13.8
33.4	–2.2	–5	48.4	–0.6	–14.2
33.9	–2.6	–5.4	49.2	–0.7	–14.5
34.2	–2.8	–5.6	49.7	–0.9	–14.8
34.6	–3.2	–5.9	50.2	–1.1	–15.1
35.1	–3.6	–6.2	50.6	–1.3	–15.4
35.4	–3.8	–6.3	51.1	–1.6	–15.7
35.8	–4.1	–6.6	51.6	–1.9	–16

Time (min)	$T(p)$ (°C)	$T(s)$ (°C)	Time (min)	$T(p)$ (°C)	$T(s)$ (°C)
52.1	–2.2	–16.2	61.2	–20	–21.5
52.3	–2.4	–16.4	61.4	–20.2	–21.6
52.8	–2.8	–16.6	61.7	–20.5	–21.8
53.0	–3	–16.8	62.1	–20.8	–22
53.3	–3.2	–16.9	62.6	–21.1	–22.2
53.5	–3.5	–17.1	63.1	–21.5	–22.4
53.8	–3.8	–17.2	63.5	–21.8	–22.7
54.0	–4.1	–17.4	64.0	–22.1	–22.9
54.3	–4.6	–17.5	64.5	–22.3	–23.1
54.5	–5.3	–17.6	64.9	–22.6	–23.4
54.8	–6.2	–17.8	65.4	–22.9	–23.6
55.0	–7.1	–18	65.8	–23.1	–23.8
55.3	–5.1	–18.1	66.3	–23.3	–23
55.5	–9	–18.3	66.8	–23.6	–24
55.8	–10	–18.4	67.3	–23.8	–24.2
56.1	–11	–18.6	67.7	–24	–24.4
56.4	–12	–18.8	68.2	–24.2	–24.6
56.6	–12.8	–18.9	68.7	–24.4	–24.9
56.9	–13.4	–19.1	69.1	–24.7	–25.1
57.1	–14.1	–19.2	69.6	–24.9	–25.3
57.4	–14.7	–19.4	70.1	–25.1	–25.5
58.1	–16.1	–19.8	70.6	–25.3	–25.7
58.4	–16.5	–19.9	71.1	–25.6	–26
58.6	–16.8	–20.1	71.6	–25.7	–26.2
58.8	–17.2	–20.2	72.0	–25.9	–26.4
59.1	–17.6	–20.3	72.5	–26.2	–26.6
59.3	–17.9	–20.4	73.0	–26.4	–26.9
59.6	–18.2	–20.9	73.5	–26.6	–27.1
59.8	–18.5	–20.7	74.0	–26.8	–27.3
60.0	–18.8	–20.8	74.5	–27	–27.5
60.3	–19	–21	75.0	–27.2	–27.8
60.5	–19.4	–21.1	75.4	–27.4	–28
61.0	–19.8	–21.4	76.0	–27.6	–28.2

Time (min)	T(p) (°C)	T(s) (°C)	Time (min)	T(p) (°C)	T(s) (°C)
76.4	-27.8	-28.4	95.4	-34.4	-35.8
76.9	-28	-28.6	96.2	-34.6	-36
77.4	-28.2	-28.8	96.9	-34.8	-36.3
78.1	-28.5	-29.4	97.6	-35	-36.5
78.6	-28.6	-29.5	98.4	-35.2	-36.8
79.4	-28.9	-29.8	98.9	-35.3	-36.9
79.8	-29	-30	99.7	-35.6	-37.2
80.3	-28.9	-30.2	100.4	-35.8	-37.4
81.1	-28.9	-30.5	101.2	-35.9	-37.7
81.6	-28.9	-30.7	102.0	-36.2	-37.9
82.0	-29	-30.9	102.7	-36.4	-38.1
82.5	-29.3	-31.1	103.5	-36.6	-38.4
83.0	-29.6	-31.2	104.2	-36.9	-38.6
83.5	-29.9	-31.5	105.0	-37.1	-38.9
84.0	-30.2	-31.7	105.8	-37.3	-39.1
84.5	-30.4	-31.9	106.5	-37.5	-39.3
85.0	-30.7	-32.1	107.3	-37.8	-39.6
85.4	-30.9	-32.3	108.1	-37.9	-39.8
85.9	-31.1	-32.5	108.8	-38.2	-40
86.4	-31.4	-32.6	109.6	-38.3	-40.2
86.9	-31.6	-32.8	110.3	-38.5	-40.5
87.4	-31.7	-33	111.4	-38.7	-40.8
87.9	-31.9	-33.2	112.1	-38.9	-40.9
88.4	-32.1	-33.4	113.1	-39.1	-41.2
89.1	-32.3	-33.6	113.9	-39.4	-41.4
89.8	-32.6	-33.8	114.6	-39.5	-41.6
90.6	-32.9	-34.1	115.4	-39.8	-41.8
91.0	-33	-34.2	116.4	-40	-42.1
92.0	-33.3	-34.6	116.9	-40	-42.2
92.8	-33.5	-34.9	118.0	-40.3	-42.5
93.3	-33.7	-35	119.0	-40.5	-42.7
94.0	-33.9	-35.3	119.7	-40.7	-43
94.7	-34.2	-35.5	120.7	-41	-43.2

Time (min)	T(p) (°C)	T(s) (°C)	Time (min)	T(p) (°C)	T(s) (°C)
121.5	–41.1	–43.4	157.3	–47.5	–50.3
122.3	–41.3	–43.6	158.3	–47.6	–50.5
123.0	–41.4	–43.8	159.3	–47.8	–50.6
124.1	–41.7	–44.1	160.5	–47.8	–50.8
125.1	–41.9	–44.3	162.1	–48	–51
126.1	–42.1	–44.6	163.3	–48.2	–51.2
127.1	–42.3	–44.8	165.3	–48.4	–51.4
128.1	–42.5	–44.9	167.4	–48.6	–51.7
129.1	–42.7	–45.2	168.9	–48.8	–51.8
130.1	–42.9	–45.4	169.5	–48.8	–51.9
131.1	–43.2	–45.6			
132.2	–43.4	–45.8			
133.2	–43.5	–46.1			
134.0	–43.8	–46.3			

Warming Tabular Data for a 0.9% NaCl Solution

Time (min)	T(p) (°C)	T(s) (°C)
135.2	–44	–46.5
132.2	–44.1	–46.7
137.2	–44.3	–46.9
138.0	–44.5	–47
139.0	–44.7	–47.3
140.3	–44.9	–47.5
141.5	–45	–47.8
142.8	–45.3	–48
143.6	–45.4	–48.1
144.8	–45.6	–48.3
145.9	–45.7	–48.5
147.1	–46	–48.7
148.4	–46.2	–49
149.4	–46.3	–49.1
150.9	–46.5	–49.3
152.2	–46.7	–49.5
153.4	–46.9	–49.8
154.5	–47	–49.9
155.7	–47.2	–50.1

Continued table (right column):

Time (min)	T(p) (°C)	T(s) (°C)
0.5	–49	–52
4.4	–49	–51.1
5.2	–48.2	–50.8
5.3	–48.6	–50.2
6.8	–48.4	–50
7.7	–48.2	–49.6
8.5	–48	–49.2
9.0	–47.8	–49
9.6	–47.5	–48.7
10.1	–47.3	–48.4
10.7	–47.1	–48.1
11.3	–46.9	–47.8
11.8	–46.7	–47.6
12.3	–46.5	–47.4
13.1	–46.2	–47
13.6	–46	–46.7
14.1	–45.8	–46.5
14.6	–45.6	–46.3
15.1	–45.4	–46.1

Time (min)	$T(p)$ (°C)	$T(s)$ (°C)	Time (min)	$T(p)$ (°C)	$T(s)$ (°C)
15.6	−45.2	−45.8	32.5	−37.8	−37.5
16.2	−44.9	−45.5	33.0	−37.6	−37.2
16.7	−44.7	−45.3	33.5	−37.3	−37
17.2	−44.5	−45	34.0	−37.2	−36.7
17.7	−44.2	−44.8	34.5	−36.9	−36.5
18.2	−44.1	−44.6	35.0	−36.6	−36.3
18.6	−43.9	−44.2	35.5	−36.4	−36
19.2	−43.6	−44	36.0	−36.2	−35.8
19.7	−43.4	−43.7	36.5	−36	−35.5
20.2	−43.2	−43.5	37.1	−35.8	−35.3
20.7	−43	−43.2	37.6	−35.5	−35.1
21.3	−42.7	−43	38.1	−35.3	−34.8
21.8	−42.5	−42.7	38.6	−35	−34.5
22.3	−42.3	−42.5	39.1	−34.9	−34.3
22.8	−42	−42.2	39.6	−34.6	−34
23.3	−41.8	−42	40.1	−34.5	−33.8
23.8	−41.6	−41.7	40.6	−34.2	−33.6
24.3	−41.4	−41.5	41.0	−34	−33.3
24.8	−41.2	−41.2	41.6	−33.8	−33.1
25.3	−40.9	−40.9	42.0	−33.5	−32.8
25.8	−40.7	−40.7	42.8	−33.2	−32.5
26.3	−40.5	−40.5	43.3	−33	−32.3
26.9	−40.2	−40.2	43.7	−32.8	−32
27.4	−40	−40	44.2	−32.6	−31.8
27.9	−39.8	−39.7	44.7	−32.3	−31.6
28.4	−39.6	−39.4	45.2	−32.2	−31.4
28.9	−39.4	−39.3	45.7	−32	−31.1
29.4	−39.1	−39	46.2	−31.7	−31.9
29.9	−38.9	−38.7	46.7	−31.5	−30.6
30.4	−38.7	−38.5	47.2	−31.3	−30.4
30.9	−38.5	−38.3	47.7	−31	−30.2
31.4	−38.2	−38	48.1	−30.8	−29.9
32.0	−38	−37.7	48.6	−30.6	−29.7

Time (min)	$T(p)$ (°C)	$T(s)$ (°C)	Time (min)	$T(p)$ (°C)	$T(s)$ (°C)
49.1	–30.4	–29.5	65.7	–23.2	–21.9
49.6	–30.2	–29.2	66.2	–23	–21.7
50.1	–30	–29	66.7	–22.8	–21.5
50.6	–29.7	–28.8	67.2	–22.5	–21.2
51.1	–29.5	–28.5	67.6	–22.3	–21
51.6	–29.3	–28.3	68.2	–22.2	–20.8
52.3	–29	–28	68.6	–22	–20.6
52.8	–28.8	–27.8	69.1	–21.8	–20.4
53.3	–28.6	–27.6	69.6	–21.5	–20.2
53.7	–28.3	–27.3	70.3	–21.3	–19.9
54.2	–28.1	–27.1	70.8	–21.1	–19.7
54.7	–27.9	–26.9	71.3	–20.9	–19.4
55.2	–27.7	–26.7	71.7	–20.8	–19.2
55.7	–27.5	–26.5	72.4	–20.6	–18.9
56.2	–27.3	–26.2	73.1	–20.4	–18.6
56.7	–27.1	–26	73.8	–20.3	–18.3
57.2	–26.9	–25.8	74.5	–20.2	–18
57.7	–26.7	–25.6	75.0	–20.2	–17.8
58.4	–26.4	–25.2	75.4	–20.1	–17.6
58.9	–26.2	–25	76.1	–19.8	–17.3
59.4	–25.9	–24.8	76.6	–19.6	–17.1
59.8	–25.7	–24.5	77.1	–19.2	–16.9
60.3	–25.5	–24.3	77.6	–18.9	–16.7
60.8	–25.3	–24.1	78.0	–18.6	–16.4
61.3	–25.1	–23.9	78.5	–18.3	–16.3
61.8	–24.9	–23.7	78.9	–18	–16.1
62.3	–24.7	–23.4	79.4	–17.8	–15.9
62.8	–24.5	–23.2	79.9	–17.6	–15.7
63.3	–24.3	–23	80.3	–17.4	–15.6
63.8	–24	–22.8	80.8	–17.1	–15.4
64.2	–23.8	–22.6	81.3	–16.9	–15.1
64.7	–23.6	–22.3	81.7	–16.7	–15
65.2	–23.4	–22.1	82.2	–16.5	–14.8

Time (min)	$T(p)$ (°C)	$T(s)$ (°C)	Time (min)	$T(p)$ (°C)	$T(s)$ (°C)
82.7	-16.2	-14.6	101.7	-8.6	-6.8
83.1	-16	-14.4	102.4	-8.3	-6.6
83.8	-15.7	-14.1	103.1	-8.1	-6.3
84.5	-15.4	-13.7	103.6	-7.9	-6.2
85.2	-15.1	-13.4	104.3	-7.7	-5.9
85.9	-14.8	-13.1	105.0	-7.4	-5.6
86.4	-14.6	-13	105.5	-7.2	-5.4
86.9	-14.4	-12.8	106.2	-7	-5.2
87.3	-14.2	-12.6	109.9	-5.8	-3.8
88.0	-13.9	-12.3	110.6	-5.6	-3.5
88.5	-13.7	-12.1	111.3	-5.3	-3.2
89.2	-13.5	-11.9	112.0	-5.1	-3
89.7	-13.3	-11.7	112.6	-4.9	-2.7
90.1	-13.1	-11.5	113.2	-4.3	-2.5
90.6	-12.9	-11.3	113.8	-4.6	-2.3
91.3	-12.6	-11	114.3	-4.5	-2.1
91.7	-12.5	-10.8	115.0	-4.4	-1.9
92.4	-12.2	-10.6	115.5	-4.2	-1.7
93.1	-11.9	-10.2	116.2	-4	-1.4
93.6	-11.7	-10	116.9	-3.9	-1.1
94.1	-11.6	-9.8	117.6	-3.7	-0.9
94.5	-11.4	-9.7	118.3	-3.6	-0.7
95.2	-11.1	-9.4	119.0	-3.4	-0.5
95.9	-10.8	-9.2	119.9	-3.2	-0.2
96.4	-10.6	-9	120.4	-3.1	-0.1
97.1	-10.3	-8.6	121.0	-2.9	0.2
97.8	-10.1	-8.4	121.7	-2.8	0.5
98.3	-9.9	-8.2	122.4	-2.7	0.7
98.9	-9.6	-8	123.6	-2.5	1
99.7	-9.4	-7.7	124.5	-2.4	-1.3
100.4	-9.1	-7.4	125.2	-2.3	-1.5
100.8	-9	-7.2	126.2	-2.2	1.7
101.3	-8.7	-7	126.8	-2.1	1.9

Time (min)	T(p) (°C)	T(s) (°C)	Time (min)	T(p) (°C)	T(s) (°C)
127.5	-2	2.2	148.4	0.2	9.7
128.5	-1.9	2.4	149.1	0.4	9.9
129.4	-1.8	2.8	149.8	0.6	10.1
130.1	-1.8	3.1	150.2	0.9	10.3
130.6	-1.7	3.2	150.7	1.1	10.4
131.0	-1.7	3.5	151.3	1.3	10.5
131.7	-1.6	3.7	152.0	1.6	10.7
132.2	-1.5	3.8	152.4	1.9	10.9
132.7	-1.5	4.1	152.9	2.1	11
133.1	-1.4	4.2	153.3	2.4	11.2
133.6	-1.4	4.4	153.8	2.7	11.3
134.1	-1.4	4.5	154.0	2.9	11.4
134.7	-1.3	4.7	154.4	3.3	11.5
135.4	-1.3	5	154.6	3.5	11.6
135.8	-1.2	5.2	154.9	3.7	11.7
136.5	-1.2	5.4	155.1	3.9	11.7
136.9	-1.2	5.6	155.3	4.4	11.8
137.6	-1.1	5.8	155.6	4.9	11.9
138.3	-1	6	155.8	5.6	12
138.9	-1	6.3	156.0	6.2	12
139.6	-0.9	6.6	156.2	6.8	12.1
140.3	-0.8	6.8	156.5	7.4	12.2
140.9	-0.8	7.1	156.7	7.9	12.2
141.6	-0.7	7.3	156.9	8.2	12.3
142.2	-0.6	7.5	157.1	8.7	12.4
142.9	-0.6	7.8	157.4	9	12.4
143.6	-0.5	8	157.6	9.4	12.4
144.2	-0.4	8.2	157.8	9.6	12.5
144.9	-0.3	8.4	158.0	9.9	12.6
145.8	-0.2	8.7	158.3	10.1	12.7
146.4	-0.1	8.9	158.5	10.3	12.8
147.1	-0.1	9.1	158.9	10.7	12.9
147.8	0.1	9.4	159.4	11.1	13

Time (min)	$T(p)$ (°C)	$T(s)$ (°C)	Time (min)	$T(p)$ (°C)	$T(s)$ (°C)
159.8	11.3	13	174.4	16.2	17
160.3	11.7	13.2	175.3	16.4	17.2
160.7	11.9	13.3	176.2	16.6	17.4
161.2	12.1	13.5	177.1	16.8	17.7
161.8	12.4	13.7	178.0	17.1	17.9
162.3	12.7	13.8	178.8	17.3	18.1
162.9	12.9	14	179.7	17.5	18.4
163.6	13.2	14.2	180.4	17.7	18.5
164.2	13.5	14.4	181.3	18	18.8
164.9	13.7	14.6	182.1	18.2	19
165.6	13.9	14.9	182.8	18.4	19.1
166.3	14.2	15	183.5	18.5	19.2
166.7	14.3	15.1	184.1	18.7	19.4
167.4	14.5	15.3	184.8	18.9	19.6
168.2	14.8	15.5	185.7	19.2	19.8
168.9	14.9	15.6	186.6	19.4	20
169.8	15.1	15.8	187.2	19.6	20.2
170.4	15.3	16	188.1	19.8	20.4
171.3	15.5	16.2	189.0	20.1	20.7
172.2	15.7	16.4	189.2	20.1	20.7
173.5	16	16.7			

APPENDIX B

Data for a D_2 and DTA Analysis of a 0.9% (wt/v) NaCl Solution

```
D2 &DTA Run: Lacsal — 04-10-1996 — TAJ — V3.8-SN009-8/31/95


D2&DTA System — RUN IDENTIFICATION — COOLING

SAMPLE NAME : PC.92.061
DATA FILE NAME    :  /DATA/NACL0.9.A/NACL.9.1.C
DATE              :  05-21-1992
OPERATOR          :
SYSTEM ID         :
FILL VOLUME       :  2 ML
DTA REFERENCE     :  METHANOL
WATER SAMPLE      :  DI
NOTEBOOK NUMBER   :  12-92
NOTEBOOK PAGE     :  11


PARAMETER TABLE

RESISTANCE READ COUNT       :  5
TEMPERATURE READ COUNT      :  10
ELAPSED TIME LIMIT (HRS)    :  5
COOLING START TEMP. (C)     :  18
COOLING STOP TEMP. (C)      :  -60
COOLING EVENT WINDOW (C)    :  .2


GRID SPACE FOR DTA GRAPHS (DEGREES (C)/DIVISION)

(T(p) - T(r))/ REF TEMP     :  2
(T(p) - T(i))/ REF TEMP     :  2
(T(i) - T(r))/ REF TEMP     :  2
(T(p) - T(r))/ EVENTS       :  2


CALIBRATION - CORRECTION TEMPERATURES (C)

ICE             SAMPLE      REF         SHELF-SURFACE
.972            .019        .186        0.791
BATH TEMP       : 0


STANDARD DEVIATION - DEGREE (C)

ICE             SAMPLE      REF         SHELF-SURFACE
.070            .055        .060        .022
```

TEMPERATURE CORRECTIONS DEGREES (C)

ICE	SAMPLE	REF	SHELF-SURFACE
-.927	-.019	-.186	-.791

RESISTANCE CHECK (KILO-OHMS)   ICE (CELL)   SAMPLE (CELL)

		ICE (CELL)	SAMPLE (CELL)
SHORT	:	$1. \times 10^{-4}$	$1. \times 10^{-4}$
OPEN	:	9.164×10^{7}	9.164×10^{7}
CELL CONSTANT	:	14.977	14.315

Cooling Tabular Data for 0.9% NaCl Solution

EVENT	Time (min)	R(i)	R(s)	D2	T(p)	T(r)	DT1	T(s)
0	10.8	206.696	2.143	96.45	17.94	18.52	-0.58	15.80
1	10.5	207.764	2.15	96.63	17.62	18.22	-0.61	15.40
2	11.3	208.834	2.155	96.91	17.30	17.88	-0.58	14.90
3	11.6	209.702	2.16	97.08	17.10	17.65	-0.55	14.60
4	11.8	210.048	2.163	97.11	16.89	17.44	-0.55	14.30
5	12.0	210.704	2.169	97.14	16.70	17.22	-0.52	14.20
6	12.5	211.988	2.182	97.15	16.30	16.83	-0.53	13.70
7	13.0	212.934	2.185	97.45	15.95	16.44	-0.49	13.20
8	13.2	213.784	2.188	97.71	15.76	16.21	-0.45	12.90
9	13.7	215.156	2.206	97.53	15.34	15.81	-0.47	12.40
10	14.0	214.218	2.2	97.37	15.06	15.50	-0.44	12.00
11	14.2	216.088	2.188	98.76	14.85	15.3	-0.45	11.80
12	14.5	217.034	2.22	97.76	14.65	15.03	-0.38	11.60
13	14.7	217.856	2.202	98.94	14.41	14.81	-0.40	11.30
14	15.0	218.516	2.239	97.60	14.15	14.58	-0.43	11.10
15	15.2	217.988	2.24	97.32	13.94	14.37	-0.43	10.80
16	15.5	219.938	2.22	99.07	13.68	14.11	-0.43	10.50
17	15.9	221.248	2.261	97.85	13.22	13.63	-0.41	10.10
18	16.2	221.906	2.226	99.69	13.00	13.44	-0.44	9.80
19	16.4	222.776	2.263	98.44	12.79	13.18	-0.39	9.60
20	16.7	223.102	2.263	98.59	12.53	12.93	-0.40	9.30
21	16.9	223.072	2.274	98.10	12.30	12.71	-0.41	9.10
22	17.2	224.614	2.253	99.70	12.08	12.45	-0.37	8.90
23	17.4	225.412	2.2284	101.15	11.84	12.21	-0.37	8.60
24	17.6	226.308	2.289	98.87	11.59	12.02	-0.43	8.40
25	17.9	227.036	2.294	98.97	11.35	11.82	-0.47	8.10
26	18.1	227.622	2.296	99.14	11.13	11.6	-0.47	7.90
27	18.4	228.212	2.306	98.96	10.90	11.34	-0.44	7.70
28	18.6	229.184	2.309	99.26	10.70	11.1	-0.40	7.40
29	18.8	229.772	2.307	99.60	10.40	10.89	-0.49	7.20
30	19.1	229.26	2.308	99.33	10.14	10.68	-0.54	6.90

EVENT	Time (min)	R(i)	R(s)	D2	T(p)	T(r)	DT1	T(s)
31	19.3	231.244	2.302	100.45	9.90	10.45	-0.55	6.70
32	19.6	232.058	2.331	99.55	9.70	10.19	-0.49	6.50
33	19.8	232.976	2.307	100.99	9.40	9.95	-0.55	6.20
34	20.1	233.862	2.343	99.81	9.20	9.68	-0.48	6.00
35	20.3	233.468	2.353	99.22	8.90	9.5	-0.60	5.70
36	20.6	235.1	2.357	99.75	8.80	9.22	-0.42	5.50
37	20.8	235.92	2.372	99.46	8.50	9.01	-0.51	5.30
38	21.0	237.052	2.38	99.60	8.30	8.8	-0.50	5.10
39	21.7	237.348	2.382	99.64	8.10	8.59	-0.49	4.90
40	21.5	238.346	2.378	100.23	7.90	8.36	-0.46	4.70
41	21.7	238.542	2.398	99.48	7.60	8.10	-0.50	4.40
42	22.0	240.348	2.386	100.73	7.40	7.89	-0.49	4.20
43	22.3	241.396	2.427	99.46	7.20	7.68	-0.48	4.00
44	22.5	242.18	2.4	100.91	6.90	7.48	-0.58	3.70
45	22.7	242.64	2.441	99.40	6.70	7.24	-0.54	3.50
46	23.0	242.994	2.454	99.02	6.40	7.02	-0.62	3.30
47	23.2	244.414	2.431	100.54	6.20	6.78	-0.59	3.10
48	23.5	245.938	2.467	99.69	6.10	6.55	-0.45	2.90
49	23.7	246.534	2.472	99.73	5.80	6.33	-0.53	2.70
50	24.0	247.244	2.471	100.06	5.60	6.07	-0.47	2.40
51	24.5	249.586	2.512	99.36	5.20	5.67	-0.47	2.00
52	24.7	250.182	2.519	99.32	5.00	5.42	-0.42	1.80
53	24.9	251.32	2.517	99.85	4.80	5.2	-0.40	1.60
54	25.2	252.17	2.525	99.87	4.60	5.02	-0.42	1.40
55	25.4	253.082	2.543	99.52	4.4	4.79	-0.39	1.20
56	25.7	254.084	2.565	99.06	4.2	4.60	-0.40	1.00
57	25.9	254.772	2.569	99.17	3.9	4.39	-0.49	0.70
58	26.1	255.388	2.57	99.37	3.8	4.14	-0.34	0.6
59	26.6	256.924	2.583	99.47	3.4	3.72	-0.32	0.2
60	26.9	256.766	2.578	99.60	3.2	3.47	-0.27	0
61	27.1	258.848	2.575	100.52	3	3.31	-0.31	-0.3
62	27.4	259.62	2.619	99.13	2.8	3.10	-0.30	-0.4
63	27.6	260.246	2.639	98.62	2.7	2.89	-0.19	-0.7

EVENT	Time (min)	R(i)	R(s)	D2	T(p)	T(r)	DT1	T(s)
64	27.8	261.176	2.632	99.23	2.5	2.67	-0.17	-0.9
65	28.3	263.266	2.645	99.53	2.2	2.25	-0.05	-1.3
66	28.8	265.044	2.62	101.16	1.8	1.86	-0.06	-1.6
67	29.0	266.222	2.676	99.49	1.6	1.62	-0.02	-1.8
68	29.5	268.166	2.676	100.21	1.2	1.25	v0.05	-2.2
69	29.8	269.436	2.673	100.80	1.1	1.04	0.06	-2.4
70	30.0	269.2	2.666	100.98	0.8	0.9	v0.10	-2.6
71	30.3	271.952	2.668	101.93	0.6	0.65	-0.05	-2.8
72	30.5	273.164	2.615	104.46	0.4	0.46	-0.06	-3
73	30.8	274.182	2.669	102.73	0.2	0.27	-0.07	-3.2
74	31.0	274.086	2.669	102.69	-0.1	0.05	-0.15	-3.3
75	31.5	277.897	2.658	104.55	-0.4	-0.32	-0.08	-3.7
76	32.0	280.358	2.664	105.24	-0.9	-0.7	-0.20	-4
77	32.2	281.556	2.681	105.02	-1.1	-0.91	-0.19	-4.2
78	32.4	282.446	2.681	105.35	-1.3	-1.09	-0.21	-4.4
79	32.7	284.05	2.678	106.07	-1.5	-1.25	-0.25	-4.6
80	32.9	283.97	2.639	107.61	-1.8	-1.45	-0.35	-4.7
81	33.2	286.216	2.661	107.56	-2	-1.61	-0.39	-4.9
82	33.4	287.998	2.677	107.58	-2.2	-1.77	-0.43	-5
83	33.9	289.866	2.693	107.64	-2.6	-2.17	-0.43	-5.4
84	34.2	291.334	2.693	108.18	-2.8	-2.29	-0.51	-5.6
85	34.6	293.616	2.71	108.35	-3.2	-2.67	-0.53	-5.9
86	35.1	296.06	2.716	109.01	-3.6	-2.97	-0.63	-6.2
87	35.4	296.388	2.715	109.17	-3.8	-3.14	-0.66	-6.3
88	35.8	297.792	2.716	109.64	-4.1	-3.47	-0.63	-6.6
89	36.1	298.446	2.723	109.60	-4.3	-3.66	-0.64	-6.8
90	36.3	297.872	2.716	109.67	-4.5	-3.81	-0.69	-7
91	36.6	299.684	2.703	110.87	-4.8	-3.99	-0.81	-7.2
92	37.1	300.424	2.729	110.09	-5.1	-4.32	-0.78	-7.5
93	37.5	301.518	2.723	110.73	-5.4	-4.63	-0.77	-7.8
94	38.0	302.252	2.711	111.49	-5.8	-4.96	-0.84	-8
95	38.5	303.508	2.716	111.75	-6.1	-5.28	-0.82	-8.4
96	40.0	304.834	2.733	111.54	-6.4	-5.60	-0.80	-8.6

EVENT	Time (min)	R(i)	R(s)	D2	T(p)	T(r)	DT1	T(s)
97	39.2	305.58	2.733	111.81	-6.6	-5.68	-0.92	-8.8
98	39.5	306.592	2.733	112.18	-6.8	-5.90	-0.90	-9
99	40.0	308.502	2.752	112.10	-7.1	-6.16	-0.94	-9.3
100	40.5	310.684	2.763	112.44	-7.5	-6.5	-1.00	-9.6
101	40.9	313.698	2.771	113.21	-7.8	-6.78	-1.01	-9.9
102	41.4	316.734	2.77	114.34	-8.1	-7.08	-1.02	-10.2
103	41.9	319.542	2.788	114.61	-8.5	-7.37	-1.13	-10.5
104	42.4	323.198	2.796	115.59	-8.8	-7.68	-1.12	-10.2
105	42.9	326.616	2.642	123.62	-9.1	-8.0	-1.10	-11.1
106	43.3	329.784	2.439	135.21	-9.4	-8.22	-1.18	-11.4
107	43.8	332.546	2.366	140.55	-9.7	-8.50	-1.19	-11.6
108	44.3	1324.58	2.297	576.66	-10	-8.7	-1.30	-11.9
109	44.8	1048.45	2.254	465.15	-10.3	-9.01	-1.29	-12.2
110	45.1	1017.96	2.231	456.28	-10.5	-9.16	-1.34	-12.3
111	45.3	1011.61	2.225	454.66	-0.7	-9.23	8.53	-12.4
112	45.5	977.876	2.195	445.50	-0.5	-9.32	8.82	-12.5
113	46.0	948.458	2.212	428.78	-0.5	-9.63	9.13	-12.8
114	46.5	923.392	2.209	418.01	-0.4	-9.86	9.46	-13
115	47.0	917.804	2.188	419.47	-0.5	-10.07	9.57	-13.3
116	47.5	889.832	2.169	410.25	-0.5	-10.37	9.87	-13.6
117	48.0	888.97	2.143	414.83	-0.5	-10.62	10.12	-13.8
118	48.4	908.332	2.125	427.45	-0.6	-10.91	10.31	-14.2
119	49.2	962.276	2.101	458.01	-0.7	-11.28	10.58	-14.5
120	49.7	1015.77	2.091	485.78	-0.9	-11.55	10.65	-14.8
121	50.2	1062.92	2.074	512.50	-1.1	-11.78	10.68	-15.1
122	50.6	969.624	2.071	468.19	-1.3	-12.02	10.72	-15.4
123	51.1	1002.95	2.063	486.16	-1.6	-12.33	10.73	-15.7
124	51.6	1053.57	2.05	513.94	-1.9	-12.56	10.66	-16
125	52.1	1078.4	2.059	523.75	-2.2	-12.83	10.63	-16.2
126	52.3	1098.85	2.063	532.65	-2.4	-12.91	10.51	-16.4
127	52.8	1095.44	2.067	529.97	-2.8	-13.19	10.39	-16.6
128	53.0	1081.82	2.076	521.11	-3	-13.37	10.37	-16.8
129	53.3	1066.06	2.091	509.83	-3.2	-13.49	10.29	-16.9

EVENT	Time (min)	R(i)	R(s)	D2	T(p)	T(r)	DT1	T(s)
130	53.5	1065.39	2.098	507.81	-3.5	-13.63	10.13	-17.1
131	53.8	1112.23	2.095	530.90	-3.8	-13.75	9.95	-17.2
132	54.0	1321.5	2.141	617.23	-4.1	-13.88	9.78	-17.4
133	54.3	2393.44	2.176	1099.93	-4.6	-14.04	9.44	-17.5
134	54.5	20665.8	2.21	9351.04	-5.3	-14.18	8.89	-17.6
135	54.8	93739	2.261	41459.09	-6.2	-14.32	8.12	-17.8
136	55.0	119233	2.319	51415.70	-7.1	-14.46	7.36	-18
137	55.3	132780	2.386	55649.62	-8.08	-14.6	6.52	-18.1
138	55.5	152294	2.451	62135.45	-9	-14.74	5.74	-18.3
139	55.8	168904	2.524	66919.18	-10	-14.89	4.89	-18.4
140	56.1	184982	2.593	71338.99	-11	-15.07	4.07	-18.6
141	56.4	199043	2.669	74575.87	-12	-15.2	3.20	-18.8
142	56.6	209232	2.741	76334.18	-12.8	-15.38	2.58	-18.9
143	56.9	218592	2.805	77929.41	-13.4	-15.55	2.15	-19.1
144	57.1	230060	2.859	80468.70	-14.1	-15.71	1.61	-19.2
145	57.4	249266	2.897	86042.80	-14.7	-15.85	1.15	-19.4
146	58.1	262410	2.997	87557.56	-16.1	-16.32	0.22	-19.8
147	58.4	273418	3.097	88284.79	-16.5	-16.36	-0.14	-19.9
148	58.6	285468	3.141	90884.43	-16.8	-16.52	-0.28	-20.1
149	58.8	294734	3.204	91989.39	-17.2	-16.66	-0.54	-20.2
150	59.1	304116	3.222	94387.34	-17.6	-16.77	-0.83	-20.3
151	59.3	312714	3.232	96755.57	-17.9	-16.94	-0.96	-20.4
152	59.6	313764	3.268	96011.02	-18.2	-17.04	-1.16	-20.9
153	59.8	322606	3.287	98146.03	-18.5	-17.15	-1.35	-20.7
154	60.0	331514	3.338	99315.16	-18.8	-17.27	-1.53	-20.8
155	60.3	339788	3.341	101702.48	-19	-17.39	-1.61	-21
156	60.5	348588	3.333	104586.86	-19.4	-17.52	-1.88	-21.1
157	61.0	359726	3.42	105183.04	-19.8	-17.84	-1.96	-21.4
158	61.2	359706	3.436	104687.43	-20	-18.00	-2.00	-21.5
159	61.4	357418	3.481	102676.82	-20.2	-18.09	-2.11	-21.6
160	61.7	369930	3.5	105694.29	-20.5	-18.22	-2.28	-21.8
161	62.1	393490	3.538	111218.20	-20.8	-18.47	-2.33	-22
162	62.6	404330	3.59	112626.74	-21.1	-18.68	-2.42	-22.2

EVENT	Time (min)	R(i)	R(s)	D2	T(p)	T(r)	DT1	T(s)
163	63.1	408118	3.632	112367.29	-21.5	-18.94	-2.56	-22.4
164	63.5	403414	3.671	109892.13	-21.8	-19.17	-2.63	-22.7
165	64.0	410674	3.718	110455.62	-22.1	-19.46	-2.64	-22.9
166	64.5	386654	3.804	101644.06	-22.3	-19.71	-2.59	-23.1
167	64.9	417942	3.873	107911.70	-22.6	-19.95	-2.65	-23.4
168	65.4	429726	3.953	108708.83	-22.9	-20.17	-2.73	-23.6
169	65.8	439638	4.161	105656.81	-23.1	-20.42	-2.68	-23.8
170	66.3	455202	4.687	97120.12	-23.3	-20.66	-2.64	-23
171	66.8	460104	5.291	86959.74	-23.6	-20.89	-2.71	-24
172	67.3	470698	6.149	76548.71	-23.8	-21.09	-2.71	-24.2
173	67.7	468964	8.185	57295.54	-24	-21.34	-2.66	-24.4
174	68.2	471718	10.917	43209.49	-24.2	-21.57	-2.63	-24.6
175	68.7	478908	12.478	38380.19	-24.4	-21.8	-2.60	-24.9
176	69.1	484694	14.909	32510.16	-24.7	-22.05	-2.65	-25.1
177	69.6	480530	21.939	21903.00	-24.9	-22.26	-2.64	-25.3
178	70.1	479540	156.029	3073.40	-25.1	-22.44	-2.67	-25.5
179	70.6	480782	215.494	2231.07	-25.3	-22.68	-2.62	-25.7
180	71.1	481024	262.822	1830.23	-25.6	-22.92	-2.68	-26
181	71.6	473450	303.238	1561.31	-25.7	-23.15	-2.55	-26.2
182	72.0	473394	318.17	1487.86	-25.9	-23.37	-2.53	-26.4
183	72.5	474512	334.128	1420.15	-26.2	-23.58	—2.62	-26.6
184	73.0	468554	351.918	1331.43	-26.4	-23.78	-2.62	-26.9
185	73.5	467508	371.932	1256.97	-26.6	-24.01	-2.59	-27.1
186	74.0	462770	392.358	1179.46	-26.8	-24.21	-2.59	-27.3
187	74.5	458694	414.082	1107.74	-27	-24.42	-2.58	-27.5
188	75.0	455992	436.9	1043.70	-27.2	-24.67	-2.53	-7.8
189	75.4	452114	460.274	982.27	-27.4	-24.85	-2.55	-28
190	76.0	450744	484.95	929.46	-27.6	-25.09	-2.51	-28.2
191	76.4	441294	510.698	864.10	-27.8	-25.29	-2.51	-28.4
192	76.9	437476	537.352	814.13	-28	-25.53	-2.47	-28.6
193	77.4	434288	563.904	770.15	-28.2	-25.76	-2.44	-28.8
194	78.1	426148	612.95	695.24	-28.5	-26.05	-2.45	-29.4
195	78.6	416942	652.808	638.69	-28.6	-26.26	-2.34	-29.5

EVENT	Time (min)	R(i)	R(s)	D2	T(p)	T(r)	DT1	T(s)
196	79.4	411820	710.354	579.74	–28.9	–26.54	–2.37	–29.8
197	79.8	405970	748.024	542.72	–29	–26.79	–2.21	–30
198	80.3	398820	783.99	508.71	–28.9	–27.01	–1.89	–30.2
199	81.1	384780	836.792	459.83	–28.9	–27.24	–1.66	–30.5
200	81.6	379332	872.03	435.00	–28.9	–27.48	–1.42	–30.7
201	82.0	373016	894.188	417.16	–29	–27.71	–1.29	–30.9
202	82.5	371218	922.124	402.57	–29.3	–27.81	–1.49	–31.1
203	83.0	366508	956.454	383.19	–29.6	–28.04	–1.56	–31.2
204	83.5	359964	991.062	363.21	–29.9	–28.24	–1.66	–31.5
205	84.0	349974	1031.17	339.40	–30.2	–28.41	–1.79	–31.7
206	84.5	346238	1074.16	322.33	–30.4	–28.58	–1.82	–31.9
207	85.0	340278	1124.26	302.67	–30.7	–28.81	–1.89	–32.1
208	85.4	336910	1187.37	283.74	–30.9	–28.95	–1.95	–32.3
209	85.9	332590	1245.89	266.95	–31.1	–29.18	–1.92	–32.5
210	86.4	325978	1298.56	251.03	–31.4	–29.35	–2.05	–32.6
211	86.9	316112	1347.02	234.68	–31.6	–29.51	–2.09	–32.8
212	87.4	293628	1396.47	210.26	–31.7	–29.75	–1.95	–33
213	87.9	285880	1449.77	197.19	–31.9	–29.91	–1.99	–33.2
214	88.4	276848	1499.07	184.68	–32.1	–30.16	–1.94	–33.4
215	89.1	269292	1577.73	170.68	–32.3	–30.42	–1.88	–33.6
216	89.8	265298	1659.73	159.84	–32.6	–30.65	–1.95	–33.8
217	90.6	257424	1743.16	147.68	–32.9	–30.9	–2.00	–34.1
218	91.0	256178	1802.06	142.16	–33	–31.11	–1.89	–34.2
219	92.0	245820	1919.9	128.04	–33.3	–31.43	–1.87	–34.6
220	92.8	241244	2013.52	119.81	–33.5	–31.66	–1.84	–34.9
221	93.3	239106	2076.88	115.13	–33.7	–31.9	–1.80	–35
222	94.0	234212	2176.16	107.63	–33.9	–32.16	–1.74	–35.3
223	94.7	222420	2271	97.94	–34.2	–32.38	–1.82	–35.5
224	95.4	218154	2367.86	92.13	–34.4	–32.68	–1.72	–35.8
225	96.2	214686	2474.84	86.75	–34.6	–32.93	–1.67	–36
226	96.9	209944	2579.26	81.40	–34.8	–33.14	–1.66	–36.3
227	97.6	206900	4372.08	47.32	–35	–33.38	–1.62	–36.5
228	98.4	202220	5460.1	37.04	–35.2	–33.63	–1.57	–36.8

EVENT	Time (min)	R(i)	R(s)	D2	T(p)	T(r)	DT1	T(s)
229	98.9	200298	5649.3	35.46	-35.3	-33.85	-1.45	-36.9
230	99.7	201260	5617.1	35.83	-35.6	-34.05	-1.55	-37.2
231	100.4	202424	5667.98	35.71	-35.8	-34.32	-1.48	-37.4
232	101.2	205278	5756.62	35.66	-35.9	-34.57	-1.33	-37.7
233	102.0	207148	5869.76	35.29	-36.2	-34.78	-1.42	-37.9
234	102.7	211306	5987.52	35.29	-36.4	-34.98	-1.42	-38.1
235	103.5	214078	6111.52	35.03	-36.6	-35.26	-1.34	-38.4
236	104.2	218276	6229.6	35.04	-36.9	-35.44	-1.46	-38.6
237	105.0	221078	6359.24	34.76	-37.1	-35.7	-1.40	-38.9
238	105.8	225090	6497.2	34.64	-37.3	-35.95	-1.35	-39.1
239	106.5	226322	6631.3	34.13	-37.5	-36.12	-1.38	-39.3
240	107.3	231770	6761.04	34.28	-37.8	-36.30	-1.50	-39.6
241	108.1	236434	6888.22	34.32	-37.9	-36.62	-1.28	-39.8
242	108.8	241326	7039	34.28	-38.2	-36.83	-1.37	-40
243	109.6	239530	7290.18	32.86	-38.3	-37.09	-1.21	-40.2
244	110.3	241298	7412.36	32.55	-38.5	-37.32	-1.18	-40.5
245	111.4	247424	7611.5	32.51	-38.7	-37.56	-1.14	-40.8
246	112.1	251472	7752.28	32.44	-38.9	-37.83	-1.07	-40.9
247	113.1	256864	7934.14	32.37	-39.1	-38.05	-1.05	-41.2
248	113.9	263506	8075.9	32.63	-39.4	-38.34	-1.06	-41.4
249	114.6	265550	8235.4	32.24	-39.5	-38.55	-0.95	-41.6
250	115.4	268626	8388.4	32.02	-39.8	-38.7	-1.10	-41.8
251	116.4	274740	8584.34	32.00	-40	-38.92	-1.08	-42.1
252	116.9	277722	8675.94	32.01	-40	-39.15	-0.85	-42.2
253	118.0	282524	8884	31.80	-40.3	-39.32	-0.97	-42.5
254	119.0	286970	9062.88	31.66	-40.5	-39.55	-0.95	-42.7
255	119.7	293712	9211.3	31.89	-40.7	-39.76	-0.94	-43
256	120.7	300720	9409.32	31.96	-41	-40.0	-1.00	-43.2
257	121.5	304608	9554.56	31.88	-41.1	-40.20	-0.90	-43.4
258	122.3	308910	9674.22	31.93	-41.3	-40.24	-1.06	-43.6
259	123.0	319356	9846.64	32.43	-41.4	-40.63	-0.77	-43.8
260	124.1	325312	10030	32.43	-41.7	-40.82	-0.88	-44.1
261	125.1	331096	10236.4	32.34	-41.9	-41.08	-0.82	-44.3

EVENT	Time (min)	R(i)	R(s)	D2	T(p)	T(r)	DT1	T(s)
262	126.1	335808	10426.4	32.21	-42.1	-41.37	-0.73	-44.6
263	127.1	344790	10619	32.47	-42.3	-41.60	-0.70	-44.8
264	128.1	351742	10782.6	32.62	-42.5	-41.81	-0.69	-44.9
265	129.1	358792	10979.5	32.68	-42.7	-42.04	-0.66	-45.2
266	130.1	368020	11181	32.91	-42.9	-42.28	-0.62	-45.4
267	131.1	374726	11370.2	32.96	-43.2	-42.46	-0.75	-45.6
268	132.2	378258	11566.5	32.70	-43.4	-42.66	-0.73	-45.8
269	133.2	389524	11755.2	33.14	-43.5	-42.91	-0.59	-46.1
270	134.0	396634	11884.8	33.37	-43.8	-43.1	-0.70	-46.3
271	135.2	407544	12120.7	33.62	-44	-43.34	-0.66	-46.5
272	132.2	413754	12302.5	33.63	-44.1	-43.55	-0.55	-46.7
273	137.2	422362	12499.5	33.79	-44.3	-43.74	-0.56	-46.9
274	138.0	424016	12640.1	33.55	-44.5	-43.94	-0.56	-47
275	139.0	437366	12806.1	34.15	-44.7	-44.15	-0.55	-47.3
276	140.3	446830	13049.7	34.24	-44.9	-44.34	-0.56	-47.5
277	141.5	458338	13258.2	34.57	-45	-44.59	-0.41	-47.8
278	142.8	469126	13474.7	34.82	-45.3	-44.81	-0.49	-48
279	143.6	471940	13620.5	34.65	-45.4	-45.02	-0.38	-48.1
280	144.8	489534	13838.6	35.37	-45.6	-45.24	-0.36	-48.3
281	145.9	498390	14024.9	35.54	-45.7	-45.45	-0.25	-48.5
282	147.1	504660	14233.5	35.46	-46	-45.64	-0.36	-48.7
283	148.4	518044	14474	35.79	-46.2	-45.85	-0.35	-49
284	149.4	526578	14622.2	36.01	-46.3	-46.09	-0.21	-49.1
285	150.9	525928	14864.9	35.38	-46.5	-46.3	-0.20	-49.3
286	152.2	528000	15080.3	35.01	-46.7	-46.44	-0.26	-49.5
287	153.4	528000	15298.2	34.51	-46.9	-46.71	-0.19	-49.8
288	154.5	528000	15462.7	34.15	-47	-46.92	-0.08	-49.9
289	155.7	528000	15663.1	33.71	-47.2	-47.06	-0.14	-50.1
290	157.3	528000	15926.2	33.15	-47.5	-47.26	-0.24	-50.3
291	158.3	528000	16064	32.87	-47.6	-47.48	-0.12	-50.5
292	159.3	528000	16231.9	32.53	-47.8	-47.57	-0.23	-50.6
293	160.5	528000	16430.8	32.13	-47.8	-47.78	-0.02	-50.8
294	162.1	528000	16631.8	31.75	-48	-47.95	-0.05	-51

EVENT	Time (min)	R(i)	R(s)	D2	T(p)	T(r)	DT1	T(s)
295	163.3	528000	16831.7	31.37	–48.2	–48.2	0.00	–51.2
296	165.3	528000	17124.9	30.83	–48.4	–48.45	0.05	–51.4
297	167.4	528000	17433.2	30.29	–48.6	–48.69	0.09	–51.7
298	168.9	528000	17637.8	29.94	–48.8	–48.92	0.12	–51.8
299	169.5	528000	17658.8	29.90	–48.8	–48.93	0.13	–51.9

D2 &DTA Run: pc.92.061 — 05-21-1992 — — V3.8-SN009-8/31/95

D2&DTA System — RUN IDENTIFICATION — WARMING

SAMPLE NAME : PC.92.061
DATA FILE NAME   : /DATA/NACL0.9.A/NACL.9.1.W
DATE   : 05-21-1992
OPERATOR   : ·
SYSTEM ID   :
FILL VOLUME   : 2 ML
DTA REFERENCE   : METHANOL
WATER SAMPLE   : DI
NOTEBOOK NUMBER   : 12-92
NOTEBOOK PAGE   : 11

PARAMETER TABLE

RESISTANCE READ COUNT   : 5
TEMPERATURE READ COUNT   : 10
ELAPSED TIME LIMIT (HRS)   : 5
COOLING START TEMP. (C)   : -60
COOLING STOP TEMP. (C)   : 20
COOLING EVENT WINDOW (C)   : .2

GRID SPACE FOR DTA GRAPHS (DEGREES (C)/DIVISION)

(T(p) - T(r))/ REF TEMP   : 2
(T(p) - T(i))/ REF TEMP   : 2
(T(i) - T(r))/ REF TEMP   : 2
(T(p) - T(r))/ EVENTS   : 2

CALIBRATION - CORRECTION TEMPERATURES (C)

ICE	SAMPLE	REF	SHELF-SURFACE
.972	.019	.186	0.791
BATH TEMP	: 0		

STANDARD DEVIATION - DEGREE (C)

ICE	SAMPLE	REF	SHELF-SURFACE
.070	.055	.060	.022

TEMPERATURE CORRECTIONS DEGREES (C)

ICE	SAMPLE	REF	SHELF-SURFACE
-.927	-.019	-.186	-.791

```
RESISTANCE CHECK (KILO-OHMS)   ICE (CELL)   SAMPLE (CELL)
```

RESISTANCE CHECK (KILO-OHMS)		ICE (CELL)	SAMPLE (CELL)
SHORT	:	$1. \times 10^{-4}$	$1. \times 10^{-4}$
OPEN	:	9.164×10^{7}	9.164×10^{7}
CELL CONSTANT	:	14.977	14.315

Warming Tabular Data for 0.9% NaCl Solution

EVENT	R(i)	R(s)	D2	T(r)	T(p)	DT1
0	528000	18057.3	29.24	-49	-48.958	0.042
1	528000	18182.2	29.04	-48.787	-48.985	-0.198
2	528000	18000.4	29.33	-48.573	-48.825	-0.252
3	528000	17781.4	29.69	-48.307	-48.566	-0.259
4	528000	17648.3	29.92	-48.058	-48.389	-0.331
5	528000	17426.2	30.30	-47.81	-48.202	-0.392
6	528000	17199.9	30.70	-47.508	-47.962	-0.454
7	528000	17064	30.94	-47.242	-47.767	-0.525
8	528000	16871.8	31.29	-47.056	-47.554	-0.498
9	528000	16708.2	31.60	-46.799	-47.323	-0.524
10	528000	16338.5	32.32	-46.542	-47.128	-0.586
11	528000	16394.8	32.21	-46.295	-46.942	-0.647
12	528000	16300.2	32.39	-46.1	-46.703	-0.603
13	528000	16086.5	32.82	-45.835	-46.481	-0.646
14	528000	15779.8	33.46	-45.526	-46.252	-0.726
15	528000	15578.3	33.89	-45.35	-45.995	-0.645
16	528000	15366.3	34.36	-45.085	-45.765	-0.680
17	528000	15181.3	34.78	-44.838	-45.562	-0.724
18	528000	14946.5	35.33	-44.6	-45.385	-0.785
19	528000	14738.7	35.82	-44.326	-45.183	-0.857
20	528000	14539	36.32	-44.133	-44.935	-0.802
21	528000	14338.7	36.82	-43.895	-44.723	-0.828
22	528000	14111.5	37.42	-43.622	-44.494	-0.872
23	528000	13892.9	38.01	-43.447	-44.248	-0.801
24	528000	13674.2	38.61	-43.166	-44.106	-0.940
25	528000	13471.9	39.19	-42.911	-43.86	-0.949
26	528000	13260.9	39.82	-42.727	-43.613	-0.886
27	528000	13030.6	40.52	-42.473	-43.438	-0.965
28	528000	12798	41.26	-42.253	-43.183	-0.930
29	528000	12568.3	42.01	-42.051	-42.954	-0.903
30	528000	12307.7	42.90	-41.736	-42.735	-0.999

EVENT	R(i)	R(s)	D2	T(r)	T(p)	DT1
31	528000	12079.4	43.71	–41.508	–42.489	–0.981
32	528000	11851.5	44.55	–41.325	–42.279	–0.954
33	528000	11591.6	45.55	–41.01	–42.041	–1.031
34	528000	11372.7	46.43	–40.809	–41.891	–1.082
35	528000	11132.1	47.43	–40.564	–41.612	–1.048
36	528000	10892.8	48.47	–40.311	–41.393	–1.082
37	528000	10649.6	49.58	–40.118	–41.166	–1.048
38	528000	10428.7	50.63	–39.866	–40.93	–1.064
39	528000	10201.6	51.76	–39.613	–40.72	–1.107
40	528000	9941.34	53.11	–39.334	–40.475	–1.141
41	528000	9716.6	54.34	–39.108	–40.248	–1.140
42	528000	9496.98	55.60	–38.881	–40.03	–1.149
43	528000	9258.84	57.03	–38.664	–39.82	–1.156
44	528000	9052.1	58.33	–38.429	–39.593	–1.164
45	528000	8818.46	59.87	–38.203	–39.376	–1.173
46	528000	8602.82	61.38	–37.968	–39.141	–1.173
47	528000	8400.46	62.85	–37.682	–38.906	–1.224
48	528000	8192.72	64.45	–37.43	–38.714	–1.284
49	528000	7991.7	66.07	–37.24	–38.505	–1.265
50	528000	7759.94	68.04	–36.997	–38.2	–1.203
51	528000	7570.54	69.74	–36.763	–38.027	–1.264
52	528000	7391.08	71.44	–36.521	–37.792	–1.271
53	528000	7193.08	73.40	–36.295	–37.558	–1.263
54	528000	7008.94	75.33	–36.053	–37.324	–1.271
55	528000	6832.56	77.28	–35.803	–37.15	–1.347
56	528000	6661.98	79.26	–35.587	–36.899	–1.312
57	528000	6491.68	81.33	–35.362	–36.656	–1.294
58	528000	6333.92	83.36	–35.095	–36.457	–1.362
59	528000	6178.02	85.46	–34.81	–36.232	–1.422
60	528000	6006.44	87.91	–34.604	–35.955	–1.351
61	528000	5876.48	89.85	–34.388	–35.773	–1.385
62	528000	5735.8	92.05	–34.173	–35.549	–1.376
63	528000	5603.26	94.23	–33.949	–35.29	–1.341

EVENT	R(i)	R(s)	D2	T(r)	T(p)	DT1
64	525928	5475.26	96.06	–33.734	–35.074	–1.340
65	525286	5343.6	98.30	–33.424	–34.901	–1.477
66	524024	3429.98	152.78	–33.209	–34.651	–1.442
67	524024	2605.4	201.13	–32.977	–34.462	–1.485
68	521982	. 2506	208.29	–32.746	–34.22	–1.474
69	521982	2406.86	216.87	–32.505	–34.014	–1.509
70	518816	2309.46	224.65	–32.308	–33.773	–1.465
71	512876	2216.38	231.40	–32.042	–33.523	–1.481
72	513136	20863.7	246.26	–31.717	–33.188	–1.471
73	506178	2000.6	253.01	–31.511	–33.007	–1.496
74	500418	1920.98	260.50	–31.272	–32.827	–1.555
75	498908	1842.75	270.74	–31.007	–32.629	–1.622
76	496252	1765.19	281.13	–30.827	–32.346	–1.519
77	489808	1695.32	288.92	–30.597	–32.158	–1.561
78	486596	1625.89	299.28	–30.417	–31.952	–1.535
79	477434	1559.25	306.19	–30.187	–31.738	–1.551
80	473732	1491.36	317.65	–29.939	–31.481	–1.542
81	473480	1431.34	330.79	–29.701	–31.275	–1.574
82	460870	1372.49	335.79	–29.479	–31.036	–1.557
83	453692	1315.09	344.99	–29.275	–30.805	–1.530
84	454116	1261.91	359.86	–29.037	–30.608	–1.571
85	448624	1210.4	370.64	–28.782	–30.395	–1.613
86	446502	1161.41	384.45	–28.518	–30.19	–1.672
87	441642	1114.1	396.41	–28.306	–29.96	–1.654
88	435926	1067.36	408.42	–28.111	–29.738	–1.627
89	429044	1023.24	419.30	–27.89	–29.506	–1.616
90	426280	978.486	435.65	–27.678	–29.261	–1.583
91	416662	918.596	453.59	–27.381	–28.989	–1.608
92	408518	879.16	464.67	–27.11	–28.776	–1.666
93	407090	840.33	484.44	–26.881	–28.563	–1.682
94	399506	802.894	497.58	–26.687	–28.333	–1.464
95	395254	767.114	515.25	–26.416	–28.138	–1.722
96	388742	731.988	531.08	–26.204	–27.934	–1.730

EVENT	R(i)	R(s)	D2	T(r)	T(p)	DT1
97	388888	696.406	558.42	–26.035	–27.713	–1.678
98	380318	664.274	572.53	–25.782	–27.544	–1.762
99	373832	631.79	591.70	–25.639	–27.289	–1.650
100	367976	599.164	614.15	–25.377	–27.069	–1.692
101	365396	568.418	642.83	–25.166	–26.866	–1.700
102	360330	539.292	668.15	–24.964	–26.688	–1.724
103	351976	496.654	708.69	–24.678	–26.358	–1.680
104	348274	469.59	741.66	–24.442	–26.172	–1.730
105	347826	443.18	784.84	–24.265	–25.91	–1.645
106	342308	418.78	817.39	–24.063	–25.724	–1.661
107	336946	395.008	853.01	–23.853	–25.521	–1.668
108	328854	371.374	885.51	–23.634	–25.31	–1.676
109	326432	350.482	931.38	–23.424	–25.063	–1.639
110	319870	329.448	970.93	–23.165	–24.889	–1.724
111	316836	310.732	1019.64	–22.972	–24.661	–1.689
112	314238	291.72	1077.19	–22.711	–24.491	–1.780
113	310308	275.088	1128.03	–22.443	–24.257	–1.814
114	303338	258.642	1172.81	–22.284	–24.03	–1.746
115	296286	243.744	1215.56	–22.083	–23.828	–1.745
116	296322	229.8	1289.48	–21.874	–23.576	–1.702
117	290502	216.092	1344.34	–21.665	–23.408	–1.743
118	287036	204.312	1404.89	–21.423	–23.24	–1.817
119	281328	191.53	1468.85	–21.197	–22.997	–1.800
120	278006	180.322	1541.72	–20.997	–22.812	–1.815
121	274480	169.787	1616.61	–20.771	–22.543	–1.772
122	271218	159.372	1701.79	–20.538	–22.334	–1.796
123	266286	148.895	1788.41	–20.329	–22.15	–1.821
124	261052	139.898	1866.02	–20.113	–21.966	–1.853
125	260088	129.617	2006.59	–19.855	–21.765	–1.910
126	254252	132.849	1913.84	–19.663	–21.522	–1.859
127	248772	22.4696	11071.49	–19.373	–21.297	–1.924
128	246190	17.871	13775.95	–19.164	–21.121	–1.957
129	243838	15.049	16202.94	–19.023	–20.862	–1.839

EVENT	R(i)	R(s)	D2	T(r)	T(p)	DT1
130	239808	12.821	18704.31	–18.741	–20.796	–2.055
131	234256	9.821	23852.56	–18.442	–20.57	–2.128
132	231126	6.998	33027.44	–18.193	–20.362	–2.169
133	225654	5.353	42154.68	–17.845	–20.312	–2.467
134	219722	. 4.474	49110.86	–17.63	–20.22	–2.590
135	217224	4.045	53701.85	–17.423	–20.161	–2.738
136	212670	3.709	57338.91	–17.208	–20.095	–2.887
137	210410	3.382	62214.67	–16.952	–19.828	–2.876
138	206502	3.246	63617.38	–16.754	–19.562	–2.808
139	202644	3.192	63484.96	–16.539	–19.213	–2.674
140	199509	3.18	62738.68	–16.341	–18.897	–2.556
141	197040	3.136	62831.63	–16.118	–18.639	–2.521
142	194288	3.088	62917.10	–15.92	–18.316	–2.396
143	191623	3.042	62992.44	–15.747	–18.058	–2.311
144	188117	2.985	63020.77	–15.558	–17.827	–2.269
145	185775	2.973	62487.39	–15.385	–17.57	–2.185
146	182977	2.944	62152.51	–15.171	–17.354	–2.183
147	179393	2.932	61184.52	–14.965	–17.139	–2.174
148	176013	2.898	60736.02	–14.727	–16.899	–2.172
149	174422	2.872	60731.89	–14.529	–16.676	–2.147
150	171136	2.818	60729.60	–14.373	–16.461	–2.088
151	168197	2.816	59729.05	–14.209	–16.236	–2.027
152	165893	2.795	59353.49	–14.029	–15.999	–1.970
153	161534	2.765	58420.98	–13.709	–15.694	–1.985
154	156793	2.728	57475.44	–13.43	–15.389	–1.959
155	154213	2.692	57285.66	–13.11	–15.117	–2.007
156	149175	2.646	56377.55	–12.849	–14.846	–1.997
157	147391	2.647	55682.28	–12.685	–14.608	–1.923
158	145301	2.621	55437.24	–12.473	–14.41	–1.937
159	141534	2.602	54394.31	–12.26	–14.213	–1.953
160	138210	2.587	53424.82	–12.064	–13.942	–1.878
161	134577	2.574	52283.22	–11.803	–13.729	–1.926
162	132854	2.524	52636.29	–11.526	–13.499	–1.973

EVENT	R(i)	R(s)	D2	T(r)	T(p)	DT1
163	131151	2.491	52649.94	−11.371	−13.295	−1.924
164	128547	2.489	51646.04	−11.184	−13.073	−1.889
165	124847	2.487	50199.84	−10.947	−12.942	−1.995
166	121494	2.441	49772.22	−10.752	−12.64	−1.888
167	118861	2.422	49075.56	−10.459	−12.468	−2.009
168	115640	2.414	47903.89	−10.232	−12.182	−1.950
169	113606	2.372	47894.60	−10.004	−11.896	−1.892
170	112202	2.378	47183.35	−9.777	−11.676	−1.899
171	109773	2.348	46751.70	−9.566	−11.569	−2.003
172	107986	2.337	46207.10	−9.339	−11.39	−2.051
173	104993	2.287	45908.61	−9.104	−11.089	−1.985
174	101121	2.287	44215.57	−8.828	−10.839	−2.011
175	99564.2	2.292	43439.88	−8.683	−10.6	−1.917
176	96339.8	2.273	42384.43	−8.424	−10.34	−1.916
177	94079.2	2.25	41812.98	−8.165	−10.036	−1.871
178	92007.4	2.229	41277.43	−7.931	−9.868	−1.937
179	89405	2.21	40454.75	−7.713	−9.633	−1.920
180	85427.4	2.2	38830.64	−7.471	−9.365	−1.894
181	83835.2	2.156	38884.60	−7.204	−9.122	−1.918
182	81523	2.136	38166.20	−7.003	−8.968	−1.965
183	79308.6	2.1127	37538.98	−6.89	−8.741	−1.851
184	77545.2	2.117	36629.76	−6.49	−8.571	−2.081
185	74945.4	2.102	35654.33	−6.41	−8.328	−1.918
186	72807.8	2.092	34802.96	−6.102	−8.085	−1.983
187	70756.4	2.071	34165.33	−5.973	−7.867	−1.894
188	97959.4	2.0375	48078.23	−5.716	−7.698	−1.982
189	65179.2	2.016	32330.95	−5.467	−7.448	−1.981
190	63794.4	2.013	31691.21	−5.282	˙−7.246	−1.964
191	61245	1.995	30699.25	−5.01	−7.028	−2.018
192	59231.2	1.986	29824.37	−4.769	−6.81	−2.041
193	58150.8	1.965	29593.28	−4.633	−6.609	−1.976
194	56679	1.939	29231.05	−4.425	−6.451	−2.026
195	55569	1.948	28526.18	−4.225	−6.335	−2.110

EVENT	R(i)	R(s)	D2	T(r)	T(p)	DT1
196	53373.6	1.954	27315.05	–3.985	–6.143	–2.158
197	52024	1.934	26899.69	–3.841	–5.917	–2.076
198	50923.4	1.915	26591.85	–3.633	–5.772	–2.139
199	48826.4	1.9089	25578.29	–3.37	–5.579	–2.209
200	46762	. 1.894	24689.55	–3.122	–5.338	–2.216
201	44515.6	1.899	23441.60	–2.888	–5.106	–2.218
202	42638.2	1.875	22740.37	–2.612	–4.945	–2.333
203	41260.2	1.866	22111.58	–2.405	–4.841	–2.436
204	39116.6	1.863	20998.17	–2.174	–4.617	–2.443
205	37505.6	1.845	20328.24	–1.968	–4.535	–2.567
206	34393.2	1.836	18732.68	–1.761	–4.353	–2.592
207	32874.8	· 1.84	17866.74	–1.539	–4.2	–2.661
208	29902.8	1.832	16322.49	–1.364	–3.984	–2.620
209	27704.2	1.824	15188.71	–1.046	–3.873	–2.827
210	26266.2	1.804	14559.98	–0.816	–3.72	–2.904
211	24008.8	1.83	13119.56	–0.547	–3.553	–3.006
212	22513.4	1.795	12542.28	–0.341	–3.369	–3.028
213	19995.5	1.821	10980.51	–0.088	–3.241	–3.153
214	17997.5	1.823	9872.46	0.133	–3.113	–3.246
215	15699	1.801	8716.82	0.362	–2.938	–3.300
216	14665.2	1.817	8071.11	0.591	–2.834	–3.425
217	12813.8	1.798	7126.70	0.82	–2.6909	–3.511
218	10231.9	1.813	5643.63	1.096	–2.5315	–3.628
219	10479.9	1.805	5806.04	1.317	–2.372	–3.689
220	9825.88	1.808	5434.67	1.522	–2.284	–3.806
221	9435.04	1.8255	5168.47	1.758	–2.2	–3.958
222	9086.5	1.823	4984.37	1.963	–2.077	–4.040
223	8781.28	1.83	4798.81	2.175	–1.982	–4.157
224	7970.84	1.847	4315.56	2.403	–1.918	–4.321
225	7102.9	1.851	3837.33	2.694	–1.823	–4.517
226	6852.62	1.875	3654.73	2.906	–1.783	–4.689
227	6654.26	1.87	3558.43	3.125	–1.727	–4.852
228	6451.32	1.894	3406.19	3.36	–1.695	–5.055

EVENT	R(i)	R(s)	D2	T(r)	T(p)	DT1
229	6165.12	1.892	3258.52	3.658	–1.592	–5.250
230	5907.52	1.9	3109.22	3.924	–1.544	–5.468
231	5702.1	1.903	2996.37	4.206	–1.497	–5.703
232	5514.5	1.903	2897.79	4.495	–1.433	–5.928
233	5163.3	1.917	2693.43	4.737	–1.409	–6.146
234	4199.1	1.922	2184.76	5.003	–1.354	–6.357
235	2460.84	1.914	1285.71	5.283	–1.33	–6.613
236	2206.52	1.934	1140.91	5.533	–1.266	–6.799
237	2040.32	1.902	1072.72	5.751	–1258	–7.009
238	1796.33	1.906	942.46	6.015	–1.211	–7.226
239	1623.18	1.912	848.94	6.226	–1.187	–7.413
240	1365.39	1.936	705.26	6.459	–1.115	–7.574
241	1158.49	1.956	592.28	6.7	–1.044	–7.744
242	999.926	1.954	511.73	6.948	–1.02	–7.968
243	878.97	1.956	449.37	7.181	–0.941	–8.122
244	484.624	1.952	248.27	7.413	–0.846	–8.259
245	703.274	1.954	359.92	7.653	–0.79	–8.443
246	649.24	1.956	331.92	7.886	–0.771	–8.597
247	601.516	1.915	314.11	8.118	–0.639	–8.757
248	560.294	1.917	292.28	8.357	–0.552	–8.909
249	529.322	1.954	270.89	8.574	–0.505	–9.079
250	510.044	1.988	256.56	8.79	–0.426	–9.216
251	487.246	1.996	244.11	9.0143	–0.315	–9.329
252	434.602	2.039	213.14	9.261	–0.196	–9.457
253	393.002	2.046	192.08	9.515	–0.117	–9.632
254	381.562	2.059	185.31	9.731	–0.069	–9.800
255	335.664	2.096	160.15	9.961	0.08	–9.881
256	314.032	2.156	145.65	10.208	0.231	–9.977
257	295.948	2.245	131.83	10.423	0.444	–9.979
258	280.954	3.055	91.91	10.661	0.642	–10.019
259	273.338	9.892	27.63	10.807	0.863	–9.944
260	264.9	8.903	29.75	10.95	1.084	–9.866
261	254.558	6.111	41.66	11.114	1.312	–9.802
262	244.734	5.7769	42.36	11.344	1.659	–9.685

EVENT	R(i)	R(s)	D2	T(r)	T(p)	DT1
263	238.666	5.729	41.66	11.497	1.896	–9.601
264	233.512	5.561	41.99	11.85	2.108	–9.742
265	228.494	4.402	42.30	11.796	2.368	–9.428
266	224.362	5.196	43.18	11.949	2.69	–9.259
267	222.394	· 4.871	45.66	11.979	2.91	–9.069
268	221.086	5.072	43.59	12.125	3.263	–8.862
269	219.926	5.037	43.66	12.178	3.467	–8.711
270	220.176	4.93	44.66	12.201	3.694	–8.507
271	221.504	4.951	44.74	12.339	3.945	–8.394
272	221.592	4.918	45.06	12.415	4.438	–7.977
276	222.752	4.872	45.72	12.491	4.985	–7.506
274	223.64	4.835	46.25	12.537	5.616	–6.921
275	224.74	4.814	46.68	12.621	6.224	–6.397
276	226.756	4.765	47.59	12.721	6.846	–5.875
277	227.216	4.783	47.50	12.797	7.389	–5.408
278	227.668	4.779	47.64	12.812	7.877	–4.935
279	226.5	4.758	47.60	12.896	8.256	–4.640
280	225.752	4.699	48.04	12.957	8.712	–4.245
281	225.908	4.685	48.22	13.026	9.037	–3.989
282	226.01	4.683	48.26	13.117	9.368	–3.749
283	225.566	4.664	48.36	13.178	9.623	–3.555
284	225.58	4.647	48.54	13.216	9.88	–3.336
285	227.498	4.614	49.31	13.3	10.139	–3.161
286	229.22	4.579	50.06	13.333	10.354	–2.979
287	225.322	4.631	48.66	13.46	10.7	–2.760
288	219.586	4.566	48.09	13.551	11.068	–2.483
289	213.11	4.553	46.81	13.704	11.329	–2.375
290	207.106	4.537	45.65	13.818	11.682	–2.136
291	200.132	4.508	44.39	13.94	11.904	–2.036
292	195.293	4.521	43.20	14.099	12.141	–1.958
293	189.352	4.478	42.28	14.312	12.432	–1.880
294	176.896	4.418	40.04	14.434	12.661	–1.773
295	165.575	4.459	37.13	14.654	12.905	–1.749
296	158.583	4.503	35.22	14.813	13.188	–1.625

EVENT	R(i)	R(s)	D2	T(r)	T(p)	DT1
297	154.961	4.527	34.23	15.026	13.47	−1.566
298	151.506	4.555	33.26	15.238	13.683	−1.555
299	150.702	4.527	33.29	15.465	13.927	−1.538
300	149.668	4.568	32.76	15.632	14.223	−1.409
301	150.575	4.619	32.60	15.836	14.337	−1.499
302	150.469	4.667	32.24	15.979	14.543	−1.436
303	151.115	4.68	32.29	16.131	14.808	−1.323
304	152.451	4.712	32.35	16.357	14.884	−1.473
305	152.77	4.751	32.16	16.516	15.119	−1.397
306	153.984	4.812	32.00	16.644	15.324	−1.320
307	155.899	4.74	32.89	16.855	15.483	−1.372
308	157.136	4.6411	33.86	17.074	15.748	−1.326
309	158.6	4.571	34.70	17.345	16.021	−1.324
310	158.977	4.594	34.61	17.751	16.187	−1.384
311	160.534	4.615	34.79	17.774	16.331	−1.443
312	160.647	4.628	34.71	17.977	16.625	−1.352
313	161.91	4.597	35.22	18.248	16.829	−1.419
314	162.607	4.4909	36.21	18.428	17.101	−1.327
315	163.466	4.444	36.78	18.675	17.304	−1.371
316	163.008	4.4157	36.92	18.923	17.537	−1.386
317	163.644	4.388	37.29	19.103	17.748	−1.355
318	164.613	4.333	37.99	19.342	17.967	−1.375
319	164.999	4.309	38.29	19.499	18.214	−1.285
320	164.469	4.213	39.04	19.746	18.357	−1.389
321	164.537	4.181	39.35	19.956	18.508	−1.448
322	165.179	4.221	39.13	20.067	18.723	−1.344
323	165.142	4.236	38.99	20.269	18.943	−1.326
324	164.85	4.236	38.92	20.53	'19.15	−1.380
325	164.391	4.254	38.64	20.724	19.407	−1.317
326	163.766	4.193	39.06	20.933	19.557	−1.376
327	162.973	4.151	39.26	21.134	19.841	−1.293
328	163.064	4.2	38.82	21.313	20.111	−1.202
329	162.141	4.121	39.35	21.387	20.11	−1.277

APPENDIX C

Data for 3% (wt/v) lactose and 0.9% NaCl (wt/v) Solution

```
D2 &DTA Run: Lacsal — 04-10-1996 — TAJ — V3.8-SN009-8/31/95


D2&DTA System — RUN IDENTIFICATION — COOLING

SAMPLE NAME : LACSAL
DATA FILE NAME    : /DATA/LACSAL.A/LACSAL.1.C
DATE              : 04-10-1996
OPERATOR          : TAJ
SYSTEM ID         : V3.8-SN009-8/31/95
FILL VOLUME       : 2 ML
DTA REFERENCE     : METHANOL
WATER SAMPLE      : DI
NOTEBOOK NUMBER   : 3-86
NOTEBOOK PAGE     : 19


PARAMETER TABLE

RESISTANCE READ COUNT           :     5
TEMPERATURE READ COUNT          :     10
ELAPSED TIME LIMIT (HRS)        :     5
COOLING START TEMP. (C)         :     22
COOLING STOP TEMP. (C)          :     -60
COOLING EVENT WINDOW (C)        :     .2


GRID SPACE FOR DTA GRAPHS (DEGREES (C)/DIVISION)

(T(p) - T(r))/ REF TEMP : 2
(T(p) - T(i))/ REF TEMP : 2
(T(i) - T(r))/ REF TEMP : 2
(T(p) - T(r))/ EVENTS   : 2


CALIBRATION - CORRECTION TEMPERATURES (C)

ICE             SAMPLE      REF         SHELF-SURFACE
.198            - .172      .752        0.044
BATH TEMP        : 0

STANDARD DEVIATION - DEGREE (C)

ICE             SAMPLE      REF         SHELF-SURFACE
.030            .036        .085        .042
```

TEMPERATURE CORRECTIONS DEGREES (C)

ICE	SAMPLE	REF	SHELF-SURFACE
-.198	.172	-.751	-.044

RESISTANCE CHECK (KILO-OHMS)   ICE (CELL)   SAMPLE (CELL)

		ICE (CELL)	SAMPLE (CELL)
SHORT	:	2.974×10^{-3}	6.447×10^{-3}
OPEN	:	8.293×10^{6}	9.164×10^{7}
CELL CONSTANT	:	9.851	10.809

Cooling Tabular Data for 3% Lactose and 0.9% NaCl Solution

EVENTS	T(i)	T(p)	T(r)	DT1	DT3
0	21.933	21.924	21.572	0.352	0.361
1	21.725	21.664	21.283	0.381	0.442
2	21.420	21.440	21.112	0.328	0.308
3	21.197	21.157	20.792	0.365	0.405
4	20.914	20.941	20.442	0.499	0.472
5	20.601	20.576	20.189	0.387	0.412
6	20.519	20.494	19.988	0.506	0.531
7	20.161	20.127	19.622	0.505	0.539
8	19.862	19.843	19.301	0.542	0.561
9	19.660	19.731	19.092	0.639	0.568
10	19.406	19.349	18.830	0.519	0.576
11	18.994	19.004	18.455	0.549	0.539
12	18.881	18.802	18.230	0.572	0.651
13	18.469	18.562	17.803	0.759	0.666
14	18.341	18.343	17.660	0.683	0.681
15	18.011	18.050	17.254	0.796	0.757
16	17.650	17.681	16.908	0.773	0.742
17	17.454	17.471	16.675	0.796	0.779
18	17.115	17.177	16.373	0.804	0.742
19	16.995	17.033	16.132	0.901	0.863
20	16.754	16.724	15.883	0.841	0.871
21	16.406	16.422	15.566	0.856	0.840
22	16.195	16.248	15.331	0.917	0.864
23	15.817	15.900	15.006	0.894	0.811
24	15.643	15.718	14.741	0.977	0.902
25	15.241	15.331	14.324	1.007	0.917
26	15.105	15.119	14.180	0.939	0.925
27	14.673	14.838	13.823	1.015	0.850
28	14.529	14.603	13.618	0.985	0.911
29	14.141	14.268	13.223	1.045	0.918
30	13.928	14.009	12.994	1.015	0.934

EVENTS	T(i)	T(p)	T(r)	DT1	DT3
31	13.708	13.796	12.842	0.954	0.866
32	13.388	13.423	12.461	0.962	0.927
33	13.205	13.202	12.316	0.886	0.889
34	12.991	12.957	12.110	0.847	0.881
35	12.778	12.751	11.881	0.870	0.897
36	12.404	12.361	11.590	0.771	0.814
37	12.174	12.200	11.384	0.816	0.790
38	12.052	11.994	11.101	0.893	0.951
39	11.845	11.741	10.940	0.801	0.905
40	11.608	11.503	10.774	0.729	0.834
41	11.440	11.312	10.564	0.748	0.876
42	11.041	10.920	10.188	0.732	0.853
43	10.880	10.782	9.974	0.809	0.907
44	10.634	10.490	9.758	0.732	0.876
45	10.427	10.305	9.528	0.777	0.899
46	10.212	10.059	9.374	0.685	0.838
47	10.027	9.905	9.174	0.732	0.853
48	9.804	9.705	8.974	0.731	0.831
49	9.489	9.312	8.580	0.731	0.908
50	9.250	9.119	8.280	0.839	0.970
51	9.057	8.856	8.110	0.746	0.947
52	8.648	8.500	7.777	0.723	0.871
53	8.401	8.299	7.546	0.754	0.855
54	8.269	8.084	7.321	0.763	0.948
55	8.084	7.850	7.104	0.746	0.979
56	7.859	7.680	6.880	0.800	0.980
57	7.511	7.401	6.562	0.839	0.949
58	7.255	7.191	6.314	0.877	.0.941
59	7.154	6.966	6.096	0.870	1.058
60	6.960	6.702	5.902	0.800	1.058
61	6.790	6.531	5.677	0.854	1.113
62	6.386	6.219	5.295	0.924	1.090
63	6.075	5.822	4.867	0.955	1.208

EVENTS	T(i)	T(p)	T(r)	DT1	DT3
64	5.678	5.487	4.516	0.971	1.162
65	5.514	5.323	4.243	1.080	1.271
66	5.241	5.034	3.970	1.065	1.272
67	5.000	4.823	3.751	1.072	1.249
68	4.836	4.589	3.524	1.065	1.312
69	4.726	4.362	3.297	1.065	1.429
70	4.390	3.994	2.913	1.081	1.470
71	4.297	3.736	2.710	1.026	1.587
72	4.039	3.532	2.514	1.018	1.520
73	3.859	3.414	2.286	1.128	1.570
74	3.577	3.014	1.886	1.128	1.691
75	3.412	2.810	1.752	1.058	1.660
76	3.279	2.574	1.587	0.987	1.692
77	2.879	2.204	1.296	0.908	1.583
78	2.683	1.984	1.060	0.924	1.623
79	2.298	1.574	0.674	0.900	1.624
80	2.062	1.321	0.461	0.861	1.601
81	1.881	1.203	0.248	0.956	1.633
82	1.676	0.990	0.066	0.924	1.610
83	1.558	0.779	–0.147	0.926	1.705
84	1.006	0.421	–0.543	0.964	1.549
85	0.620	–0.022	–0.891	0.869	1.510
86	0.169	–0.299	–1.152	0.853	1.321
87	–0.013	–0.513	–1.334	0.821	1.322
88	–0.092	–0.728	–1.588	0.860	1.496
89	–0.313	–0.918	–1.810	0.892	1.497
90	–0.527	–1.124	–1.953	0.829	1.426
91	–0.995	–1.561	–2.342	0.781	1.348
92	–1.249	–1.784	–2.573	0.789	1.324
93	–1.471	–1.943	–2.788	0.844	1.317
94	–1.638	–2.214	–2.955	0.741	1.317
95	–1.876	–2.342	–3.186	0.844	1.310
96	–2.290	–2.644	–3.513	0.868	1.223

EVENTS	T(i)	T(p)	T(r)	DT1	DT3
97	-2.544	-2.796	-3.752	0.956	1.208
98	-2.871	-3.211	-4.127	0.916	1.256
99	-3.286	-3.571	-4.518	0.948	1.233
100	-3.741	-3.858	-4.870	1.012	1.129
101	-3.973	-4.002	-5.078	1.076	1.106
102	-4.133	-4.203	-5.335	1.132	1.202
103	-4.581	-4.523	-5.695	1.173	1.115
104	-4.773	-4.707	-5.920	1.213	1.147
105	-5.182	-5.020	-6.217	1.197	1.035
106	-5.519	-5.357	-6.554	1.197	1.036
107	-5.752	-5.542	-6.772	1.230	1.020
108	-5.985	-5.679	-6.973	1.294	0.988
109	-6.129	-5.888	-7.125	1.238	0.996
110	-6.395	-6.121	-7.295	1.174	0.900
111	-6.556	-6.266	-7.520	1.254	0.964
112	-6.717	-6.363	-7.738	1.375	1.021
113	-7.160	-6.709	-8.101	1.392	0.941
114	-7.321	-6.919	-8.294	1.376	0.973
115	-4.7432	-7.120	-8.448	1.328	3.705
116	0.54	-7.451	-8.755	1.304	9.295
117	0.532	-7.677	-8.941	1.264	9.473
118	0.54	-8.106	-9.3534	1.248	9.893
119	0.619	-8.511	-9.6045	1.094	10.224
120	0.564	-8.745	-9.8152	1.070	10.379
121	0.548	-9.053	-10.163	1.110	10.711
122	0.6038	-9.427	-10.504	1.077	11.108
123	0.603	-0.51343	-10.659	10.146	11.262
124	0.635	-0.25186	-10.805	10.553	11.440
125	0.556	-0.23602	-11.114	10.878	11.670
126	0.501	-0.30733	-11.447	11.140	11.948
127	0.398	-0.31526	-11.732	11.417	12.130
128	0.256	-0.40243	-12.058	11.656	12.314
129	0.058	-0.42621	-12.336	11.910	12.394

EVENTS	*T(i)*	*T(p)*	*T(r)*	DT1	DT3
130	-0.139	-0.64823	-12.621	11.973	12.482
131	-0.186	-0.77516	-12.932	12.157	12.746
132	-0.337	-0.90212	-13.177	12.275	12.840
133	-0.606	-1.1164	-13.537	12.421	12.931
134	-0.915	-1.3548	-13.766	12.411	12.851
135	-1.32	-1.6251	-14.028	12.403	12.708
136	-0.764	-1.8955	-14.372	12.477	13.608
137	-2.067	-2.2459	-14.643	12.397	12.576
138	-2.496	-2.5487	-15.004	12.455	12.508
139	-2.687	-2.7799	-15.094	12.314	12.407
140	-2.871	-3.0193	-15.275	12.256	12.404
141	-3.421	-3.4266	-15.547	12.120	12.126
142	-4.356	-3.8503	-15.851	12.001	11.495
143	-5.0616	-4.1144	-16.057	11.943	10.995
144	-5.816	-4.3787	-16.205	11.826	10.389
145	-6.684	-4.6912	-16.386	11.695	9.702
146	-7.668	-5.1004	-16.452	11.352	8.784
147	-8.759	-5.5661	-16.617	11.051	7.858
148	-9.805	-6.1693	-16.807	10.638	7.002
149	-10.732	-6.9589	-16.947	9.988	6.215
150	-11.692	-7.8551	-17.129	9.274	5.437
151	-12.509	-8.7616	-17.294	8.532	4.785
152	-13.196	-9.5727	-17.459	7.886	4.263
153	-13.86	-10.393	-17.6	7.207	3.740
154	-14.561	-11.306	-17.707	6.401	3.146
155	-15.124	-12.065	-17.881	5.816	2.757
156	-15.668	-12.834	-18.063	5.229	2.395
157	-16.195	-13.489	-18.228	4.739	2.033
158	-16.715	-14.13	-18.369	4.239	1.654
159	-17.153	-14.787	-18.576	3.789	1.423
160	-17.716	-15.413	-18.725	3.312	1.009
161	-18.197	-15.965	-18.908	2.943	0.711
162	-18.595	-16.535	-18.974	2.439	0.379

EVENTS	T(i)	T(p)	T(r)	DT1	DT3
163	−18.952	−16.99	−19.165	2.175	0.213
164	−19.392	−17.42	−19.323	1.903	−0.069
165	−19.742	−17.909	−19.505	1.596	−0.237
166	−20.024	−18.324	−19.713	1.389	−0.311
167	−20.308	−18.631	−19.879	1.248	−0.429
168	−20.591	−19.072	−20.07	0.998	−0.521
169	−20.933	−19.446	−20.22	0.774	−0.713
170	−21.15	−19.862	−20.345	0.483	−0.805
171	−21.367	−20.137	−20.528	0.391	−0.839
172	−21.61	−20.471	−20.678	0.207	−0.932
173	−21.752	−20.755	−20.87	0.115	−0.882
174	−21.969	−21.047	−21.062	0.015	−0.907
175	−22.179	−21.373	−21.136	−0.237	−1.043
176	−22.354	−21.574	−21.387	−0.187	−0.967
177	−22.48	−21.817	−21.579	−0.238	−0.901
178	−22.664	−22.093	−21.654	−0.439	−1.010
179	−22.882	−22.369	−21.779	−0.590	−1.103
180	−23.008	−22.655	−21.938	−0.717	−1.070
181	−23.193	−22.906	−22.163	−0.743	−1.030
182	−23.36	−23.15	−22.264	−0.886	−1.096
183	−23.528	−23.36	−22.464	−0.896	−1.064
184	−23.763	−23.511	−22.707	−0.804	−1.056
185	−23.789	−23.747	−22.841	−0.906	−0.948
186	−24.016	−23.957	−23	−0.957	−1.016
187	−24.176	−24.175	−23.184	−0.991	−0.992
188	−24.428	−24.504	−23.545	−0.959	−0.883
189	−24.571	−24.748	−23.746	−1.002	−0.825
190	−24.916	−25.044	−24.073	−0.971	−0.843
191	−25.043	−25.254	−24.225	−1.029	−0.818
192	−25.262	−25.558	−24.477	−1.081	−0.785
193	−25.498	−25.863	−24.788	−1.075	−0.710
194	−25.608	−26.099	−24.956	−1.143	−0.652
195	−25.853	−26.345	−25.226	−1.119	−0.627

EVENTS	$T(i)$	$T(p)$	$T(r)$	DT1	DT3
196	–26.199	–26.602	–25.596	–1.006	–0.603
197	–26.309	–26.988	–25.765	–1.223	–0.544
198	–26.393	–27.209	–25.925	–1.284	–0.468
199	–26.571	–27.489	–26.195	–1.294	–0.376
200	–26.681	–27.794	–26.271	–1.523	–0.410
201	–26.96	–28.015	–26.575	–1.440	–0.385
202	–27.13	–28.312	–26.862	–1.450	–0.268
203	–27.392	–28.61	–27.133	–1.477	–0.259
204	–27.605	–28.823	–27.395	–1.428	–0.210
205	–27.816	–29.052	–27.666	–1.386	–0.150
206	–28.012	–29.342	–27.946	–1.396	–0.066
207	–28.301	–29.632	–28.251	–1.381	–0.050
208	–28.496	–29.888	–28.446	–1.442	–0.050
209	–28.649	–30.033	–28.659	–1.374	0.010
210	–28.946	–30.204	–28.93	–1.274	–0.016
211	–29.159	–30.434	–29.245	–1.189	0.086
212	–29.304	–30.648	–29.509	–1.139	0.205
213	–29.474	–30.836	–29.738	–1.098	0.264
214	–29.772	–31.041	–30.027	–1.014	0.255
215	–29.968	–31.272	–30.3	–0.972	0.332
216	–30.224	–31.444	–30.581	–0.863	0.357
217	–30.472	–31.581	–30.897	–0.684	0.425
218	–30.608	–31.786	–31.11	–0.676	0.502
219	–30.916	–31.726	–31.435	–0.291	0.519
220	–31.027	–31.512	–31.512	0.000	0.485
221	–31.206	–31.641	–31.811	0.170	0.605
222	–31.411	–31.864	–32.034	0.170	0.623
223	–31.583	–32.224	–32.248	0.024	0.665
224	–31.805	–32.593	–32.556	–0.037	0.751
225	–31.977	–32.842	–32.736	–0.106	0.759
226	–32.182	–33.057	–33.002	–0.055	0.820
227	–32.439	–33.418	–33.225	–0.193	0.786
228	–32.542	–33.685	–33.465	–0.220	0.923

EVENTS	T(i)	T(p)	T(r)	DT1	DT3
229	-32.868	-33.986	-33.68	-0.306	0.812
230	-33.066	-34.21	-33.912	-0.298	0.846
231	-33.212	-34.451	-34.144	-0.307	0.932
232	-33.386	-34.684	-34.385	-0.299	0.999
233	-33.745	-35.029	-34.738	-0.291	0.993
234	-34.037	-35.305	-34.987	-0.318	0.950
235	-34.15	-35.539	-35.211	-0.328	1.061
236	-34.39	-35.712	-35.479	-0.233	1.089
237	-34.58	-35.876	-35.686	-0.190	1.106
238	-34.744	-36.136	-35.945	-0.191	1.201
239	-34.934	-36.326	-36.238	-0.088	1.304
240	-35.175	-36.491	-36.506	0.015	1.331
241	-35.313	-36.69	-36.713	0.023	1.400
242	-35.564	-36.95	-36.93	-0.020	1.366
243	-35.831	-37.089	-37.19	0.101	1.359
244	-35.987	-37.254	-37.441	0.187	1.454
245	-36.169	-37.583	-37.727	0.144	1.558
246	-36.498	-37.896	-38.047	0.151	1.549
247	-36.792	-38.21	-38.429	0.219	1.637
248	-37.034	-38.435	-38.603	0.168	1.569
249	-37.216	-38.671	-38.846	0.175	1.630
250	-37.346	-38.871	-38.994	0.123	1.648
251	-37.442	-39.132	-39.212	0.080	1.770
252	-37.65	-39.446	-39.359	-0.087	1.709
253	-37.876	-39.594	-39.568	-0.026	1.692
254	-38.049	-39.803	-39.76	-0.043	1.711
255	-38.206	-40.022	-40.004	-0.018	1.798
256	-38.319	-40.223	-40.152	-0.071	· 1.833
257	-38.58	-40.406	-40.457	0.051	1.877
258	-38.745	-40.59	-40.666	0.076	1.921
259	-38.962	-40.808	-40.788	-0.020	1.826
260	-39.172	-41.115	-41.077	-0.038	1.905
261	-39.352	-41.324	-41.321	-0.003	1.969

EVENTS	T(i)	T(p)	T(r)	DT1	DT3
262	–39.537	–41.491	–41.54	0.049	2.003
263	–39.755	–41.692	–41.767	0.075	2.012
264	–39.947	–41.876	–42.021	0.145	2.074
265	–40.121	–42.148	–42.274	0.126	2.153
266	–40.244	–42.35	–42.529	0.179	2.285
267	–40.541	–42.526	–42.8	0.274	2.259
268	–40.862	–42.78	–43.037	0.257	2.175
269	–41.021	–42.895	–43.318	0.423	2.297
270	–41.275	–43.193	–43.493	0.300	2.218
271	–41.528	–43.404	–43.713	0.309	2.185
272	–41.669	–43.677	–43.88	0.203	2.211
273	–41.808	–43.923	–44.056	0.133	2.248
274	–42.028	–44.047	–44.266	0.219	2.238
275	–42.211	–44.276	–44.557	0.281	2.346
276	–42.396	–44.523	–44.663	0.140	2.267
277	–42.632	–44.814	–44.874	0.060	2.242
278	–42.878	–45.026	–45.156	0.130	2.278
279	–43.01	–45.308	–45.332	0.024	2.322
280	–43.124	–45.485	–45.579	0.094	2.455
281	–43.406	–45.679	–45.817	0.138	2.411
282	–43.59	–45.883	–46.135	0.252	2.545
283	–43.827	–46.069	–46.365	0.296	2.538
284	–44.074	–46.272	–46.621	0.349	2.547
285	–44.197	–46.502	–46.842	0.340	2.645
286	–44.409	–46.671	–47.063	0.392	2.654
287	–44.708	–46.83	–47.338	0.508	2.630
288	–44.902	–47.132	–47.621	0.489	2.719
289	–45.07	–47.344	–47.772	0.428	2.702
290	–45.273	–47.549	–47.931	0.382	2.658
291	–45.432	–47.797	–48.135	0.338	2.703
292	–45.643	–48.002	–48.384	0.382	2.741
293	–45.89	–48.179	–48.623	0.444	2.733
294	–46.006	–48.401	–48.854	0.453	2.848

EVENTS	T(i)	T(p)	T(r)	DT1	DT3
295	–46.262	–48.606	–49.076	0.470	2.814
296	–46.439	–48.828	–49.201	0.373	2.762
297	–46.616	–48.962	–49.423	0.461	2.807
298	–46.828	–49.113	–49.672	0.559	2.844
299	–47.023	–49.336	–49.904	0.568	2.881
300	–47.2	–49.479	–50.162	0.683	2.962
301	–47.413	–49.702	–50.385	0.683	2.972
302	–47.67	–49.817	–50.634	0.817	2.964
303	–47.83	–50.031	–50.786	0.755	2.956
304	–47.999	–50.263	–50.964	0.701	2.965
305	–48.221	–50.478	–51.125	0.647	2.904
306	–48.425	–50.701	–51.268	0.567	2.843
307	–48.576	–50.915	–51.527	0.612	2.951
308	–48.754	–51.085	–51.75	0.665	2.996
309	–48.941	–51.336	–51.857	0.521	2.916
310	–49.074	–51.381	–52.063	0.682	2.989
311	–49.323	–51.542	–52.278	0.736	2.955
312	–49.412	–51.757	–52.475	0.718	3.063
313	–49.697	–51.81	–52.689	0.879	2.992
314	–49.893	–51.972	–52.896	0.924	3.003
315	–50.063	–52.205	–53.173	0.968	3.110
316	–50.25	–52.339	–53.379	1.040	3.129
317	–50.429	–52.617	–53.505	0.888	3.076
318	–50.661	–52.85	–53.72	0.870	3.059
319	–50.83	–53.057	–53.846	0.789	3.016
320	–51.089	–53.272	–54.071	0.799	2.982
321	–51.286	–53.488	–54.295	0.807	3.009
322	–51.402	–53.479	–54.511	1.032	3.109
323	–51.554	–53.668	–54.754	1.086	3.200
324	–51.751	–53.839	–54.961	1.122	3.210
325	–51.984	–54.064	–55.177	1.113	3.193
326	–52.172	–54.271	–55.312	1.041	3.140
327	–52.387	–54.478	–55.456	0.978	3.069
328	–52.584	–54.703	–55.654	0.951	3.070

EVENTS	T(i)	T(p)	T(r)	DT1	DT3
329	–52.718	–54.703	–55.888	1.185	3.170
330	–52.925	–54.793	–56.105	1.312	3.180
331	–53.023	–55.01	–56.213	1.203	3.190
332	–53.185	–55.271	–56.276	1.005	3.091
333	–53.41	–55.487	–56.411	0.924	3.001
334	–53.536	–55.524	–56.682	1.158	3.146
335	–53.733	–55.677	–56.917	1.240	3.184
336	–54.012	–55.74	–57.152	1.412	3.140
337	–54.102	–55.948	–57.188	1.240	3.086
338	–54.209	–56.174	–57.234	1.060	3.025
339	–54.362	–56.418	–57.451	1.033	3.089
340	–54.596	–56.527	–57.685	1.158	3.089
341	–54.731	–56.744	–57.813	1.069	3.082
342	–55.038	–56.871	–58.057	1.186	3.019
343	–55.137	–56.925	–58.266	1.341	3.129
344	–55.281	–57.133	–58.257	1.124	2.976
345	–55.398	–57.115	–58.465	1.350	3.067
346	–55.606	–57.333	–58.646	1.313	3.040
347	–55.696	–57.541	–58.601	1.060	2.905
348	–55.804	–57.55	–58.818	1.268	3.014
349	–55.93	–57.65	–59.036	1.386	3.106
350	–56.02	–57.914	–59.045	1.131	3.025
351	–56.174	–57.677	–59.181	1.504	3.007
352	–56.273	–57.94	–59.263	1.323	2.990
353	–56.291	–58.149	–59.236	1.087	2.945
354	–56.499	–58.122	–59.463	1.341	2.964
355	–56.59	–58.339	–59.59	1.251	3.000
356	–56.78	–58.576	–59.663	1.087	2.883
357	–56.961	–58.503	–59.872	1.369	2.911
358	–57.178	–58.712	–60.063	1.351	2.885
359	–57.114	–58.839	–59.835	0.996	2.721
360	–57.051	–58.63	–59.917	1.287	2.866
361	–57.078	–58.721	–59.908	1.187	2.830

D2 &DTA Run: Lacsal — 04-10-1996 — TAJ — V3.8-SN009-8/31/95

D2&DTA System — RUN IDENTIFICATION — WARMING

```
SAMPLE NAME : LACSAL
DATA FILE NAME    : /DATA/LACSAL.A/LACSAL.1.W
DATE              : 04-10-1996
OPERATOR          : TAJ
SYSTEM ID         : V3.8-SN009-8/31/95
FILL VOLUME       : 2 ML
DTA REFERENCE     : METHANOL
WATER SAMPLE      : DI
NOTEBOOK NUMBER   : 3-86
NOTEBOOK PAGE     : 19
```

PARAMETER TABLE

```
RESISTANCE READ COUNT        :    5
TEMPERATURE READ COUNT       :    10
ELAPSED TIME LIMIT (HRS)     :    5
WARMING START TEMP. (C)      :    -60
WARMING STOP TEMP. (C)       :    20
WARMING EVENT WINDOW (C)     :    .2
```

GRID SPACE FOR DTA GRAPHS (DEGREES (C)/DIVISION)

```
(T(p) - T(r))/ REF TEMP      : 2
(T(p) - T(i))/ REF TEMP      : 2
(T(i) - T(r))/ REF TEMP      : 2
(T(p) - T(r))/ EVENTS        : 2
```

CALIBRATION — CORRECTION TEMPERATURES (C)

ICE	SAMPLE	REF	SHELF-SURFACE
.198	-.172	.752	0.044
BATH TEMP : 0			

STANDARD DEVIATION — DEGREE (C)

ICE	SAMPLE	REF	SHELF-SURFACE
.030	.036	.085	.042

TEMPERATURE CORRECTIONS DEGREES (C)

ICE	SAMPLE	REF	SHELF-SURFACE
-.198	.172	-.751	-.044

RESISTANCE CHECK (KILO-OHMS)		ICE (CELL)	SAMPLE (CELL)
SHORT	:	2.974×10^{-3}	6.447×10^{-3}
OPEN	:	8.293×10^{6}	9.164×10^{7}
CELL CONSTANT	:	9.851	10.809

Warming Tabular Data for 3% Lactose and 0.9% NaCl Solution

EVENTS	R(i)	R(s)	D2	T(i)	T(p)	T(r)	T(s)	DT1	DT3
0	528000	528000	1	–57.44	–59.021	–60.308	–65.689	1.287	2.868
1	528000	528000	1	–57.386	–58.794	–60.19	–64.429	1.396	2.804
2	528000	528000	1	–57.187	–58.685	–59.844	–63.667	1.159	2.657
3	528000	528000	1	–56.961	–58.385	–59.636	–63.034	1.251	2.675
4	528000	528000	1	–56.789	–58.194	–59.381	–62.494	1.187	2.592
5	528000	528000	1	–56.68	–58.130	–59.154	–62.237	1.024	2.474
6	528000	528000	1	–56.481	–57.922	–58.909	–61.716	0.987	2.428
7	528000	528000	1	–56.391	–57.822	–58.665	–61.442	0.843	2.274
8	528000	483194	1.09	–56.111	–57.514	–58.384	–60.986	0.870	2.273
9	528000	409632	1.29	–55.849	–57.306	–57.957	–60.421	0.651	2.108
10	528000	380694	1.39	–55.687	–57.079	–57.831	–60.147	0.752	2.144
11	528000	326776	1.62	–55.452	–56.862	–57.442	–59.665	0.580	1.99
12	528000	280422	1.88	–55.182	–56.545	–56.972	–59.164	0.427	1.79
13	528000	238164	2.22	–54.803	–56.174	–56.718	–58.547	0.544	1.915
14	528000	218058	2.42	–54.749	–55.975	–56.393	–58.311	0.418	1.644
15	528000	202292	2.61	–54.452	–55.722	–56.312	–58.029	0.590	1.86
16	528000	185736	2.84	–54.38	–55.632	–56.024	–57.749	0.392	1.644
17	528000	170887	3.09	–54.164	–55.442	–55.78	–57.431	0.338	1.616
18	528000	145506	3.63	–53.931	–55.055	–55.528	–57.06	0.473	1.597
19	528000	134729	3.92	–53.733	–54.955	–55.222	–56.762	0.267	1.489
20	528000	122692	4.30	–53.616	–54.721	–54.988	–56.499	0.267	1.372
21	528000	113596	4.65	–53.4	–54.514	–54.799	–56.228	0.285	1.399
22	528000	104602	5.05	–53.248	–54.361	–54.592	–55.93	0.231	1.344
23	528000	88194	5.99	–52.862	–53.992	–54.178	–55.389	0.186	1.316
24	528000	81319.6	6.49	–52.629	–53.830	–53.945	–55.127	0.115	1.316
25	528000	74482.6	7.09	–52.503	–53.596	–53.828	–54.884	0.232	1.325
26	528000	68585.4	7.70	–52.315	–53.380	–53.532	–54.604	0.152	1.217
27	528000	63191.4	8.36	–52.082	–53.165	–53.353	–54.334	0.188	1.271
28	528000	58017.8	9.10	–51.912	–52.994	–53.084	–54.1	0.090	1.172
29	528000	53498	9.87	–51.697	–52.770	–52.878	–53.867	0.108	1.181
30	528000	48877.8	10.80	–51.518	–52.608	–52.663	–53.552	0.055	1.145

EVENTS	R(i)	R(s)	D2	T(i)	T(p)	T(r)	T(s)	DT1	DT3
31	528000	45139	11.70	–51.304	–52.393	–52.466	–53.274	0.073	1.162
32	528000	41649	12.68	–51.17	–52.151	–52.144	–53.031	–0.007	0.974
33	528000	38471.2	13.72	–50.964	–51.981	–51.92	–52.762	–0.061	0.956
34	528000	32853.8	16.07	–50.553	–51.417	–51.679	–52.278	0.262	1.126
35	528000	30434.8 ·	17.35	–50.411	–51.238	–51.393	–52.045	0.155	0.982
36	528000	26143.2	20.20	–50.01	–50.924	–50.929	–51.472	0.005	0.919
37	528000	24302	21.73	–49.795	–50.799	–50.67	–51.257	–0.129	0.875
38	528000	22593	23.37	–49.67	–50.522	–50.474	–51.052	–0.048	0.804
39	528000	21066.8	25.06	–49.448	–50.397	–50.233	–50.685	–0.164	0.785
40	528000	19465.7	27.12	–49.199	–50.076	–50.082	–50.462	0.006	0.883
41	528000	18243.4	28.94	–49.039	–49.987	–49.815	–50.221	–0.172	0.776
42	528000	17081	30.91	–48.878	–49.737	–49.494	–49.936	–0.243	0.616
43	528000	15031.5	35.13	–48.487	–49.380	–49.183	–49.445	–0.197	0.696
44	528000	14131.8	37.36	–48.345	–49.104	–48.934	–49.196	–0.170	0.589
45	528000	12512.8	42.20	–47.866	–48.659	–48.508	–48.627	–0.151	0.642
46	528000	11801	44.74	–47.661	–48.526	–48.286	–48.44	–0.240	0.625
47	525716	11150	47.15	–47.493	–48.375	–47.949	–48.254	–0.426	0.456
48	518948	10525.5	49.30	–47.245	–48.144	–47.922	–47.933	–0.222	0.677
49	510098	9976.54	51.13	–47.121	–47.930	–47.648	–47.739	–0.282	0.527
50	502584	9360.5	53.69	–46.926	–47.726	–47.4	–47.428	–0.326	0.474
51	490656	8900.6	55.13	–46.651	–47.549	–47.134	–47.162	–0.415	0.483
52	486590	8431.8	57.71	–46.483	–47.309	–46.913	–46.985	–0.396	0.43
53	478710	8018.2	59.70	–46.306	–47.123	–46.683	–46.781	–0.440	0.377
54	471574	7607.1	61.99	–46.094	–46.883	–46.497	–46.418	–0.386	0.403
55	448362	6915.98	64.83	–45.696	–46.467	–46.17	–45.931	–0.297	0.474
56	439242	6568.5	66.87	–45.546	–46.184	–45.897	–45.719	–0.287	0.351
57	426828	6264.3	68.14	–45.237	–46.060	–45.659	–45.516	–0.401	0.422
58	419362	5985.9	70.06	–45.132	–45.812	–45.456	–45.26	–0.356	0.324
59	406726	5726.02	71.03	–44.875	–45.600	–45.191	–45.022	–0.409	0.316
60	401106	5409.98	74.14	–44.594	–45.459	–44.953	–44.757	–0.506	0.359
61	396840	2647.5	149.89	–44.506	–45.299	–44.723	–44.572	–0.576	0.217
62	384732	2535.34	151.75	–44.355	–45.114	–44.381	–44.334	–0.733	0.026
63	374308	2277.38	164.36	–43.898	–44.620	–44.047	43.894	–0.573	0.149

EVENTS	R(i)	R(s)	D2	T(i)	T(p)	T(r)	T(s)	DT1	DT3
64	365114	2160.06	169.03	–43.704	–44.514	–43.783	–43.603	–0.731	0.079
65	349480	1936.73	180.45	–43.3	–44.135	–43.406	–43.155	–0.729	0.106
66	345550	1838.18	187.98	–43.072	–43.862	–43.335	–42.936	–0.527	0.263
67	336194	1742.76	192.91	–42.922	–43.739	–43.09	–42.689	–0.649	0.168
68	331412	1657.69	199.92	–42.817	–43.563	–42.879	–42.523	–0.684	0.062
69	325686	1574.17	206.89	–42.668	–43.307	–42.642	–42.347	–0.665	–0.026
70	304070	1484.83	204.78	–42.334	–43.088	–42.494	–42.023	–0.594	0.16
71	300768	1415.65	212.46	–42.168	–42.912	–42.292	–41.821	–0.620	0.124
72	299004	1348.34	221.76	–41.966	–42.754	–42.073	–41.576	–0.681	0.107
73	287482	1286.2	223.51	–41.8	–42.508	–41.898	–41.392	–0.610	0.098
74	288358	1226.29	235.15	–41.703	–42.359	–41.566	–41.217	–0.793	–0.137
75	288760	1172.66	246.24	–41.546	–42.148	–41.391	–40.99	–0.757	–0.155
76	277012	1072.42	258.31	–41.065	–41.806	–41.059	–40.5	–0.747	–0.006
77	272510	1026.05	265.59	–40.916	–41.570	–40.718	–40.273	–0.852	–0.198
78	265554	982.696	270.23	–40.706	–41.324	–40.631	–40.168	–0.693	–0.075
79	256500	904.698	283.52	–40.261	–40.966	–40.273	–39.662	–0.693	0.012
80	249580	863.088	289.17	–40.069	–40.712	–40.108	–39.462	–0.604	0.039
81	246376	833.86	295.46	–39.955	–40.642	–39.795	–39.192	–0.847	–0.16
82	242436	803.832	301.60	–39.737	–40.406	–39.699	–39.018	–0.707	–0.038
83	239844	774.754	309.57	–39.572	–40.196	–39.411	–38.808	–0.785	–0.161
84	236666	746.896	316.87	–39.407	–39.961	–39.263	–38.634	–0.698	–0.144
85	222356	695.564	319.68	–38.98	–39.585	–38.846	–38.243	–0.739	–0.134
86	216410	673.512	321.32	–38.901	–39.358	–38.646	–37.983	–0.712	–0.255
87	211714	651.646	324.89	–38.71	–39.228	–38.429	–37.722	–0.799	–0.281
88	206606	631.118	327.37	–38.502	–39.106	–38.204	–37.531	–0.902	–0.298
89	203750	611.732	333.07	–38.293	–38.862	–38.099	–37.297	–0.763	–0.194
90	200502	588.234	340.85	–38.154	–38.688	–37.813	–37.149	–0.875	–0.341
91	195571	572.14	341.82	–37.91	–38.452	–37.666	–36.959	–0.786	–0.244
92	187680	538.136	348.76	–37.52	–38.183	–37.328	–36.491	–0.855	–0.192
93	183284	508.126	360.71	–37.26	–37.766	–36.921	–36.11	–0.845	–0.339
94	179972	495.576	363.16	–37.06	–37.618	–36.679	–35.833	–0.939	–0.381
95	173994	469.212	370.82	–36.662	–37.167	–36.411	–35.488	–0.756	–0.251
96	170805	458.224	372.75	–36.532	–37.019	–36.091	–35.272	–0.928	–0.441

EVENTS	R(i)	R(s)	D2	T(i)	T(p)	T(r)	T(s)	DT1	DT3
97	164190	436.578	376.08	-36.082	-36.759	-35.746	-34.78	-1.013	-0.336
98	154768	416.144	371.91	-35.9	-36.378	-35.332	-34.53	-1.046	-0.568
99	149194	406.62	366.91	-35.728	-36.119	-35.091	-34.366	-1.028	-0.637
100	145381	387.048	375.61	-35.27	-35.738	-34.755	-33.824	-0.983	-0.515
101	140196	370.946	377.94	-34.916	-35.331	-34.444	-33.445	-0.887	-0.472
102	135715	356.02	381.20	-34.58	-34.986	-34.007	-33.067	-0.979	-0.573
103	129715	341.94	379.35	-34.227	-34.607	-33.645	-32.654	-0.962	-0.582
104	127845	335.392	381.18	-34.055	-34.348	-33.396	-32.423	-0.952	-0.659
105	122881	322.272	381.30	-33.651	-34.124	-33.13	-31.986	-0.994	-0.521
106	118590	310.726	381.65	-33.238	-33.728	-32.762	-31.652	-0.966	-0.476
107	114872	298.974	384.22	-32.937	-33.383	-32.47	-31.258	-0.913	-0.467
108	112546	293.886	382.96	-32.791	-33.220	-32.162	-31.035	-1.058	-0.629
109	108991	283.078	385.02	-32.362	-32.850	-31.837	-30.642	-1.013	-0.525
110	106563	276.908	384.83	-32.225	-32.644	-31.666	-30.36	-0.978	-0.559
111	105072	272.568	385.49	-31.985	-32.507	-31.461	-30.267	-1.046	-0.524
112	103377	267.944	385.82	-31.891	-32.301	-31.213	-30.045	-1.088	-0.678
113	101382	263.816	384.29	-31.72	-32.052	-31.11	-29.9	-0.942	-0.61
114	99814.4	259.248	385.02	-31.625	-31.872	-30.812	-29.712	-1.060	-0.813
115	98264.4	250.135	392.85	-31.275	-31.623	-30.487	-29.26	-1.136	-0.788
116	96792.6	246.49	392.68	-31.087	-31.367	-30.368	-29.09	-0.999	-0.719
117	94973.8	241.502	393.26	-30.916	-31.195	-30.147	-29.005	-1.048	-0.769
118	92770.2	237.33	390.89	-30.659	-31.093	-29.908	-28.75	-1.185	-0.751
119	90476	233.84	386.91	-30.497	-30.896	-29.696	-28.58	-1.200	-0.801
120	88411	228.514	386.90	-30.369	-30.605	-29.594	-228.37	-1.011	-0.775
121	85258.6	221.24	385.37	-29.908	-30.264	-29.262	-27.9	-1.002	-0.646
122	84248	217.904	386.63	-29.781	-30.050	-29.108	-27.798	-0.942	-0.673
123	82765	213.596	387.48	-29.627	-29.948	-28.888	-27.544	-1.060	-0.739
124	81298.4	208.88	389.21	-29.431	-29.751	-28.642	-27.332	-1.109	-0.789
125	79803.6	204.418	390.39	-29.227	-29.487	-28.429	-27.137	-1.058	-0.798
126	77054	194.704	395.75	-28.912	-29.112	-28.192	-26.713	-0.920	-0.72
127	75974.2	190.814	398.16	-28.759	-28.967	-27.938	-26.544	-1.029	-0.821
128	74546.2	186.683	399.32	-28.546	-28.755	-27.76	-26.375	-0.995	-0.786
129	73221.2	182.58	401.04	-28.36	-28.653	-27.556	-26.18	-1.097	-0.804

EVENTS	R(i)	R(s)	D2	T(i)	T(p)	T(r)	T(s)	DT1	DT3
130	71854.6	179.184	401.01	–28.114	–28.432	–27.362	–26.054	–1.070	–0.752
131	70924	176.574	401.67	–27.944	–28.261	–27.15	–25.834	–1.111	–0.794
132	69991.4	174.677	400.69	–27.867	–28.049	–27.066	–25.606	–0.983	–0.801
133	68952.2	172.079	400.70	–27.698	–27.837	–26.863	–25.387	–0.974	–0.835
134	66714.6	176.059	378.93	–27.308	–27.438	–26.508	–25.032	–0.930	–0.8
135	65238.8	174.541	373.77	–27.104	–27.293	–26.288	–24.898	–1.005	–0.816
136	62553.6	171.626	364.48	–26.774	–26.929	–25.984	–24.417	–0.945	–0.79
137	61718.8	170.49	362.01	v26.563	–26.709	–25.757	–24.308	–0.952	–0.806
138	60059	152.888	392.83	–26.275	–26.387	–25.521	–23.946	–0.866	–0.754
139	59039.2	151.666	389.27	–26.089	–26.175	–25.352	–23.736	–0.823	–0.737
140	58117.6	122.833	473.14	–25.895	–26.006	–25.108	–23.526	–0.898	–0.787
141	56184.4	25.541	2199.77	–25.523	–25.778	–24.771	–23.207	–1.007	–0.752
142	55050	25.172	2186.95	–25.422	–25.499	–24.62	–23.056	–0.879	–0.802
143	52574.4	24.478	2147.82	–25.017	–25.229	–24.292	–22.678	–0.937	–0.725
144	50662.2	23.759	2132.34	–24.782	–24.934	–23.922	–22.326	–1.012	–0.86
145	49091.4	23.014	2133.11	–24.436	–24.656	–23.612	–22.033	–1.044	–0.824
146	47213	22.341	2113.29	–24.05	–24.293	–23.251	–21.673	–1.042	–0.799
147	45159.6	21.688	2082.24	–23.839	–23.923	–22.992	–21.339	–0.931	–0.847
148	43739.4	21.094	2073.55	–23.511	–23.662	–22.623	–20.905	–1.039	–0.888
149	43000.2	20.732	2074.10	–23.318	–23.435	–22.473	–20.888	–0.962	–0.845
150	41957	20.38	2058.73	–23.151	–23.326	–22.238	–20.738	–1.088	–0.913
151	40649.6	19.944	2038.19	–22.823	–23.066	–21.904	–20.321	–1.162	–0.919
152	40011.4	19.662	2034.96	–22.639	–22.814	–21.779	–20.179	–1.035	–0.86
153	38496	19.215	2003.43	–22.296	–22.579	–21.554	–19.804	–1.025	–0.742
154	37481.4	18.96	1976.87	–22.17	–22.344	–21.395	–19.688	–0.949	–0.775
155	36768.4	18.763	1959.62	–22.061	–22.135	–21.161	–19.546	–0.974	–0.9
156	35596	18.408	1933.72	–21.66	–21.859	–20.903	–19.205	–0.956	–0.757
157	34413.8	18.009	1910.92	–21.384	–21.474	–20.57	–18.873	–0.904	–0.814
158	33021	17.593	1876.94	–21.109	–21.239	–20.228	–18.532	–1.011	–0.881
159	32012.8	17.283	1852.27	–20.791	–20.822	–19.996	–18.267	–0.826	–0.795
160	31046	16.899	1837.15	–20.616	–20.722	–19.754	–18.076	–0.968	–0.862
161	29867.4	16.66	1792.76	–20.316	–20.488	–19.489	–17.678	–0.999	–0.827
162	28837.2	16.324	1766.55	–19.908	–20.162	–19.157	–17.364	–1.005	–0.751

EVENTS	R(i)	R(s)	D2	T(i)	T(p)	T(r)	T(s)	DT1	DT3
163	28388.6	16.149	1757.92	-19.808	-19.962	-19.04	-17.24	-0.922	-0.768
164	27387.8	15.943	1717.86	-19.55	-19.637	-18.692	-16.942	-0.945	-0.858
165	26432.8	15.702	1683.40	-19.217	-19.405	-18.361	-16.611	-1.044	-0.856
166	25473.2	15.413	1652.71	-18.918	-19.064	-18.013	-16.306	-1.051	-0.905
167	24700.6	15.21	1623.97	-18.603	-18.764	-17.674	-16	-1.090	-0.929
168	23873.4	15.074	1583.75	-18.338	-18.349	-17.517	-15.761	-0.832	-0.821
169	23043	14.701	1567.44	-18.023	-18.067	-17.228	-15.423	-0.839	-0.795
170	22545	14.378	1568.02	-17.84	-17.942	-16.988	-15.267	-0.954	-0.852
171	21992.2	14.384	1528.93	-17.501	-17.685	-16.675	-14.897	-1.010	-0.826
172	21702	14.21	1527.23	-17.377	-17.462	-16.502	-14.814	-0.960	-0.875
173	21001	13.978	1502.43	-17.054	-17.189	-16.524	-14.453	-0.665	-0.53
174	20594.8	13.779	1494.65	-16.922	-16.940	-16.09	-14.272	-0.850	-0.832
175	19913	13.557	1468.84	-16.575	-16.668	-15.769	-14.026	-0.899	-0.806
176	19203.9	13.138	1461.71	-16.286	-16.428	-15.489	-13.689	-0.939	-0.797
177	18808.4	13.029	1443.58	-16.187	-16.188	-15.284	-13.583	-0.904	-0.903
178	16828.7	12.986	1295.91	-15.849	-15.965	-14.971	-13.23	-0.994	-0.878
179	17422.8	12.66	1376.21	-15.61	-15.537	-14.741	-12.976	-0.796	-0.869
180	16861.7	12.309	1369.87	-15.305	-15.330	-14.397	-12.723	-0.933	-0.908
181	16424.2	12.045	1363.57	-14.993	-15.050	-14.151	-12.42	-0.899	-0.842
182	15862.4	11.86	1337.47	-14.672	-14.779	-13.823	-12.069	-0.956	-0.849
183	15394.2	11.58	1329.38	-14.368	-14.491	-13.528	-11.742	-0.963	-0.84
184	14913.3	11.353	1313.60	-14.106	-14.261	-13.25	-11.489	-1.011	-0.856
185	14371	11.244	1278.10	-13.81	-13.957	-12.915	-11.212	-1.042	-0.895
186	13915.3	10.9626	1269.34	-13.45	-13.670	-12.605	-10.927	-1.065	-0.845
187	13216.9	10.81	1222.65	-13.237	-13.432	-12.352	-10.569	-1.080	-0.885
188	13049.8	10.682	1221.66	-13.13	-13.219	-12.238	-10.455	-0.981	-0.892
189	12672.3	10.547	1201.51	-12.795	-12.940	-11.969	-10.228	-0.971	-0.826
190	12292.4	10.394	1182.64	-12.533	-12.694	-11.643	-9.943	-1.051	-0.89
191	11928.3	10.152	1174.97	-12.247	-12.425	-11.268	-9.586	-1.157	-0.979
192	11592.1	10.011	1157.94	-11.913	-12.114	-11.041	-9.399	-1.073	-0.872
193	11284.4	9.643	1170.22	-11.651	-11.844	-10.837	-9.124	-1.007	-0.814
194	10948.6	9.341	1172.10	-11.399	-11.510	-10.529	-8.84	-0.981	-0.87
195	10638.4	9.283	1146.01	-11.073	-11.289	-10.285	-8.832	-1.004	-0.788

EVENTS	R(i)	R(s)	D2	T(i)	T(p)	T(r)	T(s)	DT1	DT3
196	10334.3	9.046	1142.42	–10.772	–11.020	–10.009	–8.233	–1.011	–0.763
197	10035.5	8.881	1130.00	–10.52	–10.711	–9.709	–8.007	–1.002	–0.811
198	9241	8.711	1060.84	–10.292	–10.499	–9.483	–7.716	–1.016	–0.809
199	8923.9	8.54	1044.95	–10.016	–10.288	–9.183	–7.514	–1.105	–0.833
200	8594.54	8.468	1014.94	–9.699	–10.027	–8.892	–7.231	–1.135	–0.807
201	8341.32	8.236	1012.79	–9.423	–9.743	–8.601	–6.981	–1.142	–0.822
202	8122.94	8.161	995.34	–9.131	–9.499	–8.342	–6.707	–1.157	–0.789
203	7839.62	7.912	990.85	–8.872	–9.124	–8.124	–6.45	–1.000	–0.748
204	7723.54	7.839	985.27	–8.759	–9.061	–7.979	–6.353	–1.082	–0.78
205	7575.62	7.739	978.89	–8.549	–8.907	–7.745	–6.2	–1.162	–0.804
206	7299.3	7.702	947.71	–8.362	–8.745	–7.512	–5.935	–1.233	–0.85
207	7033.18	7.51	936.51	–8.047	–8.478	–7.27	–5.694	–1.208	–0.777
208	6781.7	7.499	904.35	–7.878	–8.186	–7.012	–5.405	–1.174	–0.866
209	6547	7.253	902.66	–7.514	–7.968	–6.715	–5.172	–1.253	–0.799
210	6289.08	7.203	873.12	–7.256	–7.628	–6.53	–4.963	–1.098	–0.726
211	6069.52	7.008	866.08	–7.03	–7.394	–6.168	–4.747	–1.226	–0.862
212	5859.88	6.866	853.46	–6.773	–7.241	–5.968	–4.491	–1.273	–0.805
213	5651.84	6.824	828.23	–6.563	–6.959	–5.711	–4.226	–1.248	–0.852
214	5308.2	6.815	778.90	–6.274	–6.813	–5.486	–3.963	–1.327	–0.788
215	2635.28	6.697	393.50	–5.992	–6.539	–5.23	–3.771	–1.309	–0.762
216	2515.64	6.629	379.49	–5.832	–6.411	–4.974	–3.523	–1.437	–0.858
217	2405.1	6.538	367.86	–5.551	–6.193	–4.742	–3.355	–1.451	–0.809
218	2243.8	6.321	354.98	–5.238	–5.961	–4.454	–3.028	–1.507	–0.784
219	2133.62	6.257	341.00	–5.045	–5.718	–4.206	–2.765	–1.512	–0.839
220	2023.48	6.203	326.21	–4.797	–5.526	–3.967	–2.55	–1.559	–0.83
221	1938.04	6.161	314.57	–4.58	–5.325	–3.767	–2.335	–1.558	–0.813
222	1842.56	6.046	304.76	–4.324	–5.124	–3.52	–2.12	–1.604	–0.804
223	1531.04	5.865	261.05	–3.812	–4.723	–2.907	–1.642	–1.816	–0.905
224	1499.03	5.915	253.43	–3.581	–4.595	–2.684	–1.412	–1.911	–0.897
225	1423.02	5.824	244.34	–3.349	–4.419	–2.477	–1.237	–1.942	–0.872
226	1309.84	5.856	223.67	–3.07	–4.138	–2.159	–0.888	–1.979	–0.911
227	1207.47	5.829	207.15	–2.711	–4.042	–1.881	–0.587	–2.161	–0.83
228	1142.85	5.835	195.86	–2.552	–3.802	–1.746	–0.373	–2.056	–0.806
229	1081.32	5.691	190.01	–2.393	–3.706	–1.54	–0.151	–2.166	–0.853

EVENTS	R(i)	R(s)	D2	T(i)	T(p)	T(r)	T(s)	DT1	DT3
230	993.16	5.603	177.26	-2.082	-3.546	-1.286	0.014	-2.260	-0.796
231	914.67	5.641	162.15	-1.812	-3.282	-0.961	0.212	-2.321	-0.851
232	824.898	5.623	146.70	-1.463	-3.099	-0.566	0.678	-2.533	-0.897
233	727.428	5.559	130.86	-1.137	-2.819	-0.265	1.23	-2.554	-0.872
234	676.65	5.532	·122.32	-0.899	-2.692	0.01	1.53	-2.702	-0.909
235	645.392	5.563	116.02	-0.82	-2.652	0.223	1.68	-2.875	-1.043
236	604.15	5.578	108.31	-0.685	-2.397	0.507	1.97	-2.904	v1.192
237	572.616	5.482	104.45	-0.447	-2.357	0.736	2.278	-3.093	-1.183
238	551.532	5.476	100.72	-0.384	-2.309	0.965	2.44	-3.274	-1.349
239	524.398	5.304	98.87	-0.392	-2.166	1.193	2.718	-3.359	-1.585
240	511.58	5.352	95.59	-0.281	-2.142	1.453	2.954	-3.595	-1.734
241	489.384	5.375	91.05	-0.337	-1.959	1.72	3.197	-3.679	-2.057
242	468.568	5.378	87.13	-0.273	-1.760	1.987	3.558	-3.747	-2.26
243	461.438	5.408	85.33	-0.17	-1.744	2.262	3.879	-4.006	-2.432
244	451.292	5.49	82.20	-0.155	-1.601	2.505	4.129	-4.106	-2.66
245	435.146	5.41	80.43	-0.083	-1.410	2.756	4.341	-4.166	-2.839
246	442.306	5.386	82.12	-0.091	-1.387	2.991	4.497	-4.378	-3.082
247	430.526	5.437	79.18	-0.012	-1.235	3.203	4.872	-4.438	-3.215
248	428.432	5.459	78.48	-0.02	-1.187	3.422	5.02	-4.609	-3.442
249	418.94	5.474	76.53	-0.004	-1.132	3.696	5.301	-4.828	-3.7
250	410.834	5.408	75.97	0.105	-1.093	3.938	5.527	-5.031	-3.833
251	402.92	·5.395	74.68	0.113	-0.949	4.149	5.815	-5.098	-4.036
252	394.15	5.441	72.44	0.216	-0.933	4.367	6.08	-5.300	-4.151
253	387.07	5.421	71.40	0.256	-0.791	4.609	6.314	-5.400	-4.353
254	380.726	5.42	70.24	0.264	-0.719	4.828	6.539	-5.547	-4.564
255	369.474	5.387	68.59	0.303	-0.609	5.108	6.943	-5.717	-4.805
256	360.484	5.386	66.93	0.287	-0.616	5.326	7.098	-5.942	-5.039
257	350.812	5.39	65.09	0.311	-0.529	5.567	7.277	-6.096	-5.256
258	340.572	5.351	63.65	0.303	-0.481	5.769	7.463	-6.250	-5.466
259	328.93	5.364	61.32	0.327	-0.370	5.979	7.718	-6.349	-5.652
260	317.07	5.33	59.49	0.366	-0.228	6.212	7.951	-6.440	-5.846
261	305.674	5.307	57.60	0.39	-0.085	6.437	8.144	-6.522	-6.047
262	296.108	5.274	56.14	0.414	0.025	6.701	8.407	-6.676	-6.287
263	284.684	5.269	54.03	0.485	0.120	6.941	8.755	-6.821	-6.456

EVENTS	R(i)	R(s)	D2	T(i)	T(p)	T(r)	T(s)	DT1	DT3
264	274.458	5.285	51.93	0.58	0.326	7.251	8.917	–6.925	–6.671
265	265.436	5.294	50.14	0.666	0.563	7.522	9.233	–6.959	–6.856
266	259.772	5.333	48.71	0.745	0.682	7.738	9.457	–7.056	–6.993
267	256.002	5.332	48.01	0.777	0.769	7.939	9.634	–7.170	–7.162
268	252.028	5.363	46.99	0.88	0.982	8.017	9.804	–7.035	–7.137
269	246.94	5.381	45.89	0.99	1.164	8.233	9.958	–7.069	–7.243
270	242.2	5.384	44.99	1.077	1.345	8.433	10.25	–7.088	–7.356
271	237.276	5.422	43.76	1.171	1.558	8.696	10.381	–7.138	–7.525
272	234.592	5.47	42.89	1.242	1.715	8.896	10.488	–7.181	–7.654
273	230.466	5.577	41.32	1.353	1.944	9.096	10.765	–7.152	–7.743
274	226.24	5.776	39.17	1.542	2.164	9.25	10.972	–7.086	–7.708
275	224.104	5.954	37.64	1.644	2.471	9.389	11.08	–6.918	–7.745
276	221.85	6.169	35.96	1.676	2.699	9.458	11.118	–6.759	–7.782
277	220.26	6.348	34.70	1.802	3.006	9.658	11.379	–6.652	–7.856
278	218.596	6.458	33.85	1.896	3.288	9.796	11.463	–6.508	–7.9
279	218.31	6.497	33.60	1.983	3.571	9.896	11.509	–6.325	–7.913
280	217.124	6.539	33.20	2.022	3.813	9.973	11.647	–6.160	–7.951
281	219.388	6.516	33.67	2.179	4.252	10.081	11.793	–5.829	–7.902
282	219.104	6.52	33.60	2.164	4.605	10.173	11.815	–5.569	–8.009
283	218.232	6.488	33.64	2.242	4.925	10.273	11.9	–5.348	–8.031
284	217.44	6.418	33.88	2.368	5.299	10.319	11.976	–5.020	–7.951
285	216.796	6.404	33.85	2.368	5.744	10.349	12.068	–4.605	–7.981
286	215.498	6.364	33.86	2.384	6.133	10.411	12.129	–4.278	–8.027
287	214.492	6.321	33.93	2.415	6.406	10.503	12.16	–4.097	–8.088
288	213.786	6.324	33.81	2.517	6.772	10.541	12.282	–3.770	–8.024
289	212.774	6.304	33.75	2.572	7.097	10.67	12.32	–3.573	–8.098
290	211.662	6.285	33.68	2.659	7.300	10.725	12.397	–3.425	–8.066
291	212.48	6.236	34.07	2.706	7.540	10.825	12.458	–3.285	–8.119
292	212.126	6.275	33.80	2.824	7.788	10.894	12.489	–3.106	–8.07
293	211.556	6.284	33.67	2.886	8.020	10.955	12.634	–2.935	–8.069
294	211.342	6.289	33.61	2.886	8.299	11.032	12.642	–2.733	–8.146
295	210.546	6.288	33.48	2.996	8.554	11.131	12.794	–2.577	–8.135
296	211.188	6.329	33.37	3.145	8.925	11.269	12.954	–2.344	–8.124
297	211.258	6.364	33.20	3.247	9.195	11.384	12.977	–2.189	–8.137

EVENTS	R(i)	R(s)	D2	T(i)	T(p)	T(r)	T(s)	DT1	DT3
298	210.396	6.382	32.97	3.318	9.419	11.437	12.993	-2.018	-8.119
299	209.326	6.526	32.08	3.482	9.682	11.621	13.084	-1.939	-8.139
300	207.648	6.673	31.12	3.498	9.882	11.674	13.267	-1.792	-8.176
301	207.132	7.995	25.91	3.647	10.113	11.804	13.389	-1.691	-8.157
302	206.478	10.643	·19.40	3.842	10.359	11.934	13.511	-1.575	-8.092
303	205.786	10.216	20.14	3.913	10.582	12.049	13.549	-1.467	-8.136
304	204.924	9.779	20.96	4.046	10.836	12.156	13.694	-1.320	-8.11
305	204.004	9.509	21.45	4.179	11.089	12.239	13.785	-1.150	-8.06
306	202.016	9.328	21.66	4.39	11.296	12.354	13.96	-1.058	-7.964
307	196.879	9.012	21.85	4.499	11.549	12.621	14.112	-1.072	-8.122
308	193.456	8.921	21.69	4.757	11.825	12.804	14.341	-0.979	-8.047
309	191.453	8.82	21.71	5.249	12.147	12.933	14.523	-0.786	-7.684
310	189.854	8.643	21.97	5.74	12.269	13.185	14.644	-0.916	-7.445
311	188.684	8.547	22.08	6.292	12.483	13.299	14.796	-0.816	-7.007
312	186.957	8.484	22.04	7.371	12.705	13.481	14.925	-0.776	-6.11
313'	185.449	8.469	21.90	8.709	12.942	13.641	15.115	-0.699	-4.932
314	183.937	8.161	22.54	9.742	13.209	13.846	15.41	-0.637	-4.104
315	229.064	8.544	26.81	10.627	13.453	14.028	15.471	-0.575	-3.401
316	169.404	8.339	20.31	11.087	13.667	14.15	15.653	-0.483	-3.063
317	186.808	8.174	22.85	11.692	13.712	14.415	15.789	-0.703	-2.723
318	183.638	8.359	21.97	12.083	13.941	14.574	16.024	-0.633	-2.491
319	180.983	8.315	21.77	12.74	14.154	14.771	16.258	-0.617	-2.031
320	178.685	8.217	21.75	13.228	14.451	15.006	16.417	-0.555	-1.778
321	177.734	8.193	21.69	13.586	14.595	15.21	16.636	-0.615	-1.624
322	177.185	8.137	21.78	13.943	14.876	15.346	16.764	-0.470	-1.403
323	176.712	8.06	21.92	14.263	14.960	15.596	17.021	-0.636	-1.333
324	176.44	7.987	22.09	14.597	15.164	15.762	17.194	-0.598	-1.165
325	176.236	7.912	22.27	14.84	15.354	15.989	17.329	-0.635	-1.149
326	175.64	8.072	21.76	15.166	15.627	16.14	17.488	-0.513	-0.974
327	172.943	7.412	23.33	15.446	15.862	16.373	17.751	-0.511	-0.927
328	151.175	7.273	20.79	15.84	16.233	16.705	17.932	-0.472	-0.865
329	173.54	7.28	23.84	16.051	16.391	16.938	18.037	-0.547	-0.887
330	173.43	7.098	24.43	16.187	16.573	16.984	18.037	-0.411	-0.797

APPENDIX D

Data for a D_2 and DTA Analysis of a 3% Lactose Solution

D2 &DTA Run: LACTOSE — 04-07-1996 — — V3.8-SN009-8/31/95

D2&DTA System — RUN IDENTIFICATION — WARMING

```
SAMPLE NAME : LACTOSE
DATA FILE NAME    : /DATA/LACTOSE.A/LACTOSE.1.W
DATE              : 04-07-1996
OPERATOR          : TAJ
SYSTEM ID         : V3.8-SN009-8/31/93
FILL VOLUME       : 2 ML
DTA REFERENCE     : METHANOL
WATER SAMPLE      : DI
NOTEBOOK NUMBER   : 3-86
NOTEBOOK PAGE     : 17
```

PARAMETER TABLE

```
RESISTANCE READ COUNT        : 5
TEMPERATURE READ COUNT  : 10
ELAPSED TIME LIMIT (HRS)     : 5
WARMING START TEMP. (C) : -60
WARMING STOP TEMP. (C)       : 20
WARMING EVENT WINDOW (C)     : .2
```

GRID SPACE FOR DTA GRAPHS (DEGREES (C)/DIVISION)

```
(T(p) - T(r))/ REF TEMP      : 2
(T(p) - T(i))/ REF TEMP      : 2
(T(i) - T(r))/ REF TEMP      : 2
(T(p) - T(r))/ EVENTS        : 2
```

CALIBRATION - CORRECTION TEMPERATURES (C)

ICE	SAMPLE	REF	SHELF-SURFACE
.445	-.109	.897	0.256
BATH TEMP	: 0		.

STANDARD DEVIATION - DEGREE (C)

ICE	SAMPLE	REF	SHELF-SURFACE
022	.041	.070	.050

TEMPERATURE CORRECTIONS DEGREES (C)

ICE	SAMPLE	REF	SHELF-SURFACE
-.445	.109	-.897	-256

RESISTANCE CHECK (KILO-OHMS)		ICE (CELL)	SAMPLE (CELL)
SHORT	:.	$2.73. \times 10^{-3}$	$7.44. \times 10^{-3}$
OPEN	:	9.164×10^{7}	9.164×10^{7}
CELL CONSTANT	:	14.975	14.296

Warming Tabular Data for 3% Lactose Solution

EVENT	R(i)	R(s)	D2	T(p)	EVENT	R(i)	R(s)	D2	T(p)
0	528000	528000	1	-59.947	31	528000	528000	1	-52.715
1	528000	528000	1	-59.738	32	528000	528000	1	-52.536
2	528000	528000	1	-59.638	33	528000	528000	1	-52.267
3	528000	528000	1	-59.401	34	528000	528000	1	-52.025
4	528000	528000	1	-59.092	35	528000	528000	1	-51.810
5	528000	528000	1	-58.892	36	528000	528000	1	-51.595
6	528000	528000	1	-58.574	37	528000	528000	1	-51.210
7	528000	528000	1	-58.329	38	528000	528000	1	-50.987
8	528000	528000	1	-58.029	39	528000	528000	1	-50.763
9	528000	528000	1	-57.803	40	528000	528000	1	-50.522
10	528000	528000	1	-57.613	41	528000	528000	1	-50.335
11	528000	528000	1	-57.323	42	528000	528000	1	-50.138
12	528000	528000	1	-57.042	43	528000	528000	1	-49.942
13	528000	528000	1	-56.698	44	528000	528000	1	-49.728
14	528000	528000	1	-56.481	45	528000	528000	1	-49.523
15	528000	528000	1	-56.264	46	528000	528000	1	-49.078
16	528000	528000	1	-55.893	47	528000	528000	1	-48.784
17	528000	528000	1	-55.857	48	528000	528000	1	-48.615
18	528000	528000	1	-55.631	49	528000	528000	1	-48.277
19	528000	528000	1	-55.316	50	528000	528000	1	-48.100
20	528000	528000	1	-55.171	51	528000	528000	1	-47.868
21	528000	528000	1	-54.955	52	528000	528000	1	-47.602
22	528000	528000	1	-54.748	53	528000	528000	1	-47.398
23	528000	528000	1	-54.541	54	528000	528000	1	-47.221
24	528000	528000	1	-54.333	55	528000	528000	1	-46.919
25	528000	528000	1	-54.189	56	528000	528000	1	-46.671
26	528000	528000	1	-53.758	57	528000	528000	1	-46.476
27	528000	528000	1	-53.560	58	528000	528000	1	-46.290
28	528000	528000	1	-53.371	59	528000	528000	1	-46.175
29	528000	528000	1	-53.164	60	528000	528000	1	-45.954
30	528000	528000	1	-52.904	61	528000	528000	1	-45.689

EVENT	R(i)	R(s)	D2	T(p)	EVENT	R(i)	R(s)	D2	T(p)
62	528000	528000	1	-45.565	95	364488	528000	0.69	-37.325
63	528000	528000	1	-45.344	96	357628	528000	0.68	-37.108
64	528000	528000	1	-45.115	97	347996	528000	0.66	-36.839
65	528000	528000	1	-44.868	98	338516	528000	0.64	-36.458
66	528000	528000	. 1	-44.727	99	331182	528000	0.63	-36.285
67	528000	528000	1	-44.497	100	325458	528000	0.62	-36.077
68	528000	528000	1	-44.074	101	319662	528000	0.61	-35.860
69	528000	528000	1	-43.748	102	308850	528000	0.58	-35.480
70	528000	528000	1	-43.528	103	303364	528000	0.57	-35.265
71	528000	528000	1	-43.308	104	299238	528000	0.57	-35.014
72	528000	528000	1	-43.097	105	291570	528000	0.55	-34.634
73	527902	528000	1	-42.834	106	280492	528000	0.53	-34.281
74	523886	528000	0.99	-42.527	107	277454	528000	0.53	-34.074
75	517044	528000	0.98	-42.290	108	271354	528000	0.51	-33.850
76	512820	528000	0.97	-42.018	109	266354	528000	0.50	-33.713
77	505640	528000	0.96	-41.851	110	260890	528000	0.49	-33.472
78	499184	528000	0.95	-41.720	111	256662	528000	0.49	-33.240
79	489520	528000	0.93	-41.422	112	251638	528000	0.48	-32.990
80	487286	528000	0.92	-41.273	113	242312	528000	0.46	-32.629
81	477176	528000	0.90	-41.063	114	233740	528000	0.44	-32.269
82	475474	528000	0.90	-40.844	115	230180	528000	0.44	-32.003
83	461038	528000	0.87	-40.433	116	222642	528000	0.42	-31.540
84	453392	528000	0.86	-40.215	117	213576	528000	0.40	-31.326
85	444004	528000	0.84	-40.032	118	208726	528000	0.40	-31.069
86	436082	528000	0.83	-39.630	119	202636	528000	0.38	-30.719
87	425524	528000	0.81	-39.447	120	197922	528000	0.37	-30.497
88	417888	528000	0.79	-39.238	121	191462	527236	0.36	-30.129
89	410020	528000	0.78	-39.072	122	187775	510062	0.37	-29.925
90	407600	528000	0.77	-38.820	123	184040	494254	0.37	-29.745
91	399200	528000	0.76	-38.585	124	177199	463054	0.38	-29.481
92	390654	528000	0.74	-38.193	125	173583	443970	0.39	-29.174
93	383646	528000	0.73	-37.993	126	170315	430880	0.40	-28.970
94	373658	528000	0.71	-37.567	127	163586	401544	0.41	-28.587

EVENT	R(i)	R(s)	D2	T(p)	EVENT	R(i)	R(s)	D2	T(p)
128	157305	375710	0.42	-28.247	161	59582.2	63125	0.94	-20.050
129	150980	348570	0.43	-28.052	162	58660.8	60672.8	0.97	-19.958
130	147650	336760	0.44	-27.806	163	56063.2	55979.4	1.00	-19.633
131	144951	325162	0.45	-27.593	164	53904.8	51477.4	1.05	-19.367
132	139909	301928	0.46	-27.296	165	51611.4	47413.8	1.09	-18.993
133	136667	291790	0.47	-27.084	166	49599.2	43689.8	1.14	-18.685
134	134294	281022	0.48	-26.949	167	48622.4	41900.8	1.16	-18.469
135	131327	272306	0.48	-26.839	168	46664	38633.2	1.21	-18.162
136	128797	262106	0.49	-26.627	169	44701.6	35602.2	1.26	-17.872
137	123965	243990	0.51	-26.246	170	42802	32666.8	1.31	-17.458
138	119160	227358	0.52	-26.035	171	41035	30106.4	1.36	-17.135
139	116771	218956	0.53	-25.798	172	39427	27736.4	1.42	-16.738
140	114127	211464	0.54	-25.654	173	38641.6	26636	1.45	-16.573
141	111947	203294	0.55	-25.452	174	37075	24560.2	1.51	-16.218
142	107406	189541	0.57	-25.080	175	35492.2	22663.2	1.57	-15.904
143	105431	182995	0.58	-24.853	176	34054.6	20927	1.63	-15.533
144	103055	176583	0.58	-24.701	177	32607.8	19336	1.69	-15.228
145	100977	170143	0.59	-24.448	178	31346.6	17848.1	1.76	-14.956
146	97263	158377	0.61	-24.086	179	30624	17182.7	1.78	-14.751
147	93328	146378	0.64	-23.750	180	29872.6	16404.5	1.82	-14.545
148	89835	135765	0.66	-23.422	181	28790.8	15193.4	1.89	-14.241
149	87793	130427	0.67	-23.287	182	27670.8	14073.8	1.97	-14.011
150	85869.8	125296	0.69	-23.036	183	26479.6	13008.7	2.04	-13.617
151	82737.2	116642	0.71	-22.633	184	25503	12071.6	2.11	-13.338
152	75486.8	102708	0.73	-22.080	185	24939	11636.6	2.14	-13.166
153	74033	97546.4	0.76	-21.929	186	23943.4	10780.2	2.22	-12.855
154	73219.2	94126.4	0.78	-21.787	187	22989.4	10014	2.30	-12.585
155	70770.4	87308.8	0.81	-21.477	188	22103.8	9317.56	2.37	-12.356
156	69486.8	83916.8	0.83	-21.310	189	21630	8970.98	2.41	-12.209
157	67882	80542.6	0.84	-21.068	190	20663.	48314.8	2.49	-11.899
158	66625.2	77529	0.86	-20.817	191	19796.7	7781.94	2.54	-11.645
159	63824.4	71576	0.89	-20.600	192	19057.9	7263.66	2.62	-11.327
160	62298.4	68602.6	0.91	-20.400	193	18272.8	6818.3	2.68	-11.001

EVENT	R(i)	R(s)	D2	T(p)	EVENT	R(i)	R(s)	D2	T(p)
194	17517	6416.84	2.73	–10.667	227	1646.8	336.966	4.89	–2.579
195	16735.8	6045.56	2.77	–10.415	228	1421.6	322.232	4.41	–2.308
196	16068.4	5677.1	2.83	–10.130	229	1222.4	307.614	3.97	–2.045
197	15451.7	5368.06	2.88	–9.871	230	1101.4	297.168	3.71	–1.918
198	14851.4	5060.96	2.93	–9.570	231	1005.2	290.564	3.46	–1.783
199	14493.3	4934.62	2.94	–9.367	232	852.14	277.41	3.07	–1.505
200	13943	2592.2	5.38	–9.010	233	796.1	271.092	2.94	–1.385
201	13426	2392.32	5.61	–8.800	234	729.33	263.668	2.77	–1.290
202	12938	2198.94	5.88	–8.419	235	692.03	260.33	2.66	–1.274
203	12417	2017.48	6.15	–8.192	236	663.04	258.696	2.56	–1.219
204	11913	1857.56	6.41	–7.893	237	638.17	262.698	2.43	–1.147
205	11425	1708.66	6.69	–7.602	238	618.74	280.544	2.21	–1.076
206	10995	1573.52	6.99	–7.336	239	592.31	294.77	2.01	–0.917
207	10604	1451.53	7.31	–7.094	240	573.34	291.118	1.97	–0.909
208	10182	1341.4	7.59	–6.852	241	540.91	286.364	1.89	–0.853
209	9798.7	1241.45	7.89	–6.586	242	532.05	281.48	1.89	–0.766
210	9396.4	1146.73	8.19	–6.328	243	512.71	273.634	1.87	–0.631
211	9050.5	1063.26	8.51	–6.087	244	499.64	265.81	1.88	–0.512
212	8716.4	984.282	8.86	–5.805	245	483.46	258.63	1.87	–0.417
213	8412.3	914.71	9.20	–5.709	246	466.38	251.842	1.85	–0.322
214	8075.6	853.512	9.46	–5.572	247	440.53	243.786	1.81	–0.187
215	7788.1	795.822	9.79	–5.275	248	424.21	237.516	1.79	–0.132
216	7361.7	714.306	10.31	–5.018	249	404.58	231.692	1.75	–0.005
217	7064.9	663.338	10.65	–4.842	250	385.826	225.354	1.71	0.018
218	6604.9	598.904	11.03	–4.489	251	365.22	218.422	1.67	–0.005
219	6344.1	555.324	11.42	–4.369	252	347.606	213.584	1.63	–0.092
220	6046.7	517.076	11.69	–4.129	253	334.368	209.47	1.60	–0.037
221	5752	482.704	11.92	–3.945	254	312.662	202.6	1.54	–0.021
222	5454.8	452.84	12.05	–3.753	255	289.168	194.834	1.48	0.010
223	4210.9	428.842	9.82	–3.537	256	280.06	190.58	1.47	0.003
224	2489.4	408.406	6.10	–3.257	257	269.106	186.534	1.44	–0.029
225	2213.8	383.054	5.78	–3.058	258	257.162	179.945	1.43	0.003
226	1815.5	288.35	6.30	–2.755	259	245.284	174.045	1.41	–0.021

EVENT	R(i)	R(s)	D2	T(p)	EVENT	R(i)	R(s)	D2	T(p)
260	237.57	168.173	1.41	0.090	293	163.01	152.734	1.07	7.470
261	231.428	164.359	1.41	0.185	294	162.496	152.386	1.07	7.796
262	226.734	161.09	1.41	0.264	295	162.367	151.995	1.07	8.129
263	221.24	158.737	1.39	0.319	296	161.828	151.442	1.07	8.477
264	218.17	155.773	1.40	0.382	297	161.025	151.051	1.07	8.794
265	214.26	153.014	1.40	0.446	298	160.58	150.836	1.06	9.110
266	210.054	148.632	1.41	0.548	299	159.145	150.47	1.06	9.511
267	203.684	145.483	1.40	0.588	300	158.68	149.599	1.06	9.928
268	200.83	143.347	1.40	0.651	301	158.94	149.482	1.06	10.174
269	197.526	142.148	1.39	0.746	302	159.08	149.035	1.07	10.566
270	194.063	140.515	1.38	0.912	303	158.75	148.752	1.07	10.858
271	193.346	140.205	1.38	1.085	304	158.35	148.021	1.07	11.165
272	188.574	139.282	1.35	1.299	305	156.98	147.593	1.06	11.602
273	186.068	137.845	1.35	1.464	306	156.44	146.665	1.07	12.069
274	183.113	137.211	1.33	1.716	307	154.89	146.549	1.06	12.345
275	182.68	137.401	1.33	1.945	308	154.22	146.091	1.06	12.696
276	180.521	137.042	1.32	2.268	309	153.38	145.232	1.06	13.048
277	178.121	137.361	1.30	2.590	310	152.28	144.881	1.05	13.277
278	177.416	136.351	1.30	2.865	311	152.09	144.83	1.05	13.482
279	176.67	136.311	1.30	3.195	312	151.43	144.419	1.05	13.795
280	176.265	136.361	1.29	3.493	313	150.66	143.292	1.05	14.168
281	174.929	135.041	1.30	3.759	314	150.02	142.795	1.05	14.480
282	173.081	134.483	1.29	4.018	315	150.45	137.749	1.09	14.799
283	173.49	133.598	1.30	4.370	316	149.74	137.99	1.09	15.004
284	171.775	131.896	1.30	4.652	317	148.91	135.804	1.10	15.383
285	170.702	131.017	1.30	4.886	318	148.46	134.619	1.10	15.610
286	170.557	130.945	1.30	5.206	319	146.33	113.384	1.29	15.852
287	168.37	150.583	1.12	5.573	320	146.88	122.307	1.20	16.132
288	167.637	153.138	1.09	5.799	321	146.7	24.454	6.00	16.382
289	165.925	155.093	1.07	6.266	322	146.43	23.706	6.18	16.586
290	164.349	153.518	1.07	6.523	323	146.3	22.901	6.39	16.835
291	163.956	153.293	1.07	6.795	324	146.3	22.165	6.60	16.979
292	163.684	153.108	1.07	7.136	325	146.3	21.474	6.81	17.182

EVENT	R(i)	R(s)	D2	T(p)
326	146.24	21.074	6.94	17.303
327	146.36	20.57	7.12	17.529
328	146.1	20.177	7.24	17.792
329	146.3	19.85	7.37	18.026
330	146.3	19.536	7.49	18.229
331	146.3	19.226	7.61	18.462
332	146.18	18.8251	7.77	18.462
333	139.57	19.239	7.25	18.589
334	147.64	18.135	8.14	18.747

7

The Freezing Process

INTRODUCTION

In 1909, Shackell [1] realized that in order to successfully lyophilize a formulation, it was necessary to freeze the substance prior to the start of the drying process. A cursory review of the scientific literature shows that many early and current investigators reporting results of a lyophilization process merely state that the product was first frozen to a given temperature but provide little or no information regarding the nature of the freezing process [2–14]. A number of those references cited are of investigators who have contributed to the advancement of lyophilization as a science. These references are cited to stress the point that even the most experienced investigators have at times overlooked the importance of the freezing process or have not provided the rationale for selecting a given freezing process.

While simple in concept, the freezing process will be shown to be perhaps the most complex and least understood step in the lyophilization process. Yet its effect on the drying process is paramount, and it can impact the properties of the final product. Perhaps the real problem with the freezing process lies in the fact that it is relatively easy to perform. All one needs for freezing is a cold (controlled or uncontrolled) environment to reduce the temperature of the formulation to obtain a matrix where there is separation of the solutes and solvent(s), to diminish the mobility of the water in the interstitial region of the matrix such that it approaches zero, and to provide a matrix structure that offers a minimal impedance to the flow of water vapor during the drying process.

It was shown in the previous chapter that freezing a solution of 0.9% NaCl does not produce complete separation of the NaCl salt and the water solvent. There was evidence that a portion of the salt and solvent solution formed an interstitial region whose properties were different from those of the eutectic mixture. The addition of 3% (wt/v) lactose to the isotonic solution produced an interstitial system that prevented the NaCl from reaching the concentration required to form an eutectic mixture.

It is the objective of this chapter to provide the reader with an awareness of the complexities that are involved in the formation of a frozen matrix that separates the solutes from the solvent, the nature of the interstitial region of the matrix, and various means for obtaining the optimal frozen matrix structure, which are dependent on the nature of the formulation.

FORMATION OF THE FROZEN MATRIX

The reader will be provided with general insight into the mechanisms that are involved in the formation of the frozen matrix, i.e., when the mobility of the water in the interstitial region approaches zero.

Formation of Ice Crystals

In general, the first step in the freezing process is the formation of ice crystals. The formation of ice crystals was discussed rather extensively in Chapter 3. Although Chapter 3 did touch on the effect of solutes on the ice structure (i.e., the formation of clathrates or gas hydrates) that can cause the structure of the ice to be cubic or, in the case of a substance like urea, to complex with water to such an extent that a completely amorphous system is formed, it is generally assumed by those working in the field that the ice crystals formed during the freezing process are relatively pure [15–17]. In addition, the dependency of the vapor pressure on temperature will obey the phase diagram shown in Figure 3.1. For pure ice to form, the solutes of the solution are pushed ahead of the freezing front and are not incorporated into the ice structure.

While it is reasonable to assume that relatively pure ice is formed during the freezing of a formulation, there is evidence to suggest that the ice crystals formed are not entirely pure. For example, Gross [18] showed that in an apparatus illustrated by Figure 7.1, a significant voltage was generated across a frozen portion of an electrode located just above the advancing ice front. The voltage generated is referred to as the *freezing potential*. With an apparatus like that shown in Figure 7.1, Gross was able to show that the freezing potential was not only dependent on the nature and concentration of the salt solution but also increased with the freezing rate. Freezing potentials as high as 30 V were achieved with an ammonium chloride (NH_4Cl) solution, while Calenicenco and Tacu [19] were able to achieve voltages as high as –130 V for a 2×10^{-5} N thiocyanate solution. The reason for the generation of the freezing potential is a disproportion of ions incorporated in the ice matrix that creates a solution with a nonuniform charge distribution. The importance of the freezing potential with respect to the lyophilization process is that ions can be incorporated into the ice crystals during the freezing process. The effect the incorporation of such ions would have on the vapor pressure of ice at temperatures $< 0°C$ is not known at this time. Perhaps as more calorimetric measurements are made during the primary drying process, which will be described in the next chapter, the effect of solutes on the properties of ice will be better understood.

Mushy System

I have observed at times that the final dried cake may have a *chimney-like structure* in the middle of the cake, like that shown in Figure 7.2. The formation of such a

Figure 7.1. An Apparatus Used to Determine the Freezing Potential of a Dilute Electrolytic Solution

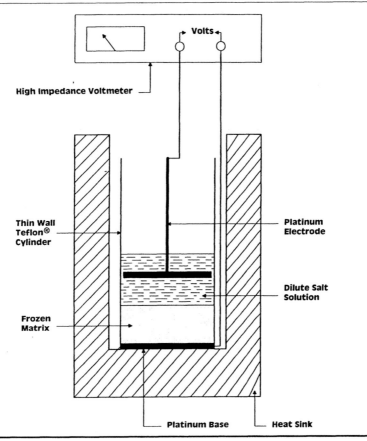

chimney occurs during the freezing process. As ice crystals grow from the bottom of the container, the system will take on what is sometimes referred to as a *mushy* state (i.e., a phase consisting of a mixture of solid and liquid phases [20–22]). The chimneys are formed as a result of two modes of convection, which are illustrated in Figure 7.3. One form of convection occurs in the liquid layer just above the mushy region, and the second occurs within the mushy region. The result of these convection currents is the formation of a chimney-like structure like that shown in Figure 7.2. While the composition of the formulation will no doubt be a factor in determining the formation of chimney-like structures, the only aqueous solution presently reported to form such structures during freezing is NH_4Cl [20–22].

Interstitial Region

In most formulations, the final frozen matrix will consist of an interstitial region composed of the solutes and uncrystallized water disbursed between the ice

Figure 7.2. Formation of a Chimney-like Structure in a Cake as a Result of the Freezing Process

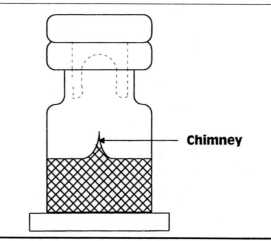

crystals. Of all the aspects of the lyophilization process, the interstitial region of the matrix is perhaps the least understood or the most unpredictable system. In the previous chapter it was shown that, from a 3% lactose solution, the concentration of lactose in the interstitial region increased to 81.5% while the NaCl in an isotonic solution increased to 13.7% in a system that contained 22% uncrystallized water. DTA analysis of the frozen matrix showed the interstitial region to be a solid solution or in a glassy state. Since such a system cannot be duplicated anywhere in nature, except as a result of the freezing process, the presence of ice crystals makes the study of such a region extremely difficult. For example, in the above interstitial region, it is not known if the composition of the glass is homogeneous or heterogeneous—is the composition of a cross-section of glass uniform or is there some variation in the composition? Examination of the dried product offers little or no insight because it is not known what effect removal of the water constituent will have on the distribution of the other constituents. It is this fundamental lack of knowledge of the nature of the interstitial region that presently prevents any accurate mathematical model form predicting the thermal properties of a formulation. We may be able to measure the thermal properties of a formulation, but one can offer little rationale as to why such thermal properties exist. The interstitial region of a frozen matrix remains one of nature's most closely guarded secrets.

The main emphasis of this text is concerned with the processes involved in the lyophilization of parenteral and biotechnology formulations. However, as I mentioned in Chapter 1, there are numerous other applications of lyophilization. Many of these systems are dependent not only on the final freezing temperature but also on the rate of freezing which will govern the size and distribution of ice crystals. The freezing of these and other systems has already been adequately described in previous publications and discussing the freezing of such a diverse range of systems is beyond the scope of this text. For those readers interested in the freezing of biological formulations, references [25–30] will provide a good introduction and offer guidelines regarding the freezing process. For those readers interested in the freezing of tissues, references [25,29–32] will prove a starting point. While there is not an abundance of information regarding the freezing of food products, the papers cited in [30] will be a good introduction. When considering the freezing of biological, tissue and food products, the reader should bear in mind that, regardless of the nature of the system, the interstitial region of the resulting matrix will tend to become elusive as the system becomes more complex. The freezing of tissue or a virus culture is far more complex than freezing an isotonic 3% lactose solution.

As a result of the freezing process, an interstitial region of a matrix may consist of one of the following systems.

Glassy State

The glassy state is the most common state found in the interstitial region as a result of freezing most pharmaceutical and biological formulations. As stated above, little is really known of the nature of the glassy state at this time; however, the one commonality is that its change from what may be called a solid solution to a liquid requires very little energy. The small amount of energy required to cause such a transition makes accurate determination of the onset temperature for the mobility of water in the interstitial a rather difficult task (see the previous chapter).

Figure 7.3. The Streamlines That Occur at the Liquid and "Mushy" Region During Freezing That Lead to the Formation of Chimneys

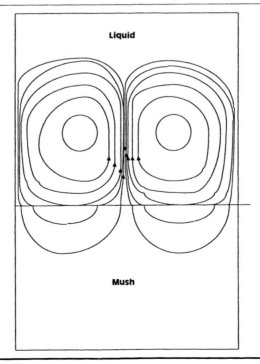

Crystalline and Glassy States

For some formulations it has been shown that it is possible to initialize crystallization by thermally treating the frozen matrix prior to the drying process [24]. It has been my experience to observe the eutectic temperature for sodium NaCl in a solution containing sucrose. However, the collapse temperature of such a system was significantly lower than the onset temperature for melting the eutectic mixture of NaCl. As the sucrose concentration was increased, the area under the eutectic peak diminished, which indicated that not all of the NaCl in the formulation was involved in the formation of the eutectic mixture. That portion not associated with the eutectic mixture was incorporated, along with the sucrose, into the glassy interstitial state. It is of interest to point out that the probable existence of an eutectic temperature for sucrose at temperatures higher than -25°C has been proposed [33]. I have not observed such an eutectic temperature for sucrose nor is it reported by any other investigator. Present information indicates that the addition of sucrose to a formulation tends to increase the formation of glassy states in the interstitial region.

Crystalline State

The data from the previous chapter indicate that dilute solutions of NaCl do not completely crystallize to form $NaCl \cdot 2H_2O$. There is some evidence, however, that some constituents, such as mannitol, regularly crystallize during the freezing process. Also, other constituents, such as inositol, have sometimes formed crystallized states [33]. From the above, the reader should be aware that complete crystallization of a constituent can occur, but such an occurrence must be considered atypical for most constituents in a lyophilized formulation.

Frozen Matrix

While it is important to obtain a completely frozen matrix prior to the start of the drying process, it is equally important that the nature of this matrix not impede the flow of water vapor during either the primary or secondary drying processes. The morphology of the frozen matrix depends not only on the composition of the formulation but also on the rate at which the formulation is frozen.

Examination of the frozen matrix is difficult because one cannot easily distinguish between ice crystals and the frozen interstitial region. If it is assumed that the degree of crystallization approaches 1 (i.e., all of the water in the formulation formed ice crystals during the freezing process), then after removal of the ice crystals by sublimation, the remaining structure will be representative of the interstitial region of the matrix when in the frozen state.

Using scanning electron micrographs, Dawson and Hockely [35] showed, that the morphology of the final dried product was dependent not only on the initial composition of the formulation but also on the freezing rate. These investigators showed that when a formulation containing 1% (wt/v) trehalose and 5 mg/mL of a protein was frozen at a slow rate of -1°C/min, a thick skin formed on the surface of the cake. This surface layer contained a relatively high concentration of the protein. It was also observed that most of the protein in the formulation was located in the upper 20% of the cake. Such a cake structure was illustrated in Figure 1.3b. Dawson and Hockely [35] found that such a thick skin impeded the flow of water vapor and that the moisture content of such dried products was relatively high. Rapid freezing (-150°C/min) of 1% (wt/v) trehalose produced a cake with a network that had a fine, filamentous structure and did not have a crust or glaze to impede the drying process.

What could be considered an ideal frozen matrix was proposed by MacKenzie [36] and is illustrated by Figure 7.4a. In this matrix, the ice crystals form channels throughout the matrix and allow for water vapor to flow from the matrix. In Figure 7.4b, the degree of crystallization is ≤ 0.5 and the ice crystals are immersed in a sea of glassy interstitial material. Under these conditions, the flow of water vapor through the resulting cake will be greatly impeded.

The preceding discussion confirms that the formation of a frozen matrix is a complex process and that the final matrix is dependent not only on the composition of the formulation but also on the freezing rate.

IMPACT OF THE FILL–VOLUME

An aspect of the freezing process that is often overlooked is the fill-volume. This section will consider the impact that the relative fill height to the diameter of the container has on the rate of freezing of a formulation. In this discussion, it will be assumed that freezing occurs in the absence of any significant degree of supercooling.

Heat-Transfer Rate

The expression for the heat-transfer rate (q) through a homogenous solid substance illustrated by Figure 7.5 can be defined as

---

Figure 7.4. A Frozen Matrix

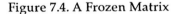

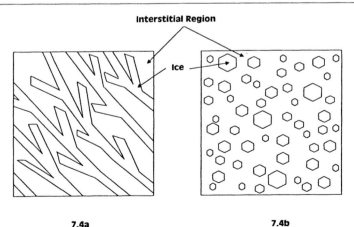

7.4a   7.4b

Figure 7.4a shows the formation of ice channels in an ideal frozen matrix. Figure 7.4b shows a matrix where the degree of crystallization is ≤ 0.5.

Figure 7.5. The Heat-Transfer Rate Through a Homogeneous Solid Substance

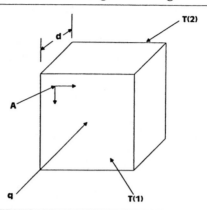

The heat-transfer rate (*q*) through a homogeneous solid substance having a cross-sectional area *A*, a thickness *d*, and a temperature differential $T(1) - T(2)$.

$$q = \frac{KA[T(1) - T(2)]}{d} \quad \frac{\text{cal}}{\text{sec}} \tag{1}$$

where *K* is the heat conductivity (cal/(cm-°C) of the substance (a bulk property that is independent the direction of the heat-transfer rate), *A* is the cross-sectional area of the solid in cm^2, $[T(1) - T(2)]$ is the temperature differential that exists between face (1) and face (2) where $T(1) > T(2)$, and *d* is the length of the conduction path in cm. Equation 1 shows that the heat-transfer rate is directly proportional to *K*, *A*, and $[T(1) - T(2)]$ but inversely proportional to *d*.

Heat Transfer During Freezing

The temperature profile during the formation of ice in a formulation is shown by Figure 7.6. Figure 7.6a shows the temperature profile across the system. $T(1)$ represents the temperature of the ice at the freezing surface, $T(2)$ is the temperature at the ice–mushy zone interface and $T(3)$ is the temperature of the liquid state. The physical system that corresponds to the temperature profile is illustrated by Figure 7.6b. In the latter figure, the cross-sectional area, *A* of the system is considered to be constant, while *d(i)* is the thickness if the ice layer, λ is the thickness of the mushy zone, and *d(l)* is the thickness of the liquid layer.

Figure 7.6a shows that the temperature profile across the ice layer varies inversely with the ice thickness, and there is a nonlinear temperature profile across the mushy zone. When $T(1) = T(2) = T(3)$, the system is in a state of equilibrium, and the rate of ice freezing will be zero. In order for ice growth to occur, a temperature profile of $T(1) < T(2) < T(3)$ must exist, and $T(1) < 0$ °C. The rate of freezing will be dependent of the rate of formation of the mushy zone.

The thickness of the mushy zone can be expressed as

Figure 7.6. Temperatures During the Freezing Process

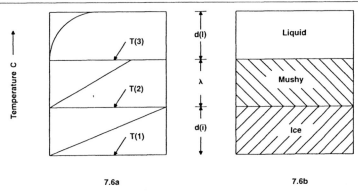

7.6a   7.6b

The temperature of system during the freezing process, where Figure 7.6a illustrates the temperature profile of the system. $T(1)$ represents the temperature of ice at the freezing surface, $T(2)$ is the temperature of the ice at the "mushy" interface, and $T(3)$ is the temperature of the liquid state. The solid lines indicate the temperature at some distance d within a given region. Figure 7.6b shows the corresponding physical states where $d(i)$ is the thickness of the ice layer, $d(l)$ is the thickness of the liquid layer, and λ is the thickness of the mushy region.

$$\lambda = \frac{\Delta H_o}{D' A \Delta H_f} \quad cm \tag{2}$$

where ΔH_o is the energy released to form the ice crystals, ΔH_f is the energy released to formed the ice crystals and is equivalent to the heat of fusion (79.7 cal/g), and D' is the density of the zone.

The rate $(d\lambda/dt)$ at which the mushy zone forms will be defined as

$$\frac{d\lambda}{dt} = \frac{1}{D' A \Delta H_f} \frac{d(\Delta H_o)}{dt} \tag{3}$$

where $d(\Delta H_o)/dt$ is the energy transport rate across the mushy zone. From equation 1, the rate of formation of the mushy zone can be expressed as

$$\frac{d\lambda}{dt} = \frac{K'[T(2) - T(3)]}{D' \lambda \Delta H_f} \quad \frac{cm}{sec} \tag{4}$$

where K' is the heat conductivity (cal/cm-°C) of the mushy zone and $d\lambda$ is the thickness of the mushy layer. Note that equation 4 shows that the rate of formation of the mushy zone is independent of the cross-sectional area of the system.

It would be reasonable to assume that $K > K'$ and that, during the freezing of a formulation, the mushy zone may extend throughout the entire fill volume prior to the formation of a completely frozen matrix. As a result, the rate of freezing will be directly related to the rate of formation of the mushy zone or

$$\frac{d\lambda}{dt} = \frac{K'[T(1) - T(3)]}{D' \lambda \Delta H_f} \quad \frac{cm}{sec} \tag{5}$$

The rate of formation of the mushy zone will decrease as the mushy zone thickness $d\lambda$ increases. From equation 5 it can be seen that the time required to freeze a formulation will be more dependent on the fill-height than the total fill-volume.

FREEZING FUNCTIONS AND METHODS

Introduction

The freezing function is defined as that process by which a formulation is converted from a homogeneous solution to a frozen matrix with thermal and physical properties suitable for lyophilization. The selection of the freezing function will be shown to be dependent on the nature of the formulation and on its thermal properties. As a result, the reader should become aware that the freezing function for one formulation may not be applicable for another formulation. In order to assist the reader in understanding the rationale for selecting a given freezing function, the freezing function for three hypothetical formulations will be described. Each of the formulations will have a 2 mL fill volume in a 5 mL glass tubing vial and freezing will be conducted on the shelf-surface of a freeze-dryer. The freezing function will be based on the observed thermal properties for each of the formulations. The thermal properties of each formulation are listed in Table 7.1. So that the reader can gain an overview of the entire lyophilization process, these same formulations will be used as examples

Table 7.1. Thermal Properties of Formulations #1, #2, and #3

Thermal Properties	#1	#2	#3
Stability of formulation at 1 atm and 4°C–8°C (hr)	≤ 2	72	96
Solid content (mg/mL)	20	35	80
Degree of supercooling (°C)	6	10	3
Equilibrium freezing temperature (°C)	–0.8	–1.3	–2.2
Freezing temperature (D2 $\geq$ 1)	–35	–40	–50
Onset temperature for interstitial phase change during freezing (°C)	none	none	–30
Collapse temperature (°C)	–23	–32	–45
Interstitial melting temperature (°C)	–12	–20	–22
Onset temperature for metastable state (°C)	–25	none	none
Onset temperature for interstitial phase change during warming (°C)	none	none	–21
Degree of crystallization	≈ 1	0.9	0.60
Activation energy (kcal/mole)	32	30	22

in later chapters that consider the primary and secondary drying processes and the properties of the lyophilized product.

Formulation #1

Formulation #1 contains a reducing agent and must be maintained under a nitrogen blanket during the filling operation. Consequently, exposure of the formulation to an atmospheric environment·after filling results in a reduction in the concentration of the reducing agent. As a result, the product stability, from Table 7.1, is less than 2 hr at temperatures ranging from 4°C to 8°C, and the product must be protected from oxidation by placing the vials directly on a cold shelf immediately after the filling operation. Since the collapse temperature (Tc) for this formulation was determined to be –23°C, a shelf-surface temperature $\leq$ –35°C would be sufficient to obtain a completely frozen matrix.

The thermal analysis of the formulation also indicated the presence of a metastable state that had an onset temperature of –25°C. It was determined that the metastable state was a result of the crystallization of the active constituent in the formulation. Increasing the matrix temperature to –20°C was necessary to maximize the crystallization of the active constituent.

The formulation matrix temperature is determined by means of a formulation matrix temperature sensor container, which is illustrated in Figure 7.7. This figure shows that the bare thermocouple junction is located in the middle of the container and approximately 1/16 in. to 1/8 in. from the bottom of the vial. The thermocouple junction consists of 3 to 4 twists of the wires coated with a thin layer of solder. It is recommended that a formulation matrix temperature sensor container be placed on each of the shelves of the dryer.

Figure 7.7. Formulation Matrix Temperature Sensor Container

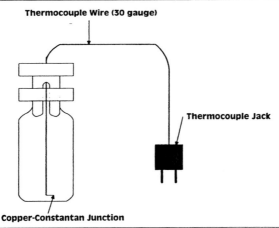

Formulation matrix temperature sensor container where the sensor is a Teflon®-insulated, bare copper-constantan thermocouple junction.

The degree of crystallization approached 1 which indicates that there is only a small fraction of uncrystallized water associated with the interstitial region of the matrix. The relatively high activation energy for conductivity in the interstitial region of the matrix and the relatively high interstitial melting temperature are indications that *Tc* is not sharp and well defined (i.e., there is relatively little change in the conductivity of the interstitial region with a temperature near the determined *Tc*). For these reasons, the formulation matrix can be considered to be in a completely frozen state at ≤ -28°C.

The freezing function for this formulation, as illustrated in Figure 7.8, must take into account the instability of the formulation at 4°C to 8°C under atmospheric pressure and the thermal properties of the formulation as listed in Table 7.1. This figure shows that the shelf surface was prechilled to ≤ -35°C. Trays of vials are loaded directly onto the shelf after filling to freeze the formulation in order to reduce the risk of oxidizing the reducing agent in the formulation. While the prechilled shelves provide a means for freezing the formulation, the shelves will tend to develop a coating of frost that can pose a problem at the onset of the primary drying process. Since 1 mm of frost is equivalent in thermal insulation to 1 cm of ice, the presence of frost on the shelf surface can affect the heat transfer between the shelf-surface and the vials and thus affect the rate of cooling of the vials.

Figure 7.8. Freezing Function for Formulation #1

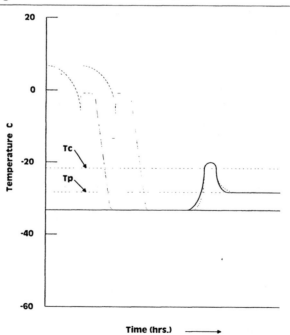

The solid line represents the shelf-surface temperature, the dashed line indicates the formulation matrix temperature, *Tc* is the collapse temperature, and *Tp* is the target product temperature during primary drying.

After all of the formulation matrix temperature sensors have reached and maintained -35°C ± 3°C for some period of time, e.g., no less than 2 hr, the shelf-surface temperature is increased to -20°C ± 2°C to remove metastable states. When all of the formulation matrix temperature sensors have reached and maintained -20°C ± 2°C for a specified period of time (1/2 hr), the shelf-surface temperature is lowered to -28°C ± 2°C, which will be the target product temperature (T_p) during the primary drying process. Thus the freezing function serves to provide not only the conditions for freezing the formulation and the conditions necessary for creating a desirable formulation matrix but also establishes the target T_p for the primary drying process.

The results of freezing for formulation #1 are shown in Figure 7.9. The heterogeneous frozen matrix of formulation #1 resulting from the freezing function, illustrated in Figure 7.8, shows the presence of three distinct regions. The fine structure at the base of the vial is a result of the rapid growth of ice crystals caused by the supercooling of the formulation. The relatively low volume of this region is the result of a low degree of supercooling of the formulation (6°C). When the vial was placed directly on the prechilled shelf-surface, only a small volume of the formulation was cooled below the equilibrium freezing temperature of -0.8°C prior to the onset of the nucleation of ice crystals.

The middle region of the matrix was that volume of liquid whose temperature was equal to or higher than that of the equilibrium freezing temperature; thus no ice crystals were formed when ice nucleation occurred in the lower region of the matrix. The coarse structure of this region was formed as a result of the growth of larger ice crystals. These ice crystals were formed near the equilibrium freezing temperature.

The thin upper region of the matrix was formed by excipients and the active ingredient being concentrated at the top of the matrix by the growth of large ice crystals in the middle region of the matrix. Such a layer is undesirable because it may cause a loss in potency in the active constituent or the possible exposure of the reducing excipient to the atmosphere prior to the commencement of primary drying. In addition, the layer may cause the formation of a glaze that may serve as a vapor

---

Figure 7.9. Frozen Matrix of Formulation #1

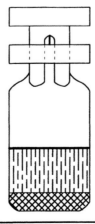

barrier and impede the vapor flow from the matrix during the primary and secondary drying processes.

Formulation #2

With a stability at 4°C to 8°C of 72 hr, it will not be necessary to prechill the shelves of the dryer to temperatures lower than 0°C. Assuming that the filling operation is performed at ambient temperature, the trays of vials can be loaded onto shelves refrigerated to maintain a surface temperature between 4°C to 8°C. At this temperature range, there will be no frost buildup to effect heat transfer to the vials or to present a problem during the start of primary drying.

Once all of the formulation matrix temperature sensors indicate that all of the formulation in the vials is at 6°C ± 2°C, then the shelf-surface temperature can be lowered at a rate that will cause the entire fill volume (2 mL) to be significantly colder than the equilibrium freezing temperature (-1.3°C) prior to the nucleation of the ice crystals. With a degree of supercooling of 10°C, the temperature of the entire volume of the formulation can be lowered below the equilibrium freezing temperature prior to the nucleation of the ice crystals. Under these conditions, there will be little or no formation of a layer of excipients and active constituent on the top surface of the matrix.

The freezing function for formulation #2 is shown by Figure 7.10. This figure shows that, because of the stability of the formulation and the thermal properties listed in Table 7.1, the freezing function is not as involved as formulation #1. The shelf-surface temperature was initially set at 6°C ± 2°C and maintained at this temperature until all of the formulation matrix temperature sensors registered between 4°C and 8°C for no less than 1 hr. The shelf-surface temperature was then lowered at a slow rate, e.g., -20°C/hr. The temperature was lowered until the shelf-surface temperature reached -42°C ± 2°C. The shelf-surface temperature was maintained at this temperature until all of the formulation matrix temperature sensors indicated temperatures of -42°C ± 2°C. The vials were maintained at this temperature for an additional period of time ($\approx$ 1 hr) to ensure complete freezing of the entire batch.

The shelf-surface temperature was then increased to -37°C ± 2°C and maintained at this temperature until commencement of the primary drying process. With a degree of crystallization of 0.9, there will be adequate separation of solutes and solvent so that there will be no need for any thermal treatment of the matrix prior to obtaining the T_p for primary drying. Since the activation energy for this formulation was 30 kcal/mole and the Tc of -32°C is not sharp and well defined, selecting a T_p of -37°C ± 3°C will ensure that the matrix will be in a completely frozen state during the primary drying process.

The frozen matrix formed by the freezing function is illustrated by Figure 7.11. Because the entire volume of the formulation was significantly below the equilibrium freezing temperature when nucleation of the ice crystals occurred, ice growth—dendritic in nature—extended throughout the entire fill volume. The formation of this ice structure served as a preliminary frozen structure that confined any unfrozen formulation to discrete regions in the fill volume. As the temperature was lowered, additional ice formed without the confined liquid phase. At temperatures approaching -40°C, the mobility of the water in the interstitial region had approached 0, and the matrix of the formulation had reached a completely frozen state. The formation

Figure 7.10. Freezing Function for Formulation #2

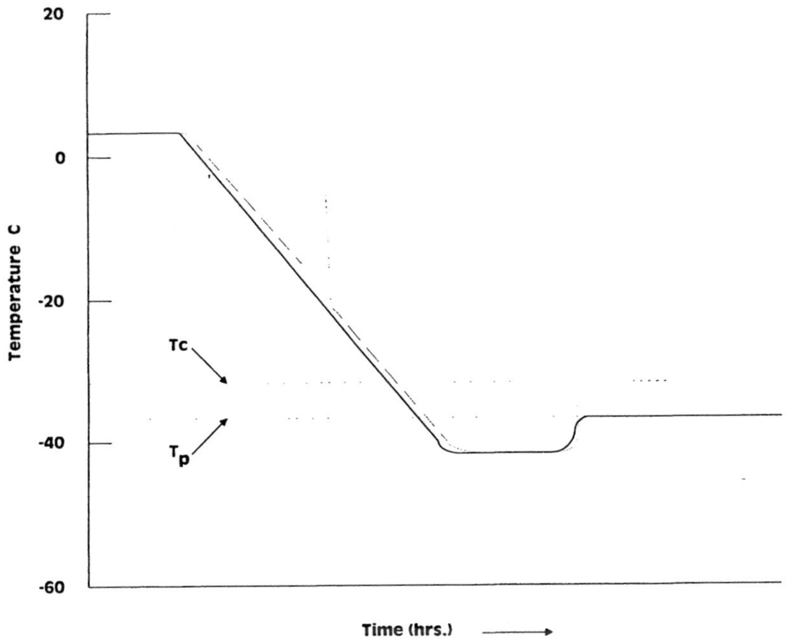

The solid line represents the shelf-surface temperature, the dashed line indicates the formulation matrix temperature, Tc is the collapse temperature, and T_p is the target product temperature during primary drying.

of the preliminary frozen structure served to form a homogeneous matrix like that illustrated in Figure 7.11, not the heterogeneous matrix shown by Figure 7.9.

This freezing function provides a means for forming an uniform frozen matrix, adequately separating the solutes and the solvent, and establishing the necessary T_p for the primary drying process.

Formulation #3

A review of the thermal properties that are listed in Table 7.1 indicates that obtaining an acceptable frozen matrix for this formulation will not be an easy task. For example, the degree of supercooling is only 3°C, while the equilibrium freezing temperature is -2.2°C. As a result of these thermal properties, the nucleation of ice crystals in the formulation will occur between -5°C and -6°C. Because of these properties, it is very unlikely that one can generate a freezing function that will result in a homogeneous matrix like that shown in Figure 7.11, and the matrix will probably be similar to that shown in Figure 7.9.

The degree of crystallization for this formulation was found to be 0.6 and the matrix structure would approach that shown in Figure 7.4b. Since there were no observed metastable states present, it is doubtful if any additional thermal treatment will increase the degree of crystallization.

Figure 7.11. Frozen Matrix of Formulation #2

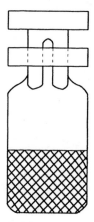

The low activation energy of 22 kcal/mole would signify that the observed Tc is rather sharp and well defined. This conclusion is supported by the relatively low interstitial melting temperature of –28°C. Although there is a phase change in the interstitial region of the matrix with an onset temperature of –21°C, which suggests the presence of a NaCl eutectic mixture, thermal analysis indicates that liquid states are already present in the interstitial region prior to the melting of the eutectic mixture. From this example, it can be seen that one must be careful in determining the thermal properties of a formulation, and one must not rely solely on a measurement of an eutectic mixture for determining the proper freezing temperature of a formulation.

The freezing function for formulation #3 is illustrated by Figure 7.12. The freezing function for this formulation is quite similar to that for formulation #2, except that a shelf-surface temperature of –55°C is required to achieve a completely frozen matrix. It should be noted that it is difficult to achieve a shelf-surface temperature of –55°C with most production freeze-dryers. For those dryers that can achieve such temperatures, the time required to reach such a temperature would be significantly longer than that for formulations #1 and #2.

Other Freezing Methods

While the above shelf-surface freezing method is perhaps the most widely used method for obtaining a frozen matrix, the reader should be aware of two other freezing techniques. Although these methods have only limited applications today, they are a viable means for freezing a formulation.

Figure 7.12. Freezing Function for Formulation #3

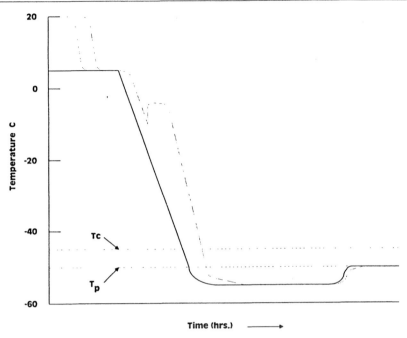

The solid line represents the shelf-surface temperature, the dashed line indicates the formulation matrix temperature, Tc is the collapse temperature, and T_p is the target product temperature during primary drying.

Immersion Freezing

Immersion freezing is a freezing technique in which the heat-transfer media, a liquid or gas, is in contact with both the bottom and the walls of the container. Perhaps the most successful application for this form of freezing was used during World War II in the drying of blood plasma.

Greaves [37] provides an excellent account of the historical development of this technique. The basic problem was the need to dry 400 mL of blood plasma contained in a 500 mL bottle. From equation 5, the rate of formation of the mushy zone ($d\lambda/dt$) is inversely proportional to the thickness of the mushy zone. Consequently, the time to freeze such a large volume was excessively long. Not only did the freezing of the plasma require a long period of time, but the fill-height also made the drying process excessively long.

The problem was overcome by spinning the bottles on their sides in the presence of a cold blast of air. With the bottles turning at 759 rpm, centrifugal force distributed the liquid around the inside periphery of the container to form a shell similar to that shown in Figure 7.13a. The high speed was necessary because at lower speeds the ice crystals formed during supercooling would form a lump, and

Figure 7.13. Frozen Matrices Formed by Immersion Freezing

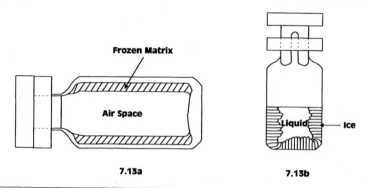

Figure 7.13a is the matrix formed by spin freezing and Figure 7.13b is the matrix formed by immersion in a cold gas or liquid environment.

the resulting matrix would become nonuniform in nature. As seen by Figure 7.13a, the high-speed spinning reduced the thickness of the mushy region, and thereby reduced the time required to obtain a completely frozen matrix. With the thinner shell and a hollow center, drying of the matrix was also faster. This technique is still used today, but, in place of the cold air, a liquid refrigerant is used, and the bottles are turned by means of a set of rollers.

When a container is immersed in a cold gas or liquid, ice will form, as shown in Figure 7.13b. The formation of ice crystals normal to the walls of the container results in the formation of layers of interstitial material perpendicular to the flow of the water vapor from the container. The presence of such layers would tend to impede vapor flow from the container during both the primary and secondary drying processes.

Snap Freezing

Snap freezing [38] is a process in which the formulation is frozen by placing the product in a chamber and then evacuating the chamber. The heat of evaporation and the heat of sublimation reduce the temperature of the formulation to form ice crystals and then reduce the temperature of the matrix to less than the T_c. From Figure 7.14, it can be seen that ice formation will occur at the top of the formulation rather than at the base of the container, as depicted in Figure 7.6.

This method of freezing may be applicable for thin layers of a formulation on a glass or ceramic surface, where there is minimal heat transfer to the film during the freezing process. The pressure should be reduced slowly to degas the formulation to prevent any frothing. Once degassing is complete, the pressure in the chamber should be reduced quickly to generate the ice phase and complete the freezing process.

Figure 7.14. "Snap Freezing."

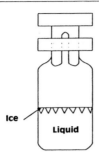

Ice

Liquid

The principal difficulty with this technique is that the evaporation of water will result in a change in the concentration of the constituents in the formulation. Such a change in concentration may lead to chemical reactions involving the active constituent and the other constituents. In order to reduce the temperature to less than the Tc, sublimation of ice must occur. Since this sublimation of ice, or primary drying, will occur at temperatures higher than the Tc, the surface layer may collapse or meltback. Any active constituent in this upper layer may be denatured and form a skin or crust that would impede the flow of water vapor from lower regions of the film.

While snap freezing may find some application in freezing thin films of a formulation with a relatively high Tc, it is my opinion that it would not be applicable for freezing formulations contained in vials like those described in formulations #1, #2, and #3. Since the shelves of the dryer are assumed to be at ambient temperature, the heat transfer to the vials would be sufficient to prevent the formation of a frozen matrix like that depicted in Figures 7.9 and 7.11. The resulting matrix would most likely exhibit a reduction in volume and a nonuniform structure containing a skin, crust, or frothy surface. Although this method of freezing has been tried from time to time, I am not aware of any successful frozen matrix obtained for fill volumes greater than 1 mL by this technique.

SYMBOLS

A	cross-sectional area of the solid (cm^2)
d	length of the thermal conduction path (cm)
$d(i)$	thickness of the ice layer (cm)
$d(l)$	thickness of the liquid layer (cm)
$d\lambda$	thickness of the mushy zone (cm)
$d(\Delta H_0)/dt$	energy transport rate across the mushy zone

$(d\lambda/dt)$	rate at which the mushy zone forms
D'	density of the ice
ΔH_f	energy released to form ice crystals and is equivalent to the heat of fusion (79.7 cal/g)
ΔH_o	energy released to form ice crystals
K	heat conductivity of a solid (cal/cm-°C)
K'	heat conductivity of the mushy layer (cal/cm-°C)
λ	thickness of the mushy zone (cm)
q	heat-transfer rate expressed in energy per unit time
$T(1)$	temperature of face 1 of a solid (°C)
$T(2)$	temperature of face 2 of a solid (°C)
$T(3)$	liquid phase temperature (°C)
Tc	collapse temperature (°C)
T_p	product temperature during primary drying (°C)

REFERENCES

1. L. F. Shackell, *Am J. Physiol.*, 24: 325 (1909).

2. P. W. Muggleton, in *Aspects Theoriques et Indusrial de la Lyophilization* (L. Rey, ed.), Herman, Paris, 1964, p. 411.

3. J. A. Bashir and K. E. Avis, *Bull Parenter Drug Assoc.*, 27: 63 (1973).

4. J. M. Barbaree, in International Symposium on freeze-drying of biological products, Washington, D.C. 1076. *Develop. Biol. Standard*, 36: S. Karger, Basel, 1977, p. 263.

5. A. Fauvel, P. Labrude, and C. Vigneron, *J. Pharm Sci.*, 73: (1984).

6. H. Seager, C. B. Taskis, M. Syrop, and T. J. Lee, *J. Parenter Sci. & Tech.*, 39: 161 (1985).

7. Y. Koyama, M. Kamat, R. J. De Angelis, R. Spinivasan, and P. P. DeLuca, loc. cit., 42: 47 (1988).

8. D. J. Korey and J. B. Schwartz, loc. cit., 43: 80 (1989).

9. P. P. DeLuca, M. S. Kamat, and Y. Koida, *Cong. Int. Technol. Pharm*, 5th, 1: 439 (1989).

10. D. Greiff, in International Symposium on Freeze-Drying and Formulation Bethesda, USA, 1990, *Develop. Biol. Standard*, 74: S. Karger, Basel, 1991, p. 85.

11. S. Vemuri, Ibid. p. 341.

12. C. Chiang, D. Fessler, R. Thirucote, and P. Tyle, *J. Parenter Sci. & Tech.*, 48: 24 (1994).

13. K. A. Kagkadis, D. M. Rekkas, P. P. Dallas, and N. H. Choulis, loc. cit., 50: 317 (1996).

14. N. Milton, M. J. Pikal, M. L. Roy, and S. L. Nail, loc. cit., 51: 7 (1997).

15. L. Rey, *Proc. R. Soc. Lond. B.*, 191: 9 (1975).

16. L. R. Rey, in International Symposium on freeze-drying of biological products, Washington, D.C. 1076. *Develop. Biol. Standard,* 36: S. Karger, Basel, 1977, p. 19.

17. T. A. Jennings, *J. Parenter Sci. & Tech.,* 42: 118 (1988).

18. G. W. Gross, *J. Colloid & Interface Sci.,* 25: 270 (1967).

19. N. Calenicenco and G. Tacu, *Bull. Inst. Poletechnic Iasi,* 6: 90 (1960).

20. M. G. Worster, *J. Fluid Mech.,* 237: 649 (1992)

21. S. Tait, K. Jahrling, and C. Jaupart, *Nature,* 359: 406 (1992).

22. S. Tait and C. Jaupart, *J. Geophys. Research,* 97: 6735 (1992).

23. A. J. Phillips, R. J. Yarwood and J. H. Collett, *Analytical Proceedings,* 23: 394 (1986).

24. L. Gatlin and P. P. DeLuca, *J. Parenter. Drug Assoc.,* 34: 398 (1980).

25. H. T. Meryman, in *Cryobiology* (H. T. Meryman, ed.), Academic Press, New York, 1966.

26. J. L. Stephenson, in *Recent Research in Freezing and Drying* (A. S. Parkes and A. Smith, eds.), Blackwell Scientific Publications, Oxford, England, 1960.

27. International Symposium on freeze-drying of biological products, Washington, D.C. 1976. *Develop. Biol. Standard,* 36: S. Karger, Basel, 1977.

28. International Symposium on Freeze-Drying and Formulation, Bethesda, USA, 1990, *Develop. Biol. Standard,* 74: S. Karger, Basel, 1991.

29. *Aspects Théoriques et Industriels de la Lyophilisation* (L. Rey, ed.), Herman, Paris, 1964.

30. *Freeze Drying and Advanced Food Technology* (S. A. Goldblith, L. Rey, and W. W. Rothmayer, eds.), Academic Press, New York, 1975.

31. L. R. Rey, in *Recent Research in Freezing and Drying* (A. S. Parkes and A. Smith, eds.) Blackwell Scientific Publications, Oxford, England, 1960, p. 40.

32. I. Gersh and J. L. Stephenson, in *Biological Applications of Freezing and Drying* (R. J. C. Harris, ed.) Academic Press Inc., New York, 1954, p. 324.

33. F. Franks, in International Symposium on Freeze-Drying and Formulation, Bethesda, USA, 1990, *Develop. Biol. Standard,* 74: S. Karger, Basel, 1991, p.9.

34. A. P. MacKenzie, in International Symposium on freeze-drying of biological products, Washington, D.C. 1076. *Develop. Biol. Standard,* 36: S. Karger, Basel, 1977, p. 51.

35. P. J. Dawson and D. J. Hockely, in International Symposium on Freeze-Drying and Formulation, Bethesda, USA, 1990, *Develop. Biol. Standard,* 74: S. Karger, Basel, 1991, p.185.

36. A. P. MacKenzie, *Bulletin Parenter. Drug Assoc.,* 20: 101 (1966).

37. R. I. N. Greaves, in *Aspects Théoriques et Industriels de la Lyophilisation* (L. Rey, ed.), Herman, Paris, 1964, p. 323.

38. L. Rey, loc. cit., p. 23.

8

The Primary Drying Process

INTRODUCTION

The Function of Primary Drying Process

The principal function of the primary drying process is to reduce a major quantity of solvent in a product while the matrix is in a frozen state (i.e., when the mobility of the water in the interstitial region approaches zero). For an aqueous formulation, the water is removed by the sublimation of ice crystals. If the mobility of the water in the interstitial is > 0 during the primary drying process, then, after sublimation of the ice crystals, there can be a lack of rigidity of the exposed interstitial region and collapse or even meltback can occur, as illustrated in Figure 1.3c. It will be shown that the chamber pressure and shelf-surface temperature necessary to complete the primary drying process will be determined by the thermal characteristics of the formulation, mainly the collapse temperature and occasionally the eutectic temperature. There is a tendency to departmentalize a given process, and lyophilization is not an exception. The reader should be aware at the start that, after removal of the ice crystals from the matrix, the secondary drying commences, and some water vapor is also removed from the exposed interstitial region of the matrix. As a result, there is no sharp demarcation between the completion of primary drying and the commencement of secondary drying.

The Perception of the Primary Drying Process

In contrast to the freezing process, which was described in the previous chapter, the primary drying process has received a great deal of attention from a number of investigators. Consequently, an understanding of the mechanisms involved in the process has not been without its share of controversy. I shall endeavor to provide the reader with an objective description of the process and to indicate those areas where dissension may prevail.

Perhaps the greatest difficulty that those entering the field of lyophilization have in understanding the primary drying process is that of the role that the phase diagram of water plays in determining the process parameters. This lack of understanding of the role of the phase diagram appears even in peer-reviewed journals would signify that even the reviewers were not aware that the data presented were *not possible to obtain in this universe.*

A basic problem with a good portion of the literature that describes the mechanisms involved in the primary drying process is that lyophilization involves a formulation whose constituents are not readily available to all investigators. Since one cannot reproduce the formulation, one also cannot reproduce the results of lyophilization processes reported in the literature. If the intent of the publication is to provide a basic understanding of the primary drying process, then one must question what intrinsic value such a description has to our general knowledge of the process.

THE SUBLIMATION PROCESS

Since the sublimation process plays such a major role in the primary drying process, it is imperative for the reader to have a fundamental understanding of this process. For this reason, a significant portion of this chapter will be devoted to the sublimation of ice. Although sublimation is defined as a solid-to-gas phase change that can occur at pressures of 1 atm, I will limit the discussion of the sublimation process to pressures lower than 4.58 Torr, or the vapor pressure of water at the triple point.

General Assumptions

To understand the sublimation process for ice crystals in a frozen formulation matrix, one must make certain assumptions regarding the nature of the system.

Matrix Temperature

Because the mass of the ice present in the formulation matrix generally exceeds the mass of the interstitial region, the temperature of the frozen matrix during primary drying will be governed by the temperature of the ice crystals at the gas-solid interface.

Disordered Ice Layer

In Chapter 3, the presence of a disordered layer on the surface of ice crystals was introduced. It was shown that there was evidence of the presence of such a layer even at temperature as low as –35°C. It was also shown that the electrical conductivity of ice can be associated with protonic conduction in the disordered layer. At a pressure of 15 Torr, Maeno and Nishimura [1] determined, especially for ice doped with hydrogen fluoride (HF), that conductance measurements showed the presence of a disordered layer on the surface of the ice during sublimation. For a rate of sublimation (χ), these authors [1] observed that the heat of sublimation was only 8.3 kcal/mole, which was significantly lower than that predicted by an Arrhenius expression for the sublimation:

$$\chi = \frac{P_{eq}}{\sqrt{2\pi MRT}} \approx \chi_o \exp^{-\frac{\Delta H_s}{RT}} \quad \text{kg}/(\text{m}^2\,\text{sec}) \qquad (1)$$

where ΔH_s is the enthalpy for sublimation of ice, R is the gas constant, and T is the ice temperature at the gas-ice surface interface expressed in K.

The observed lower enthalpy could result from the pressure of the test chamber being 15 Torr when the vapor pressure of ice at 0°C is 4.58 Torr. There are no known data that show the absence of a disordered layer on an ice surface at pressures generally used during the primary drying process, i.e., significantly lower than 4.58 Torr. It will, therefore, be assumed that sublimation of ice occurs from a disordered surface layer.

Frequency Distribution of Energy

There will be some frequency distribution of energies for the water molecules that compose the disordered layer. I [2] proposed that the nature of the frequency distribution can be expressed in terms of a *Maxwell-Boltzmann frequency distribution* (see Figure 8.1) and has the form

$$\frac{N_i}{N_o} = \sqrt{\frac{4x}{\pi}}\ \exp^{-x} \tag{2}$$

where N_i is the number of water molecules having a kinetic energy E_i; N_o is the total number of water molecules on the disordered surface; $x = E_i/(kT)$, where T is the temperature of ice surface; and k is Boltzmann's constant 3.3×10^{-24} cal/molecule-K.

For ice at a temperature of –30°C, the heat of sublimation has been given as 12,176 cal/mole or 2.02×10^{-20} cal/molecule. Since kT at –30°C has a value of

Figure 8.1. Fraction of Water Molecules in a Disordered Layer of Ice

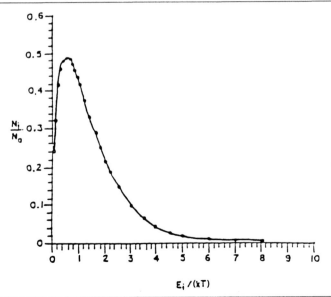

Fraction of the water molecules (N_i/N_o) in a disordered layer of ice as a function of $E_i/(kT)$ based on a Maxwell-Boltzmann–like frequency distribution.

8.02×10^{-22} cal, the fraction of molecules, at any instant, that have sufficient energy to escape the surface of the disordered layer will be 6.57×10^{-11}. The energy (E_i) where $N_i/N_o \leq 1$ molecule per mole (1.6×10^{-24}) will be about 4.4×10^{-20} cal/molecule. The total fraction of molecules having sufficient energy, at any instant, to leave the surface, can be determined from equation 2 by summing the energies between 2.02×10^{-20} cal/molecule and 4.4×10^{-20} cal/molecule. Because the difference in energy at which the number of molecules escaping the surface approaches 0 and the energy where molecules have sufficient energy to escape the surface is increased only by a factor of about 2, it would appear that the Maxwell-Boltzmann–like distribution shown by equation 2 does not provide the correct function for determining the probability of a water molecule leaving the disordered layer of an ice surface. This demonstrates the need for a better probability function to describe the sublimation process from such a complex system.

The Sublimation of Ice

Consider that the sublimation of ice occurs at a gas-ice interface, as illustrated in Figure 8.2. This figure shows that the disordered layer consists of a system of water molecules without a well-defined structure. As the gas pressure above the gas-surface interface is lowered, water molecules with sufficient energy and a proper momentum vector, indicated by the →, can leave the surface. Once in the gas phase, the water molecule may collide with residual gas molecules such as oxygen and

Figure 8.2. Sublimation at the Gas-Ice Interface

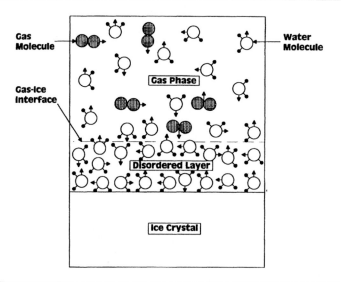

The sublimation at a gas-ice interface where the arrows indicate the momentum vector of the water and gas molecules.

nitrogen and return to the ice surface. Thus, the sublimation rate will be dependent on pressure. The following discusses the various expressions used for determining the rate of sublimation of ice.

Sublimation Model by Maeno and Nishimura

From equation 1, χ is dependent on ΔH_s and T. In Chapter 4, it was shown that for those temperatures where there is a known disordered layer present (-1°C to -30°C), ΔH_s was determined to be 12,176 cal/mole. For temperatures where there is little or no disordered layer present on the ice surface (-40°C to -90°C), ΔH_s was determined to be 12,259 cal/mole. Since ΔH_s is relatively constant for the temperature range used in the lyophilization process, χ will be strongly dependent on temperature. A plot of ln χ as a function of $1/T$, for the temperature range of -1°C to -90°C, is shown in Figure 8.3.

I [3] determined the rate of sublimation of ice from a stainless steel tray. A tray was selected because the sublimation of water vapor over the surface area would far exceed the amount of water lost as a result of sublimation that occurs along the walls of the tray. In this way, the observed total loss of water from the tray per unit time

Figure 8.3. A Plot of ln χ

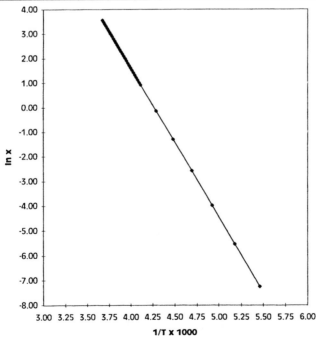

A plot of ln χ as a function of $1/T$ where the value of ΔH_s for -1°C to -40°C was 12,176 cal/mole and 12,259 cal/mole for temperatures ranging from -40°C to -90°C.

would approach the unit sublimation rate, i.e., mass/(unit area – unit time). At a shelf-surface temperature of –16°C, with the chamber pressure maintained at 55 mTorr, the sublimation rate was determined to be 1.39×10^{-4} kg/(m² sec) when the ice temperature was –47°C. For a chamber pressure maintained at 500 mTorr with a nitrogen gas bleed, the sublimation rate was determined to be 1.05×10^{-4} kg/(m² sec) for an ice temperature of –24°C.

From Figure 8.3, the estimated sublimation rate is 0.500 kg/(m² sec) at 55 mTorr and 4.357 kg/(m² sec) at 500 mTorr. A comparison of the experimental data with the calculated sublimation rates obtained from Figure 8.3 shows that the calculated values were usually significantly greater than the experimentally determined values. In addition, the experimental data show that an increase in the chamber pressure for a constant shelf-surface temperature resulted in a decrease in the sublimation rate, while the sublimation rate obtained from Figure 8.3 increased with pressure. The fact that there is a lack of agreement between experimentally determined sublimation rates and those calculated from equation 1 indicates that the measure of χ is a much more complex function and must take into account the shelf-surface temperature.

It is also of interest to examine the sublimation rates obtained from equation 1 in terms of the heat-transfer rate, i.e., energy per unit time. At temperatures lower than –40°C, the enthalpy for sublimating 1 kg of ice would be 681,055 cal. Since

$$1 \text{ cal} = 4.184 \text{ J}$$

and

$$1 \text{ J/sec} = 1 \text{ W}$$

the energy required to obtain an ice sublimation rate of 1.39×10^{-4} kg/(m² sec) in a stainless steel tray at a temperature of –47°C and a chamber pressure maintained at 55 mTorr would be 395.5 W. However, the wattage necessary to sublimate ice at a rate of 0.500 kg/(m² sec)—based on equation 1—would be 1,424,767. From this comparison of the energies of the experimentally determined sublimation rate and those calculated from equation 1, it is apparent that the rate of sublimation from an ice surface cannot be accurately calculated from equation 1 because equation 1 greatly overstates the energy needed for the sublimation process.

A further difficulty with the use of equation 1 to calculate the sublimation rate occurs with the units. The units for equation 1 should be 1/(m² sec-mole), not kg/(m² sec). Since the units of the equation are not correct, equation 1 is not applicable for determining the sublimation rate of ice at a given chamber pressure and shelf-surface temperature.

I have reported the results shown by Figure 8.4 [2]. These results show that during the initial pumpdown of a freeze-dryer containing a tray of ice, there was no decrease in the ice temperature until the pressure in the chamber approached or was lower than 170 mTorr, i.e., the equilibrium vapor pressure of ice at –35°C. The decrease in the ice temperature is an indication, based on the phase diagram of water, for the onset of the sublimation of ice from the surface. From these results, it can be concluded that for chamber pressures ≥ vapor pressure of the ice, for all practical purposes, the sublimation rate of water vapor approaches zero. Any water molecules that do manage to escape the ice surface, at pressures greater than the vapor pressure

Figure 8.4. A Plot of Chamber Pressure, Ice Temperature, and Shelf-Surface Temperature

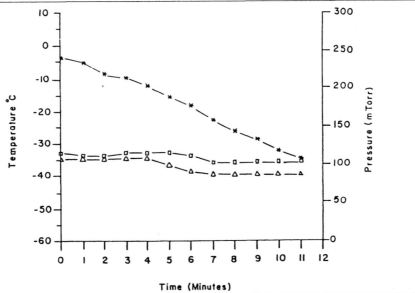

A plot of chamber pressure (denoted by -*-), ice temperature (denoted by -△-) and shelf-surface temperature (denoted by -□-) as a function of time during the initial pumpdown of a freeze-dryer. [Source *Journal of Parenteral Science and Technology*, "Discussion of Primary Drying During Lyophilization," Vol. 42, No. 4, July–Aug. 1988, Page 121, Figure 5.]

of ice, does so by diffusion, and the transfer rate of mass from the ice surface will be a relatively slow process.

Sublimation Model by Pikal, Shah, Senior, and Lang

Pikal et al. [4] reported that for pressures ranging from 0.040 Torr to 0.8 Torr, the mole fraction of air in the sample tube (X_2) was found to obey the empirical expression

$$X_2 = \exp(-0.602\dot{m}) \tag{3}$$

where $\dot{m}$ is the sublimation rate expressed in mg/min. Since these authors did not indicate the sublimation rate in terms of mass/(unit time – unit area), it is not possible to compare their data with other published sublimation rates for ice [3,5]. These authors stated that by increasing the pressure in the system to a pressure equal to that of the equilibrium vapor pressure for ice, the sublimation rate determined at low pressures decreased by a factor of 2. It is unfortunate that the authors [4] did not support their findings with unambiguous experimental data so that their data could be compared with the above experimental values [3]. In addition, their data support the results shown in Figure 8.4, which shows that the onset of sublimation occurs at pressures near or below the equilibrium vapor pressure.

Rate of Sublimation by Livesey and Rowe

Livesey and Rowe [5] proposed the following expression for the sublimation rate (R_s):

$$R_s = (P_v - P_a)\alpha \sqrt{\frac{M}{2\pi RT}} \tag{4}$$

where α is defined as the *accommodation coefficient* and is sometimes referred to as the *condensation coefficient* or *sticking coefficient*. Regardless of its designation, α is defined as the ratio of the rate per unit area that molecules actually condense on a surface to the unit rate at which they strike the surface [6].

The term P_v is given as the saturated vapor pressure (SVP) of ice expressed in Pascals (Pa) where the term SVP appears to refer to the vapor pressure of ice at a given temperature T. The pressure P_a is given as the partial pressure of water vapor above an ice surface. From the discussion presented by the authors, P_v refers to the vapor pressure at the gas-ice interface of the product, while P_a is the vapor pressure at the condenser surface. M is the molecular weight of water expressed in kg/mole and R is the gas constant. It was not made clear by the authors if the temperature T (K) refers to the temperature of the product at the gas-ice surface interface or the temperature of the condenser surface.

It is most unfortunate that Livesey and Rowe [5] implied that there is some driving force created as a result of a difference in temperature at the gas–ice surface interface of the product and the temperature of the condenser when in fact no such driving force exists. In reality, water vapor flow from a vial occurs because the pressure in the vial exceeds that of the chamber. If the sublimation rate were truly governed by equation 4, then there would be no need for evacuating the drying chamber. One would only have to refrigerate the product lower than its collapse or eutectic temperature and then reduce the temperature of the condenser; if the expression is correct, water would leave the product at rates as high as 2.4 g/(cm² hr) when the product temperature is at -47°C and the condenser is maintained at -70°C.

At a pressure of 1 atm, the drying rate will be insignificant, even if the condenser temperature is reduced to -150°C. It appears that some freeze-dryer manufacturers have, at times, been misled by the concepts defining equation 4 and have advertised that sublimation rates of a product can be increased by a lower condenser temperature. The only conditions under which the condenser temperature can affect the sublimation rate is increasing the chamber pressure by either blocking the vapor flow from the drying chamber or by reducing refrigeration to the condenser, which results in an increase in the chamber pressure.

Sublimation Model by Dushman

Using the following expression, Dushman [7] determined the vapor pressure of a metal by measuring the rate of evaporation at a given temperature.

$$\chi = 0.058 P \sqrt{\frac{M'}{T}} \quad \text{g/(cm}^2 \text{ sec)} \tag{5}$$

where M' is the molecular weight expressed in g, P is the vapor pressure expressed in Torr, and T is the ice temperature expressed in K.

From equation 5, the rate of sublimation of ice at –47°C and a pressure of 0.055 Torr was determined to be 3.24 g/(cm² hr). However, based on the measured sublimation rate of water vapor at –47°C, equation 5 should be expressed as

$$\chi = 3.22 P \sqrt{\frac{M'}{T}} \quad \text{g/(cm}^2 \text{ hr)} \tag{6}$$

When equation 6 was applied to the sublimation of ice at temperature of –24°C at a pressure of 0.5 Torr, the sublimation rate was 0.43 g/(cm² hr), whereas the measured sublimation was 0.038 g/(cm² hr) [3]. These results suggest that equation 6 is also not a general equation for determining the sublimation rate of ice as a function of chamber pressure.

Effects of Chamber Pressure and Shelf-Surface Temperature

Based on the above results and the reported sublimation rates [3] when the shelf-surface temperature was maintained at –16°C, the sublimation rate will approach 0 when the temperature of the ice is –16°C and the chamber pressure is $\geq$ 1.132 Torr. Based on the above results, equation 6 can be represented as

$$\chi = B(P,C)(P_s - P_c) \sqrt{\frac{M'}{T}} \tag{7}$$

where P_s is the vapor pressure of ice for a shelf-surface temperature (T_s) (i.e., 1.132 Torr for an ice temperature of –16°C); P_c is the pressure in the chamber expressed in Torr; and $B(P,C)$ is a nonunit term that is pressure and container dependent. For a $B(P,C)$ value of 0.2, an ice temperature of –47°C, a chamber pressure of 0.055 Torr and shelf-surface temperature of –16°C, the calculated rate of sublimation of ice, based on equation 7, was 0.060 g/(cm² hr), while the measured rate of sublimation (χ) was 0.050 g/(cm² hr). Increasing the chamber pressure to 500 mTorr while maintaining the shelf-surface temperature at –16°C and the ice temperature at –24°C resulted in an χ of 0.038 g/(cm² hr) [3]. Using equation 7, the calculated value of χ was 0.034 g/(cm² hr) for a $B(P,C)$ value of 0.2. For the same value of $B(P,C)$, a chamber pressure of 1 Torr, an ice temperature of –17°C, and a shelf-surface temperature of –16°C, the calculated value for the sublimation rate was 0.007 g/(cm² hr); thus, the calculated change in χ is consistent with experimental results.

The effect of chamber pressure on the sublimation rate of ice (χ) from a tray having a $B(P,C)$ value of 0.2 is illustrated in Figure 8.5. This figure shows that a decrease in pressure results in a marked increase in the sublimation rate. However, this figure also points out that there is a practical limit beyond which a reduction in pressure does not result in a significant increase in the sublimation rate. Between the pressures of 1031 mTorr and 29 mTorr, indicated by arrow A, there is a 13-fold increase in the sublimation rate. A further decrease in pressure from 29 mTorr to $\approx 9 \times 10^{-4}$ mTorr increased the sublimation rate by a factor of only 1.2. Thus, from Figure 8.5, a reduction of the chamber pressure to values less than 29 mTorr will not, for a given shelf temperature, substantially increase the sublimation rate.

The effect of the shelf-surface temperature on the sublimation rate of ice, from a tray having a $B(P,C)$ value of 0.2 and a chamber pressure of 30 mTorr, is illustrated

Figure 8.5. Sublimation Rate of Ice from a Tray

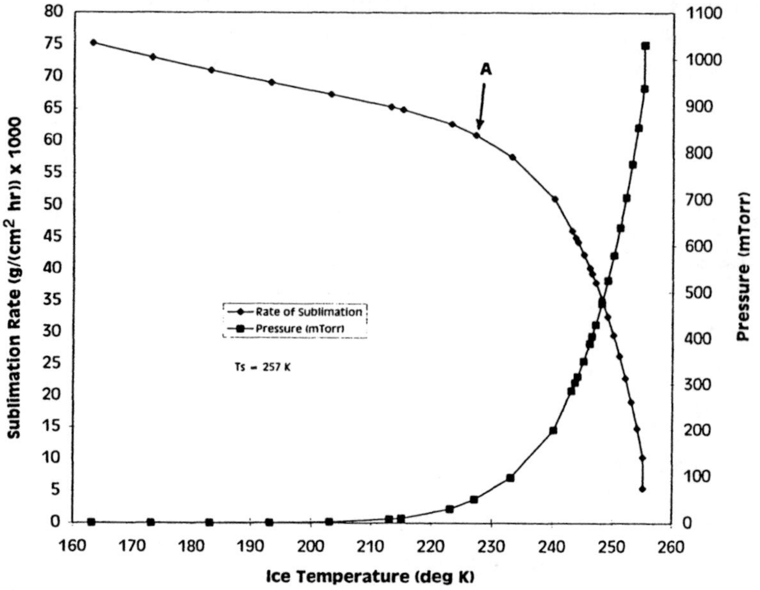

The sublimation rate of ice from a tray (denoted by -◆-), having *B(P,C)* value of 0.2, and chamber pressure (denoted by -■-) as a function of ice temperature (K) for a shelf-surface temperature (T_s) maintained at 257K.

by Figure 8.6. This figure shows a plot of the sublimation rate as a function of the shelf-surface temperature (T_s) (K) obtained from equation 7. Since the ice phase does not exist at temperatures greater than 0°C, the vapor pressure of liquid water will be used to provide a value for P_s for equation 7. Figure 8.6 shows that, for a chamber pressure of 30 mTorr, the sublimation rate increases from 1.41×10^{-3} g/(cm^2 hr) for T_s = 223K or –45°C to 1.35 g/(cm^2 hr) at a T_s = 296K or 25°C. In examining Figure 8.6, it should be noted that, as a result of the phase diagram of water, the ice-surface temperature ($T_{p,i}$) will remain constant when $T_s \geq$ –45°C (i.e., $T_{p,i}$ = 223K or –50°C) for a chamber pressure of 30 mTorr.

Effect of Container Configuration

The configuration of the container plays an important role in the sublimation process by effecting the distribution of energy into the container. This energy distribution will govern the sublimation rate at various points on the container and the configuration of the frozen structure throughout the drying process. The significance of the impact that the container can have on the drying process will become evident later in this chapter. While it is recognized that a variety of container configurations are available, this section will first consider various means for heat transfer between the shelf surface and a container. This will be followed by a description of heat transfer

Figure 8.6. Sublimation Rate of Ice from a Tray

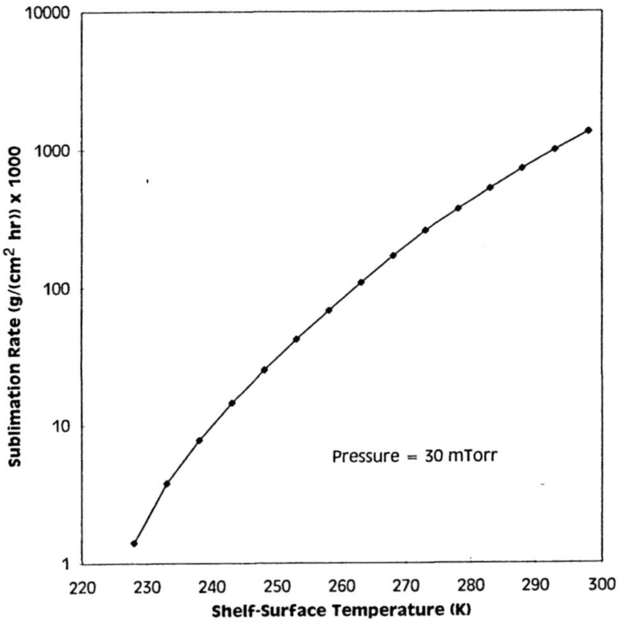

The sublimation rate of ice from a stainless steel tray having a $B(P,C)$ value of 0.2 and based on equation 7, as a function of shelf-surface temperature at a constant chamber pressure of 30 mTorr.

between the shelf surface and two of the most commonly used containers: the tray and the vial.

Heat-Transfer Rate

The heat-transfer rate (q) is defined in MKS (meter-kilogram-second) units as

$$q = 4.184 \frac{\lambda}{d}(T_s - T_{c,i}) \quad \text{W} \tag{8}$$

where λ is the thermal conductivity of the container, d is thickness of the base of the container, T_s is the shelf-surface temperature, and $T_{c,i}$ is the temperature at the ith point of the container. The total thermal conductivity (λ) between the shelf and the container is a combination of the contact between the shelf surface and the base surface of the container (see equation 8), the thermal conductivity of the gas in the chamber, and the radiant energy heat transfer.

Gas Thermal Conductivity. For *gas thermal conductivity* to be most effective, the distance between the shelf surface and the container must be equal or less than the *mean free path* (M.F.P.), i.e, the mean distance a gas molecule or atom will travel between collisions. A more detailed description of M.F.P. will be considered in Chapter 11. The energy transfer from the shelf surface to the container as a result of the thermal conductivity of gas [7] is given as

$$E_s = \frac{3}{4}\alpha\Lambda_i P\sqrt{\frac{273.2}{T_s}} \cdot (T_s - T_{c,i}) \quad W/cm^2 \tag{9}$$

where E_s is the energy transferred between two surfaces as a result of the gas thermal conductivity, Λ_i is the thermal conductivity of the *i*th gas species, and α has been previously been referred to (see equation 4) as the gas accommodation coefficient. The term α used is defined as

$$\alpha = \frac{T_r - T_i}{T_s - T_i}$$

where T_r is the temperature of the gas on leaving the shelf surface when $T_s > T_r$ and T_i is the temperature of gas striking the shelf surface. The α term will have values between 0 and 1.

The thermal conductivities of some common gases are listed in Table 8.1 [7]. This table shows that hydrogen would provide the best heat transfer between the shelf surface and a container; unfortunately, hydrogen is extremely flammable. While argon is an inert gas, the higher thermal conductivity of nitrogen would make the latter a better choice for controlling the pressure in a chamber during the drying process.

Stefan-Boltzmann Law. Using the Stefan-Boltzmann Law [8], the radiant heat transfer (E_r) between the shelf surface and the container can be expressed as

$$E_r = 5.672 \times 10^{-8}(\varepsilon_s T_s^4 - \varepsilon_c T_{c,i}^4) \quad W/m^2 \tag{10}$$

when ε_s is the *emissivity* of the shelf surface (i.e., the ratio of the emission of a given wavelength of light to that of a black body) and ε_c is the emissivity of the container.

Total Energy Transfer Between the Shelf Surface and the Container. The total energy transferred (E_t) from the shelf surface to the container is now expressed as

Table 8.1. Thermal Conductivity of Some Common Gases

Gas	W/cm²-°C × 10⁶
Hydrogen	60.7
Helium	29.3
Water vapor	26.5
Carbon dioxide	16.96
Nitrogen	16.6
Argon	9.3

$$E_t = q + E_s + E_r \; \text{cal}/(\text{cm}^2 \, \text{hr}) \qquad (11)$$

where the terms q, E_s, and E_r are defined by equations 8, 9, and 10.

Heat Transfer Between the Shelf Surface and a Tray

The stainless steel tray used to illustrate the heat transfer between the shelf surface and a tray will be assumed to have a design similar to those supplied by FTS│KINETICS (Stone Ridge, New York). This type of tray was selected because guide rails on each side of the tray that are at right angles to the door of the chamber. When such a tray is placed on the shelf, the guide rails contact a second set of rails on the sides of the shelf and generate a force that presses the bottom surface of the tray against the shelf surface. Such a tray and dryer design will maximize the physical contact between the tray and the shelf surface.

The heat transfer to a tray having the above design is illustrated by Figure 8.7. This figure shows that sublimation of the ice occurs not only along the surface area of the tray but also along the sides of the tray. While the force on the tray permits relatively uniform thermal contract between the shelf surface and the bottom of the tray, the heat transfer along the edges of the tray (illustrated by Figure 8.7a) is nonuniform because of the poor thermal conductivity of the stainless steel. As the sublimation of ice continues, the gap between the walls of the tray and the ice will increase; however, the loss of water from the tray will primarily occur as a result of sublimation from the surface of the ice rather than from the edges of the ice surface. Therefore, the rate of sublimation per unit area will closely approximate the total amount of water lost from the tray for a given time period [3]. For tray drying, the rate of sublimation will be directly dependent on the heat transfer rate for a given region of the tray. When there is uniform heat transfer throughout the tray surface area, the sublimation rate will be relatively uniform over the entire tray surface area. If, however, there is nonuniform contact between the tray and the shelf, the unit sublimation rate (χ) will also be nonuniform across the surface of the ice. The rest of this section discusses the effects that the various modes of heat transfer, represented by equations 8, 9, and 10, have on the sublimation rate of ice from a tray.

Rate of Sublimation of Ice from a Tray Resulting from Gas Thermal Conductivity.
For a tray having a uniform thickness of ice and placed on a shelf surface maintained at –16°C (257K), the effect that nitrogen gas pressure will have on the sublimation rate is shown by Figure 8.8. This plot shows that there is a maximum in the heat-transfer rate resulting from the thermal conductivity of nitrogen gas, which is reflected in an increase in the sublimation rate that occurs at a chamber pressure of 400 mTorr. For an increase in nitrogen pressure to greater than 400 mTorr (for a shelf-surface temperature of –16°C), the result will be a decrease in the heat-transfer rate (expressed in terms of the sublimation rate). It should be understood that the above merely serves to illustrate the relationship between the shelf-surface temperature and heat transfer resulting from the thermal conductivity of a gas such as nitrogen. Consequently, the shape of the plot shown by Figure 8.8 will be dependent on the shelf-surface temperature (e.g., the magnitude of the maximum sublimation rate and the range of pressures will increase as the shelf-surface temperature is increased).

Figure 8.7. Heat Transfer to a Tray

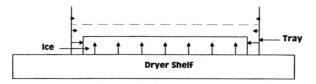

8.7a Cross-Sectional Side View

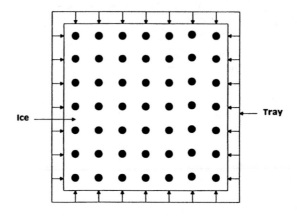

8.7b Top View

The heat transfer to a tray containing ice after sublimation has occurred for some period of time. Figure 8.7a shows the heat-transfer rate for a cross-sectional view of the tray where the arrow indicates the direction of the heat flow, and the length of the arrow its magnitude. The dotted line indicates the original fill height of the ice. Figure 8.7b illustrates the top view of tray and the • indicates the heat transfer were there is direct contact between the tray and the shelf surface.

The Effect of the Thermal Conductivity Between the Shelf Surface and the Tray.
The heat transfer resulting from the thermal conductivity between the shelf surface and the tray can be seen from Table 8.2. The total sublimation rate (χ) for a given pressure was determined from equation 7, while the heat transfer derived from gas thermal conductivity and the Stefan-Boltzmann Law, expressed as the rate of sublimation of ice, was obtained from equations 9 and 10. The thermal conductivity energy, also expressed in terms of the rate of sublimation, was obtained indirectly from equation 11. From Table 8.2, it can be seen that at low pressures, i.e., < 50 mTorr, the main source of energy for sublimation must stem from the thermal conductivity between the shelf surface and the tray. These results indicate that, in the absence of good thermal contact between the shelf surface and the tray, there could be variations in the rate of sublimation from one part of the tray to another. Such a variation in sublimation is predicted from equation 7 as a result of differences in tray temperatures. For these reasons, care must be taken to ensure good thermal contact between

Figure 8.8. Heat Transfer Between a Shelf and a Tray

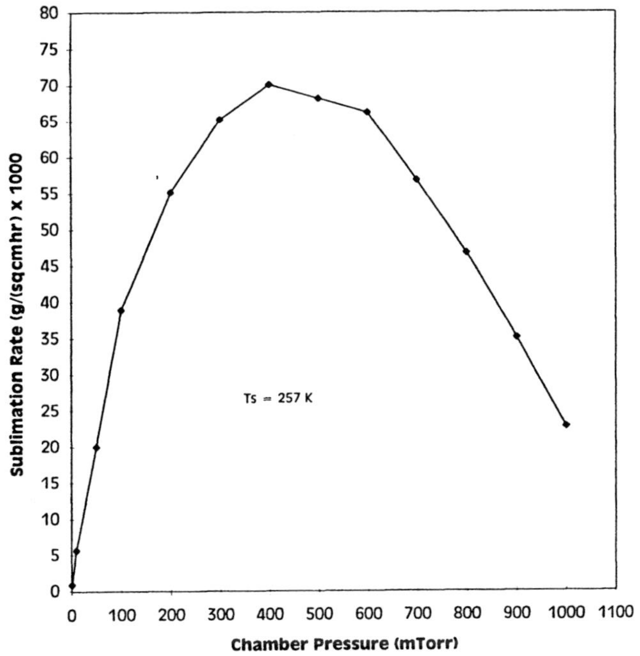

Ts = 257 K

The heat transfer between a shelf surface and stainless steel tray resulting (E_s) from the thermal conductivity of nitrogen, expressed in terms of the rate of sublimation of ice, as a function of pressure for a shelf-surface temperature maintained at 257K (–16°C).

the bottom of the tray and the shelf surface when tray drying at pressures lower than 50 mTorr.

At pressures ≥ 200 mTorr, there is sufficient energy transfer resulting from gas thermal conductivity to provide the necessary energy for the sublimation of the ice; thermal conductivity between the shelf surface and the tray is not a critical factor. A word of caution to the reader: It is important to construct a table similar to Table 8.2 for each shelf-surface temperature for the primary drying process. For example, at a shelf-surface temperature of 243K (–30°C) and chamber pressure of 200 mTorr, the heat-transfer rate determined from equation 7, where $B(P,C)$ is 0.2 and expressed in terms of the rate of sublimation, will be of the order of 4×10^{-3} g/(cm^2 hr). Therefore, the thermal contact between the shelf surface and the tray will be an important factor in achieving a uniform sublimation rate of ice across the ice surface. The impact of the shelf-surface temperature on the sublimation rate of ice for a given chamber pressure is best illustrated by Figure 8.6. As seen from this figure, increasing the shelf surface temperature from 243K (–30°C) to 268K (–5°C) will result in more than a 10-fold increase in the sublimation rate.

A third point of interest is that the thermal energy transfer, expressed in terms of the sublimation rate, resulting from E_r does not contribute significantly even when ε_s and ε_c are equal to 0.5. Thus, at pressures > 10 mTorr, the radiant heat transfer

Table 8.2. Thermal Heat Transfer Between the Shelf Surface and a Tray When the Shelf Surface Is Maintained at 257K (–16°C) and the Nitrogen Pressure Is Varied in the Drying Chamber

Pressure (mTorr)	Total Rate*	Rate (E_s)	Rate (E_r)	Rate (q)
2	67	1	1	65
10	65	6	0.8	58
50	61	20	0.6	42
100	58	39	0.5	18
200	51	55	0.4	0
300	45	65	0.3	0
400	39	70	0.2	0
526	32	68		
640	26	66		
705	23	58		
854	15			
939	10			
1,031	5			

*Sublimation rate based on equation 7 and expressed in g/(cm^2 hr) $\times$ 1,000

energy does not contribute significantly to the energy necessary to sublimate the ice from the tray at shelf-surface temperatures typically used during the lyophilization process, i.e., $T_s \leq 25°C$.

Heat Transfer Between the Shelf Surface and a Vial

Consider the heat transfer that occurs, during the sublimation of ice, between the shelf surface and a vial, as illustrated in Figure 8.9. In Figure 8.9a, it is apparent that only a small area (A) makes thermal contact with the shelf surface. This area may constitute only 2% to 5% of the total cross-sectional area of the vial, and the remaining cross-sectional area (A'') does not come into direct contact with the surface. As shown in Figure 8.9b, the heat transfer from the shelf surface at low pressures, will be defined by A in Figure 8.9a.

Preferential Sublimation of Ice from the Walls of the Vial. Pikal et al. [9] investigated the sublimation of ice from a vial. These authors used pure water and a fill height of $\approx$ 2 cm. As a result of the sublimation, a *redistribution of ice within the vial occurred such that ice near the vial wall was preferentially removed*. Preferential removal of ice along the walls of the vial was later explained [10] to be a result of increased heat transfer at the vial edge. The ice sublimated from the region of the walls was said to

Figure 8.9. Heat Transfer to a Vial

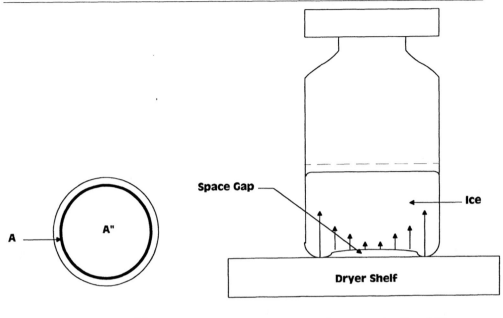

8.9a Bottom View　　　　　　　**8.9b Side Cross-Sectional View**

The heat transfer to a vial containing ice after sublimation has occurred for some period of time. Figure 8.9a shows the bottom profile of the vial where the **A** denotes the contact area between the vial and the surface and **A″** signifies the area of the gap between the vial and the shelf. Figure 8.9b shows the heat-transfer rate for a cross-sectional view of the vial where the arrows indicates the direction of the heat flow and the length of the arrow its magnitude. The dotted line indicates the original fill height of the ice. The dotted line denotes the original volume of the ice.

recondense near the middle of the vial, where the surface temperature of the ice was *slightly lower*. In support of their findings, Pikal et al. [10] state that Livesey and Rowe [5] had arrived at the same conclusion regarding the redistribution of ice during the sublimation process; however, I was unable to find any such supporting statement [5].

Pikal et al. further state that because of this phenomenon (i.e., preferential removal of the ice along the walls), temperature measurements *had to be completed before 15% of the ice was removed before the assumption of a planar ice-surface was seriously violated*. For a full load of 60 vials having an outside diameter (o.d.) of 2.4 cm, the sublimation rate of water for a shelf temperature of –16°C and chamber pressures of 68 mTorr and 490 mTorr were reported to be 0.23 g/hr and 0.26 g/hr, respectively. These results imply that the increase in the sublimation rate resulted from an increase in the chamber pressure from 68 mTorr to 490 mTorr.

The above results do not agree with those I obtained using a metal tray. The discrepancy on the effect of pressure on the sublimation rate of ice may be attributed to differences in the types of containers. The wall of the vial appears to have a greater impact on the results than that of a tray. It would have been helpful if the authors [10] had reported the increase in fill height of the ice near the center of the vial as a result of the redistribution process. According to the phase diagram of water shown by Figure 3.1, a temperature variation across an ice surface would also require a pressure gradient. The question that needs to be addressed with regard to the sublimation of an ice surface in a vial is, *Does a pressure gradient exist?*

Planar Sublimation of Ice from a Vial. I observed the sublimation of ice from weighed, stoppered, 15 mL, clear glass tubing vials having an o.d. of 2.665 cm. A random sample lot of 30 vials were marked, and the weight recorded with a 20 mm closure in place. A fill volume of 11.5 mL of deionized water was then added to the preweighed containers, and the weights of the vials, closures, and water were recorded. The vials and closures were always handled wearing plastic gloves in order to avoid weighing errors.

To minimize any effects of radiant energy heat transfer from the walls of the tray, the marked vials were placed in the middle of the tray and were surrounded by similar vials having 20 mm closures and the same fill volume of water. A copper-constantan thermocouple (36 gauge) was positioned in one of the surrounding vials so that its junction was located in the middle of the vials and about 1/8 in. from the bottom of the vial. An insulated copper-constantan (36 gauge) thermocouple was taped to the shelf surface prior to loading of the tray. The chamber pressure was determined and controlled from an MKS (MKS Instruments Inc., Andover, Mass.) capacitance manometer pressure gauge. The data were collected and stored using a modified Drying Process Monitor (DPM) computer program [11].

The shelf-surface temperature was maintained at –16°C during both the freezing and the primary drying processes. A chamber pressure of 500 mTorr was selected for the first test and was maintained by means of a nitrogen gas bleed into the drying chamber. After 5 hr at a shelf-surface temperature of –16.2°C ± 0.5°C and an ice temperature of –21.8°C ± 0.3°C, the sublimation process was stopped by backfilling the chamber to 1 atm with nitrogen gas, and the vials were stoppered. The vials were allowed to warm to ambient temperature and then reweighed to determine the loss in water as a result of the sublimation of ice.

The sublimation rate was found to be 0.0645 g/(cm² hr) ± 0.013 g/(cm² hr) or, based on total loss, 14% of the mass of water was sublimated from the vial. The frequency distribution for the sublimation rate from the vials is shown by Figure 8.10. This figure shows that there were no vials that had unit sublimation rates lower than 0.045 g/(cm² hr), while there were 8 vials with sublimation rates that ranged from 0.050 g/(cm² hr) to 0.055 g/(cm² hr). In addition, Figure 8.10 shows that the frequency distribution for the sublimation of ice is skewed to the right.

There was no evidence of the sublimation of ice along the walls of the vials. Some of the vials did have a small peak on the surface, but the presence of such a peak was explained in the previous chapter as a result of convection currents at the liquid–mushy zone interface. Some of the vials had ice surfaces that were flat, while others were sloped to indicate nonuniform heat transfer to the surface.

With the same vials and experimental setup used in the previous experiment, the average unit sublimation rate of the ice for a shelf-surface temperature of –16.1°C ± 0.7°C, an ice temperature of –40.2°C ± 0.8°C and a chamber pressure of 50 mTorr

Figure 8.10. Frequency Distribution for the Unit Sublimation Rate of Ice from a Set of 30 Vials at a Shelf-Surface Temperature of –16°C and a Chamber Pressure of 500 mTorr

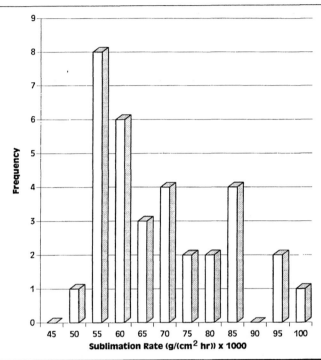

was 0.070 g/(cm^2 hr) ± 0.018 g/(cm^2 hr). The frequency distribution for these results is shown in Figure 8.11. This figure shows a more normal distribution of unit sublimation rates than that in Figure 8.10 and the predominant sublimation rate ranged between 0.065 g/(cm^2 hr) and 0.070 g/(cm^2 hr). A comparison of Figures 8.10 and 8.11 clearly shows that the average unit sublimation rate for vials at 50 mTorr is greater than that at 500 mTorr. In addition, the observed sublimation rates for the vials exceeded that from a tray [3]. In this test, an average of 18% of the water (based on amount of water measured prior to the test) was removed from the vials. Some vials did show some indication of ice sublimating from the walls of the vial, but sublimation mostly occurred from a planar surface.

The total sublimation rates at 50 mTorr and 500 mTorr are 0.396 g/hr and 0.360 g/hr, while for similar diameter vials Pikal et al. [9] reported sublimation rates of 0.23 g/hr at 68 mTorr and 0.26 g/hr at 490 mTorr. The reason for the discrepancy between these results is probably that the vials used by Pikal et al. [9] were contained in trays, whereas the results shown above were conducted with the vials in direct contact with the shelf surface. For vials in trays, the heat transfer to the vials will be dependent on the surface temperature of the tray. As pointed out earlier in this chapter, unless there is good physical contact between the tray and the shelf surface, the

Figure 8.11. Frequency Distribution for the Unit Sublimation Rate of Ice from a Set of 29 Vials at a Shelf-Surface Temperature of –16°C and a Chamber Pressure of 50 mTorr

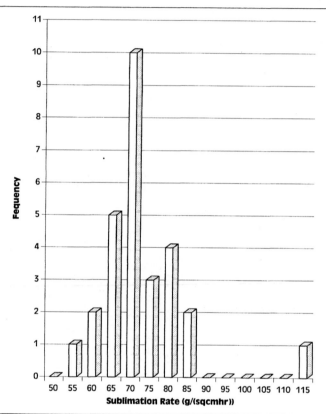

temperature of the tray surface will be dependent on pressure (i.e., an increase in pressure could result in an increase in the surface temperature of the tray). In addition, many investigators will often state only the shelf temperature, and the reader does not know if the reported shelf temperature refers to the shelf-fluid temperature or the actual shelf-surface temperature.

For the sublimation of the ice to occur, the pressure in the vial must be greater than the pressure in the drying chamber. Assuming that the observed temperature of ice is an estimate of the pressure in the vials, the value of $B(P,C)$, obtained from equation 7, for the sublimation of the ice at chamber pressures of 500 mTorr and 50 mTorr were 0.50 and 0.24, respectively. Since the $B(P,C)$ values for the trays were about 0.2 for both chamber pressures, the effect of pressure on $B(P,C)$ values of vials may stem from not having an accurate measurement of the vial pressure during the sublimation process.

The basic concept that the reader should understand is the distinction between the unit sublimation rate (g/(cm² hr) and the total sublimation rate (g/hr). For a given shelf-surface or tray-surface temperature, the unit sublimation rate of ice from either a vial or a tray will vary inversely with an increase in chamber pressure; the total sublimation rate will be directly dependent on the unit sublimation rate and the total sublimating surface area.

Change in the Configuration of Ice During Sublimation from a Vial. With the remaining water in vials from the tests conducted in the previous section, a change in the configuration of the ice was observed after sublimating at a pressure of 50 mTorr and a shelf-surface temperature of –16.4°C ± 0.9°C. Based on the observed ice configuration during the previous tests at 50 mTorr and 500 mTorr and the various ice configurations observed after 19 hr of drying, the successive changes in the ice configuration during the sublimation process were deduced. These sequential changes in configuration are illustrated by Figure 8.12. This figure shows that sublimation occurs first from a planar surface (Figure 8.12a). As the sublimation process commences, sublimation occurs preferentially with respect to the walls of the vial, and the ice takes on a mound-like configuration (Figure 8.12b). The formation of the mound is an indication that the rate of sublimation in this region is lower than near the walls of the vials. It is important to realize that, from the phase diagram of water, because the pressure above the ice will be the same throughout the vials, there must be a uniform ice surface temperature. The reason for the lower sublimation rate in the middle of the vial is because this region, as illustrated in Figure 8.7b, obtained less energy for the sublimation process.

With further sublimation, the mound becomes less pronounced (Figure 8.12c), and the ice configuration becomes cylindrical in nature. Near the completion of the sublimation process, a small mound is formed in the vial (presumably located in the center of the vial [Figure 8.12d]). The change in the configuration of ice with sublimation is expected to vary with fill volume. For example, with a low fill volume, one would anticipate that the formation of the mound will be less pronounced than for a higher fill volume. One would also expect the final ice formation to be like that illustrated in Figure 8.12d. This differs significantly from that of sublimation from a tray when the heat-transfer rate is relatively uniform over the entire surface area of the tray.

The vials were stoppered again, allowed to warm to ambient temperature, and reweighed. The observed frequency distribution of the remaining water in the vials is shown by Figure 8.13. From this figure, the mean and standard deviation of the mass of the remaining water were found to be 2.1 g ± 0.6 g. The significance of the frequency distribution in Figure 8.13 is that there were a significant number of vials having a higher mass of water than the mean, a α value of 2.5, or having an odds against 1 greater than 60. The importance of knowing the frequency distribution with regard to the primary drying process will become more apparent to the reader later in this chapter.

Other Means for Improving the Heat Transfer Between the Shelf Surface and the Vial. Patel et al. [12] reported a novel method for increasing the thermal heat transfer between a heating surface (tray or shelf) and a vial. These authors found that placement of the vials on a multilayer corrugated aluminum quilt resulted in an increase in the drying rate. The drying was performed using 3 mL of a 10% (wt/v) mannitol solution in a 10 cm^3 molded glass vial.

For a test, half of the vials on the shelf were in contact with the corrugated aluminum quilt, while the other half were placed directly on the shelf surface.

Figure 8.12. Changes in the Configuration of Ice

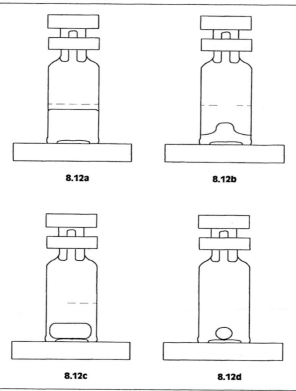

The change in the configuration of ice during the course of sublimation from a glass tubing vial, where the dotted line indicates the original fill height. Figure 8.12a shows sublimation occurring from a planar surface; Figure 8.12b shows the formation of a dome or peak; Figure 8.12c illustrates a later formation of a cylinder-like configuration, while Figure 8.12d shows the final ice to have a small dome-shaped configuration located in the middle of the vial.

Thermistors were used to measure the shelf-surface and product temperatures. The product temperature was determined by placing the temperature sensor directly on the center inside wall of the vial. After freezing the formulation at –40°C for 4 hr, the pressure in the chamber was reduced to a predetermined pressure, and the shelf temperature was increased to +5°C. The effect of the corrugated aluminum quilt on the drying rate was studied at a pressure of 20 mTorr, while the drying rate without the device was determined at pressures ranging from 20 mTorr to 2,000 mTorr. The chamber pressure was maintained at a desired value by bleeding dry air into the chamber through a controlled leak valve.

These authors [12] showed that with the same chamber pressure (20 mTorr) and shelf temperature, the use of the corrugated aluminum quilt between the vial and the shelf resulted in a substantial increase in the drying rate (i.e., 10.5 hr using the device and 17 hr without the device). It is unfortunate that, while the authors studied the

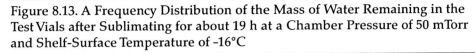

Figure 8.13. A Frequency Distribution of the Mass of Water Remaining in the Test Vials after Sublimating for about 19 h at a Chamber Pressure of 50 mTorr and Shelf-Surface Temperature of -16°C

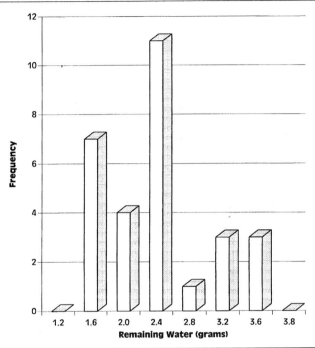

rate of drying as a function of pressure for vials dried without the device, there were no data showing the effect that an increase in pressure had on the drying rate for vials with the device.

The authors [12] state that *As the chamber pressure is increased the product attains higher temperature faster; also the drying time is decreased.* It is regrettable that their published data from the primary drying process do not support this statement. For example, in their published Figure 6, the product temperature at 20 mTorr and 3 hr after the start of the drying process was about -20°C; however, this was also the same product temperature for a chamber pressure of 165 mTorr that was shown in their Figure 7. At the 8 hr mark, as seen in their Figure 7; the product temperature was about -10°C for 550 mTorr, while, after this same time period, the product temperature shown in their Figure 6 was also -10°C for a chamber pressure of 1,000 mTorr. The question that must then be answered is, *How can one, during the primary drying process, obtain for the same shelf temperature the same product temperature for such a large difference in chamber pressures?*

In addition, the time required to complete primary drying for a chamber pressure of 270 mTorr was about 16 hr, which was also the same time required to complete primary drying for a chamber pressure of 165 mTorr. While I do not doubt the integrity of Patel and his associates, the discrepancies in the data must lie in the pressure measurements (i.e., the pressure gauge reading may not have been representative of the actual chamber pressure). Based on this, I would like to stress to the reader the need for diligence and for knowing not only the accuracy of the measurements but also the location of the pressure or temperature transducer.

Other Means for Describing the Heat Transfer in a Vial During the Primary Drying Process. It is worth noting that Nail [13] expressed the heat transfer (q) to the vial in terms of thermal resistance:

$$q = \frac{\Delta T_t}{R_t} \quad \text{cal/(cm}^2 \text{ hr)} \tag{12}$$

where R_t represents the sum of thermal resistances resulting from temperature drops (ΔT_t) from the shelf surface to the sublimation surface. Nail [13] argues that the thermal heat flux can be determined from the difference between the temperatures and the thermal resistance values. The resistance values can be controlled by pressure in the chamber. For a given shelf temperature, q is increased as a result of the decrease in R_t generated by an increase in gas thermal conductivity resulting from an increase in the chamber pressure. I have showed in Figure 8.8 that an increase in pressure to 400 mTorr results in a significant increase in the water vapor flux [g/(cm^2 hr)] leaving a tray. At pressures greater than 400 mTorr, the slope of the graph of water vapor flux as a function of pressure was negative.

In placing these results in their proper perspective, one should be aware that the vials in Nail's study [13] were contained in a tray in which there was no apparent provision made to maximize the thermal contact between the tray and the shelf surface. In addition, the shelf was maintained at +45°C. Nail did not specify if the reported shelf temperature was the shelf-fluid or shelf-surface temperature. Using only the fluid temperature can provide misleading results, which may be specific to one design and manufacturer of the dryer and possibly prevent others from reproducing the results.

In a test conducted on the sublimation of water from vials at a constant chamber pressure of 100 mTorr and having direct contact with the shelf surface, I found that, in a research size dryer, a shelf-fluid temperature of +30°C was required to maintain a shelf-surface temperature of 0°C, while a production dryer required temperatures of about 10°C to obtain similar results. In order for others to be able to repeat these results, investigators should be careful to indicate the actual shelf-surface temperature or tray-surface temperature when reporting their results. This will be particularly important when transferring a primary drying process from an R&D dryer to a large-scale production dryer.

Primary Drying of a Formulation

The above placed a great deal of emphasis on gaining some insight regarding the sublimation of ice from both a tray and glass vial. While the primary drying of a formulation will differ from that of pure ice, the principles of the sublimation process for the ice in the matrix will be the same as that for a formulation. The similarities between the two systems are that the sublimation of water vapor from ice crystals in the frozen matrix is governed by the same principles that were considered in the previous section (i.e., the effect that pressure at the gas-ice interface has on the sublimation rate and the impact that pressure plays on the heat transfer rate). The principal difference between the sublimation of ice and the formulation is the presence of the interstitial region. In the case of the sublimation of ice, the region vacated by the sublimated water offered no impedance to the vapor flow from the receding ice surface. For a formulation, however, the presence of the interstitial region can, depending on the nature of the freezing process, impede, or as preferred by some authors, represent a resistance to, water vapor flow from ice sublimating in the interior of the matrix [6,9]. This section will first consider the configuration of the frozen matrix at various times during the primary drying process and then the effect that the interstitial region has on the primary drying process.

Configuration of the Frozen Matrix

Because of the presence of the interstitial region, which eventually becomes the dried cake of the formulation, the configuration of the frozen matrix is obscured at various times during the drying process. This obscurity of the frozen matrix has led to disagreement between investigators in which I have had an active role. It is encouraging, however, that there now seems to be a more general consensus that the change in the configuration of the frozen matrix follows closely that illustrated by Figure 8.12. This section will first consider the impact that the configuration has on ascertaining the sublimation rate of ice from a frozen formulation and then discuss the current model for changing the configuration of the frozen matrix during the primary drying process.

Prior to 1980, little attention was given to the configuration of the frozen matrix during the primary drying process. For example, both Rowe [14] and Seager [15] illustrated the frozen matrix during the primary drying process as initially planar in nature. As the primary drying process proceeded, the frozen matrix took on a more convex configuration. Neither of these investigators offered a description of the change in the matrix configuration as the primary drying process neared completion.

In 1980, a paper published by Nail [6] investigated the effects that an increase in chamber pressure would have on the heat-transfer properties of a vial during the primary drying process for frozen 0.18 M solution of methylprednisolone. He determined the drying rate by measuring the change in weight of a vial that was removed from the chamber during the drying process by a special sampling mechanism (*thief*) controlled from the top of the chamber. The weight loss for a given chamber pressure was plotted as a function of time. Assuming a planar surface area, Nail [6] found that an increase in chamber pressure from 40 mTorr to 1,300 mTorr, with a constant shelf temperature of 45°C, resulted in an increase in the unit sublimation rate from 0.081 $g/(cm^2 \, hr)$ to 0.21 $g/(cm^2 \, hr)$. The vials during both tests were contained in an aluminum tray. The increase in the sublimation rate, at a given shelf temperature, with an increase in chamber pressure does not agree with the results I obtained for the sublimation of ice from a tray and a vial. I proposed in 1986 [3] that the configuration of the frozen matrix did not remain planar throughout primary drying but at some point took on a cylindrical configuration.

Similar increases in the sublimation rate with increase chamber pressure were reported by Pikal et al. [9]. These authors also used a tray to retain the vials; however, these authors used a shelf-surface temperature of –21°C for sublimating ice from a frozen 50 mg/mL solution of mannitol at chamber pressures of 45 mTorr and 228 mTorr. The sublimation rates were calculated based on differences in the vial and chamber pressures. For a chamber pressure of 45 mTorr and shelf-surface temperature of –21°C, the sublimation rate was 0.16 g/(cm^2 hr), while for a chamber pressure of 228 mTorr and same shelf-surface temperature, the sublimation rate was 0.33 g/(cm^2 hr). These results also do not agree with those I observed for the sublimation of ice, at a shelf-surface temperature of –16°C and chamber pressures of 50 mTorr and 500 mTorr. The issue that needs to be addressed is not an emotional one (i.e., whose results are right and whose results are wrong) but an explanation as to why in one instance an increase in pressure would cause an increase in the sublimation rate, while other results show that an increase in chamber pressure results in a decrease in the sublimation rate.

First one must differentiate between a *steady state* system with respect to the shelf-surface temperature and *nonsteady state* system. A steady state system is one in which the shelf-surface temperature (T_s) is maintained at a given value and the product temperature (T_p) is directly dependent on the chamber pressure (P_c). If the pressure were to be increased, then less energy would be dissipated by the sublimation of the ice and result in an increase in T_p. For a nonsteady state system, the T_s and Pc are both varied. The resulting value of T_p will be dependent on T_s and P_c but not relative to the former value of T_s. Consequently, an increase in T_p need not be associated with a decrease in sublimation rate.

Discussion of Nail's [6] Results. The vials used by Nail [6] were housed in an aluminum tray. The temperature of the shelf on which the tray rested was maintained at 45°C. Unless the tray has a special feature that forces mechanical contact between the tray and the shelf surface, there will be relatively nonuniform and generally poor thermal contact between the tray and the shelf surface. At lower pressures, the inside tray-surface temperature will be reduced by the energy loss resulting from the ice in the product and the poor heat transfer between the shelf surface and the tray. As the pressure is increased, there is greater heat transfer between the shelf surface and the tray as a result of the thermal conductivity of the chamber gas. The result is a higher inside tray-surface temperature. While the shelf temperature remains constant, the temperature of the inside tray-surface will vary with pressure, hence a nonsteady state system with respect to the inside tray surface temperature. An increase in the inside tray-surface temperature, generated by an increase in chamber pressure, resulted in a higher sublimation rate. This effect of pressure on the sublimation of ice resulting from a tray can be seen from Figure 8.8. A change in the configuration of the frozen matrix similar to that seen for ice (see Figure 8.12c) could also be a factor [2,3] in increasing the overall sublimation rate.

Discussion of Pikal et al. [9] Results. In order to avoid errors resulting from placement of the closure, Pikal et al. [9] used a fully seated closure that contained a stainless steel tube having an inside diameter (i.d.) of 0.217 cm and a length of 1.497 cm. The rate of sublimation for a given chamber pressure was obtained from the following expression for the tube resistance R_s:

$$R_s = \frac{(P_v' - P_c)}{\dot{m}} \qquad (13)$$

where P'_v is the pressure in the vial and P_c is the chamber pressure expressed in units of mm Hg. The tube resistance R_{TB} is given as

$$R_{TB}^{-1} = 0.2478 + 1.944\overline{P} \tag{14}$$

where $\overline{P}$ is equal to $(P'_v - P_c)/2$. The authors provide values for the chamber pressure but no data, such as product temperature, from which the value of P'_v can be determined.

The sublimation rate calculated by Pikal et al. [9] provides little insight regarding the method or the accuracy of the calculations. Consider the following: for a shelf-surface temperature of –16°C, the calculated sublimation rate for pure ice at a chamber pressure of 68 mTorr was 0.23 g/hr, while at a chamber pressure of 490 mTorr the rate increased only to 0.26 g/hr. However, for a frozen solution containing 50 mg/mL of mannitol, the sublimation rate, for a lower shelf-surface temperature of –21°C, was 0.16 g/hr for a chamber pressure of 45 mTorr and 0.33 g/hr at a chamber pressure of 228 mTorr. The authors offer no explanation as to why the sublimation rate of ice from a frozen mannitol solution should be 1.26 times greater than pure ice when the shelf-surface temperature and chamber pressure are lower. It is regrettable that I am unable, based on the data provided by Pikal and his associates [9], to resolve the observed differences in the effect of chamber pressure on the sublimation rate that I determined and with those discussed above. It is my belief that nature is truly an impartial enmity and does not favor one investigator over another.

The present concept regarding the change in the configuration of the frozen matrix during primary drying has evolved from a planar configuration throughout the primary drying process to a change in configuration as illustrated by Figure 8.12. Pikal and Shah [16] proposed Figure 8.14 to show that the frozen matrix remaining planar throughout the entire primary drying process as *ideal*, whereas, the *actual*, which was based on the results of their study of the configuration of the frozen matrix, followed the change in configuration of ice as seen in Figure 8.12. In further support for the configurations shown by Figure 8.12 and the *actual* in Figure 8.14, those experienced in the field of lyophilization have, at one time or another, observed partial meltback in one or more of the vials in a batch. Such partial meltback results from incomplete primary drying prior to the start of the secondary drying process. The meltback appears as a small defect in the middle of the cake and never—in my experience—as a thin layer across the entire bottom of the vial. The location of the partial meltback in the center of the cake further supports the assertion that this region is the last to undergo the primary drying process.

The Effect of the Interstitial Region on the Primary Drying Process

The primary drying of a formulation, unlike the sublimation of pure water, is dependent on the nature of the frozen matrix. For example, Figure 7.4a shows a frozen matrix of a formulation in which the ice takes the form of branches. As the ice sublimates during primary drying, these branches will serve as gas conductance channels for the flow of water vapor. The *flow rate* (Q) [7] through these channels is defined as

$$Q = F(P_1 - P'_c) \text{ Torr-L/sec} \tag{15}$$

Figure 8.14. Suggested Geometry of the Ice-Vapor Interface

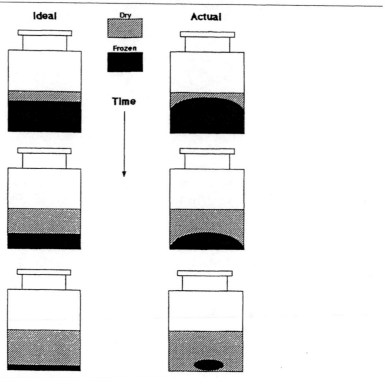

A comparison of the "ideal" planar geometry with the curved interface geometry proposed for materials studied in this research. This schematic is only intended to be qualitative. [Source: *PDA Journal*, "Intravial Distribution of Moisture . . . Freeze Drying," Vol. 51, No. 1, Jan.–Feb. 1997, Page 17, Figure 8.]

where F is the *conductance* of the channel, expressed in L/sec, and $(P_1 - P'_c)$ is the pressure differential from the surface of the sublimating ice to the end of the gas conductance channel. As will be explained in Chapter 11, the expression for F will be dependent on the nature of the gas flow through the channel. When the flow of the gas through the channels results in more gas-gas collisions than gas-wall collisions, the nature of the flow will be *viscous* in nature. When gas-wall collisions far exceed gas-gas collisions, the flow will be *molecular* in nature. For a gas conductance channel having the same pressure differential $(P_1 - P'_c)$, Q for viscous flow will be greater than Q under molecular flow conditions [7].

In addition, it is important to understand that regardless of the nature of the gas flow (molecular, transition, or viscous), F will be inversely proportional to the length (l_p) of the conduction path, which is directly proportional to the cake thickness (h_c). Therefore, as primary drying proceeds, the decrease in F will result in a decrease in Q and an increase in $(P_1 - P'_c)$. Provided the pressure above the cake P'_c, remains constant throughout the primary drying process, the pressure at the gas-ice interface

(P_1) will tend to increase as conductance decreases. From the phase diagram for water and equation 7, an increase in P_1 will, provided the vapor pressure of ice (P_s) for a T_s is greater than P'_c, cause an increase in the temperature of the frozen matrix below the gas-ice interface.

Since equations 7 and 15 both relate to the flow of gas from the cake during the primary drying process, we can now understand the effect that a change in the conductance of a cake will have on the temperature of the frozen matrix by equating equations 7 and 15 to give·

$$B(P,C)A(P_s - P_c)\sqrt{\frac{M'}{T}} = F(P_1 - P'_c) \tag{16}$$

where A is defined as the area of the sublimating surface.

Because F is inversely proportional to l_p, the increase in T_p resulting from a decrease in F will be a function of $(l_p)^2$. The dependency of an increase in T_p resulting from an increase in l_p is illustrated by Figure 8.15. Note that in this figure, the *zero* increase in temperature represents that T_p which is 5°C below the collapse temperature of the formulation. Although simple in concept, this latter figure reveals several important factors regarding the effect of the interstitial region on the primary drying process.

Figure 8.15. Product Temperature as a Function of Gas Conductance

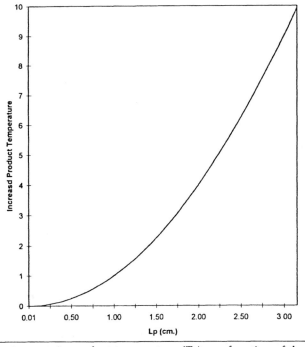

A plot of the increase in product temperature (T_p) as a function of the gas conductance path (l_p) of a cake generated by the interstitial region of the matrix. Note that the zero increase in temperature represents the T_p that is 5°C below the collapse temperature.

Effect of Fill Height. In selecting a container for a lyophilized product, the number of containers that a freeze-dryer can accommodate will often be determined by the fill height of the formulation. Since, for a given formulation, conductance will be dependent on the nature and concentration of the constituents, a fill height that produces a low gas conductance cake system can become counterproductive. Not only will the drying rate decrease significantly as the cake thickness increases, but the chances of producing defective product, e.g., partial meltback, will also increase. The detection of defective product can be an indication that the process is not reproducible. A larger diameter container may produce less product per batch, but the process time will be considerably shorter and the defective units will tend to be a result of structural defects of the container rather than related to the manufacturing process. In order to determine the optimized container diameter and fill volume dimensions, an investment in a simple development study of the effect that container diameter and fill volume has on process time and reproducibility of the lyophilized product to meet the specifications of the product will return large dividends over the product's commercial life cycle.

Effect of Cake Density. The density of the cake formed during primary drying will be dependent on the following two factors: the concentration of the constituents in the formulation and the degree of supercooling of the formulation. For a formulation having a solid content ≥ 200 mg/mL, the high solid content could result in a decrease in the short range conductance path of the cake. Such a cake structure will tend to decrease the overall conductance of the cake as primary drying proceeds. This decrease in cake conductance will be reflected in an increase in T_p and a decrease in the drying rate as the drying process proceeds. Therefore, in order to enhance the primary drying rate by increasing cake conductance, one would increase the conductance by diluting the formulation.

In the case where, as a result of the nature of water or some interaction with the constituents in the formulation, there is a significant increase in the degree of supercooling, the ice crystals will tend to be smaller, and the resulting dried cake will take on a *pressed pill* appearance, as seen in Figure 3.6. The dense structure of the *pressed pill* cake will impede the gas flow from the cake or decrease the conductance and cause an increase in T_p as the drying process proceeds. One way to avoid such a structure is to place the product directly onto a cold shelf surface ($< -40°C$). In this way, only a small portion of the volume of the formulation will undergo supercooling, and the remaining cake structure will be more conductive in nature. While the use of a cold shelf surface may solve the problem of a slow drying *pressed pill* structure, the resulting cake may contain a crust or glaze that could also impede vapor flow from the cake. One must therefore always weigh the options between a uniform slow drying cake structure and faster drying heterogeneous cake structure.

Impact of a Higher T_p. A typical target value of T_p for the primary drying process would be 5°C lower than the collapse temperature. As seen by Figure 8.15, an increase in l_p by a factor of 2.25 would, for a constant P_c and T_s, results in an increase in the T_p and the onset of mobile water in the interstitial region of the matrix. If l_p increases by a factor of 3, then T_p may increased by as much as 9°C, which could result, depending on how sharp and well defined the collapse temperature is, in a cake that would exhibit significant collapse or even meltback, as illustrated by Figure 1.3.

Effect of the Degree of Crystallization. Figure 7.4b shows a frozen matrix in which the degree of crystallization is ≤ 0.5 and there are no well-defined gas conductance channels present in the frozen matrix. Under these conditions, gas conductance will be predominantly by diffusion rather than by gas flow, as defined by equation 15. The diffusion of the gas will be through a series of unassociated conductance tubes. As a result, the overall F of the system will be low, and the T_p will tend to increase significantly during the drying process. In this example, one can see how the selection of the constituents in a formulation can have a substantial impact not only on the process time but also on the quality of the final product (e.g., low gas conductivity can result in undesirable product-product variations, such as residual moisture content).

KEY PROCESS PARAMETERS

General Relationship Between P_c, T_p, and T_s

Although an interrelationship between the process parameters (P_c, T_p, and T_s) was eluded to in the previous section, this section will describe the interrelationship and how one can use it to develop a lyophilization process.

While not obvious at first, the basis for the rate of sublimation in equation 7 reveals an important interrelationship between P_c, T_p, and T_s during the primary drying process when vials are in direct contact with the shelf surface. This relationship is stated as follows:

If any two parameters (P_c, T_p, or T_s) are fixed, the third parameter becomes invariant. For example, if the target T_p is some temperature below the collapse temperature, or in an atypical case the eutectic temperature of the formulation, then the selected value of P_c will be lower than the vapor pressure of ice at T_p. By fixing T_p and P_c, there will be only one temperature value, within limits, for T_s. As an illustration of this interoperating parameter relationship, consider formulation #1 from Table 7.1. The collapse temperature of this formulation was determined to be $-23°C$. By selecting a value of $T_p = -28°C \pm 3°C$, the matrix will remain in a frozen state throughout the entire primary drying process. Since the vapor pressure of ice at $-28°C$ is about 720 mTorr, any P_c value below 720 mTorr would be sufficient to result in the sublimation of ice crystals. By selecting a desired chamber pressure, the value of T_s necessary to maintain T_p at $-28°C \pm 3°C$ is defined. The selection of values for P_c and T_p will not only define the value of T_s but will also, assuming that the conductance of the matrix is not an impeding factor, determine the drying rate. Unfortunately, with knowledge of only P_c and T_p and not the rate of drying for such a system, it is not possible to calculate T_s from equation 7. This is a case of having one equation and two unknowns (χ and T_s); thus, it will not be possible to solve for T_s without prior knowledge of χ. The value of T_s must, therefore, be experimentally determined.

If the vials are placed in full-bottom trays, then the above relationship will still be valid, but instead of the T_s, the tray-surface temperature (T_t) will be the operating parameter. The principal difficulty in using full-bottom trays is that, because of warping of the tray as a result of handling, there will be a greater tray-to-tray variation in T_t on a given shelf than T_s across the shelf. For trays making poor thermal contact with the shelf surface, the lower T_t will result in a lower T_p and, hence, a reduced primary drying rate. If the secondary drying process were to commence before the vials in such a tray had time to complete the primary drying process, then this tray could contain defective product, such as partial meltback. Whenever possible, the vials should be in direct contact with the shelf surface during the lyophilization process.

Once T_s or T_t is determined for a given P_c and T_p, using either full-bottom or bottomless trays, the interrelationship between these parameters provides two unique advantages.

Primary Drying Independent of the Batch Size

The first important advantage of establishing primary drying by defining P_c and T_p from the collapse temperature and experimentally establishing T_s is that the vials cannot count how many other vials are in the batch. By controlling P_c and T_s, which determines T_p, the primary drying process will be the same regardless of the size of the batch. This means that the primary drying for a single shelf of product is the same as that for a fully loaded dryer.

Size and Manufacturer of the Freeze-Dryer

The second advantage of defining primary drying based on P_c and T_p is that, assuming that the emissivity of the shelf surfaces of the dryers are similar—which will be explained in Chapter 13—then the drying process will be independent of the dryer size and manufacturer. Thus, transfer from one dryer to another becomes possible because of equivalent processing parameters, i.e., P_c, T_s, and T_p. Equivalent drying environments therefore produce equivalent primary drying processes.

But many readers are aware from the experiences of others or their own experiences that the transfer of the primary drying process from one dryer to another is not as simple as described above. In addition, in order to reduce biological contamination of the product during the filling and loading operation, equipment manufacturers have introduced automatic loading equipment. While such equipment reduces the required number of personnel in the clean room, it does create its own dilemma of not being able to load product temperature monitors on each of the shelves to measure T_p. Indirect means for determining T_p during primary drying have been proposed [17] and will be discussed later in this chapter. The basic problem one encounters by not measuring T_p is that process control and its verification is dependent on a single pressure gauge and, perhaps in some instances, a single shelf-fluid temperature. In view of the relationship between P_c, T_s, and T_p, one cannot avoid expressing concern that an error in the pressure measurement will cause an error in T_p. If the T_p is in error, then there could be a substantial change in the primary drying process that could lead to a defective end product, with no data to document the cause of the defect. In light of the importance of P_c, T_s, and T_p, the role that each of these parameters plays in the primary drying process will be considered individually. In addition, the role that the condenser temperature (T_c) plays in the primary drying will also be discussed.

Chamber Pressure

In order to achieve the advantages of the interrelationship between P_c, T_p, and T_s, an accurate measurement of P_c is not only required throughout the primary drying but also the means must be employed to maintain P_c within defined limits. So that the reader may appreciate the important role that chamber pressure plays in the primary drying process, the following will consider the effect that the lack of pressure control can have on the primary drying process, the impact that the nature of the pressure gauge will have on the process, and the effect that gauge location and design features of the dryer can have on the drying process.

Effect of Lack of Pressure Control on Primary Drying

Consider a hypothetical case of a large, clean, dry, 200 ft² freeze-dryer having 10 usable shelves that can attain, when the external condenser surfaces are at -70°C, a chamber pressure of 20 mTorr. Let 1 shelf of the dryer be loaded with 2,400 vials (10 mL), each having a 2 mL fill volume of a formulation. Assume further that the formulation has a collapse temperature of -30°C.

After freezing the formulation to a temperature of -40°C and maintaining this temperature for 2 hr, the T_c is reduced to -70°C, and the chamber evacuated to its lowest pressure of 40 mTorr. The shelf temperature is increased to 0°C, which increases T_p to -35°C. The pressure in the chamber during primary drying reaches a maximum value of 45 mTorr, and the final P_c at the end of primary drying is 30 mTorr. The total primary drying process requires about 18 hr.

The dryer is then filled with 24,000 vials, each containing 2 mL of the formulation, which requires using every shelf in the dryer. T_p is again reduced to -40°C and maintained there for about 2 hr. The external condenser surfaces are refrigerated to -70°C, and the drying chamber is evacuated to 50 mTorr. The shelf temperature is again increased to 0°C, which results in the chamber pressure increasing to a maximum value of 90 mTorr. At these P_c and T_s values, T_p increases to -25°C, which exceeds the collapse temperature, and the formulation is now being *vacuum dried* rather than lyophilized (i.e., T_p exceeds the collapse temperature during primary drying). In order to lyophilize a full batch of this formulation, the shelf temperature would have to be reduced to -10°C. However, setting T_s at -10°C presents a dilemma for smaller batches, such as the 2,400 vials on one shelf, because T_p would be lowered to < -35°C, and the primary drying time would be greatly extended. As a result of the lack of control of P_c, the primary drying process becomes dependent on the number of vials in the batch and, in a sense, the vials can count.

The dilemma is resolved by freezing the formulation to -40°C, maintaining the T_p at this temperature for about 2 hr, and then refrigerating the condenser surfaces to -70°C. The chamber is then evacuated to 100 mTorr and maintained at this pressure by means of a nitrogen gas bleed. When the shelf temperature is increased to -15°C, T_p will be about -35°C, regardless if one or all of the shelves of the dryer are loaded with the formulation. Although the use of 100 mTorr will increase the primary drying process to 22 hr, the advantage that the primary drying process is independent of the batch size more than outweighs the increase in time for completing the drying process. In essence, by controlling the pressure, one establishes primary drying conditions that are independent of the number of vials in the dryer.

Impact of the Nature of the Pressure Gauge

There are basically two types of gauges currently being used to monitor and control pressure in the drying process. These gauges will be described in greater detail in

Chapter 11; a brief description of these gauges will suffice for our discussion of the primary drying process.

Thermal Conductivity Gauge. The first type of gauge determines the pressure in the chamber based on the *thermal conductivity* of gases. The presence of gases in the gauge removes energy from a hot metal filament. Low pressures are signified by a high filament temperature, and a decrease in filament temperature is indicative of the presence of a higher pressure. However, as seen from Table 8.1, the thermal conductivity of a gas will be dependent on its composition. Since calibration of these gauges (thermocouple or Pirani) is based on nitrogen, gases with higher conductivities than nitrogen will indicate false high pressures, while the presence of gases with lower conductivities will indicate false low pressures. For example, a pressure of 100 mTorr of water vapor will produce a gauge reading of about 160 mTorr, while for the same pressure of argon, the gauge would indicate that the pressure was only 56 mTorr. Since the partial pressures of gases such as nitrogen, oxygen, and water vapor in the chamber will vary during a drying process, one can only approximate the true pressure in the chamber [18] by means of a thermal conductivity gauge. Equation 17 can be used to obtain the correct P_c but requires the use of a mass spectrometer.

$$G\Lambda(N_2) = P\sum(f_i\Lambda_i) \tag{17}$$

where G is the gauge reading of the thermal conductivity gauge, $\Lambda(N_2)$ is the thermal conductivity of nitrogen, P is the total pressure in the chamber, and f_i is the fraction of the gas pressure comprised of the ith gas component having a thermal conductivity of Λ_i.

The change in the partial pressures of nitrogen and water vapor, which constitutes the major gas constituents in the chamber during a lyophilization process monitored by a thermocouple gauge, is illustrated by Figure 8.16 [18]. As shown by this figure, the partial pressures of water vapor and nitrogen are approximately the same at the start of the drying process; however, the sum of the partial pressures does not equal the observed thermocouple gauge reading. As the shelf temperature was increased, the partial pressure of water vapor in the chamber increased, while the true total pressure in the chamber, the sum of the partial pressures, decreased. The end of secondary drying is signified by a marked increase in the partial pressure of nitrogen and a reduction in the partial pressure of water vapor. Only at the end of the drying process did the sum of the partial pressures approach the observed gauge reading. Figure 8.16 shows that the thermocouple gauge indicated a false high pressure throughout most of the drying process.

As a further illustration of the difficulties one can experience in using a thermal conductivity gauge for monitoring and controlling chamber pressure during the primary drying process, consider the example of primary drying of a single shelf of product and a full dryer of product described in the previous section. To maintain the chamber pressure at 100 mTorr for the one shelf of product, the composition of the gases in chamber will be primarily nitrogen, there will be a relatively small error in the thermal conductivity gauge reading. For a batch using the maximum shelf surface, however, the partial pressure of water vapor will be the predominant gas in the dryer, and the chamber will be controlled at a false low pressure.

Figure 8.16. Thermocouple Gauge Readings and Partial Pressures

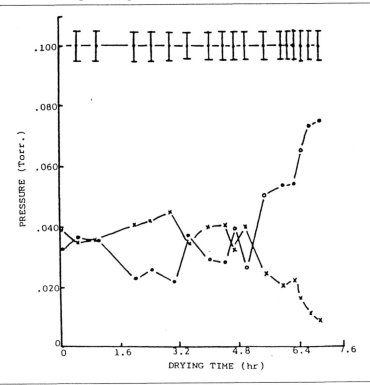

Thermocouple gauge readings and calculated partial pressures of nitrogen (denoted by o) and water vapor (denoted by x) as a function of time during the primary and secondary drying of 250 mL of a 2.5% KCl solution contained in each of four stainless steel trays. The total gauge reading is denoted by Ⱦ which signifies the variation in the gauge readings during the drying process [18].

Capacitance Manometer Gauge. The second type of gauge that is commonly used in determining chamber pressure is a capacitance manometer. This gauge determines pressure by sensing the deflection of a metal diaphragm by means of two capacitor plates. This gauge determines chamber pressure by the force per unit area exerted on the diaphragm and, for that reason, the pressure readings are independent of the composition of the gases in the chamber. This gauge will also be described in greater detail in Chapter 11.

For a full batch of product, a thermal conductivity gauge will indicate a maximum chamber pressure of 144 mTorr when the pressure in the chamber is 90 mTorr and is mainly water vapor. To reduce chamber pressure, one would have to again resort to decreasing the shelf temperature. However, when the thermal conductivity gauge finally does indicate a chamber pressure of 100 mTorr, the actual chamber pressure would be only 62 mTorr. If a capacitance manometer gauge were used in place of thermal conductivity gauge, the capacitance manometer would sense that total pressure was less than 100 mTorr and bleed dry nitrogen gas into the chamber to maintain the chamber pressure at 100 mTorr throughout the drying process.

Although thermal conductivity gauges are suitable for determining the pressure on the piping to the vacuum pump (foreline) where nitrogen is the primary gas, this type of gauge will sometimes indicate false high pressures. For this reason, this type of gauge is not suitable for monitoring and controlling the pressure in a chamber during the primary drying process.

Gauge Location and Design Features of the Dryer

The nature of the gauge is an important consideration in establishing the primary drying process, but the location of the gauge can also impact the primary drying process as can the other design features of the dryer.

Gauge Location. The obvious location for connection of the gauge pressure transducer for determining chamber pressure would be directly on a wall of the drying chamber. However, it has been my experience to find gauges used to control the process connected to the foreline of the system, not to the chamber. Since the foreline to the vacuum pump is after the condenser chamber (as illustrated in Figure 1.2), the pressure reading will not include the partial pressure of water vapor, and the gauge reading will not be representative of the chamber pressure regardless of the nature of the gauge.

In other cases, gauges for both monitoring and controlling chamber pressure may be positioned on piping from a gas supply. Depending on the length and diameter of the piping, the chamber pressure could be significantly lower than that of the piping.

Other Design Features. Another design feature that can impact primary drying is the distance between the shelves. If the shelf spacing is too close, the gas flow from the shelves can be impeded. Such impedance would cause a pressure differential across the shelf. The location of highest pressure will be dependent on other design features that govern the direction of gas flow. Regardless of where the highest pressure occurs on the shelf, at that point the product temperature will be higher and the drying rate will be lower. For a properly designed dryer, there will be no significant difference in pressure across the shelf as a result of impedance to the gas flow to the condenser.

In some dryers, the gas bleed inlet to control the chamber pressure is positioned on the foreline. The introduction of gas at this location will disturb the natural flow of gases from the drying chamber and enhance the possibility that any pump oil on the walls of the foreline may be carried back to the drying chamber, thus providing a potential source of product contamination during the primary drying process.

Summary

It is not enough for one to understand the importance that pressure control plays in the primary drying process. It is equally important to understand how the nature of the pressure transducer, its location, and the other design features can impact not only the primary drying process but also the quality of the final product.

Shelf Temperature

Shelf temperature has been attributed as playing a major role in the lyophilization process. The first issue that must be resolved is the meaning of the term *shelf temperature*. The ambiguity is illustrated by Figure 8.17. As shown in this figure, shelf temperature may be defined as either the temperature of the heat-transfer fluid (point A), the temperature of the shelf surface (point B), or the temperature of the inlet and outlet connections (C_{inlet} and C_{outlet}). For a given P_c and T_p, T_s will vary according the location of the temperature measurement.

Fluid-Shelf Temperature

The fluid-shelf temperature is determined at point A in Figure 8.17. The fluid temperature required for conducting primary drying depends on the thickness of the *barrier layer*.

The barrier layer is a region of heat-transfer fluid that is adjacent to the shelf surface and is relatively immobile (as designated by the short length of the arrows in Figure 8.17). Because of the lack of mobility of the fluid, a temperature gradient will exist across the barrier when a thermal load is exerted on the shelf surface by the sublimation process. The magnitude of the thermal gradient will be inversely proportional to the flow of the heat-transfer fluid through the shelf (i.e., as the flow diminishes, the thickness of the barrier layer increases).

The flow of fluid through the shelf, illustrated by the arrows in Figure 8.17, depends on a several factors. The first is the viscosity of the fluid at the fluid temperature. The viscosity of the fluid will vary depending on the nature of the fluid, i.e., the *viscosity index* or the change in fluid viscosity with temperature. As the fluid temperature is lowered, viscosity tends to increase. An increase in viscosity will cause a decrease in the flow of the heat-transfer fluid and an increase in the thickness of the barrier layer.

The flow through the shelf will also be dependent on the size and capacity of the heat-transfer pump. The faster the pump can force the fluid through the shelf, the smaller the barrier layer. The flow of the heat-transfer fluid through the shelf will also be dependent on design and construction of the shelf. For example, as the overall thickness of the shelf decreases, the impedance to the flow of the heat-transfer fluid, for a given temperature and heat-transfer pump size, will increase, and the barrier layer will increase.

At this point, the reader may ask, *Why so much concern over the thickness of the barrier layer?* One could always compensate for the temperature gradient across the barrier by simply increasing the temperature of the heat-transfer fluid. Consider the following example of a test conducted between my small research dryer and a large production dryer.

One shelf of each dryer was loaded with the same type of vials, each containing about 2 mL of water. The water was frozen to a temperature of $-20°C$ and allowed to remain at this temperature for about 1 hr. The condensers were then refrigerated to $\leq -60°C$, and the pressure in the chamber was lowered to 100 mTorr and maintained with a nitrogen bleed. The shelf-surface temperature in each of the dryers was increased to $0°C$, and after 1 hr the ice temperature was measured. The ice temperature in both dryers was about $-33°C \pm 3°C$. However, the fluid temperature in the

Figure 8.17. Product Undergoing Lyophilization on a Hollow Shelf

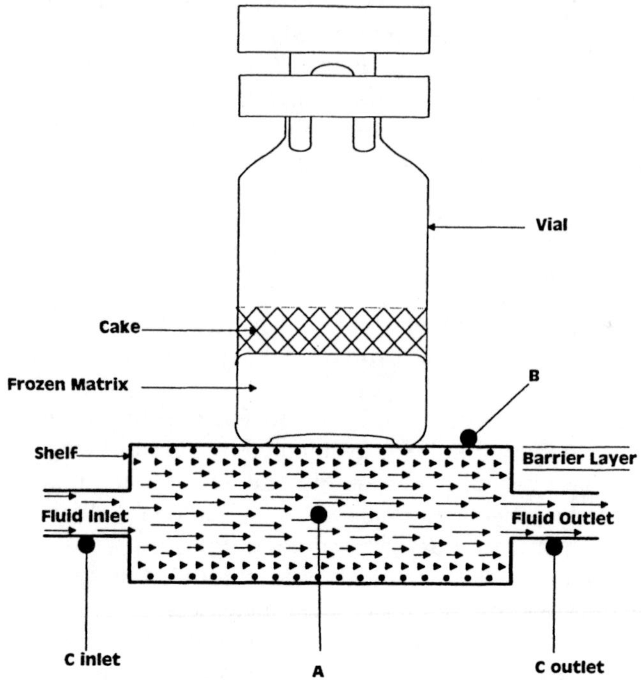

The product in vial undergoing a lyophilization process on a hollow shelf of a freeze-dryer with shelf temperature being defined as either the temperature of the heat-transfer fluid (designated by point A); the shelf-surface temperature (given as point B) or the fluid inlet and outlet temperature (C_{inlet} and C_{outlet}). The arrows denote the direction of the liquid flow and the length of the arrow its velocity.

production dryer was about 10°C, while the fluid temperature in my research dryer was about 30°C.

This example illustrates the first pitfall of defining shelf temperature by fluid temperature: One must exercise caution in transferring a primary drying process from one dryer to another because of differences in the thickness of the barrier layer. In addition, if the fluid flow though the shelves is partially blocked, a measure of the temperature of the heat-transfer fluid would not alert the operator to the increase in the barrier layer. The observed decrease in T_p may be attributed to an error in the temperature or pressure measurement and not to a malfunction of the heat-transfer fluid system.

The second pitfall is that when the thermal load is reduced as the primary drying process nears completion, there is additional heat input to the product. In the above case of the research dryer, the temperature of the product will be increased to 30°C. The product temperature in those vials that have not completed the primary drying process will increase sharply, and any remaining frozen matrix could reach temperatures that would induce severe collapse or even meltback.

Inlet and Outlet Fluid Temperatures

Attaching a temperature transducer to the fluid inlet and outlet of a shelf will also measure fluid temperature. Therefore, the pitfalls associated with the fluid-shelf temperature will also be applicable to the inlet and outlet fluid temperatures. One possible advantage of such a measurement is that the sublimation of ice will remove the energy from the heat-transfer fluid. The resulting difference between the inlet and outlet temperatures will provide some qualitative insight as to the primary drying process (i.e., as the difference in the temperatures approaches zero, the primary drying is nearing completion). Such monitoring may be helpful for a lyophilization process in which the sublimation rate is rather high. In the case where there is a low collapse temperature ($\leq$ –40°C), however, the difference between the inlet and outlet temperatures may fall within the accuracy and precision of the temperature measuring system, and the measurements would offer no real advantage over just monitoring the fluid temperature.

Shelf-Surface Temperature

I have used and highly recommend using the T_s as a means for monitoring and controlling the primary drying process. The heat transfer to the product depends only on the temperature of the shelf surface. As long as the T_s is uniform, whether the shelf is heated by the passage of a heat-transfer fluid, by heated gas or by electrical heaters does not affect the primary drying process. Using T_s overcomes two of the main pitfalls encountered with fluid-shelf temperature and the difference between the inlet and outlet fluid temperatures.

Using the T_s will eliminate the effect that the size, design, and manufacturer of the dryer could have on the transport of thermal energy to the product during primary drying. It makes no difference in the primary drying process if the fluid temperature is 10°C or 30°C, provided that the T_s values are equal. While monitoring and controlling primary drying by the fluid temperature needs only one or two temperature sensors, using T_s requires that sensors be positioned on each of the usable shelves of the dryer.

Controlling the T_s rather than the fluid temperature can prevent a significant increase in T_p as primary drying nears completion. This would permit any remaining frozen matrix to complete primary drying without being subjected to an increase in T_s that could cause partial collapse or meltback of the product.

The fluid temperature is certainly much easier to monitor and control than the T_s. I have used 30-gauge Teflon® insulated thermocouple wire taped to the shelf surface; while others have used 30-gauge to 36-gauge thermocouple wire embedded in brass disks attached to the shelf surface by means of high vacuum grease [19]. Although neither method would be suitable for use in the manufacture of aseptic injectable lyophilized formulations, other means for determining the T_s can be developed. For example, a point temperature transducer, such as a thermocouple contained in a stainless steel tube, can be attached to the outside of a full bottom tray or the retaining section of a bottomless tray. The device could be spring loaded so that the thermocouple tube is forced to make good thermal contact with the shelf surface or inside tray-surface. The location of the sensor could be positioned by movement of the tray or the retaining section. There is no doubt that other means could be found to monitor and control the T_s if those purchasing freeze-drying equipment would simply insist on having such a process control and monitoring feature as an integral feature of the freeze-dryer.

Product Temperature

The Importance of Monitoring T_p

T_p during primary drying, as defined earlier in this section, is directly dependent on T_s and P_c. Knowledge of T_p serves two important functions. The first and foremost is that monitoring of T_p throughout primary drying verifies that the primary drying was conducted at temperatures below the collapse temperature or, in a relatively few cases, the eutectic temperature. Since P_c will be lower than the vapor pressure of ice at T_p, the temperature of the frozen matrix at the gas-ice interface will be lower than T_p. Because T_p will be 5°C lower than the collapse temperature and T_p is measured 1/8 in. to 1/16 in. from the surface of the vial or tray, no point between T_p and the gas-surface interface can exceed the collapse temperature.

The second, and equally important, function is to verify that for a range of T_p values, T_s for each of the shelves, and P_c are within their defined limits, and the quality of the final product will have properties that fall within a specified range of values [20]. The verification of T_s and P_c is accomplished by placing at least 1 product sensor monitor on each shelf like that shown in Figure 7.7. By placing a product monitoring sensor on each usable shelf of the dryer, it is possible to detect changes in the accuracy of the pressure sensor or a malfunction in the shelf heat-transfer system before such changes adversely affect the quality of the final product.

Should T_p on each of the shelves show a trend toward higher temperatures, then the general increase in T_p can first be attributed to false low P_c values. The pressure gauge and its electronics should be recertified as described in Chapter 11. If the accuracy of the pressure gauge is determined to be within the manufacturer's specified limits then the T_s monitoring and control system must be examined.

One situation I have experienced is that for a given shelf, the values of T_p and T_s were lower than those observed for the other shelves of the dryer. One cannot dismiss such temperature readings as being erroneous but must recognize the possibility that there is a decrease in the flow of the heat-transfer fluid through the shelf in question. The reader should realize that monitoring the T_p and T_s throughout the primary drying process may require additional effort on the part of the operating personnel; such effort is well taken, however, if it prevents the loss of just one batch of product each year.

Indirect Determination of T_p

In an effort to reduce the effects that personnel in a clean room can have on the integrity of the product and to foster greater efficiency in loading and unloading freeze-dryers with large shelf areas, automatic vial loading equipment has been developed. Although these automatic loading and unloading devices have increased the efficiency of the manufacturing process and reduced the probability of product contamination, their use has introduced a problem regarding the monitoring of the lyophilization process. It has been reported that such systems present difficulties with the placement of thermocouples in the dryers [21]. The present systems are not designed to perform periodic loading of product temperature monitors, as described above, nor does the present equipment appear to lend itself to the placement of sensors to monitor and control the T_s. As a consequence, the use of such equipment has presented a unique set of problems in monitoring and verifying the lyophilization process. It is important to show that the primary drying process, for a given batch of a formulation, was performed within an established range of temperatures and pressures [20]. In the absence of monitoring T_p and T_s, one is forced to rely heavily on perhaps a single sensor measuring shelf-fluid temperature and P_c. In the many years of my experience in the field of vacuum technology, the sensor most likely to drift in accuracy over a period of time will be the pressure transducer. In the absence of the measurement of T_p at various locations in the dryer, there will be no effective check that the pressure transducer is operating within the limits set forth by the gauge manufacturer.

In response to the difficulties imposed by automatic loading and unloading equipment in measuring T_p, Milton et al. [17] have proposed a method by which T_p can be calculated from pressure rise in the chamber when the chamber is isolated from the condenser chamber. Before describing the method proposed by Milton et al. [17], I feel it necessary to explain some difficulties I have found with this paper so that the reader can be aware of these issues during the discussion of the paper. In the following analysis, I have made every effort to be objective in my comments.

Mean Product Temperature and Standard Deviation. In collecting product temperature data, the authors [17] used four product monitoring vials to determine the mean product temperature and standard deviation (α). Data from any monitoring vial that were *5°C or greater from the mean were discarded*. A minimum of two product vials was used to claculate the mean product temperature. The American Society for Testing and Materials (ASTM) stipulates that a frequency distribution is of little value when the number of observations (product temperature sensor vials) is less than 25 [22]. Consequently, the calculation of a mean product temperature and α for four or fewer product vials has no statistical merit. Presenting the product temperatures for each of the product-monitoring vials would have been more preferable.

Effect of the Thermocouple Transducer on the Degree of Supercooling. Roy and Pikal [17] have reported that the bare thermocouple junction served as an ice nucleation site and reduced the degree of supercooling. Because of the presence of the bare metal, the frozen matrix in which the thermocouple is located had a significantly different structure than that of the other vials. Because the matrix containing the thermocouple had larger pores, the product in the temperature sensor vials dried faster than the other vials in the system. While not in any way doubting the findings of these authors, I have not observed such differences in cake structure between vials containing the bare thermocouple junctions and other vials in the batch during the many lyophilization processes I conducted on a variety of formulations. Nor have I observed that the drying rate of vials containing the thermocouple sensors was significantly faster than other vials so that only these vials completed the primary drying process. It should be noted that Roy and Pikal [19] stated (and were cited in Milton et al. [17]) that the effect of the thermocouple on the structure of the matrix in the monitoring vials was based on a detailed study of three products, moxalactam, dobutamine hydrochloride/mannitol, and vancomycin hydrochloride, and that such

. . . measurable freezing bias may not occur for all products. What I am concerned about is the phrase *may not occur for all products* [19] has given way to the general statement that . . . *monitored vials are not representative of the entire batch . . .* [17], and what was first reported as a sound scientific observation has now become dogma.

Composition of Residual Gases in the Chamber During Primary Drying. Based on the results of Nail and Johnson [23], Milton et al. [17] assume the following basic principle of the lyophilization process: During the primary drying process, . . . *the composition of the vapor phase in the drying chamber is nearly all water vapor . . . .* Nail and Johnson showed [23] that for shelf temperatures of 40°C ± 1°C and a chamber pressure of 400 mTorr ± 5 mTorr determined by a capacitance manometer gauge, the isolation of the drying chamber resulted in an increase in chamber pressure of about 630 mTorr as determined by a (Pirani) thermal conductivity gauge during the primary drying of a formulation identified as Sol-Medrol 49 mg. At the completion of primary drying, there was a decrease in chamber pressure indicated by the Pirani gauge, which signified a decrease in the partial pressure of the water vapor in the chamber. The total drying time was about 9 hr. The high partial pressure of water vapor in the chamber was also determined by residual gas analysis (RGA) and a comparison plot of the Pirani gauge readings near the end of primary drying process. The partial pressure of water vapor determined by RGA indicated a partial pressure of water vapor similar to that indicated by the Pirani gauge. The above primary drying process was repeated for a capacitance manometer–controlled chamber pressure of 100 mTorr; unfortunately, the pressure indicated by the Pirani gauge during the primary drying was not reported.

I certainly agree that a Pirani gauge reading of about 630 mTorr or a capacitance manometer gauge reading of 400 mTorr would signify that water vapor was the predominant gas in the drying chamber. However, I am troubled when the results of RGA also indicate water vapor pressures greater than 500 mTorr when the total pressure in the chamber is only 400 mTorr. The operation of RGA will be examined in greater detail in Chapter 11; for the present, the reader should be aware that the relative ionization sensitivity for water vapor and nitrogen during an RGA are about the same [24], and the RGA measurement is certainly not dependent on the thermal conductivity of the gases. This would mean that the probability of ionization of a water molecule is the same as that for a nitrogen molecule. Because the pressures in the freeze-dryer will exceed those required by RGA, it is not possible to obtain a direct measurement of the partial pressures of gases in the drying chamber. To obtain the partial pressures of the gases in the chamber, one must know the relative ionization sensitivity for each gas, the total ion current of the RGA, and the actual pressure in the drying chamber. Since the pressure in the chamber was controlled at 400 mTorr ± 5 mTorr by a capacitance manometer gauge, it is simply not possible for the output of RGA to indicate partial pressures of water approaching 600 mTorr as seen in Figure 10 of the Nail and Johnson paper [23].

Another area of concern is that Nail and Johnson [23] make reference to my publication [18], the results of which are shown by Figure 8.17 and indicate that, under the given primary drying conditions [18], water vapor was a major but by no means the predominant gas in the drying chamber. It appears that Milton et al. [17] have once again taken the results of one or two drying processes and, in spite of published results to the contrary, made the sweeping general statement that . . . *the composition of the vapor phase in the drying chamber is nearly all water vapor* . . . during the primary drying process. There is no doubt that such a statement will be true in some instances but not necessarily in every case.

I hope that the reader will bear in mind my concerns regarding this paper: (a) the use of two to four measurements to determine a mean product temperature and standard deviation, (b) the general implication that the structure of the matrix in temperature-monitoring vials will be significantly different from the matrix of other product in the dryer, and (c) the assertion that the predominant gas in the drying chamber during the primary drying process is water vapor. The following review uses manometric measurements in determining the T_p during a primary drying process.

Manometric Measurements of T_p

The objective of the study by Milton et al. [17] was to determine the accuracy and sensitivity of manometric measurements with respect to actual T_p data, or, in other words, to determine whether or not it is possible to calculate the initial T_p from a knowledge of the change in the chamber pressure. The measurements were performed in the following manner.

The pressure and temperature data are collected by first running the data collection program for a Model DT2801 series I/O card contained in an IBM®-compatible computer and then closing the chamber isolation valve between the drying and condenser chambers. The data are then entered into a mathematical expression developed by the authors, and the temperature at the gas-ice interface is calculated.

The manometric temperature measurements were conducted on water, a 5% wt/v mannitol solution, 5% wt/v potassium chloride (KCl) solution, and a 24% wt/v KCl solution. Figure 8.18 shows the mean T_p as a function of time for the freeze-drying of a 5% wt/v mannitol solution performed at a shelf temperature of 0°C and a chamber pressure of 100 mTorr [17]. The change in the chamber pressure with respect to time, at the times indicated by the arrows, is shown by Figure 8.19. Figure 8.20 shows the manometric temperature measurements and T_p for water and the above concentrations of mannitol and KCl as determined by Milton et al. [17]. The water and the KCl drying processes were performed at the same shelf temperature and pressure as that used for mannitol in Figure 8.18 (i.e., shelf temperature of 0°C and a chamber pressure of 100 mTorr). However, without offering any explanation, the authors maintained the same chamber pressure of 100 mTorr for the drying of the 5% wt/v mannitol solution as used in the drying process illustrated by Figure 8.18, but for Figure 8.20 the authors chose to change the shelf temperature from 0°C to 20°C.

With the data provided by Figures 8.18 through 8.20 and expression A-15 from Milton et al. [17], the temperature at the gas-ice interface (T_i'') was determined from

$$T_i'' = \frac{-6144.96}{\ln P_i - 24.01849} - 273.15 \qquad (18)$$

where P_i is the pressure at the gas-ice interface.

Table 8.3 shows a comparison of the measured mean T_p values that I obtained directly from Figures 8.18 and 8.20 and the calculated T_p, at various times, based on

Figure 8.18. Mean Product Temperature and Shelf Temperature

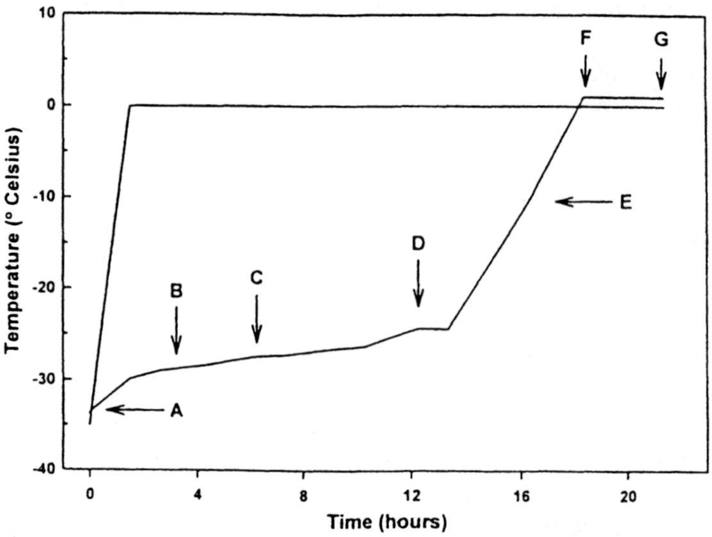

The mean product temperature (lower curve) and shelf temperature (upper curve) during the freeze-drying of a 5% wt/v mannitol solution (shelf temperature 0°C, chamber pressure 100 mTorr). Arrows indicate times at which a manometric temperature measurement was made. [Source *PDA Journal,* "Evaluation of Manometriuc Temperature During Lyophilization," Vol. 51, No. 1, Fan.–Feb. 1997, Page 10, Figure 3.]

data obtained from Figure 8.19 and the calculated T_p values taken directly from Figure 8.20. Table 8.3 reveals the following:

Measured T_p and Calculated T_p Values Based on Manometric Measurements. These data in Table 8.3 show that T_p values obtained from manometric measurements (see Figure 8.20) during the drying of the 5% wt/v mannitol solution at a shelf temperature of 20°C and chamber pressure of 100 mTorr are consistently lower than the mean product thermocouple readings. Since Milton et al. [17] provide T_p data and manometric plots for the drying of a 5% wt/v mannitol solution at a shelf temperature of 0°C, as shown in Figures 8.18 and 8.19, one may wonder why the authors chose to report product and gas-ice interface temperatures for a process conducted at a shelf temperature of 20°C, especially in view of the fact that product and gas-ice interface temperatures for manometric measurements made of the water and the two concentrations of KCl were determined for a drying process using a shelf temperature of 0°C. A second question that arises from Figure 8.20 is, *Why is there a similarity in product temperatures as a function of time for the mannitol and the 24% KCl wt/v solutions when the drying was performed at the same pressure but the mannitol drying was performed using a 20°C higher shelf temperature?* This ought to raise concerns, at least in my mind, regarding the validity of the data shown in Figure 8.20.

Willemer [25] also used barometric temperature measurements (BTMs) to ascertain the ice temperature during the primary drying process for which the

Figure 8.19. Transient Pressure Response

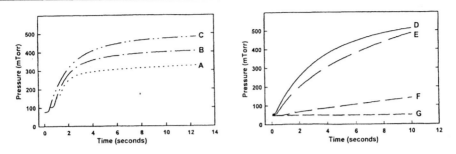

The transient pressure response during freeze drying of a 5% mannitol wt/v mannitol. A-G refer to data points indicated in Figure 8.18. [Source *PDA Journal*, "Evaluation of Manometriuc Temperature During Lyophilization," Vol. 51, No. 1, Fan.–Feb. 1997, Page 10, Figure 4.]

chamber pressure was maintained at 250 mTorr. Willemer's results showed that there was reasonable agreement between the temperatures indicated by the thermocouple sensors and those determined by the saturated water vapor pressure. The partial pressure of water vapor was determined by means of a mass spectrum based on the total chamber pressure. In addition, Willemer [25] stressed that *The time for measuring the pressure rise must be short enough to avoid a temperature increase in the product temperature.* Consequently, Willemer limited the closing time of the valve to the drying chamber to between 3 and 4 sec for what was described as *normal conditions.*

Measured T_p from Figure 8.18 and Calculated T_p from Figure 8.19. The basic difference between data provided by Milton et al. [17] for the drying of a 5% wt/v mannitol solution and that described in the previous section (Figure 8.20) is that the former was conducted at a shelf temperature of 20°C (see Table 8.3), while the data (which I determined from Figures 8.18 and 8.19) was obtained for a shelf temperature of 0°C. These results are inconsistent with the general results shown in Figure 8.20, i.e., the manometric determined T_p at points A, B, and C from Figure 8.19 are higher than the measured mean T_p values for the same elapsed time in Figure 8.18. These results would indicate that data taken 10 sec after closing the valve showed an increase in the temperature of the gas-ice interface. If there was an increase in the gas-ice interface temperature, then there also must have been an increase in mean measured T_p values. Although the manometric measurements shown in Figure 8.18 did not exceed 14 sec, Milton et al. [17] report that extending the duration of the measurement to 20 sec increases T_p by less 1°C. However, they offer no data to support such a statement.

Milton et al. [17] state that the drying process was performed using a modulated leak valve at a chamber pressure of 100 mTorr, as shown by Figure 8.18. Figure 8.19 shows that only the manometric measurement at the elapsed times indicated by arrows A and B was the initial chamber pressure of 100 mTorr. The initial chamber

Figure 8.20. Manometric Temperature and Mean Thermocouple Temperature Measurements

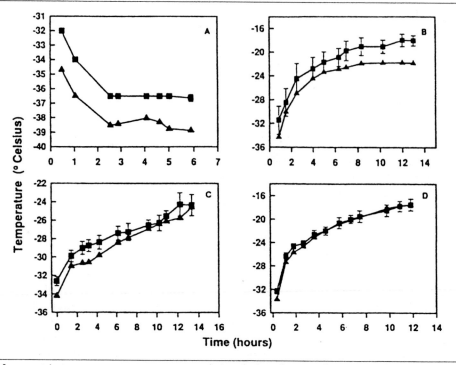

Manometric temperature measurements (triangles) and mean thermocouple temperature measurements (squares) during freeze drying of water (panel A), 5% mannitol (panel B), 5% wt/v potassium chloride (panel C), and 24% wt/v potassium chloride (panel D) solutions. [Source *PDA Journal*, "Evaluation of Manometriuc Temperature During Lyophilization," Vol. 51, No. 1, Fan.–Feb. 1997, Page 13, Figure 5.]

pressures at all other points were conducted at pressures significantly lower than 100 mTorr. As a result of these lower initial pressures, T_p values would be lower than that shown in Figure 8.18. In addition, the gas-surface interface temperature would also be lowered, producing a false low manometric determination of T_p values with respect to Figure 8.18. However, the manometric plots shown in Figure 8.19 indicate that the temperature at the gas-ice interface is greater than the T_p values shown in Figure 8.18. As stated in the previous paragraph, higher gas-ice interface temperatures must be associated with even higher T_p values. Such data would suggest that determination of T_p by manometric measurements, at least as described by these authors [17], is not a nonintrusive determination of the product temperature during the primary drying process.

The use of pressure control with a gas such as nitrogen would further complicate the manometric measurement. As shown by Willemer [25], each freeze-dryer would have to be equipped with a mass spectrometer to obtain the correct measured pressure-time plots for water vapor.

Table 8.3. A Comparison of the Measured Mean Product Temperatures and Manometric Determined Product Temperature for a 5% Mannitol Solution Based on Data Obtained from Figures 8.18, 8.19, and 8.20 [17]

Elapsed Drying Time (hr) (Letters used in Figure 8.18)	Measure of mean T_p (°C) based on Figure 8.18 ($T_s = 0$°C)	Measure of mean T_p (°C) based on Figure 8.20 ($T_s = 20$°C)
0 (A)	–33.4	–32.5*
3.2 (B)	–28.8	–24.0
6.3 (C)	–27.2	–20.9
12.3 (D)	–24.2	–18.2
16.4 (E)	–10.8	—
18.5 (F)	0.4	—
21.5 (G)	0.4	—

Elapsed Drying Time (hr) (Letters used in Figure 8.18)	Calculated T_p (°C) based on Figure 8.19‡ ($T_s = 0$°C)	Manometric Values of T_p (°C) based on Figure 8.20 ($T_s = 20$°C)
0 (A)	–28.9	–34.4*
3.2 (B)	–26.7	–26.2
6.3 (C)	–25.0	–23.1
12.3 (D)	–24.2	–21.9
16.4 (E)	–24.7	—
18.5 (F)	–36.7	—
21.5 (G)	–45.7	—
Total Primary Drying Time (h)	16	13

*Elapsed time is 0.8 hr

‡Calculated T_p values based on manometric pressures determined at the 10 sec mark.

Duration of the Primary Drying Process. The estimated completion of the primary drying process for the mannitol at a shelf temperature of 0°C was taken as the elapsed time indicated by the letter E (16 hr), while primary drying conducted at a shelf temperature of 20°C required 13 hr. Although the reported mean T_p values for the 20°C shelf temperatures were greater than those measured for a shelf temperature of 0°C, there is reasonable agreement between the respective manometric-determined temperatures. Such agreements between manometrically determined temperatures performed at significantly different shelf temperatures can only add to the confusion regarding an explanation of these experimental results.

Manometric Measurements During Secondary Drying. From Figure 8.19, the value of T_p would indicate that a gas-ice interface was no longer present and that the drying process had entered into the secondary drying stage. Although the pressure-time trace denoted by the letter G would suggest that significant quantities of water vapor were not leaving the product cake surface, the manometric measurements offer no means for determining the value of T_p. If T_p is not consistent when the desorption of water vapor is in process, then termination of the drying process could result in variations in the moisture content.

Summary. I have devoted a significant portion of this chapter to the manometric determination of T_p as reported by Milton et al. [17]. Such discussion was considered necessary because some of the general statements made by Milton et al. [17] about the primary drying process were not universally accurate.

The use the manometric determination of T_p for a steady state system will cause an increase in T_p during the measurement. Consequently, the only means for determining T_p is with the use of a temperature sensor.

While pressure-rise tests can prove useful in determining the completion of both primary drying and secondary drying [25], the absence of product-temperature sensors to monitor the process is not, in my opinion, consistent with the spirit of current Good Manufacturing Practice (cGMP) and the production of the highest quality pharmaceutical products by the conscientious monitoring of all key process parameters, i.e., P_c, T_s, and T_p.

Condenser Temperature

The above discussion of key process control parameters did not include the condenser temperature (T_c). The rate of primary drying is related to the key process parameters P_c, T_s, and T_p and does not include T_c. But T_c does play an important role in the primary drying process—the removal of water vapor from the gases entering into the condenser chamber. There are several ways in which T_c can not only impact the primary drying process but also affect the quality of the final product.

High T_c Values

If a malfunction of the condenser refrigeration system or a leak in the condenser chamber results in an increase in P_c, then for a given T_s, there will be an increase in T_p. If the T_p increase were to equal or exceed the collapse temperature, then the volume of the dried cake will not be equal to the frozen matrix. If the amount of collapse is significant, it could impact the quality of the final product, such as the amount of residual moisture. As long as the T_c is sufficient to remove the water vapor (i.e., $T_p - T_c \geq 20\ °C$), then, in essence, the product vial will not sense the presence of the condenser in the dryer.

Low Operating P_c Values

As the collapse temperature decreases, the required operating P_c values will also decrease. Should the pressure in the condenser chamber be reduced to pressures that are lower than the vapor pressure of water at T_c, then moisture will not be deposited onto the condenser surfaces but will enter the vacuum pumping system. Vacuum pumps will be considered in greater detail in Chapter 11; vacuum pumps must compress the gas entering the pump to pressures greater than 1 atm. If water vapor is present in sufficient quantities, the partial pressure of water may reach equilibrium vapor pressure for the gas-liquid system. Instead of the water vapor being discharged into the atmosphere, a portion of the water will remain in the pump and mix with the oil that is often used in the operation of the mechanical vacuum pump. As the water content in the pump increases, the pumping speed of the pump will decrease, and the pressure in the dryer will increase. The increase in P_c for a given T_s will result in an undesirable increase in T_p.

Design and Construction of the Condenser

The design and construction of the condenser can also be a factor that can impact the primary drying process. If the design of the condenser permits water vapor to enter the vacuum pump, then there can be an increase in P_c, as described in the last section. The design should not permit the buildup of ice on the condenser surface, such as reducing the flow of gases from the chamber or, in an extreme case, blocking the vapor flow from the drying chamber. In either case, a reduction or blockage of the vapor flow from the drying chamber can cause an increase in P_c, with an associated undesirable increase in T_p.

EXAMPLES OF PRIMARY DRYING PROCESSES

The following will describe the rationale for selecting the operating parameters for the primary drying process for formulations #1, #2, and #3. The thermal properties of the formulations were listed in Table 7.1. Figures 7.8, 7.10, and 7.12 illustrated the freezing functions, based on their respective thermal properties, that were used not only to obtain a completely frozen matrix for each of the formulations but also to establish the target T_p for the primary drying process.

In each of the following primary drying processes, bottomless trays, similar in design as that illustrated in Figure 8.21, were used so that vials were in direct contact with the shelf surface. As shown in Figure 8.21, the bottomless tray consists of two sections, a vial retainer and a sliding bottom section. In using the tray, the bottom section is slid under the vial retainer so that it forms a complete tray. The filled vials are then placed in the tray, and the tray is loaded onto the dryer shelf. The sliding bottom section is then pulled out, and vials rest directly on the shelf surface. The vial retainer allows the vials to be moved while on the shelf surface. The use of bottomless trays is preferred because it will eliminate vial-to-vial variations in the primary drying rate as a result of differences in the thermal heat-transfer properties of the trays. Once the drying process is completed, the bottom section is slid back under the vial retainer, and the vials can be safely removed from the dryer shelf. Some care must be exercised in sliding back the bottom section in order to prevent vial breakage.

Figure 8.21. A Bottomless Tray

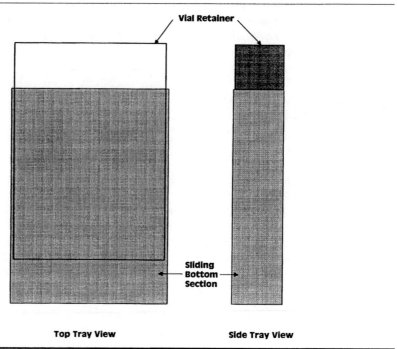

Formulation #1

Determination of the Operating Parameters

The following are guidelines for selecting or determining the primary drying parameters for formulation #1.

Target Product Temperature. The target T_p during the primary drying was set, based on a collapse temperature of –23°C (see Table 7.1), at –28°C ± 3°C. At this temperature, there is a high degree of confidence that the mobility of the water in the interstitial region of the matrix is zero, i.e., the matrix is in a completely frozen state.

Selection of a Chamber Pressure. The vapor pressure of ice at –28°C [8] is about 350 mTorr. Although any P_c lower than 350 mTorr will result in the sublimation of ice, the objective in determining the primary drying parameters is to select a practical P_c that will provide a reproducible sublimation rate based on equation 7.

Since T_p has been designated as being –28°C ± 3°C, the selection of a P_c will define T_s or the P_s as given in equation 7. In essence, the lower the selected P_c, the higher T_s or P_s will be in order to maintain T_p within its defined range of temperatures. A P_c value of 30 mTorr would certainly result in the need for a high value of T_s. However, at such a low P_c, the throughput (expressed in Torr-L/sec and discussed further in Chapter 11) of the dryer may be exceeded for the latter pressure, and P_c will vary dependent on the number of vials or trays in the dryer. An uncontrolled increase in P_c will result, for a given T_s, in an uncontrolled value of T_p. If T_p exceeds the collapse temperature of –23°C, then the product cannot be considered lyophilized; this may be exhibited by signs of collapse or meltback in the final dried product.

In this example, an arbitrary operating P_c of 100 mTorr ± 5 mTorr was selected. Such a pressure was selected because most freeze-dryers can readily attain and maintain such a pressure with dry nitrogen gas and a gas bleed system. In addition, at this pressure, there is little backstreaming of hydrocarbon vapors from the vacuum pumping system. The reader may ask at this point, *If a chamber pressure of 75 mTorr or 125 mTorr was selected as* P$_c$, *provided the throughput of the dryer was sufficient for that pressure, what effect would the selection of either pressure have on the duration of the primary drying process?*

My response to the first part of the question would be that either P_c would be acceptable. The response to the second part of the question would be that the drying process at 125 mTorr will be slightly longer than at 75 mTorr. But if one takes into account the additional time that must be included in the primary drying process to allow for the vial-to-vial variation in the drying rate, as illustrated in Figure 8.13 for the sublimation of ice from a random set of vials, then the selection of other chamber pressures will have only a slight effect on the duration of the primary drying process. But selecting a P_c of 275 mTorr or 300 mTorr would greatly reduce the required T_s to maintain T_p within its prescribed limits and result in a major increase the time necessary to complete the primary drying process.

Selection of a Shelf-Surface Temperature. With the value of T_p defined by the collapse temperature and a practical value of P_c selected, the T_s was experimentally determined. For example, when P_c decreases below 350 mTorr, T_p will decrease. After P_c has reached and is maintained at 100 mTorr, the T_s necessary to increase T_p to its designated target temperature was determined to be 10°C ± 2°C. T_s and P_c are maintained at their respective values until $T_p \approx T_s$ for a specified period of time.

Primary Drying Process for Formulation #1

The primary drying process for formulation #1 is illustrated by Figure 8.22. This figure shows that when the P_c was ≤ 350 mTorr, there was a decrease in T_p. Thus, there was sublimation of ice from the product and, since the decrease in T_p occurred when the pressure gauge indicated a P_c of about 350 mTorr, the accuracy of the pressure gauge was within acceptable limits. Had the decrease in T_p occurred at 300 mTorr, the pressure gauge would be reading false low values and, in the case of a capacitance manometer, the gauge would have to be zeroed at a pressure of ≤ 0.01 mTorr. It should be noted that the sensitivity of detecting the change in T_p with respect to P_c will decrease as the fill volume and fill height of the formulation increase.

The maximum decrease in T_p occurred when the P_c reached and was maintained at 100 mTorr ± 5 mTorr. At this point, the temperature of the heat-transfer fluid was increased until the T_s attained and was maintained at 10°C. The increase in T_s resulted in an increase in T_p until T_p reached the target temperature of –28°C ± 3°C. Under these primary drying conditions, T_p remained within the specified limits until the elapsed time approached 4 hr. At this point, the gas-ice interface fell below

Figure 8.22. The Primary Drying Process for Formulation #1

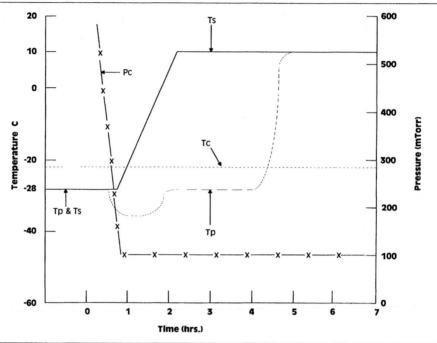

The solid line represents the shelf-surface temperature (T_s); the long dashed line indicates the target product temperature (T_p) during primary drying; the short dashed line represents the collapse temperature (Tc). The chamber pressure (P_c) is denoted by (x-x-x). The elapsed time 0 indicates the starting time of the evacuation of the dryer.

the thermocouple junction (see Figure 7.7), and the temperature of the cake started to increase. Primary drying is considered to be complete when T_p approaches T_s for a specified period of time; primary drying is also determined to be complete through the use of one of the monitoring methods that will be described later in this chapter. In this example, the primary drying ($\approx$ 4 hr) was relatively rapid, and the average energy heat-transfer rate was 47 mcal/sec during the sublimation process.

Formulation #2

Determination of the Operating Parameters

Following are guidelines for selecting or determining the operating parameters for the primary drying of formulation #2.

Target Product Temperature. From Figure 7.10, the target T_p, based on a collapse temperature of –32°C (see Table 7.1), was –37°C ± 3°C. When the target T_p is within the above range of temperatures, there is high degree of confidence that the mobility of the water in the interstitial region of the matrix approaches 0, and the product matrix is considered to be in a frozen state.

Selection of a Chamber Pressure. The vapor pressure of ice at −37°C is about 135 mTorr. While sublimation for this T_p will occur at 100 mTorr, the limits for T_p are given at −34°C and −40°C. Should the value of T_p approach −40 °C the drying rate at 100 mTorr will also approach 0. At a pressure of 50 mTorr, the gas-ice interface will attain a temperature of about −46°C. Although most freeze-dryers can maintain a P_c at 50 mTorr ± 5 mTorr throughout the drying process, for some freeze-dryers there could be insufficient vapor throughput to the condenser to prevent an increase in P_c and a resulting loss of control over the primary drying process.

Selection of a Shelf-Surface Temperature. For a T_p value of −37°C ± 3°C and a P_c value of 50 mTorr, the T_s was experimentally determined to be −20°C ± 2°C. The vapor pressure of ice at −20°C is 776 mTorr, which will give a pressure differential (P_s − P_c) of 726 mTorr, considerably less than the 9,200 mTorr of the previous primary drying process and, based on equation 7, the sublimation rate (χ) for formulation #1 will exceed that of formulation #2.

Primary Drying Process for Formulation #2

Figure 8.23 illustrates the primary drying process for formulation #2, which was found to have a collapse temperature of −32°C. This figure shows that the 12°C temperature differential between T_s and T_p remained relatively constant throughout the entire process, until the gas-ice interface fell below the thermocouple junction of the T_p monitor vial. This suggests that the matrix resulting from the 10°C supercooling of the formulation did not impede the transport of water vapor from the product. The longer primary drying time was a result of a lower heat-transfer rate to the product to provide the necessary energy for sublimation. Based on a primary drying time of about 8 hr, the heat-transfer rate for sublimation of the ice crystals in the formulation was 24 mcal/sec. This primary drying process is typical for formulations having the thermal properties that are listed for formulation #2 in Table 7.1.

Formulation #3

If one were to divide formulations into two distinct categories based on the impact that thermal properties have on the primary drying process, then I would suggest the terms *soup* and *cocktail*. The thermal properties of formulation #1 resulted in relatively rapid primary drying and permitted the use of operating parameters (P_c and T_s) that are readily available in freeze-drying equipment. While the thermal properties of formulation #2 resulted in an extended primary drying time, the process parameters were still within the operating range of most commercial freeze-drying equipment. Because the primary drying processes of formulations #1 and #2 can be readily reproduced from batch to batch, they represent a *healthy* manufacturing process, and thus can be classified as a *soup*. Formulation #3 has some undesirable thermal properties (degree of crystallization of 0.6 and a collapse temperature of −45°C) and presents operating conditions that most commercial freeze-dryers would find difficult to meet without expensive, major modifications. Because of the

Figure 8.23. The Primary Drying Process for Formulation #2

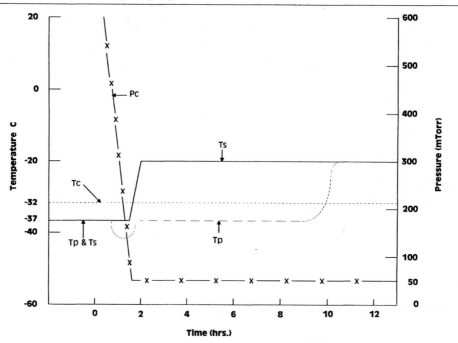

The solid line represents the shelf-surface temperature (T_s); the long dashed line indicates the target product temperature (T_p) during primary drying; the short dashed line represents the collapse temperature (Tc). The chamber pressure (P_c) is denoted by (x-x-x). The elapsed time 0 indicates the starting time of the evacuation of the dryer.

headaches that such a formulation poses on the drying equipment during the primary drying process, the term *cocktail* appears to be quite appropriate.

Determination of the Operating Parameters

Guidelines for selecting or determining the parameters for the primary drying of formulation #3 are as follows.

Target Product Temperature. In the previous formulations, T_p was selected as being $T_c - (5°C \pm 3°C)$ because T_p remained relatively stable (see Figures 8.22 and 8.23) from the time the operating T_s was attained until the gas-ice interface fell below the thermocouple junction. Figure 8.24 shows that T_p slowly increases during the primary drying process, and this increase in T_p must be taken into account to ensure that the frozen matrix remains below the collapse temperature of the formulation throughout the drying process.

Since the collapse temperature of formulation #3 was determined to be –45°C, the target temperature for the primary drying was defined as –51°C (–3°C to +2°C) to allow for the anticipated increase in T_p. The tolerances of –3°C to +2°C for the T_p was found necessary to ensure that T_p did not exceed –45°C.

Selection of a Chamber Pressure. From the above, the lowest value of T_p during the primary drying is given as –54°C. The vapor pressure of ice at this temperature is about 18 mTorr. For sublimation to occur, the pressure in the chamber must be lower than 18 mTorr (e.g., 10 mTorr ± 5 mTorr). Needless to say, at such pressures, back-streaming of hydrocarbon vapors from oil-lubricated vacuum pumps into the dryer is a major concern, and attaining such operating pressures with current freeze-drying equipment without major modifications of the vacuum system is highly unlikely.

Selection of a Shelf-Surface Temperature. The required T_s was found to be –40°C ± 3°C. For most commercial freeze-drying production equipment, the lowest T_s that can be maintained during primary drying is –35°C; therefore, maintaining a T_s of –40°C ± 3°C would require additional refrigeration equipment, which would increase the cost of the drying equipment.

Figure 8.24. The Primary Drying Process for Formulation #3

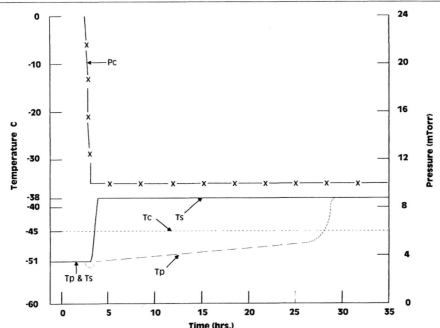

The solid line represents the shelf-surface temperature (T_s); the long dashed line indicates the target product temperature (Tp) during primary drying; the short dashed line represents the collapse temperature (Tc). The chamber pressure (P_c) is denoted by (x-x-x). The elapsed time 0 indicates the starting time of the evacuation of the dryer.

Primary Drying Process for Formulation #3

The primary drying process for formulation #3 is illustrated in Figure 8.24. This figure shows the impact of a *cocktail* formulation.

Pumpdown Time. The first effect of formulation is the extended pumpdown time. The time required to reach the necessary P_c of 10 mTorr ± 5 mTorr is considerably longer than the pumpdown time required for the primary drying of formulations #1 and #2.

Duration of Primary Drying. The primary drying of formulation #3 also takes considerably longer than the previous formulations. The primary drying time for this formulation was five times longer than for formulation #1 and three times longer than the primary drying time for formulation #2.

Increase in T_p Prior to Completion of the Primary Drying Process. The gradual increase in T_p is a result of the relatively low degree of crystallization (0.6) of the formulation (see Table 7.1). As the gas-ice interface descends through the matrix, the flow of gas from the cake is impeded by the lack of conducting paths that occurs because 40% of the water in the formulation did not undergo crystallization during the freezing process, and the cake that was produced had a poorly defined cake structure. The impedance to gas flow not only results in an increase in T_p but also extends the drying time.

Extension of the End of the Drying Process. In order to ensure that the formulation in all of the vials has completed primary drying, it is necessary to greatly extend the time period when the value of T_p shown on all of the monitors is within the limits defined by T_s.

Low Heat-Transfer Rate. The heat-transfer rate during the sublimation of the ice from the product is estimated to be about 7 mcal/sec. Such a low heat-transfer rate is directly related to the excessive time required to complete the primary drying process for this formulation.

Summary

The above three examples of primary drying provided the reader with guidelines for selecting the necessary T_p and P_c and determining the value for T_s. In addition, and perhaps of equal importance, the examples show the major impact that the thermal properties of the formulation have on the operating process parameters (P_c and T_s) and the required design and operation of the freeze-drying equipment. The reader should bear in mind that there are additional pitfalls that await the unwary who select lyophilization to stabilize a formulation. For example, in formulation #2, the degree of supercooling of 10°C had little impact on the drying process. However, the use of a fill volume significantly greater than 1 mL could result in a matrix that could impede the flow of water vapor from the product, and an increase in T_p could occur while an ice matrix was still present in the vial.

PROCESS MONITORING TECHNIQUES

The end of primary drying for formulations #1 through #3 was empirically determined when the T_p approached that the T_s for a specified time interval. The basic problem with such a method is that T_p measures only a single vial on the shelf and other vials may not have completed primary drying. If secondary drying commences before all of the containers completed primary drying, then some containers may show defects from collapse or partial meltback. Collapse will generally affect a large region of the cake; partial meltback is generally limited to a relatively small volume of the product and is located in the middle and bottom of the vial. In response to this problem, a number of other techniques for determining the completion of the primary drying process have been proposed.

Pressure Rise Test

The use of the pressure rise test for determining T_p has already been discussed in rather lengthy detail. While its use in an isolated chamber for determining the T_p during primary drying was quantitative in nature, the pressure rise rate for ascertaining the completion of the primary drying process is qualitative in nature.

As in the manometric method for determining T_p [17], determining the end of primary drying requires that the drying chamber be isolated from the condenser chamber by means of a valve. Such a test method would not be applicable in those dryers in which the condenser surfaces are located in the drying chamber. Another criterion for the use of such a method is that the leak rate of the chamber must be small and inconsequential for the test time period [23,25]. The rate of the rise in pressure will be dependent on the shelf temperature. For a shelf temperature of 40°C, the pressure rose from 400 mTorr to 760 mTorr in 3 sec [23]. Such a rapid increase in pressure indicates that ice is still present in the chamber. If the pressure does not reach a predetermined pressure within the time restraints of the test, then primary drying is considered to be complete, and one can commence with the secondary drying process. It would make sense that one would not perform the pressure rise test until all of the T_p thermocouple sensors were within the specified temperature limits of the shelf surface. Performing the pressure rise test during the main portion of the primary drying process would result in an increase in T_p, and there is always a concern that the T_p could exceed the collapse temperature of the frozen matrix [19].

Comparative Pressure Measurement

In an earlier section of this chapter, it was pointed out that Nail and Johnson [23] monitored the primary drying process by determining the difference between the output of a thermal conductivity gauge (Pirani gauge) and the output of a capacitance manometer. The capacitance manometer determines pressure by sensing the deflection of a diaphragm by the pressure in the chamber. For a given pressure, the output of the Pirani gauge will be dependent on the composition of the gases. In the presence of water vapor (see Table 8.1), the output of the thermal conductivity gauge will indicate a false high pressure. The difference between pressure measured by a thermal conductivity gauge and that measured by a capacitance manometer gauge will depend on the partial pressure of water vapor. Results of tests conducted by Nail and Johnson [23] indicate that there is a substantial quantity of water vapor in the residual atmosphere of the drying chamber at the end of the primary drying process. In view of the presence of such water vapor in the drying chamber, this monitoring technique is not always applicable for ascertaining the completion of the primary drying process, as suggested by Bardat, et al. [26].

Dielectric Moisture Sensor

When a voltage is applied across an electrical insulator, equal and opposite charges will build up until the voltage across the insulator is equal to the applied voltage. The amount of charge required to raise its potential one unit is referred to as the *capacitance* of the substance. The capacitance (C) of a substance is directly related to the *dielectric constant* (ε) by the expression

$$C = \frac{\varepsilon A}{4\pi d'} \tag{19}$$

where d' is the distance between the capacitance plates. ε is defined as the ratio of the capacity of the insulator to the capacity of a vacuum. The value of ε will be > 1.

As an alternating voltage is applied to the capacitor, some of the energy is stored, while another portion of the energy, known as the *loss*, is dissipated in the form of infrared heat (IR). The IR is generated as a result of the resistance properties of the capacitor and the current through the capacitor. The reciprocal of the loss is referred to as the quality factor (Q'). As the *loss* decreases, the Q' of the capacitor will increase.

At a frequency of 1 MHz, aluminum oxide (Al_2O_3) [8] has a ε ranging from 4.5 to 8.4. The Q' for Al_2O_3 is moisture dependent and ranges from 100 to 5,000. Because the Q' of Al_2O_3 is so sensitive to moisture, its measurement provides an excellent means for determining the relative humidity in a chamber.

An illustration of an Al_2O_3 sensor is shown in Figure 8.25. As shown in this figure, the moisture sensor consists of an aluminum metal base containing a hole and a top surface on which Al_2O_3 film is formed [19,26]. A thin gold film is vapor deposited over the Al_2O_3. The gold film is thick enough to be electrically conductive but thin enough to permit the diffusion of water vapor. Electrical contact with the gold film is made by the insertion of a metal pin to serve as a counterelectrode. Because the counterelectrode is insulated from the main aluminum block, the gold film, Al_2O_3 film, and the aluminum block form a capacitor. Since the oxide is a thin film, an equilibrium between the partial pressure of water vapor above the gold film and the absorbed water in the oxide is quickly established. The sensors are calibrated to an accuracy of ±2°C of the frost point or ±23% in vapor pressure [19].

Roy and Pikal [19] also compared the output of the electronic sensor with respect to the *trap method*. The trap method measures the difference in pressure of a small-volume system with the trap at ambient temperature and the observed pressure when the trap is chilled with dry ice (-78°C). The observed difference in pressure is a result of the change in the partial pressure of water vapor in the chamber [9]. When the results of the measurement of the partial pressure of water vapor in a chamber with the electronic capacitor sensor are compared to those of the trap method, there was agreement between the two measurement methods at pressures > 60 mTorr. At pressures less than 60 mTorr, however, the capacitor sensor tended to

Figure 8.25. An Aluminum Oxide Moisture Sensor

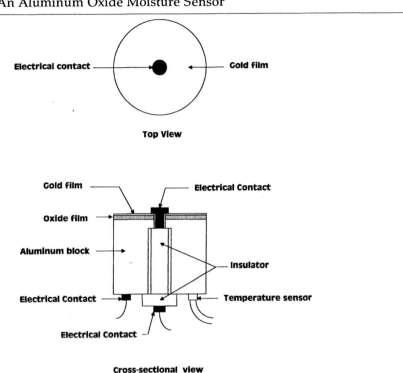

indicate a higher partial pressure. Roy and Pikal [19] attributed this higher reading to the relatively long time required for the oxide film to equilibrate with the lower partial pressure of water vapor. Since the output of the capacitor sensor is also temperature dependent, maintaining calibration in a freeze-dryer could be difficult. For this reason, Roy and Pikal [19] described the capacitor sensor as being *semiquantitative* in nature. In spite of this limitation, Roy and Pikal found the capacitance moisture sensor to be useful in determining the endpoint of the primary drying process. The sensor is reported to be capable of sensing that as few as 0.3% of the vials in a batch have not completed the primary drying process.

Bardat et al. [26] made a comparative study of the electronic capacitance sensor and the comparative pressure measurement using the pressure difference between a capacitance manometer and a Pirani gauge. They found that the electronic capacitance sensor indicated that primary drying was still occurring in the dryer when the comparative pressure measurement indicated a completion of the drying process. The chief advantage of the capacitance sensor, as reported by Bardat et al. [26], is its reliability in sensing the completion of the primary drying process. Such a feature is most important during the scale-up of a primary drying process from a pilot dryer to a production-size dryer or when changing the batch load of a dryer.

Windmill Device

Another device for monitoring the primary drying process that never found wide acceptance but because of its uniqueness deserves to be mentioned, is the windmill [27]. This device consists of a cover that fits over a tray containing the vials. At either end of the tray are two turbines or *windmills*. These windmills have a one-inch diameter and consist of a polished aluminum propeller supported on a metal rod. In an effort to reduce friction, the end of the rod is sharpened to a fine point and makes contact with a sapphire bearing. Because of the reduced friction, the windmills require relatively little difference in pressure between the closed container and the drying chamber to cause them to spin. The use of the windmills requires that they be visible to the operator. Since the revolutions per minute (rpm) of the windmill are not measured or recorded, the use of such a device depends on frequent visual checks by the operator. In spite of this limitation, it was observed that there was often a correlation between the time that the windmills stopped spinning and when the T_p approached the T_s. In other tests, the windmills were observed to continue to spin for some time after the T_p had reached the T_s. These results would indicate that there were variations in the sublimation rate. For those Don Quixotes who may attack the idea of the use of windmills, this device did serve to stimulate others to seek new means for determining the completion of the primary drying process. Thus, the science of lyophilization, like the tide, advances one wave at a time.

Summary

The previous sections have reviewed various means that have been employed to monitor the primary drying process and, most importantly, to ascertain the end of the primary drying process. The description of the windmill by Couriel [27] reflects the need for determining the end of the primary drying process. In 1977, prior to the abundant availability of computers, the physical measurement of operating parameters was still in vogue (e.g., the thermocouple, Pirani, and capacitance manometer gauges had not yet replaced the mechanical measurement of the chamber pressure via a mercury McLeod gauge).

By using more sophisticated, selective pressure measurement techniques, such as the electronic capacitance humidity sensor, Roy and Pikal [19] started to put quantitative values to their measurements. It is unfortunate that these authors did not provide us with the data or the method they used to determined that the electronic capacitance humidity sensor could sense that as few as 0.3% of the vials in a batch had not completed the primary drying process. This was a very important step because it now placed the completion of the primary drying process on a statistical basis. If one were using the electronic capacitance humidity sensor to monitor the primary drying of 100 vials of product, there would be little chance of finding a vial that was defective because its temperature was increased for secondary drying before the contents of the vial had completed primary drying. However, for a production batch of 100,000 vials (and such batch sizes do exist), the number of defective vials would increase to 300. Not only must such defective vials be found and removed by some visual, nondestructive test; the harm that may befall some individual who is depending on the quality of such a product must be considered. We owe it, not to any governmental agency, such as the FDA, but to those individuals who depend on us, as scientists or engineers, that the product that they use will be free of any defects generated by the manufacturing process.

CALORIMETRIC MEASUREMENT TECHNIQUES

Introduction

Discussion of the primary drying process has thus far been defined, controlled and monitored using two or more of the key parameters, i.e., P_c, T_s, and T_p). But these represent *indirect*, interrelated parameters. As illustrated in Figures 8.22 through 8.24, during the primary drying process, one cannot use indirect parameters to obtain any information that would serve as a basis for forecasting the termination of the drying process. The only comfort one can have when using these parameters is reliance on the consistency of nature (e.g., that vapor pressure of pure water at -40°C will be 96 mTorr not only today but also tomorrow).

The true driving force of primary drying is not the key parameters listed above but the energy necessary for the drying process. This energy includes not only the heat of sublimation but also the energy necessary to sublimate the uncrystallized water that with the other constituents in the formulation composes the glassy interstitial region of the frozen matrix. If one could establish the total energy (H_m) necessary to conduct the primary drying process under the conditions defined by the key parameters, then by determining the rate of heat transport (Q_m) into a container like a glass vial, one would have a scientific means for forecasting the duration of the primary drying process. In addition, if one also had knowledge of the frequency distribution of the heat-transfer coefficients (C_o) of the vials, one would have the rationale for extending the primary drying process such that the number of process-generated defective vials in a given lot would be beyond normal chance. If a lot size of the glass vials was 10 million vials and the product batches consisted of 100,000 vials, then the primary drying process would be extended such that the probability of finding a process-defective vial in any of the 100 lyophilized batches would be about 6σ, or about 1 in 50 million. This would not mean that there could not be a process-defective vial in a batch, but the probability of such an occurrence is beyond normal chance.

In this section, the method for making such calorimetric measurements will be described, along with the means for determining the C_o of a given vial [11]. The application of Q_m and H_m measurements during a lyophilization process will be described. The final portion of this section will address the issue of developing a primary drying process in which the number of process-defective vials will be considered beyond normal chance.

Method for Making Calorimetric Measurements

Determination of Q_m

The heat-transfer rate (q) into a container has been previously defined by equation 8; when expressed in terms of cal/sec, however, the general expression for q' during the primary drying process becomes

$$q' = \frac{KA}{d}(T_s - T_{c,i}) \quad \text{cal/sec} \tag{20}$$

where K is the thermal conductivity as a function of E_t (see equation 11).

If there are two containers in which KA/d are approximately equal, then the difference in Q_m between these containers can be expressed as

$$Q_m = (q'_s - q'_r) \quad \text{cal/sec} \tag{21}$$

were the containers are distinguished by the notations r and s. The Q_m of these containers can only be approximately equal when their C_o values are also approximately equal. The C_o term is defined as

$$C_o = \frac{KA}{d} \quad \frac{\text{mcal}}{\text{cm}^2 \text{sec}} \tag{22}$$

where the energy is expressed in mcal.

If the temperatures of both containers are measured by attaching a 32-gauge thermocouple to the middle of each container, as shown in Figure 8.26, then substituting equations 20 and 22 into equation 21, one obtains

$$Q_m = C_o(T'_r - T'_s) \quad \text{mcal/sec} \tag{23}$$

where T_r' is the temperature of the reference container and T_s' is the temperature of the sample container. For $Q_m > 0$, the reaction will be endothermic, e.g., sublimation process. $Q_m < 0$ will be an indication of an exothermic reaction, e.g. the formation of ice crystals. When $Q_m \to 0$, no phase change is occurring in the sample vial.

Determination of $H_{m,t}$

The total energy at any time t during the primary drying process is defined as

$$H_{m,t} = 1 \times 10^{-3} \int_o^t Q_m dt \quad \text{cal} \tag{24}$$

where dt is the time differential.

For the primary drying process, the total measured energy ($H_{m,t}$) at any time t will be equal to the heat of sublimation and the energy necessary to sublimate the uncrystallized water from the glassy interstitial region. Since the uncrystallized water in the interstitial region of the matrix is immobile during the primary drying portion of the lyophilization process, one can assume that it has formed a solid solution with the other constituents in the formulation. If such a solid solution were to follow Raoult's Law, then the vapor pressure of the water (P_A) above the interstitial region would be related to the equation

$$P_A = X_A P_{eq} \quad \text{mTorr} \tag{25}$$

where X_A is the mole fraction of water in the solid solution. For this reason, the vapor pressure of the water in the interstitial region will be lower than P_{eq} for a given temperature. This would imply that those formulations having low degrees of crystallization will have significant quantities of uncrystallized water in the interstitial region of the matrix. The sublimation of the uncrystallized water will tend to require more energy than the sublimation of the ice crystals.

Determination of C_o

In order to make calorimetric measurements during the primary drying process, it will first be necessary to make an accurate determination of the heat transfer coefficient (C_o) of the vials. The following will be a general description of the preparation of the vial, the measurement method and the determination of C_o.

Preparation of the Sensor Vial

The thermocouples are first calibrated at –78°C (dry ice), 0°C (ice–water slush bath), and 100°C (boiling water). The thermocouple is then attached to the bottom of a marked vial with a small quantity of a conductive epoxy, as illustrated by Figure 8.26. The epoxy is cured at about 250°C. A small known quantity of water (≈ 1 g), just sufficient to cover the epoxy-covered thermocouple, is added to the vial. The thermocouple wires are then passed through a closure. The closure is then crimp sealed to the vial. The last step is necessary in order to prevent water vapor from leaking from the vial during the measurement.

Measurement Method

The prepared vials are placed on the shelf of a small laboratory-scale freeze-dryer. Special precautions are taken to cover the walls and door of the dryer with aluminum foil to protect the vials from stray radiant energy. The thermocouple leads are

---

Figure 8.26. The Configuration of the Vial Sensors Used in Making Calometric Measurements During the Primary Drying Process

---

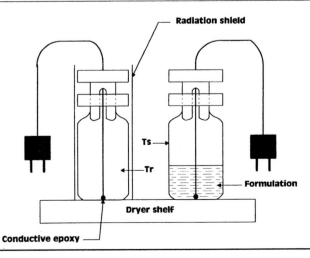

then connected to a computerized data acquisition system so the temperature can be recorded at discrete time intervals (e.g., 1 min).

The temperature of the shelf surface is refrigerated to ≤ -20°C to ensure complete freezing of the water. The condenser is chilled to < -60°C, and the drying chamber is evacuated to the pressure that will be used during the primary drying process.

The T_s is then increased and maintained at -4°C ± 1°C until all of the vials are within ±1°C of the T_s for no less than 30 min.

The temperature of the shelf surface is set at 10°C, and the temperatures of the vials are measured at discrete time intervals. The test is continued until all of the vial thermocouples are within ±1°C of the T_s for no less than 30 min. The vials are removed from the dryer, and the water is removed from the vials. The vials are then stored in a clean, dry environment.

Determination of C_o Value

The C_o is determined from

$$C_o = \frac{1}{area}(m_w C_i \Delta T_i + m_w \Delta H_f + m_w C_w \Delta T_s + m_v C_c \Delta T_s +$$

$$\tag{27}$$

$$m_s C_s \Delta T_s + m_a C_a \Delta T_s + m_{TC} C_{TC} \Delta T_{TC} \quad \frac{cal}{(°C\ sec)}$$

where the term *area* is the area under the plot of $(T_s - T_v)$ as a function of time where T_v is the vial temperature, m_w denotes the mass of water in the vial expressed in grams, C_i is the heat capacity of ice, ΔT_i represents the difference in temperature $(0°C - T_i')$ where T_i' is the temperature of the ice, ΔH_f is the heat of fusion of water (79.72 cal/g), m_v denotes the mass of the vial expressed in grams, C_c is the heat capacity of type I glass, ΔT_s represents the difference in temperature $(T_i' - T_s')$ where T_s' is the temperature of the sample vial, m_{TC} indicates the mass of the thermocouple wire in the vial, C_{TC} denotes the heat capacity of the thermocouple wire, m_s denotes the mass of the closure, m_a the mass of the crimp seal, and C_s and C_a represent the heat capacities for the closure and the crimp seal, respectively.

Values of C_o measured thus far for various types and size vials range from about 0.001 cal/(cm² sec) to about 0.01 cal/(cm² sec). As more C_o measurements are made, this range of C_o values may be expanded.

Measurement of Q_m and H_m During Primary Drying

In order to understand the impact that calorimetric measurements can have on monitoring the primary drying process, and especially on ascertaining the completion of the process, consider the primary drying of a formulation using the parameters P_c, T_p, and T_s. Figure 8.27 shows two primary drying processes for the same formulation. Both drying processes were performed at a pressure of 30 mTorr ± 5 mTorr. In Figure 8.27a, the completion of primary drying was based on when the T_p approached the T_s. Under these drying conditions, it required about 136 hr before T_p approached T_s. However, Figure 8.27b shows that the drying was terminated after

Figure 8.27. The Primary Drying of a Formulation

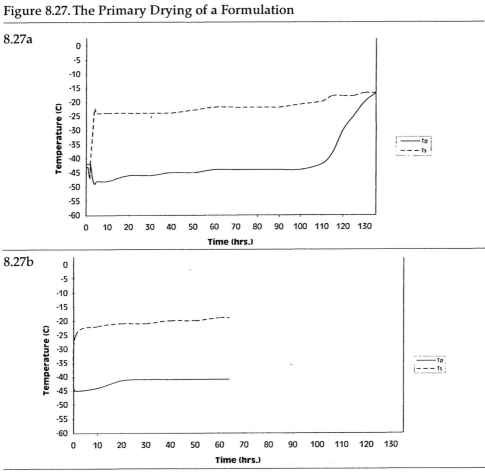

8.27a

8.27b

The primary drying of a formulation at a chamber pressure of 30 mTorr ± 5 mTorr. Figure 8.27a shows the primary drying process when using $T_p \rightarrow T_s$ to signify the completion of the primary drying process. Figure 8.27b is the primary drying of the same formulation and using the same parameters as in Figure 8.27a; however, primary drying was terminated based on calorimetric measurements.

only 64 hr and $T_p \ll T_s$. The reason for terminating the primary drying at 64 hr was based on Figure 8.28, which is a plot of Q_m and H_m as a function of time.

Figure 8.28 shows that $Q_m \rightarrow 0$ after 55 hr, signifying the completion of the primary drying process. It is of interest that there was a slight decrease in Q_m during the drying process. This decrease could be associated with an increase in the pressure at the gas-ice interface. The increase in Q_m near the end of the drying process could stem from a decrease in T_s', as a result of the approach of the gas-ice interface, as illustrated by Figure 8.12d. Unlike the slow increase in T_p to indicate the completion of primary drying, as seen in Figures 8.22 through 8.24, the completion of primary drying in the calorimetric thermograph seen in Figure 8.28 is quite sharp and well defined.

Figure 8.28. A Plot of Q_m and H_m as a Function of Time for the Primary Drying Process in Figure 8.27b

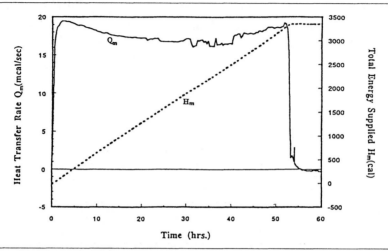

[Source: *PDA Journal*, "Calorimetric Monitoring of Lyophilization," Vol. 49, No. 6, Nov.–Dec. 1995, Page 280, Figure 5c.]

The question that now should be addressed is, *Why should the completion of the primary drying process be so sharp for calorimetric measurements while being undetectable for conventional P_c, T_p, and T_s measurements?* One possible explanation for the apparent discrepancy is that the formulation had a low solid content and, for that reason, there was poor thermal conductivity to the product thermocouple used to determine T_p. A second possibility is that secondary drying of the cake had begun, and the cake was at a significantly lower temperature than the bottom of the vials. I would like to point out that the approximately 50% reduction in the primary drying time should not be expected for all formulations. While some other formulations showed significant reduction in the time to complete primary drying, there was little or no change in the primary drying time for other formulations.

Applications of the Frequency Distribution of C_o

In an earlier section, a method for measuring and calculating the C_o of a container such as a glass vial was described. If the C_o values for a random sample of 25–30 vials of a given manufacturing lot are determined, one has a statistically adequate number of observations of a system (vials) to obtain a frequency distribution of the C_o of the lot. From such a frequency distribution, one can obtain the following:

Sensors for Calorimetric Measurements

From the frequency distribution, one can select those vials that are representative of the mean $\overline{C}_o$ value. Because these vials will represent the mean $\overline{C}_o$ value of the vials in the lot, one can now obtain a calorimetric thermograph that will be representative of the vast majority of the vials in the lot. In addition, the C_o values of these vials will be quite similar and, therefore, meet the requirements for determining Q_m based on equation 23.

Standard Deviation

From the measurements of C_o, one can calculate the standard deviation (α). A knowledge of α will enable one to make two important determinations. The first is how long the primary drying must be extended to ensure that the odds of finding a product-defective vial in the lot beyond normal chance.

Assume that the mean value of a distribution of C_o values for a random sample of vials is 3.50×10^{-3} cal/(°C sec) and has a standard deviation of $\pm 0.05 \times 10^{-3}$ cal/(°C sec). Our concern is not necessarily with C_o values $> 3.50 \times 10^{-3}$ cal/(°C sec), unless the distribution is exceptionally broad, but with vials having C_o values $< 3.50 \times 10^{-3}$ cal/(°C sec). These latter vials will require more time to complete the primary drying process. The additional time (t_o) necessary to complete primary drying can be estimated from the following equation:

$$t_o = \frac{\overline{C_o}}{\overline{C_o} - n\sigma} \times t_p \quad \text{hr} \tag{28}$$

where n is a coefficient of σ and t_p is the primary drying time for vials having the mean value of $\overline{C_o}$.

Assume that the sensor vials used to generate the thermogram shown in Figure 8.28 have a mean value of C_o. Since primary drying required 55 hr to complete, the additional time necessary to ensure completion of primary drying for a product batch containing 350 vials or 3σ is estimated, from equation 29 to be 57 hr. For a batch containing 100,000 vials, one may want to be certain that all of the vials completed primary drying so $n = 5$ or the odds of having a vial not complete primary drying would be 1 in 1 million. In order to achieve such a confidence, the t_o would have to be extended to 59 hr.

If the frequency distribution of $\overline{C_{o,b}}$ for a second lot of vials is found to have a mean value of 3.40×10^{-3} cal/(°C sec) $\pm 0.04 \times 10^{-3}$ cal/(°C sec), then two questions need to be answered: (1) Is there a reasonable chance that a process-defective product vial will appear in a given batch (100,000 vials/batch)? (2) What is the number of expected process-defective product vials that will appear in the entire lot of glass (lot size is 2 million vials) when the duration of the primary drying process is 84 hr?

In order to answer the above two questions, equation 28 can be expressed in terms of n as

$$n = \frac{1}{\sigma_b} \left(\overline{C_{o,b}} - \frac{\overline{C_o} t_p}{t_o} \right)$$

here σ_b has a value of 0.08×10^{-3} cal/(°C sec). With the values given above, n is calculated to have a value of 3.4; thus the odds of a process-defective vial in the batch is about 1 in 1,500. For a batch of 100,000 vials, one would anticipate about 66 (0.07%) process-defective vials in the batch. In the entire lot of vials, the number of cake defects would be expected to approach 1,300.

Summary

Calorimetric monitoring of the primary drying process offers several important advantages over the conventional means for monitoring the primary drying process, i.e., P_c, T_p, and T_s.

The first advantage is that it monitors the basic process parameter, which is the energy for sublimation of ice and the glassy ice in the interstitial region of the matrix. From a knowledge of Q_m and the total sublimation energy, one can almost immediately ascertain the duration of the primary drying process.

The second and perhaps more important advantage of this technique is that it provides a means for determining the frequency distribution of C_0 for a lot of vials. From such information, one has a means for predicting the number of process-defective vials that may be generated during a given batch and for a given lot of vials.

SYMBOLS

A area

A'' cross-sectional area defining the gap between the vial and the surface

Area   area under the plot of $(T_s - T_v)$ as a function of time

α ratio of the rate per unit area that molecules actually condense on a surface to the unit rate at which they strike the surface

α' the accommodation coefficient for a gas

$B(P,C)$ a nonunit term that is pressure and container dependent

C capacitance of a substance

C_a heat capacity of the crimp seal

C_c heat capacity of type I glass

C_i heat capacity of ice

C_o heat-transfer coefficient of a container

$\overline{C_{o,b}}$ mean heat-transfer coefficient for a given lot of vials

C_s heat capacity for the closure

C_{TC} heat capacity of the thermocouple wire

d' distance between two capacitance plates

d thickness of the base of the container

ΔH_f heat of fusion of water (79.72 cal/g)

ΔT_i difference in temperature $(0°C - T_i)$

ΔT_s difference in temperature $(T_i' - T_s)$

dt time differential

ϵ dielectric constant of a substance

ϵ_s emissivity of shelf surface

ϵ_c emissivity of the container

E_i kinetic energy of ith number of molecules

E_s energy transferred between two surfaces as a result of gas thermal conductivity

E_r energy transfer resulting from Stefan-Boltzmann Law

E_t total heat transfer from a surface to a container

f_i fraction of the gas pressure comprised of the ith gas component

F gas conductance of the channel (L/sec)

G gauge reading of the thermal conductivity gauge

h_c cake thickness

ΔH_s enthalpy for sublimation of ice

H_m total energy for primary drying process

I electric current expressed in amps

k Boltzmann's constant, 3.3×10^{-24} cal/(molecule-K)

l_p length of the conduction path

λ thermal conductivity of a substance

Λ_i thermal conductivity of the ith gas species

$\dot{m}$ rate of sublimation expressed in g/hr

m_a mass of the crimp seal in g

m_s mass of closure in g

m_{TC} mass of the thermocouple wire in the vial in g

m_v mass of the vial expressed in g

m_w mass of water in g

mfp the mean distance a gas molecule or atom will travel between collisions

M molecular weight expressed in kg

M' molecular weight expressed in g

n coefficient of $\alpha \geq 1$

N_i number of water molecules having an energy E_i

N_o total number of water molecules on a disordered surface

χ rate of sublimation expressed in kg/(m^2 sec)

χ_o preexponential constant

P pressure expressed in either Torr or mTorr

P_a vapor pressure at the condenser surface

P_A vapor pressure of water vapor over a solid solution having a mole fraction of X_A

P_c chamber pressure

P'_c pressure above the cake

P_{eq} equilibrium vapor pressure of ice expressed in Pascals

P_i manometric pressure expressed in Torr

P_s vapor pressure of ice at the shelf-surface temperature

P_v saturated vapor pressure (SVP) of ice expressed in Pascals

P'_v pressure in the vial

P_1 pressure at the gas-ice interface

q heat-transfer rate resulting from thermal conductivity

q' total heat transfer between the shelf surface and a container

q_x' the heat transfer of a container denoted by the letter x

Q gas conductance of a system (Torr-L/sec)

Q' quality factor of a capacitor

Q_m rate of heat transport into a container

R gas constant [8.3144 J/(K mole)]

R' electrical resistance of a substance

R_s sublimation rate

R_t sum of the thermal resistances resulting from temperature drops from the shelf surface to the sublimation surface

σ standard deviation

σ_b standard deviation for the frequency distribution for a lot of vials having a mean value of $\overline{C}_{o,b}$

t_o time required for completion of the primary drying for vials having a mean value of $\overline{C}_o$

t_p primary drying time for vials having the mean value of $\overline{C}_o$

T temperature in °C unless specified in K

Tc collapse temperature of the frozen formulation

T_c condenser temperature

$T_{c,i}$ temperature at the ith point of the container

T_i temperature of gas striking the shelf surface

T_i' temperature of the ice

T_i'' calculated temperature at the gas-ice interface

T_p product temperature (frozen or dried cake)

$T_{p,i}$ gas-ice interface temperature

T_r temperature of the gas upon leaving the shelf surface

T_r' temperature of the reference container

T_s shelf-surface temperature

T_s' temperature of the sample container

T_t tray-surface temperature

T_v temperature of the vial

X_2 mole fraction of air in the sample tube

REFERENCES

1. N. Maeno and H. Nishimura, *J. of Glaciology*, 21: 193 (1978).

2. T. A. Jennings, *J. Parenter. Sci. & Tech.*, 42: 118 (1988).

3. T. A. Jennings, loc. cit., 40: 95 (1986).

4. M. J. Pikal, S. Shah, D. Senior, and J. E. Lang, *J. Pharm. Sci.*, 72: 635 (1983).

5. R. G. Livesey and T. W. G. Rowe, *J. Parenter. Sci. & Tech.*, 41: 169 (1987).

6. S. L. Nail, *J. Parenter. Drug Assoc.*, 34: 358 (1980).

7. S. Dushman, in *Scientific Foundations of Vacuum Technique* (J. M. Lafferty, ed.), John Wiley & Sons, Inc., New York (1962).

8. *CRC Handbook of Chemistry and Physics* (E. C. Weast, ed.), 65th ed., CRC Press, Inc., Boca Raton, Florida (1984).

9. J. Pikal, M. L. Roy, and S. Shah, *J. Pharm. Sci.*, 73: 1224 (1984).

10. M. J. Pikal and S. Shah, *PDA J. Parenter. Sci. & Tech.*, 51: 17 (1997).

11. T. A. Jennings and H. Duran, *PDA J. Parenter. Sci. & Tech.*, 49: 272 (1995).

12. S. D. Patel, B. Gupta, and S. H. Yalkowsky, *J. Parenter. Sci. & Tech.*, 43: 8 (1989).

13. S. L. Nail, *J. Parenter. Drug Assoc.*, 34: 358 (1980).

14. T. W. G. Rowe, in International Symposium on freeze-drying of biological products, Washington, D.C. 1076. *Develop. Biol. Standard*, 36: S. Karger, Basel, 1977, p. 79.

15. H. Seager, *Man. Chem. & Aerosol News*, Dec. 1978, p. 59.

16. M. J. Pikal and S. Shah, *PDA J. Parenter. Sci. & Tech.* 51: 17 (1997).

17. N. Milton, M. J. Pikal, M. L. Roy, and S. L. Nail, *PDA J. Parenter. Sci. & Tech.*, 51: 7 (1997).

18. T. A. Jennings, *J. Parenter. Drug Assoc.*, 34: 62 (1980).

19. M. L. Roy and M. J. Pikal, *J. Parenter. Sci. & Tech.*, 43: 60 (1989).

20. T. A. Jennings, *J. Valid. Tech.*, 3: 386 (1997).

21. D. E. McVean, Notes from PDA Lyophilization Interest Group held at the 1996 Annual Meeting in Philadelphia, PA.

22. *Manual on Presentation of Data and Control Chart Analysis*, 6th ed., prepared by Committee E-11 on Quality and Statistics, ASTM, Philadelphia, Penn., 1990.

23. S. L. Nail and W. Johnson, in International Symposium on Biological Product Freeze-Drying and Formulation, Bethesda, USA 1990, *Develop. Biol. Standard.*, 74: Karger, Basel, 1991, p 137.

24. *Vacscan Manual*, revision 1.00 dated December 14, 1990, by Spectra Instruments, Morgan Hill, Calif. 95037.

25. H. Willemer, in International Symposium on Biological Product Freeze-Drying and Formulation, Bethesda, USA 1990, *Develop. Biol. Standard.*, 74: Karger, Basel, 1991, p 123.

26. A. Bardat, J. Buguet, E. Chatenet, and F. Courteille, *J. Parenter. Sci. & Tech.*, 47: 293 (1993).

27. B. Couriel, *Bull. Parenter. Drug Assoc.*, 31: 227 (1977).

9

Secondary Drying Processes

INTRODUCTION

While the principal objective of the primary drying is to sublimate the solvent (generally water) from the matrix, the primary function of the secondary drying is to reduce the residual moisture content of the product to levels that will no longer support biological growth or chemical reactions. It is this stage of the lyophilization process that serves as a means for slowing the kinetic clock of the active constituent. This slowing of the kinetic clock accounts for the unusually long stability of a lyophilized product. A formulation such as a vaccine that can be stored effectively at 4–8°C for a few days can now be stored for months or years at ambient temperature.

While it was pointed out in the previous chapter that there is no sharp line of demarcation between the primary and secondary drying processes, the quantity of water remaining absorbed in the cake upon completion of the primary drying process is generally too high for achieving the desired product stability. It is, therefore, the main function of the secondary drying process to reduce the *residual moisture* content to acceptable values.

This chapter will first consider the basic role of the secondary drying process. It will then consider the importance that desorption isotherms play in determining the final residual moisture in the product and the importance that *bound water* has to the stability of the final product. Examples of the secondary drying processes for previously described formulations #1, #2, and #3 will provide the reader with various means for completing the primary drying process. The final section of the chapter will consider the various methods for determining the completion of the secondary drying process.

ROLE OF THE DRYING PROCESS

Historical Background

It is always of interest when first examining a particular topic to have some understanding of its evolution. Secondary drying was certainly present when Shackell [1] first dried frozen specimens in a vacuum system, where sulfuric acid served as a condenser system, but it is doubtful if Shackell in 1909 was concerned with defining the drying process into two distinct segments, namely, primary and secondary. Hill and Pfeiffer [2], as late as 1940, still referred to what we now call lyophilization as *vacuum desiccation from a frozen state.* The earliest reference, that I could locate that divided the drying process into two segments and in which the second segment was referred to as secondary drying was by Greaves [3].

Greaves [3] found that the stability of dried blood serum at temperatures near 100°C was inversely related to the moisture content when the blood was dried in the presence of phosphorous pentoxide (P_2O_5). It should be noted that the equilibrium vapor pressure of water above a fresh surface of P_2O_5 at ambient temperature is considered too low to be measured. For blood serum, Greaves [3] recommended that laboratory samples be perfectly dry and that the minimum time for secondary drying in the presence of the above desiccant should be no less than 3 weeks. While Greaves may have been justified in making such a recommendation for dried blood serum, it will be shown later in this chapter that the duration of the secondary drying process can be considerably shorter than 3 weeks and that the required moisture content depends on the product (i.e., some products may require residual moisture values > 1% wt/wt for maximum stability).

General Description of Secondary Drying

The most common definition of the secondary drying process is the removal (desorption) of water vapor from the product after completion of the primary drying process [4–8]. For the most part little if any consideration is given to the rate at which the product temperature (T_p) is increased from the final primary drying temperature—see Figures 8.22 through 8.24—to the terminal secondary temperature. The final secondary shelf-surface temperature (T_s) is often selected arbitrarily or from the lyophilization of another formulation that may or may not have the same constituents. This absence of rationale also holds true for the selection of the chamber pressure (P_c) and the duration of secondary drying once $T_p \rightarrow T_s$. While the primary drying parameters were based on the thermal properties of the formulation, the phase diagram of water, and the frequency distribution of the heat-transfer coefficients (C_0) of the containers (vials), the basis for selecting the operating parameters for the secondary drying is far more complex. Because of the complexity of the secondary drying process, it is not uncommon to come across processes whose duration is a matter of days; if we were to follow the advice of Greaves [3], secondary drying could be extended for a matter of weeks.

CRITERIA FOR SELECTING SECONDARY DRYING PARAMETERS

Here we will consider the basis for selecting the secondary drying process parameters. These parameters will be based on (a) the quantity and nature of the residual water in the product; (b) the absorption, adsorption and desorption processes; and (c) a model for the secondary drying process of a lyophilized formulation.

Residual Moisture and Stability

Our general definition of lyophilization requires a stabilizing process in which there was a reduction in the quantity of the solvent (generally water) to levels that will no longer support biological growth or chemical reactions. Greaves [3] pointed out that merely reducing the moisture content of the final product to values less than 1% wt/wt would not necessarily ensure the desired storage life or stability of the dried product. How true those words were, yet Greaves did not realize that in the years to come the stability for some active substances would require moisture values greater than 1% wt/wt. Let us now consider the two categories of water one can encounter in a lyophilized formulation and the impact that such waters have on the final stability of the product.

Residual Moisture

Contrary to some [7], the residual moisture or *free water* in the final product is water absorbed by the cake or adsorbed on the cake surface. This water can stem either directly from the interstitial region (final cake) or from the flow of water vapor that occurs during the primary drying process. The residual moisture, as anticipated, will not be uniformly distributed throughout the cake. Pikal and Shah [9] demonstrated that during the secondary drying process, a large variation in moisture content can exist throughout cakes comprised solely of human serum albumin (HSA), dextran 40, and bovine somatotropin (BST). These authors found the lowest amounts of moisture existed at the top of the cake; however, they also found that even lower moisture content existed near the walls of a vial. The low moisture near the walls of a vial is attributed to some shrinkage of the cake away from the walls. This cake shrinkage provides a gas conduction path for the water vapor to leave the vial. Storage in a sealed container for no less than 24 hr eliminates the gradient in the moisture content and distributes the moisture uniformly throughout the cake [9].

The reader should exercise caution and not generalize that during the drying process the lowest moisture content will occur near the walls of a vial. This condition will occur only when there is shrinkage of the cake away from the wall of a vial. It has also been my experience that vials with cake shrinkage along the walls can be readily seen by inverting the vials; if shrinkage has occurred, the cake falls to the top of the vial. Yet some lyophilized cakes do not separate from the walls of a vial, even when the walls are tapped while the vial is in an inverted position. For these cakes the top of the cake may represent the region of the lowest moisture content during the secondary drying process.

It is generally agreed that the presence of excess quantities of residual moisture is responsible for the instability of dried formulations at given temperatures. The reader should note that the amount of residual moisture in the cake can arise from other sources during or after the completion of the lyophilization process; these sources of residual moisture will be addressed in subsequent chapters.

Bound Water

Bound water is water associated with the stability of an active constituent or an excipient. For example, bound water may constitute the hydrate form of a compound or may be involved in defining the configuration of a protein molecule.

Water of Hydration. Kovalcik and Guillory [10] found that amorphous cakes of cyclophosphamide underwent rapid degradation regardless of the presence and nature of other excipients. Crystalline cyclophosphamide, whether in an anhydride or monohydrate form, has been found to be stable at ambient temperatures. Kovalcik and Guillory [10] set out to determine if the amorphous cyclophosphamide formed by a lyophilization process could be converted to a crystalline state, thereby enhancing its stability.

The authors lyophilized solutions that consisted of 25 mg cyclophosphamide and 100 mg of various excipients per mL. The excipients that were selected were mannitol, lactose monohydrate, and sodium bicarbonate. The same lyophilization process was used for each of the formulations regardless of the excipient. X-ray diffraction patterns of the cakes immediately after lyophilization showed some crystallization of the cyclophosphamide in the mannitol and sodium bicarbonate cakes but no crystallization in the cake containing the lactose excipient. Macro sample tubes containing 5 μL to 20 μL of water were then added to the vials. The vials were protected from light and stored at 24°C ± 2°C. The initial moisture of cyclophosphamide containing 100 mg of mannitol was < 1% wt/wt, and the initial residual moisture in the lactose-containing vials was > 2%. After 13 days of exposure to 20 μL of water, the moisture content in the cyclophosphamide containing the mannitol increased to 3.3%, while the moisture in the product containing the lactose increased to 7.4%. X-ray diffraction of the cyclophosphamide-mannitol samples showed that the addition of water significantly increased the crystallization of the mannitol and also increased the crystallization and stability of the cyclophosphamide. However, for the cyclophosphamide-lactose samples, the addition of water vapor resulted only in the formation of crystalline α-lactose monohydrate.

The above study showed that amorphous cakes of lyophilized formulations can be crystallized by treatment with the addition of water vapor. However, it also showed that the addition of water vapor resulted in a decrease in stability because of the preferential formation of the α-lactose monohydrate.

Water of Stability. *Surface water* is another form of bound water and is identified with the stability of protein molecules. The presence of surface water is often a factor in determining the configuration of the protein and, hence, its activity. The role that surface water plays in the stability of protein molecules has been reported by Teeter [11]. This surface water is important in stabilizing the configuration of the protein molecule. Teeter studied the surface water associated with the protein crambin and found that there were two basic types of water groups associated with the protein surface. The first is pentagonal rings that were found to be associated with the hydrophobic regions; the second is chain-like arrays of the surface water in the hydrophilic regions. It is these chain-like water arrays that appear to be influenced by the nature of the protein. These results and the works of others [9] show that the loss of activity of a key substance, such as a protein, by overdrying is possible.

Volkin and Klibanov [12] studied various means by which proteins can be irreversibly thermally inactivated. Such inactivation of the protein was attributed to either partial unfolding of the protein or to disruption of the structure leading to chemical, generally covalent, bonding. Lyophilized protein cakes that were hydrated during storage formed aggregates that decreased the solubility of the protein in an aqueous solution.

Thus water and the selection of excipients play important parts in determining the stability of various compounds. The role that water plays in the stability of the final product is quite complex, and there appears to be, at least at this time, no hard and fast rules for accurately predicting the product's behavior. That brings us once again to a fundamental truth regarding the lyophilization process: the formulation is the most important step in the lyophilization process.

Absorption, Adsorption, and Desorption of Gases

I consider it prudent for the reader to have some general familiarity with the interaction that gases have with a surface. The description of such interactions that follows will not involve detailed mathematical derivations to show the interaction of gases with surfaces; such a detailed description is beyond the scope of this text. Those readers wishing to obtain a more rigorous understanding of these processes are referred to a text by Dushman [13].

Absorption

Absorption is a process that occurs when a gas striking a surface remains on the surface long enough to diffuse into the material. In order to remove the gas, the gas would have to diffuse back to the surface of the material. In the case of a lyophilized product, it was shown in previous chapters that there is good reason to believe, at least for most formulations, that the interstitial region is glassy in nature and that one of the constituents of this glass is water. After removal of the ice crystals, as a result of the primary drying process, a major portion of the remaining water will be on the surface of the cake, while some water may be absorbed in the cake. It has been my experience that it is often difficult to attain low moisture values for formulations containing a relatively high percentage of sucrose. My hypotheses, which has no experimental data to support it, is that the difficultly in attaining low moisture values for formulations containing sucrose may be a result of absorbed water in the cake and the low diffusion rate of such water to the surface. This may account for some processes requiring an increase in T_p values (> 30°C) in order to achieve desired moisture values.

Adsorption

Adsorption is the process in which gas molecules strike a surface and remain there for some finite period of time. The molecules that remain on the surface are bound mainly by polarization forces between the surface and the gas molecule or by van der Waals forces, which are typically < 10 kcal/mole. With regard to lyophilization, there are two concerns: (1) How much water will be adsorbed on the surface at the start of the secondary drying? (2) What is the maximum amount of adsorbed water to achieve the necessary stability of the product?

The first model for the adsorption of gas molecules on a surface was proposed by Langmuir [13]. In this model, Langmuir hypothesized that, for a given clean surface, molecules impinging the surface do not instantly rebound from the surface but, depending on the nature of the surface, remain on the surface for some finite period of time. If these adsorbed molecules also interact such that there is a reasonable probability of forming a pair or group of molecules, it will be less likely that these paired molecules will leave the surface as compared to a single molecule. In time, with the formation of pairs or groups, the surface will be covered with a monolayer of gas molecules. According to the Langmuir model, once the monolayer is formed, the forces between the surface and molecules striking the monolayer are not sufficient to form a second layer. If we were to plot the volume of gas adsorbed by a surface at a given temperature and function of pressure, the Langmuir model would produce a plot as illustrated by Figure 9.1.

In examining Figure 9.1, the reader should keep the following in mind.

Langmuir's model assumes that there is no absorption of the gas by the material. It is highly unlikely that this assumption is generally applicable for the surface of a lyophilized cake.

It should be realized that the process is reversible (i.e., the desorption of the gas from the surface will follow the same function as that for adsorption). As a consequence, low moisture quantities of adsorbed water will require low partial pressures of the gas.

Figure 9.1. The Langmuir Monolayer Model

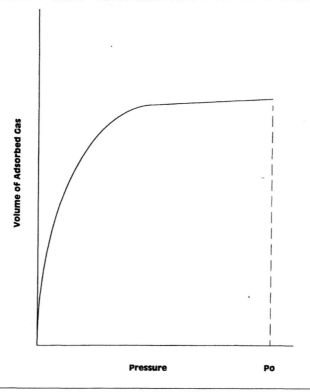

The Langmuir monolayer model for the volume of gas adsorbed on a surface at a given temperature as a function of pressure, where *Po* is the saturation pressure.

The *saturation pressure* (*Po*) will be directly dependent on temperature. As the temperature of the surface increases, *Po* will have to increase in order to maintain a monolayer of adsorbed gas on the surface. In addition, increasing the pressure of the gas to values greater than *Po* will not increase the quantity of gas adsorbed on the surface.

Using Langmuir's basic adsorption model, Brunauer, Emmett, and Teller devised the BET theory for the adsorption of gases on a surface. The BET theory expands the Langmuir model to include the adsorption of multilayers of gas on the surface. The results of the BET theory will be the same as that of Langmuir's model at low pressures when there is sufficient gas present to support a monolayer of gas.

A multilayer adsorption isotherm based on the BET theory can take the form shown in Figure 9.2. In this figure, the pressure *PL* at a given temperature is that pressure where a monolayer of gas is formed on the surface, while *Po* denotes the saturation pressure. The formation of a multilayer of adsorbed gases will be dependent on the nature of the surface and that of the adsorbing gas. If the surface material is hydrophilic in nature, then the first water layer will have the highest bond

Figure 9.2. The Adsorption of Multilayers of Gas

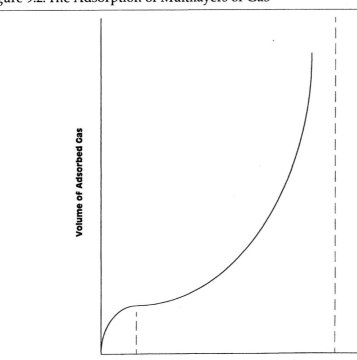

The adsorption of multilayers of gas based on the BET theory, where *Po* is the saturation pressure and *PL* represents the pressure for the formation of a monolayer of gas.

energy ($\geq$ 10 kcal/mole), while subsequent layers will be dependent on hydrogen bonding ($\approx$5 kcal/mole). The number of water layers on the surface will also be temperature dependent. It has been my experience that removing water from a stainless steel surface required that the temperature of the surface be $\geq$ 200°C for a number of hours and that the pressure in the chamber be $< 1 \times 10^{-6}$ Torr. For a clean and dry freeze-dryer chamber, with the shelves and condenser at ambient temperatures, the major gas constituent ($< 90\%$) during a leak test of the system was found to be water vapor. Such a high partial pressure of water vapor was obtained in spite of the fact that the system had been purged with dry nitrogen gas for a period of > 18 hr. The number of layers of water vapor adsorbed on a cake surface at the completion of primary drying, generally $<< 0$°C, must be substantial.

While it is not stated explicitly, a fundamental assumption of Langmuir's model and the BET theory is that there is only adsorbed gas on the surface of the material and that there is no absorption of the gas by the material. In the case of a lyophilized formulation, where absorbed water may be present, this water will have to diffuse to the surface before desorption can occur. With this statement in mind, the adsorption isotherm of a gas involving absorption would have a pressure dependence like that shown in Figure 9.3.

Figure 9.3 shows that absorption is the principal mechanism for the initial increase in gas volume retained by the solid. The quantity of gas absorbed (Q_a) will be dependent on the following [13]:

$$Q_a = \frac{P_{cm}K_o}{d}\exp^{\frac{-E_a}{RT}} \quad \frac{cm^3(STP)}{cm^2\,sec} \tag{1}$$

where P_{cm} is the pressure above the surface expressed in cm. of Hg, K_o is the preexponential function for the permeability of the gas through the substance at a temperature T and requiring an activation energy of E_a, and d is the thickness of the layer and R is the gas constant. From equation 1, the rate of diffusion of gas into a substance will be directly proportional to the pressure and exponentially dependent on temperature, i.e, the diffusion will increase rapidly with an increase in temperature.

Equation 1 also indicates that the absorption of the gas by the material is the predominant mechanism for increasing the gas volume in a substance. When the gas reaches a saturation level at the surface, then the increase in gas volume, as illustrated by Figure 9.3, is first by the formation of a monolayer. Depending on the nature of the gas and the surface, additional gas adsorption may occur by the formation of multilayers.

Desorption

Desorption is the reverse of adsorption: As the pressure of the gas above the surface is lowered, more gas molecules will leave the surface than will strike the surface. Because of this the gas contained by the substance, as illustrated by Figures 9.1 through 9.3, will decrease in volume.

The desorption process can be expressed by

$$R_i = \frac{dn_i}{dt} = k_{d,i}\exp^{\frac{-\Delta H_{d,i}}{RT}} \tag{2}$$

Figure 9.3. A Multilayer Adsoprtion Isotherm

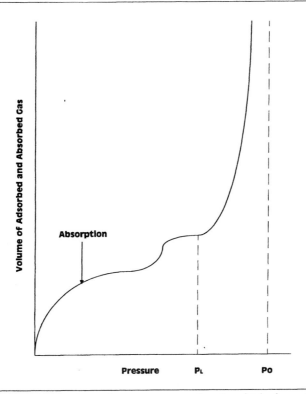

A multilayer adsorption isotherm for a gas on a surface in which absorption of the gas also occurs, where *Po* is the saturation pressure and *PL* represents the pressure for the formation of a monolayer of gas.

where n_i is the number of ith gas species leaving the surface per unit time per unit area, dn_i/dt is the rate of desorption per unit area, $k_{d,i}$ is the preexponential function for the *i*th gas specie, and $\Delta H_{d,i}$ is the heat of desorption for the ith gas species from a given surface.

The total rate of desorption (*D*) for a given temperature *T* is given as

$$D = \sum_{i=1}^{n} R_i$$

(3)

where R_i is determined from equation 2.

Model for Secondary Drying

Based on the above discussion, we can now develop a model for the secondary drying process. From this model, we can then determine the operating parameters necessary to complete the secondary drying process.

Model

At the end of the primary drying process, there is no longer any ice remaining in the matrix. The moisture in the cake formed by the constituents, as a result of their confinement in the interstitial region of the matrix during the freezing process, can contain absorbed water in the interstitial regions and adsorbed water on the surface. Such adsorbed water will most likely be in the form of multilayers.

Since there will be a limit to the P_c that one can attain in a freeze-dryer chamber, the only parameter that one can effectively use to reduce the moisture in a lyophilized cake is T_p. The value of T_p is increased by an increase in the T_s. The first water that will be removed will be the outermost layers of adsorbed water. The remaining water on the surface will be in equilibrium with the absorbed water in the cake. As T_p is increased, the rate of desorption will also increase. When the monolayer of gas is attained, further desorption of water includes the diffusion of water vapor to the surface. During the desorption process for the monolayer, there is a reduction in the volume of absorbed water in the cake. The diffusion of the absorbed water to the surface provides a source of adsorbed water that can leave the cake surface. For a given T_p and partial pressure of water vapor (P_w), a steady state is reached when

$$N_i + N_{ad} \rightleftharpoons N_d + N_{ab} \tag{4}$$

where N_i represents the number of water molecules that impinge the surface per unit time per unit area, N_{ad} is the number of N_i molecules that are adsorbed on the surface, N_d symbolizes the number of water molecules that are desorbed from the surface per unit time per unit area, and N_{ab} is the number of absorbed molecules in the cake.

Equation 4 is important because it provides an insight to how one can control the residual moisture in a final product. Following are several scenarios where equation 4 can have a major impact on the final moisture content of the product. An understanding of these scenarios will prove useful when discussing the examples of secondary drying processes.

Increase in P_w at a Constant T_p. For an increase in P_w, at given T_p, there will be an increase in N_i, which will result in an increase in N_{ad}. The increase in N_{ad} will cause, even though N_d remains constant, an increase in N_{ab}, or an increase in the moisture content of the cake.

Decrease in P_w at a Constant T_p. A decrease in P_w, for a given T_p, will cause a decrease in N_i. Since N_d will for some time remain constant, there will be a depletion of N_{ad} on the surface and a subsequent decrease in N_{ab}, or a decrease in the moisture content of the cake.

Increase in T_p at a Constant P_w. An increase in T_p, for a given P_w, will generate an increase N_d, even though N_i will remain constant. Because N_d increases, there will be a decrease in N_{ad}, which will lead to a decrease in N_{ab}, or a decrease in the moisture content of the cake.

Decrease in T_p at a Constant P_w. A decrease in T_p, for a given P_w, will result in a reduction in N_d and, with N_i remaining constant, there will be an increase in N_{ad}. With the increase in N_{ad}, there will be an increase in N_{ab} and an increase in the moisture content of the cake.

Increase in T_p and P_w or Decrease in T_p and P_w. When T_p and P_w are both simultaneously increased or decreased, it will be the magnitude of the increase of the respective variables that will determine the final moisture content of the product. In general terms, significant increases in T_p will tend to reduce the moisture content, and major increases in P_w will result in an increase of moisture in the cake.

Examples of Adsorption or Desorption Isotherms of Lyophilized Products

In order to gain further insight regarding the interaction between water vapor and a lyophilized product, the following are some published examples of adsorption or desorption isotherms.

Desorption Isotherm for Blood Plasma. The desorption isotherm for blood plasma [14] is shown by Figure 9.4. The latter figure shows the desorption isotherms for blood plasma at 20°C and 40°C. An examination of these isotherms suggests that at 20°C, the isotherm is best represented by the multilayer desorption isotherm illustrated in Figure 9.2. The thermogram indicates that the reduction in the percent moisture is primarily a result of desorption from the surface of the cake, and residual moisture stems from absorbed water in the product.

The desorption isotherm at 40°C appears to be more indicative of the isotherm shown in Figure 9.3, where diffusion of water vapor to the surface may be an important factor. Note that the results at a pressure of 0.1 mbar (75 mTorr) are best described by the above third scenario, where the diffusion of the absorbed water to the surface results in a reduction of moisture in the cake.

Figure 9.4. Desorption Isotherms of Blood Plasma [14]

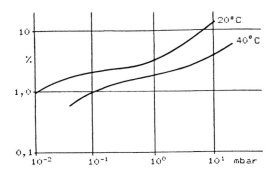

[Source: *Developments in Biological Standardization*, Vol. 74, Page 133, Figure 9. Courtesy of the International Association of Biological Standardization, Geneva, Switzerland.]

Adsorption and Desorption Isotherms of a Lyophilized, Excipient-Free, met-hGH Molecule. The adsorption and desorption isotherms for the lyophilized, excipient-free, met-hGH molecule in the temperature range of 22°C to 24°C was reported by Hsu et al. [15], and the results are shown in Figure 9.5. The adsorption and desorption isotherms appear to follow a BET isotherm for the formation of multilayers of adsorbed moisture. The isotherms do not suggest the presence of any absorbed moisture present in the lyophilized product. As a result, the water present must be in an adsorbed state on the surface of the compound. It is also of interest to note that as the relative humidity approaches zero, there is a significant amount of adsorbed water on the surface (i.e., 5.36 ± 0.08 g per 100 g of product [15]). This would suggest that the monolayer of water on the surface has a bond energy greater than that generally found for van der Waals forces. Such water may be necessary for the stability of the molecule. From the structure of the molecule, the authors [15] suggest that the water may be adsorbed in clusters at the strong polar regions of the molecule.

Pikal et al. [16] also examined the sorption isotherms for the hGH molecule with and without excipients at temperatures of 5°C and 25°C. The sorption isotherms (not shown) obtained by these authors, for the hGH molecule without excipients also show multilayer adsorption of water. However, based on the BET

Figure 9.5. Adsorption and Desorption Isotherms of Lyophilized, Excipient-Free, met-hGH at 22°C to 24°C

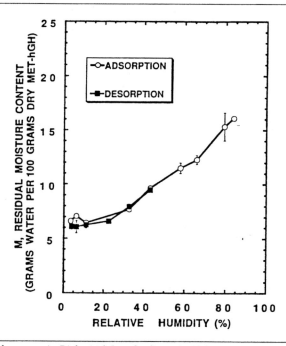

[Source: *Developments in Biological Standardization*, Vol. 74, Page 261, Figure 3. Courtesy of the International Association of Biological Standardization, Geneva, Switzerland.]

isotherm shown in Figure 9.2 and the results obtained by Hsu et al. [15], it would appear that the monolayer of water occurs at less than 1 g per 100 g of the compound. Pikal et al. [16] assess the monolayer to contain less than 2 g of water per 100 g of the hGH molecule.

The discrepancy between the sorption isotherms may lie in the lyophilization processes. Pikal et al. [16] do not provide any information in their paper or in a previous publication regarding the freezing or primary drying segments used to prepare their samples. They do state that secondary drying was conducted at temperatures below 30°C. In addition, the samples were allowed to adsorb water at the completion of the drying process by increasing the condenser temperature and reducing the product to subambient temperatures. The initial residual moisture in the lyophilized product was not reported.

Hsu et al [15] indicate that during primary drying the samples were maintained at a T_p of –35°C. for 48 hr. Secondary drying was conducted at 30°C for 35 hr. P_c was not reported for either drying process. The above drying process produced samples having 5% residual moisture.

It would appear that the discrepancy in the sorption isotherms for presumably the same compound may be attributed to differences in the lyophilization process. I feel that the above is an excellent example of the importance that the parameters play in the lyophilization process. It also points out how critical it is for detailed description of the entire lyophilization process in order that others may reproduce the results. For example, both groups of authors [15,16] expressed their isotherms in terms of percent relative humidity. The unanswered question is, *At a relative humidity of 10%, was the total chamber pressure 2.37 Torr and consisted of mainly water vapor or was the partial pressure of water vapor in the chamber 2.37 Torr for a higher total chamber pressure?*

Perhaps if the authors had provided more detailed information regarding their processes, the reason for the discrepancy in the sorption isotherms may have been resolved, and thereby provided one more step in advancing the science of lyophilization.

Use of Activity Rather than Water Vapor Pressure. In order to study chemical reactions, cleavage, oxidation, deamidation, denaturation and aggregation, some investigators have reported the results of their sorption studies in terms of activity [15]. The activity of a gas such as water (a_w) is defined as

$$a_w = \frac{f_w}{f_w^o} \tag{5}$$

where f_w is the *fugacity* of the water vapor and f_w^o is the fugacity of the water in some *standard state*. The term *fugacity* refers to an *idealized partial pressure of a gas or partial vapor pressure* [17]. Because a_w (see equation 5) does not have units, it is important to define the standard state. The standard state for the f_w^o of water vapor would be its vapor pressure at a given temperature, while f_w is some partial pressure water vapor lower than f_w^o. It should also be noted that one can express sorption isotherms in terms of activity (a_w) only if, at the given pressure, the gas behaves as an ideal gas ($PV = nRT$).

A desorption isotherm expressed in terms of activity is shown in Figure 9.6. This figure shows the desorption isotherms for PVP (polyvinyl pyrrolidone) for temperatures ranging from –40°C to 22°C as reported by MacKenzie [18]. These data suggest that for temperatures < 0°C, the desorption isotherms appear to have not formed a monolayer; for temperatures > 0°C, the desorption isotherms suggest the presence of multilayers for $a_w \geq 0.7$. It is unfortunate that MacKenzie [18] did not describe or reference his experimental method so that others can make use of it and apply it to the desorption studies of other systems. Based on the results of PVP, it would be of particular interest to compare the desorption isotherms for such systems like lyophilized mannitol and sucrose. Mannitol has been shown to crystallize in a frozen matrix [19] and in the presence of water vapor [10], while it has been my experience that the cakes of lyophilized sucrose remain amorphous.

Secondary Drying Parameters

Based on previous discussions, this section will consider the *paramount* key operating parameters for establishing a reproducible secondary drying process. The term *paramount* is used because it is not uncommon to hear the cry *everything during the drying process was the same but the product is different* (e.g., the residual moisture content is too high). The truth of the matter is that everything was not the same and the difference, although often obscure, had a significant impact on the final product. For that reason, although the following may appear to be repetitious, it is necessary to make the following assumptions before considering the key secondary parameters.

The first and, as previously stated, most important assumption is that the thermal properties of the process water and the final formulation are within prescribed limits. For example, a major increase or decrease in the degree of supercooling would affect the cake structure. An increase in the degree of supercooling could produce a dense cake structure that could impede the transport of water vapor from the product, while a decrease in supercooling may result in the formation of a crust or glaze on the top surface which could also impede the secondary drying process.

Figure 9.6. Desorption Isotherms

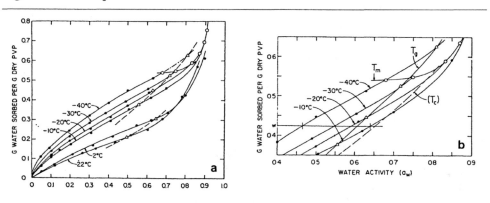

Sorption isotherms describing the desorption of water from PVP at temperatures ranging from –40°C to 22°C. [Source: *Freeze Drying and Advanced Food Technology* (Goldblith et al.), Chapter 19 ("Collapse During Freezing Drying . . . Aspects" by A. P. Mackenzie), Page 300, Figure 13.]

The second assumption is that the performance of the freeze-dryer and its associated instrumentation are functioning within the manufacturer's specifications. Perhaps one of my chief concerns with the operation of the freeze-dryer is the pressure gauge. Most of these gauges are complex electronic instruments. It is because of this complexity that one must be diligent in verifying the accuracy of pressure measurements.

The last assumption, and perhaps the most difficult to detect, is that the Standard Operating Procedures (SOPs) used in the entire lyophilization process do not contain flaws that would impact the process. In one case, the mere adjustment of *p*H during formulation determined the success or failure of the final product. When the *p*H adjustment was made with a base, an acceptable product was obtained; use of an acid, however, resulted in a major decrease in the collapse temperature. Thus the success or failure of the batch was decided prior to even opening the dryer door.

Standard Operating Secondary Drying Parameters

The secondary drying parameters are similar to those defined for primary drying (i.e., T_p, T_s, and P_c). However, unlike primary drying, when $T_p \rightarrow T_s$, as shown by the sorption isotherms shown in Figures 9.4 through 9.6, there is no indication that the secondary drying process is complete. For this reason, a fourth key parameter of time (t_s) is added. In some processes, yet another parameter must be added—the secondary drying ramp rate (R_s) of T_p to reach T_s, where R_s is $\Delta T_p / \Delta t$.

Product Temperature. As with the primary drying process, T_p plays an important role as a key parameter of the secondary drying process. While the sorption isotherms provide information as to the quantity of water (adsorbed and absorbed) by the product for a given P_w, the secondary drying is conducted under conditions where neither isothermal or isobaric conditions may exist. Since the role of temperature in the desorption process has been clearly defined, it is imperative for one to have knowledge of T_p throughout this portion of the drying process. The selection of the final T_p is a key element in the secondary drying process. In making such a selection, one should be aware of two factors.

The first is that stability of the product may be dependent on the duration that the product is maintained at T_p in order to attain the desired level of residual moisture. As a guideline, the duration of the secondary drying should be based on a small fraction of the time necessary to cause a significant change in the concentration of the active constituent.

The second factor is the time necessary to achieve the desired range of residual moisture values. It will be shown later in this section that, for a $T_p = -15°C$, the time necessary to completely remove a monolayer of gas from a surface could take a life time while at elevated temperatures, it would be a matter of days. I am careful not to suggest any hard and fast catchall temperature because the value of T_p must be based on the individual properties of each formulation.

Shelf-Surface Temperature. A measure of T_s is likewise important for several reasons. First, it provides an insight as to the energy that is being supplied to the product at any given time. A second, and perhaps more important, reason is that the drying process becomes independent of the design and construction of the freeze-dryer. For this reason, considerable time and material can be saved when scaling up the secondary drying process by eliminating the *trial and error* method that can arise when using the shelf-fluid temperature.

Chamber Pressure. While the desorption isotherms provide an important relationship between the residual water content of a product, P_w, and T_p, the pressure in the chamber during secondary drying will consist mainly of water vapor and nitrogen, assuming that the P_c is maintained at a constant pressure via a nitrogen bleed. Unlike the primary drying process when, as a result of significant pressure difference between the pressure in the vial and in the drying chamber, there is significant flow of gas from the container, the pressure differential during secondary drying will be considerably lower. This is a result of the presence of less water in the product and the water vapor's having to make its way through the cake by a pressure differential system that consists of series of adsorption and desorption processes. As P_c is increased, the transport of water vapor from the container may no longer be dependent on flow but on diffusion. Under flow conditions, the drying rate will be directly proportional to time; in the case of diffusion, however, the transport of water vapor will be dependent on the $\sqrt{\text{time}}$. A secondary process under flow conditions that requires 10 hr to complete may require as long as a 100 hr if the water must diffuse from the container. Thus, P_c is an important consideration in defining and controlling the secondary drying process.

Time. t_s plays an important role in the secondary drying process. Unlike primary drying when $T_p \rightarrow T_s$ and time was used to ensure that all of the containers had completed the primary drying process, the t_s for secondary drying will determine when the conditions defined by the sorption properties have been reached. Termination of secondary drying prior to reaching the desired sorption condition will result in higher residual moisture in the final product. Although there may be some merit in providing a guideline covering the duration of secondary drying (e.g., . . . *drying times are usually 30% of the primary drying* [7]), it is preferable that the secondary drying process be defined by the nature of the formulation.

Secondary Drying Ramp Rate. There are some formulations where the R_s of the product temperature, defined by

$$R_s = \frac{\Delta T_p}{\Delta t} \quad \frac{°C}{\text{unit time}} \tag{6}$$

where ΔT_p is the change in the product temperature and Δt is the change per unit time.

R_s could be a significant factor when the cake has a high absorbed moisture content at the conclusion of the primary drying process. A relatively low value of R_s would expose the active constituent and the excipients of the formulation to an increase in T_p in the presence of absorbed water. The chemical kinetics under these conditions may favor the formation of an insoluble substance by an excipient-excipient interaction or loss of activity by aggregation of the active constituent. With an increase in the R_s, the desorption rate and the rate of diffusion of the absorbed water to the surface are increased (see equation 1) so that the time that the constituents are exposed to the relatively high concentrations of absorbed moisture is limited. Once again, the reader is cautioned not to generalize, because a high R_s can result in physical damage to the cake. The properties of the individual formulation must be considered when deciding on the R_s for the secondary drying process [19].

Calorimetric Monitoring of Secondary Drying

Equations 1 and 2 for the gas adsorption and desorption processes involve energy, namely E_a and $\Delta H_{d,i}$. Assuming that the absorbed water in the cake is not significant, the rate of desorption of water molecules (dn_w/dt) from a surface, based on equation 2 can be represented by

$$\frac{dn_w}{dt} = k_{d,w} \exp^{\frac{-\Delta H_{d,w}}{RT_p}} \tag{7}$$

where $\Delta H_{d,w}$ is the heat of desorption of water vapor from a cake surface at a temperature of T_p and $k_{d,w}$ is the preexponential term for the desorption of water vapor for a surface. When $RT_p >> -\Delta H_{d,w}$, then $dn_w/dt = k_{d,w}$, which would approach the number of molecules on the surface.

Before considering the use of calorimetric measurements with regard to the desorption of water from a cake surface during secondary drying, it is considered best to review previous studies in this area [13].

First of all, the heat of adsorption of water vapor on a charcoal surface, prepared from sucrose, for temperatures ranging from –15°C to 40°C is 11,000 cal/mole to 9,100 cal/mole, respectively.

The number of water molecules (N_w) adsorbed on a cake having a surface area of 1 m^2, assuming closest packing, is defined by

$$N_w = \frac{1}{S} \tag{8}$$

where S is the packing factor and is defined as $0.866\xi^2$, where ξ is the diameter of the water molecule (4.68×10^{-10} m). The value of N_w, from equation 8 is 4.89×10^{18} molecules/m^2 or 8.75×10^{-5} moles, or 1.57 mg. The total energy necessary to desorb this water would be 0.095 cal at –15°C. From equation 7, with $k_{d,w} = 4.89 \times 10^{18}$ molecules/sec, the rate of desorption of the water molecules from the surface would be 2.19×10^9 molecules/sec and require 4×10^{-11} cal/sec. At this desorption rate, the time necessary to desorbe all of the water from the surface would be 70 years.

By increasing the cake surface temperature to 40°C, the rate of desorption of the water molecules is increased to 2.06×10^{12} molecules/sec or an energy of 3.1×10^{-8} cal/sec. The time to completely remove all of the water from the cake surface at 40°C would be 27.5 days.

It would appear that, because of the low quantities of energy involved, determining the rate of desorption from the surface of a cake by calorimetric measurement would require highly sensitive measuring instruments and that stray thermal energy could affect the accuracy of the measurements. It will, however, be shown later in this chapter that calorimetric measurements can be used to observe the desorption of water vapor from a cake [20]. The latter water will most likely be associated with multilayers of adsorbed water. It has been reported that as many as 18 monolayers of water molecules can exist on a surface [13].

The preceding discussion has pointed out not only the limitations of calorimetric measurements in ascertaining the desorption of a monolayer of water vapor from a surface but also the importance of selecting T_p. Rothmayr [21] recognized that the water adsorbed in the lyophilized product could be in a number of states with varying bonding energies, which makes the secondary drying process highly complex.

PROCESS MONITORING TECHNIQUES

The fundamental question that one is faced with during the development of a lyophilization process is determining the end point of the process (i.e., When is the residual moisture content within a given set of specifications?). In the absence of any guidelines, a prolonged secondary drying process is not uncommon. That secondary drying can be 30% of the duration of primary drying has been proposed as a general guideline [7]. From the above, it can be seen that once P_w and T_p have been attained and an equilibrium between P_w and the water content has been reached, for a given T_p, extending secondary drying serves no useful purpose. In fact, it will be shown in Chapter 13 that the exposure of the product to low pressures could lead to contamination of the product by the *backstreaming* of hydrocarbon vapors from the pumping system. Perhaps Rey [22] summed it up nicely: ... *this determination, which looks simple enough at first glance, is very difficult and might lead to odd and conflicting results depending upon the way it is done.* The following will outline various techniques for ascertaining the completion of the secondary process.

Pressure Rise Test

Description of the Test Method

The pressure rise test is conducted in much the same fashion as completion of the primary drying process [7,14] (i.e., the chamber is isolated from the condenser, and the rate of pressure rise is determined). If the rate of pressure rise (ROR) exceeds a predetermined value, then secondary drying is allowed to continue. If the ROR is lower than a prescribed rate, then the moisture content is assumed to be within acceptable limits. Those who advocate the use of the pressure rise test are quick to stress that the leak rate of the chamber should not in any way influence the ROR [14].

The chief advantage of this test method is that it is relatively easy to perform. The operators of the freeze-drying equipment can easily conduct the test, and the results are not dependent on the subjectivity of the operator. The test method uses the same pressure gauges that monitor the drying process, and one is not faced with additional instruments or equipment that may be expensive and/or require additional training or validation.

Commentary

The pressure rise method for determining the completion of the secondary drying process is certainly simple enough to perform, but as Rey [22] points out, it ... *might lead to odd and conflicting results depending upon the way it is done.* While I have myself used this technique to ascertain the completion of the secondary drying process, there are pitfalls that should be taken into consideration.

Effective Volume of the Dryer. The vapor pressure of water in the area surrounding the dryer is generally ≥ 20 Torr at the T_p selected for the primary drying process. The water vapor leaking into the chamber can be treated as an ideal gas. The ROR of the drying chamber (dP/dt) can be expressed as

$$\frac{dp}{dt} = \text{ROR} = \frac{Q}{V_e} \quad \frac{\text{pressure liters}}{\text{time}} \tag{9}$$

where Q is the *leak rate* expressed in units of pressure liters per unit time, and V_e is the *effective volume* of the chamber (in L) [23].

The effective volume (in L) of a drying chamber (V_e) is defined as

$$V_e = V - V_h \tag{10}$$

where V is the total volume of the chamber and V_h is the internal volume occupied by the hardware in the system, e.g., shelves and stoppering mechanism [23].

From equations 9 and 10, the observed ROR for a product will be inversely proportional to V_e (i.e., as V_e increases, ROR decreases for a given value of Q). Thus it will be no simple task, considering that real or virtual leaks in either system do not complicate matters by indicating a false ROR, to transfer the ROR for the termination of the secondary drying process from a relatively small development dryer (e.g., ≤ 24 ft^2 of shelf area), to a dryer having shelf area ≥ 200 ft^2. Even the transfer of the drying process from one dryer to another of equal shelf area may not be without problems if the dryers come from different manufacturers. The effects of competition have, at least for the moment, prevented the dryer-manufacturing industry from agreeing on any common set of standards that would simplify the transfer of drying processes from one dryer to another.

Batch Size. Another possible pitfall is batch size. In the preceding discussion, it may appear that a small batch completed the secondary drying process prior to that process being used on a full batch. What is really taking place is that the Q for the small batch differs significantly from the Q for a large batch. From equation 9, a change in Q for a given V_e will result in a change in ROR.

Comparative Pressure Sensors

Description of the Method

As described in the previous chapter, Nail and Johnson [24] investigated the difference between the output of a thermal conductivity gauge (Pirani) with that of a capacitance manometer to determine if the difference in gauge readings could be used to determine the completion of the primary and secondary drying processes. The basis for such an investigation is that the output of thermal conductivity gauges is dependent on the composition of the gases [25]. Since the thermal conductivity gauges are calibrated against a known pressure of air or nitrogen, gases such as hydrogen, helium and water vapor will, because of their higher thermal conductivity (see Table 8.1), indicate a false high pressure. For a drying process in which P_c is controlled by bleeding in dry nitrogen, the completion of secondary drying will be signified by the pressure indicated by the thermal conductivity gauge approaching that indicated by the capacitance manometer gauge. The output of the latter gauge, which will be discussed in Chapter 11, is independent of the composition of the gases in the chamber.

In their investigation of the differences in the output of the above two gauges to sense the completion of secondary drying, Nail and Johnson [24] performed primary and secondary drying at a pressure of 400 mTorr as determined and controlled by means of a capacitance manometer gauge. For a portion of the vials in the chamber, the drying process was terminated by activating the stoppering mechanism of the dryer in such a way that only one shelf was stoppered at a time. Their results showed that prior to the Pirani gauge reaching its lowest value, the observed moisture content was about 8% wt/wt. As the pressure indicated by the Pirani gauge decreased, so did the measured moisture content in the vials. After the Pirani gauge had reached its lowest value for no less than 2 hr, the observed moisture content decreased to less than 2% wt/wt. It should be noted that at the completion of the secondary drying process the output of the Pirani gauge was still significantly greater than that of the capacitance manometer gauge. The authors attributed this discrepancy between the two gauges to small differences in calibration and contend that the . . . *absolute value of the pressure reading from the thermal conductivity gauge is not important, since the method only depends on changes in response arising from changes in the composition of the gas phase in within the chamber.*

When using this method for monitoring the secondary drying process, Nail and Johnson [24] caution that one should have precise control over P_c. This is necessary in order to observe any significant change in the output of the thermal conductivity gauge, especially if the pressures are lower than 100 mTorr. They also claim that this method senses the last vials to complete the secondary drying process.

Commentary

I view the above means of monitoring the secondary drying process as a *rough* estimate of the completion of the secondary drying process. Because of the relatively low flow of gas from the vial as a result of the desorption process, a few vials still undergoing secondary drying will not have a significant impact on the composition of the gases in a dynamic system (i.e, one in which there is a flow of gas into the system to equal the pumping speed of the vacuum system and in which the chamber is maintained at a constant pressure). Once again, the relative size of the drying chamber and batch size could affect the determination of the end point for the secondary drying process. The sensitivity for determining the endpoint in a development dryer having a shelf-surface area ≤ 25 ft^2 could be significantly different from that of a production dryer having a shelf-surface ≥ 200 ft^2.

Nail and Johnson [24] do show that, when compared to the observed partial pressure of water vapor in the chamber as measured by a residual gas mass spectrometer, the output of the Pirani gauge does approach that of the capacitance manometer. As a result, one is left in a bit of a quandary. Is the difference in the Pirani gauge reading a result of errors in the calibration of the gauges, as previously suggested by the authors, or is there a significantly high partial pressure of water vapor in the chamber? In addition, if the absolute value of the output of the Pirani gauge is not important in ascertaining the endpoint of secondary drying, then how does one know when secondary drying is complete, and how does one validate a drying process when there are inconsistences in the final output of the Pirani gauge?

Humidity Sensor

Description of the Method

The electronic humidity sensor was described in detail in the previous chapter. While this device was first introduced as a means for monitoring and determining the completion of the primary drying process [26], Bardat et al. [27] extended the use of the sensor to include monitoring of and the determination of the end point of the secondary drying process.

The output of the electronic sensor was compared to that of a Pirani gauge by monitoring the moisture in samples removed by means of a *thief*, a special device that allows the stoppering and removal of a single vial from the dryer without altering the drying process. The results showed that the output of the sensor was closely related to the residual moisture in the sample for values down to 1.9% wt/wt. The output of the electronic sensor at that moisture value was 0.10%.

Commentary

The monitoring of the partial pressure of water vapor in the chamber with an electronic sensor has been found to have higher sensitivity for monitoring secondary drying than either the pressure rise test or the Pirani gauge. However, sensitivity will be dependent on variations in the batch size or the dimensions of the dryer. The humidity sensor is most helpful when scaling up a lyophilization process from a development dryer to a production dryer or transferring the lyophilization process between production dryers. A major drawback to using this device for determining the endpoint of secondary drying is that it cannot withstand the effects of steam sterilization [24]. As a result, its use is limited to the development of lyophilization processes and the manufacture of nonsterile lyophilized products.

Residual Gas Analysis

Description of the Analytical Method

The residual gas analyzer (RGA) has been used to determine the endpoint of the secondary drying process. Completion of the secondary drying process is determined by RGA by measuring the partial pressures of the gases in a freeze-drying chamber. The description and operation of RGA will be considered in Chapter 11. This analysis is performed best at pressures ≤ 0.001 mTorr; therefore, the sensor used in RGA cannot be attached directly to the freeze-dryer chamber when the pressure is > 1 mTorr. For this reason, the sensor must be contained in a separate vacuum system that can attain pressures approaching 1×10^{-8} Torr. At its lowest operating pressure, a blank mass spectrum of the gases is recorded. Gases from the drying chamber are then bled into the RGA system by means of a micrometer valve. With the pumping system of the RGA system in full operation, the pressure is increased to about 1×10^{-6} Torr by adjustment of the micrometer valve. The blank current for each of the observed gas species is subtracted from the measured value, and the partial pressure of a given gas (P_i) is

$$P_i = \frac{C_i - C_{o,i}}{S_i \sum_{j=1}^{n}(C_j - C_{o,j})} \times P_c \quad \text{mTorr}$$

(11)

where C_i is the ion current of the *ith* gas species measured during the flow of gases from the chamber, $C_{o,i}$ is the ion current of the *ith* gas species obtained from the blank mass spectra, $\Sigma(C_j - C_{o,j})$ is the sum of the ion currents of all of the gas species, and S_i is the ion sensitivity of the *ith* gas species. The S_i of gases such as oxygen, nitrogen and water vapor is 1, while hydrogen and carbon dioxide are 0.44 and 1.4, respectively.

RGA measures the partial pressures of the gases in the chamber. By assuming that the source of the water vapor is the lyophilized cake, the moisture content of the cake will be related to P_w as determined by RGA. The sensitivity of RGA for determining the partial pressure of water vapor in a chamber is far greater than that of the comparative pressure sensors or the electronic humidity sensor.

Commentary

In spite of its greater sensitivity, RGA is not widely used either in the development of lyophilization processes or in production equipment. While the cost of an RGA system (> $30,000) is certainly a factor, perhaps the greatest difficultly is that making the measurements and interpreting the data requires a certain degree of expertise. For that reason, RGA does not lend itself well in a production environment. A small mistake in performing the measurements can have a major effect on the operation of the instrument. Since there are limited data regarding the use of RGA in determining the endpoint of secondary drying, RGA will be used in determining the completion of secondary drying in the forthcoming examples of secondary drying processes.

Purge Method

I have used and recommended the purge method for terminating the secondary drying process but have been unable to find any reference to this method in the literature. The basic principle of this method is drying by purging the cake repeatedly with a gas such as dry nitrogen, although other gases such as dry helium or argon would also be applicable.

When using the purge method, the temperature of the product is first increased to its terminal temperature under what would be considered to be a typical P_c (e.g., 100 mTorr). When the $T_p \rightarrow T_s$, the P_c is increased, with the vacuum system still in operation, to a higher pressure (e.g., 2,000 mTorr). The latter pressure is accomplished by bleeding the appropriate dry gas into the drying chamber. The chamber is maintained at the higher pressure for about 2 min, the gas bleed is turned off, and the drying chamber is evacuated to the initial secondary pressure. This process is repeated from 7 to 10 times, depending on the residual moisture desired in the cake.

The purge method reduces the moisture in the cake to a given range of moisture values by the following means. The increase in P_c forces the dry gas into each of the vials and throughout each cake. During the time that the purge gas is in the cake, water vapor will desorb from the cake surface. The high rate at which the drying gas strikes the surface of the cake prevents any desorbed moisture vapor from readsorbing on the cake surface. The gas phase in the cake then consists of desorbed water vapor and the drying gas. When the gas bleed is turned off, the mixture of drying gas and water vapor flows out of the cake and vial. Under these flow conditions, the gas flow will tend to be viscous in nature for a portion of the time. As a result, the water vapor is swept out without contacting the cake. The moisture is transported to the condenser where it is adsorbed. When the chamber pressure returns to its initial value (100 mTorr), more moisture is desorbed from the walls of the cake and is removed by another purge of the dry gas.

The effectiveness of the purge can be monitored by the rate of pumpdown of the chamber from the purge pressure (2,000 mTorr) to the initial pressure (100 mTorr). When the purge removes sufficient quantities of water from the product, water vapor will be removed by the condenser, and the pumpdown time will be less than that for a clean dry chamber. As the number of purges is increased, less water vapor will be included in the pumpdown gases, and the pumpdown time will be extended.

Another means for monitoring the drying progress of the purge method is to monitor the partial pressure of water vapor in the chamber when the pressure is again at its initial pressure. While I prefer determining the composition of the chamber gases by means of RGA, other methods, such as the electronic humidity sensor, could prove useful. Once the effective number of purges has been established for a given formulation, the process can be repeated regardless of the size of the batch or the design and construction of the dryer.

The purge method for conducting and monitoring the secondary drying will be particularly useful when the cake results from a relatively high fill volume and for cakes having a dense structure. Care should be exercised when using the purge method for cakes prepared from formulations with a solid content less than 2% wt/vol. Cakes prepared from formulations having low solid content may have poor supporting structures, and the cake may break apart during the purging, which not only would result in a loss of product but also would represent a potential source of contamination of the dryer and a hazard to operating personnel. An example of the use of the purge method during secondary drying will be given later in this chapter.

Calorimetric Measurements

Results of Calorimetric Measurements During Secondary Drying

Calorimetric measurements described in the previous chapter, have been used to monitor the secondary drying process [20]. A typical calorimetric desorption peak that occurs during the increase in T_p is illustrated in Figure 9.7. The thermogram shows that the desorption process occurred over a period of nearly 4 hr, with the maximum desorption rate having a heat-transfer energy rate (Q_m) approaching 20 mcal/sec. The total energy (H_m) for the secondary desorption of water vapor from the cake was about 140 cal.

Estimating the cake surface area to be about 6 m^2 and assuming the presence of a monolayer of adsorbed water vapor and a heat of desorption of 9,100 cal/mole,

Figure 9.7. A Typical DPM Thermogram

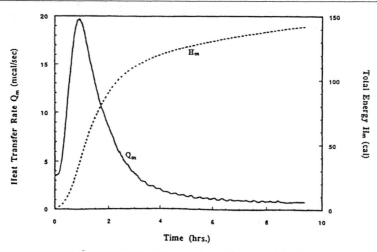

A typical DPM thermogram for secondary drying of a formulation where Q_m is the heat-transfer rate (mcal/sec) and H_m the total energy of desorption (cal). [Source: *PDA Journal*, "Calorimetric Monitoring of Lyophilization," Vol. 49, No. 6, Nov.–Dec. 1995, Page 280, Figure 6.]

the total quantity of water desorbed from the cake would be 278 mg. Since the total solid content of the cake was only about 70 mg, the residual moisture content of the cake would approach a highly unlikely value of 400%. If the actual desorbed moisture from the cake was of the order of 14 mg or 20% wt/wt, then, based on the above calorimetric measurements, the heat of desorption for the water vapor from the cake would have to be about 180 kcal/mole or 20 times the anticipated heat of desorption of water vapor. Since such a high heat of desorption of water vapor would also appear to be highly unlikely, the observed endothermic peak could be related to some physical-chemical phenomena associated with the cake (e.g., the activation energy necessary for converting one or more of the cake constituents from an amorphous to a crystalline state).

Another thermogram obtained during calorimetric measurements is shown in Figure 9.8. In this calorimetric thermogram, there is a small endothermic peak followed by a relatively broad exothermic peak. The exothermic peak was later associated with a reaction between two of the excipients forming an insoluble compound in the cake.

Commentary

The secondary drying process has often been viewed as merely the step in the lyophilization process in which the final residual moisture is removed from the cake in order to achieve the desired long-term stability. However, the introduction of

Figure 9.8. A Calorimetric Thermogram During Secondary Drying Showing the Pesence of an Exothermic Peak

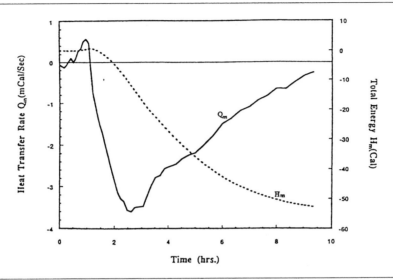

[Source: *PDA Journal*, "Calorimetric Monitoring of Lyophilization," Vol. 49, No. 6, Nov.– Dec. 1995, Page 281, Figure 7.]

calorimetric measurements during secondary drying has revealed that unusual endothermic and exothermic reactions are occurring during this portion of the lyophilization process. In addition, these measurements permit one to observe excipient-excipient, excipient-active, and active-active interactions that can result in an increase in the turbidity of the reconstituted cake or prolong the reconstitution time. I hope that others will use calorimetric measurements to monitor the lyophilization process and perhaps add to our general understanding of the highly complex secondary drying process.

EXAMPLES OF THE SECONDARY DRYING PROCESS

The examples for the secondary drying process are a continuation of the lyophilization of the three formulations whose thermal properties were listed in Table 7.1. The freezing functions for these formulations were illustrated by Figures 7.8, 7.10, and 7.12 and the primary drying processes were shown by Figures 8.22, 8.23, and 8.24. This chapter will complete the lyophilization process by illustrating the completion of the secondary drying process for these formulations. The format for describing each secondary drying process will be selecting the proper operating parameters (i.e., T_p, T_s, and P_c), an illustration of the process, and the means used to ascertain the completion of the process. Because it is not possible to cover every variation one might encounter in a secondary drying process, a short commentary section will follow each of the illustrated processes to alert the reader to any potential pitfalls or areas of possible misconceptions.

Formulation #1

Selection of the Operating Parameters

The terminal value for T_p was selected as 20°C to limit any product degradation that might result from exposure to higher temperatures. P_c was maintained at 100 mTorr to reduce or eliminate any backstreaming of hydrocarbons from the vacuum pumping system. If the use of 100 mTorr results in a prolonged secondary drying process, then one would have to either increase the terminal value for T_p or reduce P_c.

Secondary Drying for Formulation #1

The primary and secondary drying processes for formulation #1 are illustrated by Figure 9.9. Since T_p at the completion of the primary drying process was already at 10°C, T_s had to be increased only another 10°C in order to provide the product with sufficient energy to complete the desorption of residual moisture to less than 2% wt/wt. It was determined that there was no lower limit for the residual moisture value for this product, so there was no danger of product degradation as a result of overdrying.

Figure 9.9 shows that when T_s was increased to 20°C, there was a slight lag in the increase in T_p. It is possible that such a temperature lag in T_p may be a result of the desorption of water vapor from the surface; since the product consisted of a 2 mL fill volume in a 5 mL glass tubing vial (see Chapter 7), the surface area of the cake was ≤ 2 m^2. From the discussion of the calorimetric measurements of the secondary drying process, the lag in T_p is most likely associated with the heat conductivity of the cake rather than the energy associated with the desorption of water vapor from the cake surface.

Determination of Completion of Secondary Drying

In determining the completion of secondary drying, especially when in the development and validation stages of a lyophilization process, I prefer the use of an RGA system (mass spectrometer) for determining the partial pressure of water vapor in the drying chamber. Although there is no general direct relationship between the partial pressure of water vapor and the residual moisture content of a cake, knowledge of the partial pressure of the water in the chamber at termination of the secondary drying process will serve as a benchmark for determining when the residual moisture content of a product will be within the desired product specifications.

Since nitrogen and water vapor will be the primary gas species in the chamber at the end of the primary drying process, a comparison of the relative height of the current peaks for these gases can provide a guide regarding the completion of the secondary drying process. The relative water vapor and nitrogen peaks for the time period when $T_p \rightarrow T_s$ is illustrated by Figure 9.10. Note that in this figure the

Figure 9.9. The Primary and Secondary Drying Processes for Formulation #1

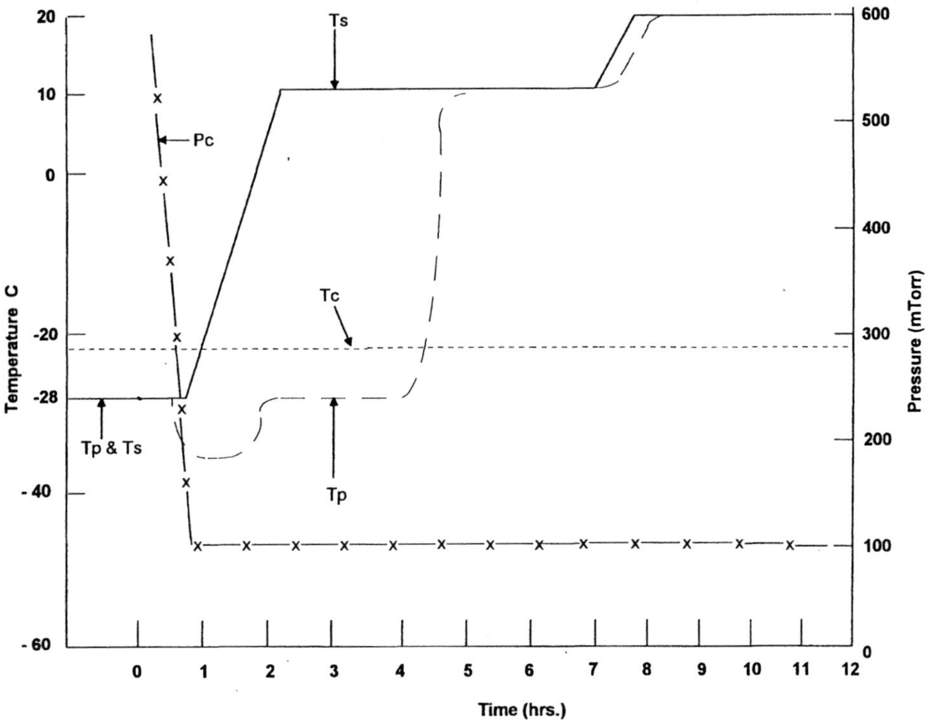

The solid line represents the T_s, the long dashed line indicates the target T_p during primary drying, and the short dashed line represents the collapse temperature (Tc). P_c is denoted by x-x-x. The elapsed time 0 indicates the starting time of evacuation of the dryer.

mass/charge (m/e) ratio increases from right to left for each set of peaks. A comparison of Figures 9.9 and 9.10 shows that after 8 hr of drying, the major gas species in the drying chamber was water vapor. The major water peak has a m/e = 18 (H_2O^+), and the smaller peak of m/e = 17 (OH^+) is quite prevalent. In hour 9, the partial pressure of nitrogen exceeded that of the water vapor. By hour 12, the current peak for the water vapor having an m/e = 18 is hardly detectable.

Stoppering of the Vials

Secondary drying was stopped by increasing the pressure in the chamber with dry nitrogen pressure to about 600 Torr. The stoppering mechanism of the dryer was then activated, and the fluted stoppers were seated on the vials to form a seal. The pressure of 600 Torr was selected so that the stoppered vial would be under a slight vacuum when removed from the chamber, but the pressure differential between the

Figure 9.10. The Relative Partial Pressure Peaks of Water Vapor and Nitrogen During the Secondary Drying Process of Formulation #1

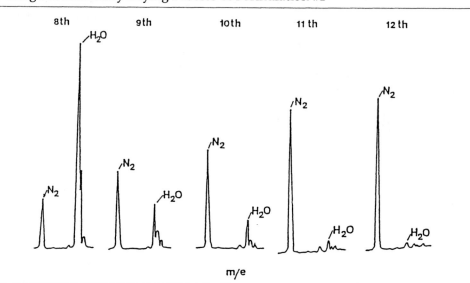

atmosphere and the internal vial pressure would not be sufficient to cause the closure to become deformed in the region designated for the penetration of a syringe needle to reconstitute the lyophilized product. Stressed closure material in the region used for the penetration of the needle will increase the possibility of coring of the elastomer material.

Commentary

Because there was no lower limit specification of the residual moisture content for this product, there was no concern regarding overdrying of the product. From Figure 9.10, the use of RGA permits detection of partial pressures of water vapor well beyond the sensitivity of either the electronic humidity sensor or the comparison of pressure gauges. By using a detection system with a higher sensitivity, there is a higher degree of confidence that duplicating the drying process will lead to reproducible results. This is not to say that the other methods for sensing the partial pressure of water vapor cannot be used in determining the completion of the secondary drying process. However, if low partial pressures of water vapor in the drying chamber are necessary to achieve consistent moisture values in the product, then these latter analytical methods may not be adequate.

Coring with the insertion of a needle through the elastomer closure is always a possibility. A closure that is stressed by a large pressure differential can only serve to increase the probability of coring. As will be seen in the secondary drying of the next formulation, there are advantages to stoppering the vials while under low pressures, i.e., < 100 mTorr.

Formulation #2

Selection of the Operating Parameters

Of the three formulations, the drying parameters for formulation #2 are perhaps the most typical. The pressure for the primary and secondary drying processes is set at 50 mTorr, and the T_p is once again set at 20°C. The one concern about this process is that at 50 mTorr, backstreaming of the hydrocarbon vapors from the vacuum pumping system can occur. These hydrocarbon vapors will enter the condenser system and, in time, will find their way into the drying chamber and onto the shelves of the dryer.

Secondary Drying for Formulation #2

The primary and secondary drying processes for formulation #2 are illustrated in Figure 9.11. The increase in T_s during secondary drying results in a major difference in T_s and T_p. Although the cake occupies a volume of 2 cm^3 it has a mass of only 70 mg or density of 0.035 g/cm^3. If the main solute constituent in the formulation was lactose (which would agree with the observed collapse temperature), then the ratio of the bulk density of lactose (β-anomer), 1.58 g/cm^3 [28] to the density of the cake would be 45.2, or 2 cm^3 of bulk lactose would occupy 90.4 cm^3 in the lyophilized form. Thus, the lyophilized cake consists mainly of voids, and at a pressure of about 50 mTorr, heat transfer from the walls of the vial to the junction of the sensor is limited by the thermal conductivity of thin layers of lactose that were formed during the freezing process. As a consequence, the significant temperature differential of $T_s - T_p$ is attributed to the poor heat-transfer properties of the lyophilized cake.

Determination of Completion of Secondary Drying

The specification for the range of the residual moisture in the lyophilized cake of formulation #2 was established, from stability studies, to be from 1% wt/wt to 6% wt/wt. The stability studies showed that residual moisture values, in order to achieve the desired shelf life of the product, had to be maintained at values less than 6% wt/wt. It was also found that there was a significant decrease in product stability when the residual moisture fell below 1% wt/wt. For this formulation, it was possible to overdry the cake. The basic problem in determining the end of the secondary drying process was not in staying within the specification for the upper limit of residual moisture but in establishing a means by which the product was not overdried. There are several possible means for solving this problem.

Sampling with a Thief. Without breaking vacuum, samples are removed from the dryer, and the residual moisture content is determined. If the residual moisture is within the prescribed specifications, then the product is stoppered and removed from the dryer. The advantage of this procedure is that if the moisture value exceeds 6% wt/wt, one can continue the secondary drying process. There are, however, three principal disadvantages to using this technique, even during the development stage of the lyophilization process.

Figure 9.11. The Prinary and Secondary Drying Processes for Formulation #2

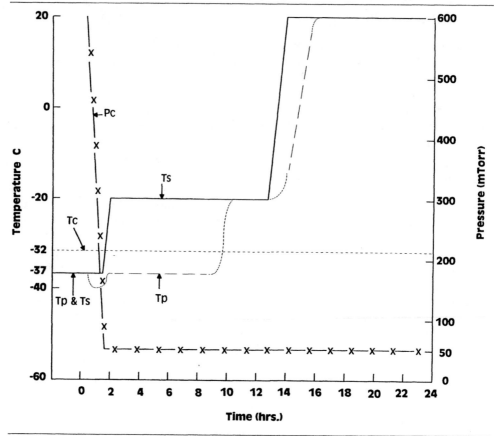

The solid line represents the T_s, the long dashed line indicates the target T_p during primary drying, and the short dashed line represents the collapse temperature (Tc). P_c is denoted by x-x-x. The elapsed time 0 indicates the starting time evacuation of the dryer.

The first is that the vials accessible to the thief for sampling are often limited. If the thief is located on the door of the dryer, then only those containers near the front of the dryer would be accessible. Should the door of the dryer not be thermally insulated, then the moisture values in these cakes will tend to be lower than the average moisture content of the product.

Another difficulty in using this technique, especially when there is a lower limit to the moisture content, is that completing the moisture analysis requires some time. To stop the drying process while the moisture analysis is being completed, the valve between the condenser and the drying chamber would have to be closed, or the vacuum pumping system would have to be turned off and P_c increased to > 2 Torr. If the results of the residual moisture analysis indicate that the moisture value exceeds the upper limit, then one would have to evacuate the chamber to the pressure selected for secondary drying and continue the secondary drying process.

The act of reducing the pressure in the chamber, as described in the above *purge method* will remove moisture from the cake, and the moisture content of the product when the drying is continued will be less than when the drying process was temporarily terminated.

There will be a finite time period to determine the residual moisture in the cake and, for some products, the samples must be pooled in order to obtain a significant sample mass. In the worst case, the moisture analysis is performed using the loss-in-weight method, it is days before the results are known.

The last drawback to using the thief technique is that the results of the moisture analysis may indicate that the product is over dried. One must then have a means for increasing the moisture content of the product. One method that has been proposed for increasing the moisture content is to increase the temperature of the condenser [16]. Before increasing the condenser temperature to water vapor pressures that will exceed P_c, one should isolate the vacuum system so that water does not enter those pumping systems that use oil-lubricated vacuum pumps. Many models of freeze-dryers do not permit the setting of the condenser temperature and require the operator to manually control the condenser temperature by turning the refrigeration system to the condenser *on* and *off*. Equipment manufacturers of freeze-dryers would probably disapprove of such a practice.

Post Drying Treatment. The second possible solution is to perform secondary drying like that shown in Figure 9.11. Then, without stoppering the vials, back fill the drying chamber to 1 atm with dry nitrogen gas. It has already been shown that the addition of water vapor to a dried cake can be beneficial [10], but that method of increasing the moisture content of a product presents its own set of problems, especially when the product must be maintained under Class 100 and aseptic conditions. There are a number of chemical compounds (see Table 9.1) that can provide a range

Table 9.1. The Percent Humidity and Vapor Pressure of Water for Various Solid Compounds at 20°C

Solid Phase	Percent Humidity	Pressure (Torr)
$LiCl \cdot H_2O$	15	2.60
$KC_2H_3O_2$	20	3.47
$CaCl_2 \cdot 6H_2O$	32.3	5.61
KNO_2	45	7.81
$NaHSO_2 \cdot H_2O$	52	9.03
$NaNO_2$	66	11.5
$H_2C_2O_4 \cdot 2H_2O$	76	13.2
$(NH_4)_2SO_4$	81	14.1
$ZnSO_4 \cdot 7H_2O$	90	15.6
$CuSO_2 \cdot 5H_2O$	98	17.0

of relative humidity values at 20°C [28] but just how one would introduce such compounds into aseptic clean room could be a problem. One possible means could be packaging the compound in a covered plastic suitable container and then sterilizing by gamma radiation. The outside of the package would then be sterilized before the package was transferred into the manufacturing area. Once inside the dryer, the covering could be removed, thereby exposing the compound to a closed dryer chamber. The length of time required for all of the product in the chamber to reach a suitable moisture content and be ready for stoppering will depend on the volume of the dryer and the quantity of the compound present. The humidity could be determined using an electronic humidity sensor as describe above.

Pressure Rise Method. The use of the pressure rise method [14] is not considered suitable for monitoring the secondary drying of this formulation. My objection to using this method is that it would be difficult to correlate the rate of rise to a given moisture content in a product. This method would be more applicable for products where there is no lower limit on the moisture content.

Other Methods. The comparative pressure method [24] is not considered suitable for this formulation because it lacks the sensitivity. The electronic humidity sensor [26,27] is considered applicable for this formulation because the moisture content in the final product can be related to a range of humidity values.

Residual Gas Analysis. Because of its relatively high sensitivity, I prefer using RGA [23] to determine the endpoint for the secondary drying of a formulation having moisture specifications like those for formulation #2. The RGA measurements taken during the secondary drying process for formulation #2, at a pressure of 50 mTorr, are shown in Figure 9.12, where the change in the partial pressure of water vapor is illustrated by the relative amplitude of the ion currents for nitrogen and water vapor. This figure shows that at the hour 16, the major gas component in the dryer was water vapor, and the moisture content in the cake would exceed the residual moisture content upper limit of 6% wt/wt. By hour 22, there was a significant decrease in the amplitude of the water peak and the average residual moisture content in the cake had decreased to about 2% wt/wt.

The reader should bear in mind that, regardless of the method selected for determining the completion of the secondary drying process, the relationship between the measurement and the residual moisture content must be carefully established. For example, does the relative amplitude of the ion current for water vapor at hour 21 represent a higher but acceptable average moisture content for the cake? If the moisture content at that time was greater than 6% wt/wt, then one would have to be concerned about the reliability of using RGA to establish the termination of the secondary drying process.

Formulation #3

Selection of the Operating Parameters

From Table 7.1, the formulation had a solid content of 80 mg/mL or 8% wt/v. In addition, Table 7.1 indicated that the degree of crystallization was only 0.60, which

Figure 9.12. The Relative Partial Pressure Peaks of Water Vapor and Nitrogen During the Secondary Drying Process of Formulation #2

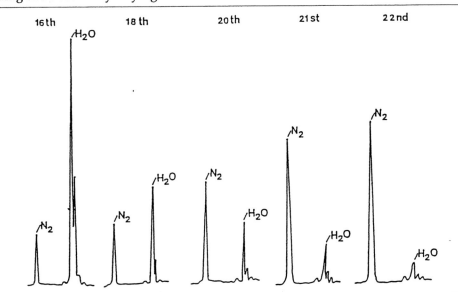

means that 40% of the water present in the formulation did not form ice crystals during the freezing process. The effect of the low degree of crystallization is the formation of a matrix that can impede the flow of water vapor from the cake during secondary drying. Such an impedance becomes troublesome when the specifications for the residual moisture in the cake is limited to $\leq 1\%$ wt/wt.

In order to maximize the flow of water vapor from the cake, the lowest P_c was set at approximately 50 mTorr. The use of such a low P_c increases the probability of backstreaming of the hydrocarbon vapors from the vacuum pumping system.

Since this product was also found to be quite heat liable, the terminal T_p was limited to approximately 20°C. In order to limit T_p, the upper value of T_s was also set at 20°C. Because of the poor heat-transfer properties of the cake, the R_s for the T_s was set at about 1°C/min.

Secondary Drying for Formulation #3

The primary and secondary drying processes for formulation #3 are illustrated in Figure 9.13. This figure shows a large temperature lag between T_s and T_p as a result of the cake's poor heat transfer properties. In spite of a low P_c, poor structural properties—resulting from the low degree of crystallization—serve to impede the flow of water vapor and could greatly extend the completion of secondary drying. However, the 8% solid content produces a relatively rigid cake structure that can withstand the periodic purging of the cake with dry nitrogen gas.

The nitrogen gas purge consisted of increasing P_c with the valves to the condenser and the vacuum system in an open position. P_c was increased to 2,000 mTorr and maintained at that pressure for about 2 min. The gas bleed was then turned off, and P_c was once again reduced to about 50 mTorr. The purging process was repeated a total of 10 times.

Determination of Completion of Secondary Drying

The effectiveness of the purge method was determined by measuring the partial pressures of the residual gases in the chamber when P_c again reached 50 mTorr. A terminal pressure of 10 mTorr could also have been selected; however, the pump-down time from 50 mTorr to 10 mTorr would have greatly extended the time to complete secondary drying without significantly lowering the residual moisture values in the product.

Figure 9.13. The Primary and Secondary Drying Processes for Formulation #3

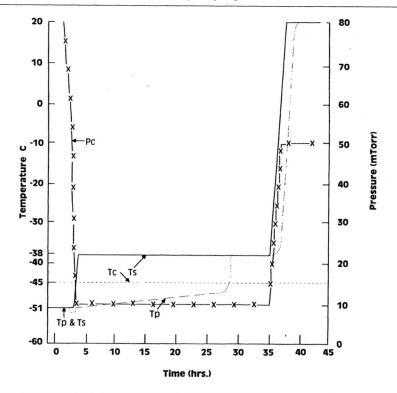

The solid line represents the T_s, the long dashed line indicates the target T_p during primary drying, and the short dashed line represents the collapse temperature (Tc). P_c is denoted by x-x-x. The elapsed time 0 indicates the starting time of evacuation of the dryer.

The partial pressures of water vapor, nitrogen, and the other gases (oxygen, carbon dioxide, etc.) after each of the purges is illustrated by Figure 9.14. This figure shows that even after 5 purges, the partial pressure in the chamber was slightly less than 10 mTorr. A total of 9 purges were required to reduce the partial pressure of water vapor to 2 mTorr and obtain a residual moisture content of less than 1% in the lyophilized cake. The chief advantage of the purge method is that the endpoint will be independent of the batch size and of the volume and of the dryer. In addition, the process can be readily transferred from a development dryer to a production dryer. The only serious limitation imposed by the purge method is that the cake must have a self-supporting structure sufficient to withstand the fluctuations in pressure that will occur within the container. If the cake is a weak self-supporting structure, then the purge method can cause a loss in product from the containers, and the presence of product dust in the dryer could represent a health hazard to operating personnel. As with any method, one must always exercise caution and common sense.

Figure 9.14. Partial Pressure of Water Vapor (denoted by –◆–), Nitrogen (denoted by –■–), and Other Gases (denoted by –▲–) After Each Nitrogen Purge

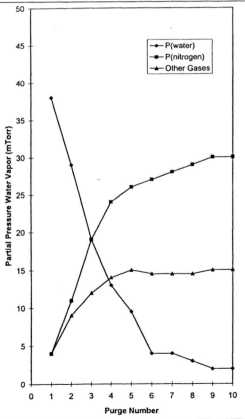

SYMBOLS

a_w	chemical activity of a gas such as water
C_i	ion current of the ith gas species measured by RGA
$C_{o,i}$	blank ion current of the ith gas species
C_o	heat-transfer coefficient (cal/°C sec)
d	thickness of the layer
dn_i/dt	rate of desorption per unit area
dn_w/dt	rate of desorption of water molecules from a surface
dP/dt	rate of pressure rise in the drying chamber
D	total rate of the desorption of gases from a surface
e	charge equal to that of 1 electron
$\Delta H_{d,i}$	heat of desorption for the ith gas molecule from a surface
$\Delta H_{d,w}$	heat of desorption of water from a surface
$\Sigma(C_j - C_{o,j})$	sum of the ion currents of all of the gas species
ΔT_p	change in product temperature
Δt	change per unit time
E_a	activation energy for the diffusion of a gas through a material
f_w	fugacity of water vapor at a given temperature
$f_w°$	fugacity of water in some standard state
H_m	total energy (cal)
$k_{d,i}$	preexponential function for the ith gas specie
$k_{d,w}$	preexponential desorption of water function from a surface
K_o	preexponential function for the permeability of the gas through the substance
m	mass of the ion species
n_i	number of ith gas species leaving the surface per unit time per unit area
N_{ab}	number of absorbed molecules in the cake
N_{ad}	number of N_i molecules that are adsorbed on the surface
N_d	number of water molecules that desorb from the surface per unit time per unit area
N_i	number of water molecules that impinge the surface per unit time per unit area

N_w	number of water molecules adsorbed on a cake having a surface area of 1 m^2
ξ	diameter of the water molecule
P_c	chamber pressure
P_{cm}	pressure above the surface expressed in cm of Hg
P_L	the pressure at which a monolayer is formed on the surface of a material
Po	saturation pressure
P_w	partial pressure of water above the cake surface
Q	leak rate expressed in units of pressure liters per unit time
Q_a	quantity of gas absorbed
Q_m	heat-transfer energy rate (mcal/sec)
R	gas constant
R_i	rate at which the ith gas molecule will desorb from a surface
R_s	secondary drying ramp rate ($\Delta T_p / \Delta t$)
S	packing factor
S_i	ion sensitivity of the ith gas species
t_s	duration of the secondary drying process
T_p	product temperature in the container (°C)
T_s	shelf-surface temperature (°C)
V	total volume of the chamber
V_e	effective volume of the chamber (L)
V_h	internal volume occupied by the hardware in the system, e.g., shelves and stoppering mechanism

REFERENCES

1. L. F. Shackell, *Am J. Physiol.*, 24: 325 (1909).

2. J. M. Hill and D. C. Pfeiffer, *Annals of Internal Medicine*, 14, 201 (1940).

3. R. I. N. Greaves, in *Biological Applications of Freezing and Drying* (R. J. C. Harris, ed.), Academic Press, Inc., New York, 1952, p. 87.

4. N. A. Williams and G. P. Polli, *J. Parenter. Sci. and Technol.*, 38: 48 (1984).

5. J. W. Snowman, *Pharm. Eng.*, November/December (1993) p. 26.

6. L. Rey, in *Aspects Théoriques et Industriels de la Lyophilisation* (L. Rey, ed.), Herman, Paris, 1964, p. 23.

7.   K. Murgatroyd, in *Good Pharmaceutical Freeze-Drying Practice* (P. Cameron, ed.), Inter-pharm Press, Inc., Buffalo Grove, Ill., 1997, p. 1.

8.   A. T. P. Skrabanja, A. L. J. De Meere, R. A. De Ruiter, and P. J. M. Den Oetelaar, *PDA J. Pharm. Sci. & Tech.*, 48: 311 (1994).

9.   M. J. Pikal and S. Shah, *PDA J. Pharm. Sci. & Tech.*, 51: 17 (1997).

10.   T. R. Kovalcik and J. K. Guillory, *J. Parenter. Sci. and Technol.*, 42: 29 (1988).

11.   M. M. Teeter, International Symposium on Biological Product Freeze-Drying and For-mulation, Bethesda, USA, 1990, *Develop Biol. Standard.*, 74: p.63.

12.   D. B. Volkin and A. M. Klibanov, ibid., p. 73.

13.   S. Dushman, *Scientific Foundations of Vacuum Technique* (J. M. Lafferty, ed.), John Whiley & Sons, Inc., New York (1962).

14.   H. Willemer, International Symposium on Biological Product Freeze-Drying and For-mulation, Bethesda, USA, 1990, *Develop Biol. Standard.*, 74: p. 123.

15.   C. C. Hsu, C. A. Ward, R. Pealman, H. M. Nguyen, D. A. Yeung, and J. G. Curley, loc. cit., p. 255.

16.   M. J. Pikal, K. Dellerman and M. L. Roy, loc. cit., p. 21.

17.   W. J. Moore, *Physical Chemistry*, Prentice-Hall, Inc., Englewood Cliffs, New Jersey, 1964.

18.   A. P. MacKenzie in *Freeze Drying and Advanced Food Technology* (S. A. Goldblith, L. Rey, and W. W. Rothmayr, eds.), Academic Press, New York, 1975, p. 277.

19.   N. A. Williams, Y. Lee, G. P. Polli, and T. A. Jennings, *J. Parenter. Sci. & Tech.*, 40: 135 (1986).

20.   T. A. Jennings and H. Duan, *PDA J. Pharm. Sci. & Tech.*, 49: 272 (1995).

21.   W. W. Rothmayr, in *Freeze Drying and Advanced Food Technology* (S. A. Goldblith, L. Rey, and W. W. Rothmayr, eds.), Academic Press, New York, 1975, p. 203.

22.   L. R. Rey, International Symposium on Freeze-Drying of Biological Products, Washing-ton, USA, 1977, *Develop Biol. Standard.*, 36: p.19.

23.   T. A. Jennings, *J. Parenter. Sci. and Technol.*, 44: 22 (1990).

24.   S. J. Nail and W. Johnson, International Symposium on Biological Product Freeze-Dry-ing and Formulation, Bethesda, USA, 1990, *Develop Biol. Standard.*, 74: p. 137.

25.   T. A. Jennings, *J. Parenter. Drug Assoc.*, 34: 62, 1980.

26.   M. L. Roy and M. J. Pikal, *J. Parenter. Sci. and Technol.*, 43: 60 (1989).

27.   A Bardat, J. Biguet, E. Chatenet, and F. Courteille, *J. Parenter. Sci. and Technol.*, 47: 293 (1993).

28.   *CRC Handbook of Chemistry and Physics* (E. C. Weast, ed.), 65th Edition, CRC Press, Inc., Boca Raton, Florida (1984).

10

Product Properties

INTRODUCTION

Need for Evaluation of the Lyophilized Product

An evaluation of the final lyophilized product serves to characterize its various physical and chemical properties and the effect that the storage conditions will have on such properties. The evaluation of these properties is, without question, of paramount importance. However, the results of such evaluation should also provide insight as to what effect the initial formulation and the various steps in the lyophilization process may have had on the observed properties of the product. It is not merely a question of what the properties of the final product are but also their source. The relationship between the observed properties of the final product and the materials and the processes that were involved (as described in the previous chapters) can be evident only as a result of careful documentation of the analytical methods and their results. This documentation of the analytical methods and nature of the final product not only verifies batch-to-batch consistency but also provides a vital benchmark that can be an invaluable source of information should a significant change occur in the nature of the lyophilized product.

General Overview of the Chapter

Before embarking on reading this chapter, it will be helpful to the reader to have an idea of why I chose this particular arrangement of the various topics. I feel this chapter should provide the reader not only with methods for product evaluation but also with the benefits derived from understanding the work of other investigators. The reader should understand that I will at times describe and disagree with, rather than ignore, the methods and/or the results of some investigators. The purpose for the critiques of the works of others is not to belittle their contribution to the field of lyophilization but to familiarize the reader with possible pitfalls that may be encountered in selecting an evaluation method or interpreting the results. I am keenly aware that without the works of other investigators, this book would not have been possible, and for that reason I feel deeply indebted. Consequently, this chapter will consider four major topic areas: physical properties, moisture, reconstitution, and product stability.

General Physical Properties of the Cake

The first section will provide the reader with a description of the general physical properties of the final product. Wherever possible, special attention will be given to the relationship between the observed physical properties of the cake and the composition of the original formulation. It will be shown how selecting the constituents of the formulation can affect a wide range of cake properties (e.g., from temperature effects to physical characteristics that include color, porosity, and structure). It will be shown that even vial breakage has been found to be related to the composition of the formulation.

Moisture

It has been established that the moisture content in a lyophilized product is an important factor that often relates to its physical and chemical properties, i.e., its stability at a given temperature. This section will first consider the different roles that water plays in the final product and will expand on the roles of *residual moisture* or *free* water and the *bound* water that were introduced in the previous chapter. The basic difference between these two forms of water is not in their chemical configuration but in their interaction with the constituents of the formulation. Various means for determining the moisture content of a product will be presented and examined.

Reconstitution

In most applications of lyophilization, with the exception of some food products such as freeze-dried ice cream, a prescribed diluent must be added to the dried product in order to obtain the desired formulation. This process of dissolving the cake by the addition of a diluent is generally referred to as *reconstitution*. The rate and properties of reconstitution of a lyophilized product will be dependent not only on the nature of the diluent but also on the surface area of the cake that results from both the nature of the original formulation and the lyophilization process. Specifications for and means for determining particulate matter in the reconstituted product will be considered, as will the effects of storage conditions. While reconstitution is a destructive test of a lyophilized cake, it represents the key criterion for acceptability of the final product.

Stability

The first part of the definition of the lyophilization process states that . . . *it is a stabilizing process. . .* [1]. For that reason, the last section of this chapter will be devoted exclusively to identifying those properties and contents of the cake that play a key role in determining the stability of the final product. The two basic methods for determining the stability of a formulation will be described and compared. With the increased emphasis on the lyophilization of biotechnology products, special attention will be given to introducing the reader to aspects that can affect the stability of lyophilized products containing proteins.

PHYSICAL PROPERTIES

The physical properties of a lyophilized product can provide key information regarding the nature of the formulation, e.g., nature and concentration of the constituents and the impact of the lyophilization process. I have, for the sake of clarity, broken down this section into two discrete sections: those cake properties that are mainly associated with the formulation and those that are directly related to the process. The reader is cautioned that observed physical properties may result from a combination of formulation and lyophilization process parameters that may prove difficult to separate and the two topic areas may overlap under some circumstances. I would also like to stress that some of the following physical characteristics, while not appearing as important as potency or activity of the product, should be viewed as fingerprints of the final product and that documenting such fingerprints could be vital in troubleshooting the formulation and the lyophilization process at some later time. One should not ask what can go wrong but rather what steps do we take when it does.

Effects of the Formulation

Transition Temperature

When the temperature of a lyophilized cake is raised, the cake can undergo physical transformation, i.e., shrinkage or liquefaction. The onset temperature for such a transformation is referred to as the *transition temperature* $T_g(w_g)$ [2]. What can often lead to some confusion is that the *glass transition temperature* associated with the freezing process is denoted as T_g' [2], which is for all practical purposes equivalent to the *collapse temperature* (*Tc*) defined in the previous chapters. Because of the possible confusion between the terms $T_g(w_g)$ and T_g', the transition temperature will be referred to as *incipient melting* and be symbolized as T_a [3]. The onset temperature for T_a has been found to be dependent on the residual moisture content. As the residual moisture in the cake increases, the onset temperature for T_a will tend to decrease [2]. A lowering of T_a can account for the disappearance of the cake during storage, as illustrated in Figure 10.1a. While T_a has been found to be moisture dependent, the reader should not overlook that T_a is also dependent on the nature and quantity of the constituents that make up the lyophilized cake. For that reason, as it was in the thermal analysis of the frozen matrix, there is no simple way to predict and determine the value of T_a for a given residual moisture content. In order to ensure that the cake does not undergo a physical transformation, the value of T_a for the upper limit moisture limit of the product should be established.

Differential scanning calorimetry (DSC) has been used to determine the T_a of a dried product [4]. The difficulty with using DSC to ascertain T_a for a given lyophilized cake is that one must transfer a small sample from the protected environment of the product vial to the DSC pan. In order to prevent an error in the measurement, one would have to know the relative humidity in the product container

Figure 10.1. Product Defects

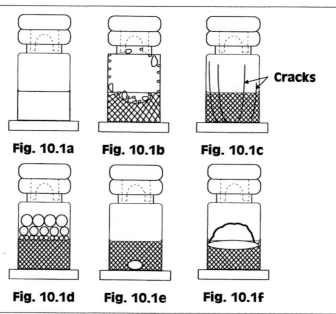

Additional product defects resulting from the formulation and/or the lyophilization process. Figure 10.1a is an illustration of the disappearing cake resulting from the temperature of the cake exceeding the transition temperature T_a; Figure 10.1b is an illustration of a cake with a poor self-supporting structure; Figure 10.1c is an illustration of vial breakage by the use of mannitol as an excipient; Figure 10.1d is an illustration of frothing or foaming; Figure 10.1e represents an example of partial meltback and Figure 10.1f is an illustration of puffing of the cake.

and match it with the relative humidity in the system used to conduct the product transfer. If the humidities are not equivalent, then a higher transfer environment humidity will result in a false low value of T_a because the moisture content in the sample will be increased. A lower humidity in the transfer environment will result in a lower moisture in the sample and a false high T_a value.

There appears to be, at least at this time, no accurate means for determining the value of T_a while the sample is still confined in the protective environment of the container-closure system. It will be shown in Chapter 12 that as the temperature is increased, there will be considerable outgassing of water vapor from the closure. If the outgassing of water vapor from the closure substantially increases the relative humidity in the container-closure system, then the value of T_a is not a true representation of the dried material obtained from the lyophilization process but some combination of the residual moisture in the product and that moisture outgassed from the closure. T_a is an important property of the lyophilized product; however, accurate means for its measurement awaits development.

Cake Volume

The volume of the cake can be an indicator that the cake was a result of a lyophilization process. For the typical formulation, the volume of the cake will be equal to that of the frozen matrix, like that illustrated in Figure 1.1a [5–8]. It has been my experience that the use of some excipients, such as polyvinylpyrrolidone (PVP) can provide a cake with a proper volume even when primary drying of the lyophilization process was performed at temperatures that exceeded the Tc (see Figure 1.3c). For that reason, if a product was produced by lyophilization, one must exercise caution on making an assessment based solely on appearance (i.e., appearances can be deceptive).

Color

The *color* of the cake, especially in the case of food products, is relative to the structure of the cake. As was emphasized in Chapter 7, the formation of ice crystals will affect the structure of the lyophilized cake.

If frozen rapidly, some products, especially foods, tend to produce a light-colored cake. When the product is frozen at a slower rate, the resulting dried product will usually have a darker color [9].

The color of the final product can also·be affected by the conditions during the primary drying process. Lyophilized products tend to be lighter in color than products that haven been vacuum dried temperatures above the *Tc* [9].

That most lyophilized formulations are colorless does not diminish the importance of observing the shade of the final product. A change in the shade of the product can be a result of a change in the freezing process of the formulation. One should note any relationship between the color or shade of the product and other key properties such as potency of the product, especially during the development stage of the product.

I have occasionally been involved in the lyophilization of colored solutions. In all cases, the color of the initial formulation differed significantly from that of the lyophilized cake. I was often surprised to find that at times little or no attention was given to documenting the color of the initial formulation let alone the resulting lyophilized cake. Such documentation may be quite useful should a problem occur with the quality of the final product. It is true that one can always go back and make a comparison with the retention samples from previous batches, but there would always be the question whether or not the color of the cake had changed during storage.

Texture

The ideal *texture* of a lyophilized cake will be highly porous, and the cake will have a sponge-like appearance. Because of the individual characteristics of each formulation, there is no general rule for an acceptable texture of the cake and each product must be examined with regard to its individual characteristics [10].

The importance of identifying the general characteristics of the texture of the cake is that the texture may be altered either by the freezing process or the primary drying stage [10]. Placement of the formulation on a prechilled shelf (e.g., -40°C) could alter the texture from Figure 1.3a to that of Figure 1.3b. Conducting the drying process when the product temperature (T_p) > Tc can also result in a change in the texture of the cake [10], as illustrated in Figure 1.3c. It is for these reasons that the observed texture of the cake is an important physical characteristic to document in order to ensure that there is continued quality of the product from batch to batch during manufacturing.

Cake Density

The *cake density* is, in the absence of collapse or meltback during the primary drying process, an indication of the self-supporting structure of the cake. This physical characteristic of the cake will, therefore, be primarily related to the freezing process. For lyophilized products, the cake density is surprisingly low. One means for determining the bulk density of a lyophilized product was described by Fink [12]. This method involved adding a known mass of lyophilized product to a graduated cylinder. The cylinder was then dropped 10 times from a height of 4 cm. The density of the cake was determined from the ratio of the known mass of the cake and the final volume of the cake in the graduated cylinder.

Shrinkage or Collapse

Cake *shrinkage*, more commonly referred to as *collapse*, is apparent when the cake volume is less than that of the frozen matrix, as illustrated in Figure 1.3c. This figure shows that the cake volume is less than that of the original frozen matrix, which is represented by the upper solid line. Collapse, not merely a lack of adherence of the cake to the walls of the container, is an indication of a possible change in the composition of the formulation, improper operating parameters during primary drying, or the presence of excessive moisture.

A simplified pictorial explanation of shrinkage or collapse is illustrated in Figure 10.2 using the analogy of a brick wall. In Figure 10.2a, the frozen matrix is illustrated by a brick wall. The bricks are the dark rectangles, while the clear area represents the mortar, or the region occupied by the constituents and any uncrystallized water. In the lyophilization process, shown by Figure 10.2b, the ice (bricks) is removed by sublimation and the remaining interstitial region (mortar) forms a highly porous system that retains the original volume of wall. If the bricks (ice) are removed while there is still mobile water in the interstitial region or mortar, the result is shrinkage or collapse, as shown by Figure 10.2c.

Shrinkage or collapse can occur as a result of a change in one of the constituents in the formulation (e.g., the active constituent), especially if such a constituent stems from nature or from improper preparation of the formulation. A change in the composition of the formulation can result in a significant change in the thermal properties. Should the Tc of the formulation occur at a lower temperature than that previously determined, then the T_p will be conducted at a temperature, during primary drying, that exceeds the Tc of the frozen matrix. Because T_p > Tc, shrinkage or collapse of the cake can occur, as illustrated in Figure 10.2c. Generally when shrinkage or collapse does occur as a result of a change in the thermal properties of the formulation, the shrinkage or collapse of the cake will be apparent throughout the entire batch. Shrinkage or collapse of the cake as a result of a change in composition of one of the constituents in the formulation can be easily prevented by determining the thermal properties of the formulation when first using a new lot

Figure 10.2. The Interstitial Region During Primary Drying

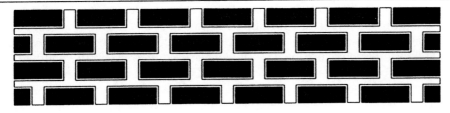

Figure 10.2a

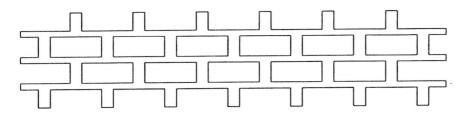

Figure 10.2b

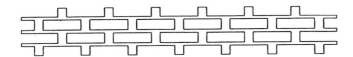

Figure 10.2c

Figure 10.2d

Pictorial illustration of the function of the interstitial region during primary drying from the perspective of a brick wall. Figure 10.2a shows the matrix of ice crystals (bricks), indicated by the dark rectangles, which are surrounded by the constituents and any uncrystallized water (mortar). The wall with the bricks removed (by sublimation) is illustrated by Figure 10.2b and is used to illustrate a lyophilized cake. Figure 10.2c illustrates the effect of removing the bricks while mobile water is present in the mortar, i.e., shrinkage or collapse. Figure 10.2d shows meltback as a result of removing the bricks when the mortar is "wet" and not self-supporting.

of the constituent. If a major change in the thermal properties is observed, then one has the choice of either using a more acceptable lot of the constituent or altering the primary drying parameters to accommodate the change in thermal properties.

The addition of PVP to the formulation can be used to prevent the collapse of the cake. However, shrinkage or collapse can still occur if the concentration of the PVP in the formulation is too low [5]. The increase in the concentration of the PVP in the formulation may prevent shrinkage or collapse of the cake; however, primary drying may be conducted in the presence of mobile water that may compromise the quality or potency of the active constituent. In this sense, the presence of shrinkage or collapse is a warning that the product may not have undergone a lyophilization process. Masking the collapse process with a compound such as PVP may prove counterproductive in the sense that it removes a natural indicator of the state of the drying process.

Shrinkage or collapse of the cake can occur in only a few of the containers or vials that are randomly scattered throughout the batch. The shrinkage is probably not generated by a major change in the thermal properties of the formulation but is a result of a change in the frequency distribution of the heat-transfer coefficient (C_o) of the containers [1]. The collapse is a result of the presence of some containers, because of a broad frequency distribution of C_o values, having significantly higher heat-transfer coefficients than the mean heat-transfer coefficient ($\overline{C_o}$). During the primary drying process, the T_p of these latter products will exceed the Tc, resulting in shrinkage, collapse, or meltback of the cakes. This topic will be considered later in this chapter; but for now the reader is referred to Figure 1.3d and Figure 10.2.

Excessive residual moisture in the cake can also cause shrinkage to occur. In this instance, the shrinkage or collapse may not be apparent immediately upon removal of the vials from the dryer but occur over a period of time. As previously mentioned, the source of the moisture may be a result of outgassing of water vapor from the closures during storage at an elevated temperature.

Pores

As described in Chapter 6, the freezing process generates a matrix of ice crystals and an interstitial region that consists of the other constituents in the formulation. The removal of ice crystals during the primary drying process will then generate a system consisting of pores. While it should be understood by the reader that such a pore system is highly complex and strongly dependent on both the composition of the formulation and the drying parameters, for this discussion the pore will be represented as cylinders like that shown in Figure 10.3. It is further assumed that the cylinders or pores are open on both ends. For pore diameters of the order of 100 Å, an increase in the ratio of P/P_o' (where P is the pressure of the *adsorbate* [e.g., water vapor] above the cake and P_o' is the saturation pressure for the adsorbate at a given temperature T) to values ≥ 0.5 will result in a condensed state of the *adsorbent* [13].

In Figure 10.3, when the R_1 of the adsorbed gas layer becomes equal to the radius of the meniscus of the liquid adsorbate (R_2), the pore will be filled with adsorbate. The total radius (R) will be equal to

Figure 10.3. The Condensation of a Liquid Phase of an Adsorbate in a Pore Having a Radius of R_1 When $R_1 = R_2$, the Radius of the Meniscus of the Liquid Absorbate

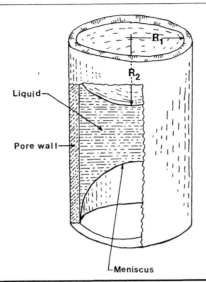

$$R = R_1 + t' \tag{1}$$

where the term t' is related to the thickness of the adsorbed layer.

The *critical volume of liquid* (V_L) in the pores of a cake to induce incipient melting can be expressed as

$$V_L = \frac{P V_a V_m}{R_o T} \tag{2}$$

where V_a is the volume of gas adsorbed on the surface and V_m is the molar volume of the liquid adsorbate.

If it is assumed that the macro pores (i.e., those pores that do not become filled with liquid adsorbate) will have little contribution to the formation of V_L, then the average pore radius $R_{p,T}$ for a pressure P and temperature T to induce incipient melting can be expressed as

$$R_{p,T} = \frac{2V_L}{A_s} \tag{3}$$

where A_s is the surface area of the cake determined by the BET theory (see Chapter 9). A knowledge of A_s will provide a measure of $R_{p,T}$ for a given V_L but will not reveal the frequency distribution of R. If the frequency distribution is relatively broad, the formation of the liquid states, for a given temperature, will occur over a relatively large range of pressures. For a broad frequency distribution of R, the partial pressure of the adsorbate in the system may not be sufficient for the cake to achieve V_L. However, if the frequency distribution of the pores is rather narrow then the total number of pores containing liquid states may result in a phase transition for the system (i.e., from a solid state to a liquid state). Although this statement offers a plausible explanation for the occurrence of the *disappearing cake* (i.e., a change from a solid to a liquid) in all of the vials of a particular batch of a lyophilized product, there is a clear need for further research in this area.

Crystallization

X-ray diffraction is an excellent analytical method for ascertaining the crystalline nature of a material [10,14,15]. So that others may duplicate the results for lyophilized materials, it is recommended that the wavelength of the radiation used with the diffractometer and the means used for calibration of the instrument be described [15]. Simatos and Blond [10] examined the effects that lyophilization had on various excipients. For example, they were able to show that for a casein-NaCl solution, the NaCl could exist in three states (bound to the casein, free and amorphous, and free and crystalline [10]. After the casein-NaCl cakes were exposed to water vapor an X-ray diffraction pattern showed that the crystalline NaCl was not hygroscopic. The free amorphous NaCl, however, when exposed to the water vapor, was hygroscopic and continued to show a diffused X-ray pattern.

In addition to X-ray diffraction, electron diffraction has been used to characterize the structure of lyophilized cakes [10]. The cake resulting from the lyophilization of a sucrose solution was observed to be mostly amorphous. Results of electron diffraction studies of the sucrose cake did reveal some organization. In some spots of the cake, there were diffraction patterns associated with single crystal, i.e., Laue pattern. The presence of the Laue pattern was never observed in the vitreous portion of the sucrose cake, and the presence of such sucrose crystal nuclei in the cake is not thought to have been induced by the energy of the electron beam. Such nuclei will, upon warming of the cake, promote complete crystallization of the sucrose cake [10].

Structure

The self-supporting structure of a cake is intuitively associated with the solid content of the formulation. Although I did not perform any comprehensive study of the relationship between the solid content of a formulation and the self-supporting structure of the lyophilized cake, experience has shown that, when the solid content becomes less than 2% wt/v, there is a greater risk that the resulting cake structure will not have sufficient strength to prevent the physical breakup of the cake during the drying process. A cake having a poor self-supporting structure is illustrated by Figure 10.1b. The poor self-supporting cake structure appears to be aggravated by the quantity of the fill volume. Large fill volumes of low solid content formulations, perhaps a result of the low solubility of the active constituents, will have a greater tendency to exhibit a poor self-supporting structure than the same formulation with a lower fill volume.

In some instances, the loss of product from the container, as a result of a poor self-supporting structure can create serious problems involving not only a loss of product but also a hazard to the operating personnel and the facilities. The portion of the cake that leaves the vial is often in the form of a fine powder that is not always visible in the chamber when the door of the dryer is opened. This fine powder can present several problems.

Operating Personnel. In case there is an apparent loss of cake from the containers, operating personnel should be equipped with the appropriate gowns and breathing apparatus, in compliance with the requirements specified by the Materials Safety Data Sheet (MSDS) for each of the solid constituents that make up the formulation. Special procedures and training of personnel will be required to ensure personnel are not exposed to the dust during removal of the gown. Gowns that may be contaminated should be disposed of in accordance with the instructions specified in the MSDS. My major concern is that the information contained in the MSDS may not have been completely documented. This will be of particular concern for those individuals engaged in the development of the lyophilization process. Because these personnel are working with relatively small amounts of product, personnel safety issues can often be overlooked or ignored. It is unfortunate but often true that *familiarity breeds contempt.*

Facilities. In a facility that contains a clean-air handling system, an increase in particle count could be an indication of the entry of dust particles into the manufacturing area. However, in those facilities not requiring such monitoring, as in the case of some diagnostic products, there may be a buildup of dust in the room that could be harmful not only to personnel but in some cases to the electronic equipment.

Environmental. If it is known that significant quantities of the product are entering the dryer, then the condensate from the condenser and the water used to rinse the dryer may require treatment prior to being discharged into the public sewage system. Such a need for expensive wastewater treatment could have been avoided had adequate consideration been given to the manufacturing of the product during the formulation development stage.

Product. Although each of the above effects of poor cake structure is of concern, perhaps the greatest concern is with the product itself. A key problem is that any loss of cake from the container will reduce the potency of the drug in that container. With no supporting data, it is my opinion that the amount of loss from product will vary from vial-to-vial. One may be tempted to compensate for the loss in potency by increasing the potency to the upper-limit allowed by the USP for the particular drug, a concept similar to the rationale of keeping extra sheep in the pen to compensate for those that may escape through a hole in the fence.

Perhaps the most difficult problem associated with dust generated by a poor self-supporting cake structure is the dust that will be on the outside of the container. Then the question arises, *How does one ensure, especially if the product is one that should not come in contact with the skin, that such product containers can be safely handled once they are in the public domain?* One possible solution would be to caution the user to handle only while wearing gloves. Such a label would not instill confidence in manufacturing practices.

The solution to these problems is relatively simple: during the initial formation of the product, make certain that the solid content is $\geq 2\%$ wt/v.

Cake Strength

The strength of a cake is a measure of its ability to withstand stress and is a gauge of its durability. This physical property of the cake is related to the nature and the concentration of the excipient(s) in the formulation [5]. Except in the case of an electrolyte such as NaCl, PVP is an excellent excipient to improve the strength of a lyophilized cake [5]. PVP's capability to increase the strength of a cake is attributed to its elastic and thread-like binding properties arising from polymeric characteristics [5]. It is not certain if such binding by PVP occurs during the freezing or the primary drying process.

A means for determining the strength of a cake has been devised [12]. In this method, the cake is transferred to a closed glass container. The container is then placed on a spring-suspended table and shaken by means of an eccentric drive mechanism. The drive mechanism generates 213 horizonal movements per minute and the same number of vertical movements per minute with an amplitude of 4 cm. After shaking for 30 min, the density of the cake was determined in the same manner as described above [12]. The percentage in the reduction in the density of the cake is considered to be proportional to the original volume of the specimen [12] and is associated with the mechanical strength of the cake.

Fractureability

The *fractureability* of a cake is a measure of the force necessary to break or crack the cake. A method by Bashir and Avis for determining the fractureability of a cake has already been described in detail in Chapter 2 [5]. These authors found that cakes made with excipients like mannitol and PVP required a considerable force to break. For example, cakes formed by the lyophilization of 0.2 M mannitol solution had a fracturability of 88 g ± 19.6 g. Such a large standard deviation would suggest that the fracturability of a cake is best viewed as a qualitative rather than a quantitative analysis of the cake structure. Forming a suitable cake with mannitol required the addition of NaCl and even those cakes were rather fragile.

Vial Breakage

The composition of a formulation, in combination with the lyophilization process, can result in vial breakage like that shown in Figure 10.1c. The breakage of vials as a result of using mannitol as an excipient was first reported by Jennings [16]. Williams et al. [17] first demonstrated that actual vial breakage occurred during the warming of a frozen mannitol matrix. When vial breakage did occur, these authors observed an increase in the electrical resistance of the system. The increased electrical resistance was found to be associated with an ice recrystallization process. Rapid cooling of the mannitol formulation produced a matrix that formed a system that could result in ice recrystallization. When the mannitol formulation was cooled slowly, a matrix that did not result in ice recrystallization was formed and vial breakage was not observed. It was later established that the breakage was associated with an exotherm resulting from the recrystallization of the mannitol [18].

Effects of the Lyophilization Parameters

The properties of the cake that are a direct result of the lyophilization process or storage conditions will be considered here. The reason for separating those properties that are primarily associated with the formulation and those associated with the lyophilization process is to provide the reader with a basis for identifying a possible source of a defective product. The reader is again advised, as demonstrated above, that the observed properties may be related to either the formulation, the lyophilization process, or a combination of both.

Potency

For a loss in potency in the final product, one of the first processes to consider is the freezing process. The reason, as stressed in Chapter 7, is that during the freezing process, ice crystals are formed, and there are major changes in the concentration of the excipients. In the case of biological products such as vaccines, the freezing process may cause a loss in the titer of the vaccine. This loss in titer can be a result of a change in pH, a major change in the concentration of gases that could prove harmful to the virus, or the removal of water that results in the denaturation of the proteins associated with the virus [19,20].

Matrix Structure

Nei and Fujkawa [21] used optical microscopy to gain an insight into the mechanisms involved during the freezing process of a formulation. Thin specimens of the formulation were sandwiched between two cover glasses, and the temperature was then lowered. As freezing of the formulation occurs, one obtains a pictorial account of the growth of the frozen matrix. While such pictorial accounts are often dramatic to observe, these authors expressed some reservations about whether the freezing in the thin layer specimen was truly representative of what actually occurred during the freezing of bulk materials [21].

Bilayer Structure

Situations may occur where certain constituents in the formulation are found to be incompatible because of possible chemical interactions. Such chemical interactions can result in a loss in potency before and during the freezing process.

For example, Haby et al. [22] reported that a product required the nitroimidazole ligand known as BMS 181321 to react with technetium-99m. The BMS 181321 ligand was obtained from the reduction of a ligand identified as BMS 181032 by a reduced form of technetium in the presence of stannous chloride ($SnCl_2$). The problem was that the ligand BMS 181032 was found not to be stable in the presence of stannous chloride. These investigators solved this dilemma by forming a bilayer matrix.

The bilayer matrix was formed by first freezing a given volume of the ligand BMS 181032 formulation containing sulfated β-cyclodextrin solution to -50°C and maintaining this temperature for a period of 1 hr. A given volume of $SnCl_2$ was added to the container, and the entire system was again frozen to -50°C.

The bilayer frozen matrix was then lyophilized, and the result was a cake having two separate layers. Lyophilization of the formulation frozen in a bilayer configuration resulted in a final product in which the ligand BMS 181032 retained > 90% of its initial concentration. The ligand still retained 85% of its original potency even after 4 weeks of storage at 40°C [22].

Froth or Foam

The final product cake may contain a *froth or foam* on the upper surface like that shown in Figure 10.1d. In general, frothing or foaming is undesirable because it provides a system that may have a different composition than that of the main formulation and could lead to interactions, such as the aggregation of proteins. Even if the frothing or foaming is found to have no deteriorating effect on the quality of the lyophilized product, its presence does not instill the same degree of confidence in the user as does a well-formed matrix, like that illustrated in Figure 1.3a.

The formation of this froth or foam on the upper surface of the cake can stem from two possible sources, one of which is the filling of the vials. In some formulations, notably those formulations containing proteins, the rapid dispensing of the fill volume into the vial can produce a foam on the surface. The foam will tend to persist during the freezing process and will be present in the final lyophilized cake.

A second source of foaming or frothing could stem directly from the freezing process. If the formulation contains significant quantities of dissolved gases, particularly CO_2, the gas will be concentrated in the interstitial region where it can reach saturation. If the interstitial region is still in a liquid state, the increased gas pressure can deform the matrix and provide a path for release of the gas. The release of the gas results in the formation of a surface that is characteristic of frothing or foaming. Frothing or foaming will tend to be more prevalent during the freezing of natural food products than in pharmaceutical or biotechnology formulations. If foaming or frothing is a problem that occurs during the freezing process, the gas concentration can be decreased by *sparging* the formulation with a gas that has a low solubility, such as nitrogen or helium. Even with sparging, one must exercise caution that excessive foaming does not result in the denaturation of the protein constituents in the formulation.

Cake Strength

The freezing process can also impact the cake strength. It has been found that quickly frozen products will tend to have a higher cake strength, as described in the previous section, than products that are frozen slowly [12]. While cake strength can be affected by the freezing process, the solid content of the formulation (i.e., wt/v) will still be the prevailing factor for the cake structure.

Crust or Glaze

The formation of a *crust* or *glaze* on the surface of the cake is strongly dependent on the nature of the freezing process. As illustrated in Figure 1.3b, the formation of a heterogenous matrix is a major factor in forming a crust or glaze during the freezing process. In particular, a crust is generated by the formation of large ice crystals that tend to concentrate the unfrozen formulation and push the solutes to the top surface.

Such cake surfaces can affect the transport of water vapor from the product and may contain key constituents of the formulation. Dawson and Hockley [23] used scanning electron microscopy to examine the glaze or crust on the surface of the cake. These authors were able not only to examine the structure of the surface layer but also to determine its composition. For example, in one case they found that the crust contained a protein constituent. They reported that the compound was not uniformly distributed throughout the crust but was more concentrated in the upper one-fifth of the crust. These results are of major importance because the composition of the crust may vary significantly from the composition of a homogeneous cake. The ability to identify the composition of a particular region of the cake using scanning electron microscopy offers an important analytical tool for determining not only the structure but also the composition of a lyophilized cake. It is unfortunate that these authors did not elaborate on the analytical technique that was used to ascertain the composition of the crust.

Collapse

The most prevalent impact of the primary drying process is on the nature of the cake structure of the final product. While the thermal properties of the formulation are an important factor, T_p with respect to Tc will have a major impact on cake structure. If $T_p > Tc$, then the presence of mobile water in the interstitial region can lead to *collapse* of the cake, as illustrated in Figure 1.3c or in Figure 10.2c. The degree of collapse will be related to the magnitude of the difference $(T_p - Tc)$. As $T_p - Tc$ increases, the volume of the cake will decrease, as illustrated by comparing Figure 10.2b and 10.2c. With the assumption that Tc remains fairly constant, the magnitude of $T_p - Tc$ will be directly related to the chamber pressure (P_c) and the shelf-surface temperature (Ts).

The reader should be aware that collapse can not only cause a cosmetic change in the appearance of the cake but also can be responsible for a loss in potency. By using scanning electron microscopy of the cake after the completion of the primary drying process, Dawson and Hockley [23] were able to observe aggregation of a protein constituent in a cake exhibiting a collapse structure. Therefore, collapse should not be viewed as merely a cosmetic defect but as a condition that can lead to a reduction in the potency of the final product.

It is hoped that the reader will realize the danger of not measuring T_p during the lyophilization process. By determining that the T_p was approaching the known Tc of the frozen matrix, one could assume that there was an error in the measurement of the T_s or the P_c. If the temperature readings were correct during the freezing process, one would be justified in believing that such measurements are still within specification, and it is the pressure gauge that, for whatever reason, is reading a false low. By adjusting the pressure control system to bring the T_p within the specified temperature range for the primary drying process, one could prevent the possible loss of a batch of product and have the pressure gauge calibrated before attempting to lyophilize another batch of the formulation.

Meltback

Meltback can be viewed as an extreme case of collapse in which liquid states exist in the interstitial region during the primary drying process. In this instance, the interstitial region is no longer self-supporting, and collapse occurs to a point where there is little cake structure, as illustrated by Figure 10.2d or no cake structure as shown by Figure 1.3d. As with collapse, meltback not only produces a cosmetically unacceptable product but can result in aggregation of the constituents to lower the potency to unacceptable levels. There are basically two forms of meltback.

Partial Meltback. *Partial meltback* occurs when the temperature of the shelf surface is increased to commence the secondary drying process prior to completion of the primary drying process. With the increase in the shelf temperature, the result is that $T_p > Tc$, a small quantity of frozen matrix located near the center bottom of the vial forms a liquid state, and the small region of the cake, like that shown by Figure 10.1e, undergoes meltback.

If partial meltback occurs throughout the entire batch, then the problem is associated with an improper lyophilization process or an error in one of the major process control transducers. An improper lyophilization process will be evident by a T_p not approaching the T_s for a specified period, which indicates an incomplete primary drying process. The problem can be corrected by allowing T_p to approach T_s for some specified period of time. If the frequency distribution of the C_o values for the vials is known, then, from the standard deviation, one can obtain the estimated additional time for maintaining T_p at approximately T_s in order to ensure that all of the product in containers have completed the primary drying process [1].

Should partial meltback occur in a small number of vials scattered throughout the entire batch, then the problem is again associated with the frequency distribution of the C_o values of the vials. In this instance, however, it will be the vials in the frequency distribution that have C_o values that are significantly lower than the mean $\overline{C_o}$ value of the vials. To correct this problem, the primary drying process will have to be extended to include those vials with the lower C_o values.

Total Meltback. When *total meltback* occurs throughout the entire batch, then all of the vials have a T_p that exceeds Tc throughout the major portion of the drying process. This condition can be corrected by either (1) lowering the P_c while maintaining the same T_s; (2) decreasing the T_s while maintaining the same P_c; or (3) decreasing both the T_s and the P_c so that T_p will be lower than Tc during the period of time when frozen matrix is present in the vials. Except for option 1, revising the process parameters (T_s and P_c) to lower the operating T_p will tend to prolong the primary drying process.

Should total meltback occur in a small number of vials scattered throughout the entire batch, the problem is once again associated with the frequency distribution of the C_o values of the vials. In this instance, however, it will be the vials in the frequency distribution that have C_o values that greatly exceed the mean $\overline{C_o}$ value of the vials. To correct this problem, the drying process will have to be slowed down so that the temperatures of the product in the vials having the higher C_o values will not generate T_p values that greatly exceed the Tc.

Puffing

Puffing is a cake defect (illustrated by Figure 10.1f) in which the upper surface of cake expands during the initial portion of the primary drying process. The reason for the expansion of the cake surface is incomplete freezing of the formulation prior to lowering P_c to commence the primary drying process [24,25]. Since the puffed portion of the cake stemmed from an unfrozen portion of the matrix, this material should be considered as being a result of partial meltback, even though it is associated with the top of the cake rather than the bottom. If the formulation contains protein constituents, then one should be concerned with aggregation and the possible formation of particulate matter. In addition, aggregation of the protein can result, depending on the fill volume, in a significant reduction in the potency of the final product. The loss in potency resulting from puffing will be more prevalent with formulations involving low fill volumes.

Browning

When cakes having carbohydrates and ascorbic acid as constituents are exposed to excessive heat during the secondary drying process the results can be a discoloration or *browning* of the cake [9]. Although browning is dependent on time and temperature, it is an irreversible reaction. While discoloration of the cake is cosmetic, browning can affect other product properties, such as the potency of the active constituent.

Effects of Storage

The underlying function of lyophilization is to serve as a stabilizing process; nevertheless, lyophilized products can undergo physical and chemical changes during storage. Some of these changes can be beneficial, while others can have a deteriorating effect on the potency of the active constituent.

Effect of Crystallization. Sucrose is often used as a major constituent in a formulation. When such a formulation is lyophilized, the sucrose will tend to be in an amorphous state. Amorphous sucrose has a high affinity for moisture. Because of sucrose's high affinity for moisture, it is often difficult to achieve low moisture content in the formulation. Yet when the amorphous sucrose is stored at an elevated temperature for a specified period of time (e.g., 60°C for 1 month), crystallized sucrose that has a relatively low affinity for moisture is formed [19]. Therefore, the conversion of sucrose from an amorphous to crystalline state will limit any further increase in moisture resulting from the presence of sucrose.

Crystalline mannitol can be formed during the lyophilization process and requires no thermal treatment after lyophilization. But the presence of crystalline mannitol in the lyophilized cake causes a major reduction in the potency of the protein erythropoeitin (EPO) after storage at 4°C for only 6 months [19].

Browning Occurring During Storage. While the previous two examples illustrate how crystallization of an excipient may prove useful in some instances and harmful to active constituents in others, other constituents may, during product storage, produce reactions that can have an impact on the active constituent. Consider a cake whose composition consists of a protein, sucrose, citric acid, and dibasic sodium phosphate. If this system were stored at 60°C for 1 month, there would be a decrease in *p*H upon reconstitution. At lower *p*H values, sucrose hydrolyses into fructose and glucose. The resulting glucose reacts chemically (browning reaction) with the protein constituent [19].

A similar reaction occurs during the storage of dried blood plasma. A chemical reaction involving glucose and protein amino groups, which results in the formation of glucose amides, gives the dried blood plasma a brown color [26]. Even dried blood plasma stored at 20°C and at relative humidities ≥ 7% can, after some period of time, produce a brown color [26].

The key point for the reader to realize is that just because a formulation has undergone the lyophilization, it does not mean that physical and chemical changes cannot occur during the storage of the product. By understanding the nature of these changes and how they impact the active constituent, one can take appropriate steps to protect the active constituent by changing either the constituents in the formulation or the conditions under which the product is stored.

MOISTURE

The moisture content has been long recognized as important because of the major role that it plays in determining the stability of a lyophilized product. In 1949, Flosdorf [27] reported that when a lyophilized formulation containing diphtheria antitoxin had a moisture content of about 0.5% by wt, there was negligible loss in the potency after 3 years of storage. Yet when the moisture content of the latter formulation was increased to between 5% and 8% there was an appreciable loss in potency after just 6 months of storage. In 1950, Lea and Hannan [28] found that the moisture content was responsible for the chemical reaction involving the amino group of casein and glucose. At low moisture levels in the cake, the interaction between casein and glucose was found to be negligible. Thus, it is important to be able to equate the moisture content of a lyophilized formulation with its stability.

The remaining portion of this section will be devoted to a discussion of the nature of the various roles that water performs in the lyophilized cake and a description of analytical methods for determining the moisture content.

Residual and Bound Water

Residual Moisture and Free Water

Free water, sometimes referred to as surface water, freezable water, or residual water [29–31], is defined as water that obeys the desorption isotherms that were illustrated by Figures 9.4 and 9.5. For clarity sake, I will use the term *residual moisture* instead of free water. The principal effect of the residual moisture content on the lyophilized product, as stated above, is to alter the product's stability. Excessive residual moisture will tend to reduce the shelf life of the product [27,28], however, the addition of water has been found to enhance the stability of some lyophilized products [32]. The actual residual moisture necessary to achieve stability will vary from product to product depending on the nature of the individual constituents [25].

Dependence on Surface Area

The adsorption isotherms for water vapor on the surface of a lyophilized cake have already been covered in the previous chapter. The quantity of moisture on the surface was shown to be dependent on the key factors of the lyophilized product.

Cake Properties. The nature of a cake surface and the heat of desorption will be key factors affecting the quantity of adsorbed water. The heat of desorption of water will be related to the bond energy between the surface of the cake and the adsorbed water. For a given temperature and partial pressure of water vapor, as the bond energy increases, the weight of adsorbed water increases.

Partial Pressure of Water. The quantity of water in a product will be dependent on the vapor pressure of water above the cake surface. At a given cake temperature, the amount of adsorbed water will increase as the vapor pressure is increased. Means for determining the partial pressure of water vapor in the chamber (e.g., an electronic humidity sensor) were described in the previous chapter.

Cake Temperature. As seen in the previous chapter, the cake temperature for a given partial pressure of water vapor has a major impact on the quantity of adsorbed moisture. An increase in the cake temperature will decrease the residual moisture content of the cake. It is important to realize that, in order to prevent readsorption of moisture on the surface as temperature is reduced, stoppering must be performed at the higher temperature before reducing the temperature of the product. Lowering the T_p, while holding the partial pressure water vapor constant, will result in an increase in the residual moisture content of the product.

Change in the Residual Moisture Content

The following are some examples of the increase in the residual moisture in the lyophilized product and the potential source of such moisture.

The residual moisture in live bovine brucellosis vaccine stored at 4°C showed a relatively rapid increase in moisture during the first 6 months, but the rate of moisture increase in the vaccine declined in subsequent months [33]. The cause for the increase in the moisture content of the vaccine was attributed to the closures. These authors felt that the composition of the closures was an important factor in determining the stability of the vaccine [33]. It should be noted that an increase in the moisture content of the vaccine stored at 4°C and associated with water vapors from the closures is contrary to my observations. In outgassing studies of closures at about 4°C, I found that at the rate at which water is released from closures at these temperatures would require years to increase the moisture content of 1 mg of material by 1%, even if the sticking coefficient were assumed to be approaching 1. The *sticking coefficient* is defined as the number of gas molecules remaining on the surface to the number of gas molecules striking the surface.

The moisture of lyophilized moxalactam disodium was found to increase from 0.68% to 0.98% over the first 6 months of storage at 25°C. Pikal and Shah [34] contend that, since the moisture content remains constant after 6 months, the source of the moisture must stem from the closure rather than transmission (permeation) through the closure. The outgassing of the closure will be considered in greater detail in Chapter 12, but for the present, the reader should be aware that the residual moisture content of a lyophilized product may be altered as a result of the container-closure system.

The mass of lyophilized pediatric vaccines such as Hemophilus b Conjugate Vaccine (Meningococcal Protein Conjugate) is only 3 mg. For this product, relatively small quantities of moisture entering the vial can result in large increases in the moisture content of the cake. The introduction of such moisture can cause collapse of the cake. Once again, the source of the moisture was attributed to the closures. The moisture in the closures can be effectively reduced by heating them in a dry heat oven at 110°C for 22 hr or at 143°C for 4 hr. The residual moisture remaining in the closure was the same for each of the drying methods [35].

Bound Water

Bound water has been previously defined as water that is an intrinsic part of the active constituent [36]. It is often this form of water that is associated with maintaining the unique configuration of a protein molecule. The removal of such water from a protein molecule can cause a change in configuration or the protein to become denatured. A change in configuration could impact the specific activity of the molecule and possibly result in cross-linking with other molecules in the system.

The desorption of bound water from a cake will follow the same basic criteria as that for residual moisture except that the desorption energy will usually be significantly greater. The relationship between the bound water and the active constituent can be expressed as

$$A \cdot n\mathrm{H_2O} \xrightarrow{k_d} A + n\mathrm{H_2O} \qquad (4)$$

where n represents the moles of bound water associated with a mole of the active constituent A and k_d is the dissociation constant, which is defined as

$$k_d = \frac{[A][\mathrm{H_2O}]^n}{[A \cdot n\mathrm{H_2O}]} \qquad (5)$$

For active compounds (A) that have relatively low values of k_d, e.g., 0.1, the stability of the product can be maintained only when there is excess water (residual moisture) in the system to prevent the dissociation. This is represented by equation 4 for the compound $A \cdot n\mathrm{H_2O}$. The stability of the compound $A \cdot n\mathrm{H_2O}$ is maintained by the presence of residual moisture. This residual moisture can be referred to as *associated water* because of its role in maintaining the stability of the active constituent. From this discussion, it can be seen that it is possible for some active constituents containing bound water to be dissociated or denatured by over-drying the product.

It should also be realized that, in spite of the presence of associated water, the dissociation reaction may still occur as a result of the presence of an excipient such as lactose in the system. While lyophilized lactose will initially be anhydrous, its natural state is a monohydrate, and reverting to its natural state could be at the expense of the active constituent. By competing for the associated water in the system, a constituent such as lactose could cause a loss in potency by allowing the reaction shown in equation 4 to proceed. Once again, the reader should recognize the importance of the proper selection of excipients for a formulation, in the above.

Analytical Methods

The following will be a description and discussion of the various analytical methods used to determine the moisture content in a lyophilized product. It is important for the reader not only to understand how the analysis is conducted but also to be aware of the limitations and possible pitfalls that could have a serious impact on the accuracy and precision of the measurement.

Gravimetric Analysis of Bulk Materials

The gravimetric method is the oldest analytical technique for determining the moisture content in a lyophilized product. In 1909, Shackell [37] used the loss-in-weight method to determine the total moisture content in natural substances such as milk, honey, salt, and other substances. He accomplished this by weighing the material before and after about 12 hr of desiccation in a vacuum chamber. The pressure in the chamber during the desiccation process was ≥ 1 Torr. Shackell used concentrated sulfuric acid as a desiccant. He repeated the drying process until a constant weight was obtained. By using this technique, Shackell found that milk consisted of 84.46% water and the moisture in honey and a salt (presumably NaCl) was 11.18% and 2.39%, respectively.

In this method for determining the moisture content of these substances, there was apparently no provision made for controlling the temperature of the sample. For example, in the case of milk, if the pressure of 1 Torr was maintained throughout the entire drying process, then, as a result of the phase diagram for water, the temperature during the primary drying process must have approached -20°C. At this temperature, one would have expected ice crystals to be present. If the milk samples were removed from the drying apparatus prior to completion of the sublimation of the ice crystals, the ice would have melted and, because of the relatively low temperature, the sample would have adsorbed moisture from the atmosphere. It may be a coincidence, but there are manufacturers of freeze-dryers who test the performance of their dryer by freeze-drying milk in trays. Although there is certainly nothing wrong with testing the performance of a dryer by the drying of milk, such a practice may reveal a rudimentary aspect of the field of lyophilization, i.e., that old practices are continued with little consideration of their basis or rationale.

The use of sulfuric acid as a desiccant can also affect the results of a gravimetric determination of the residual moisture. Vapors such as sulfur dioxide (SO_2) and sulfur trioxide (SO_3) from the acid, especially near the end of the drying process when there in little or no flow of vapors from the sample, may backstream into the drying chamber and react with the product. Such reactions would increase the final weight of the sample and cause the result of the moisture determination to be a false low.

Gravimetric Analysis of Lyophilized Products

The gravimetric method presently being used to determine the moisture content in lyophilized products is similar in principle to that employed by Shackell [37]. The basic underlying principle is to ascertain the residual moisture content by determining the maximum loss in weight of a sample. However, the present methods determine the moisture content when the sample reaches a constant weight in a closed system at a pressure of not greater than 1 Torr and using anhydrous phosphorous pentoxide (P_2O_5) as a desiccant. Since the vapor pressure of water over a fresh surface of P_2O_5 is so low that it remains unmeasured, such a property makes this compound an excellent substitute for the original concentrated sulfuric acid desiccant. The reader is cautioned that P_2O_5 is a very strong oxidizing reagent and will cause burns, so it should be handled with extreme care. Prior to the start of each drying segment, one should remove any surface film so that any water vapor present will be exposed to a fresh surface of the desiccant.

The present gravimetric method calls for the analysis to be conducted at sample temperatures ranging from 20°C to 30°C [25,36,38–40]. The optimum sample size is about 200 mg and, in many cases, the sample must be obtained from a pool that consists of multiple samples [38]. Sample sizes ranging from 300 mg to 500 mg have been proposed [39], however, an increase in the sample size could necessitate an even larger pool of samples.

May et al. [38] prepared the samples in a low humidity glove box. In spite of using P_2O_5 as a desiccant, attaining a constant weight can be rather time consuming, i.e., from 2 to 5 days. In an age when one can become impatient if a computer does not respond in a matter of seconds, the use of such an analytical method could prove frustrating at times.

The use of a pooled sample and a transfer environment presents several problems with regard to the confidence that one can have in the measurement. First of all, the sample would have to be transferred from the original sealed container to the weighing vessel. A difference between the relative humidity in the original container and that of the environment in which the transfer was performed could result either in false low or false high moisture values. If the relative humidity in the container is greater than that in the transfer environment, then after equilibration with the transfer environment, there will be a reduction of moisture in the sample, and the results of the analysis will be a false low. Should the relative humidity in the transfer environment be higher than that in the container, then the sample will gain moisture and the results of the moisture analysis will be a false high. In order to transfer the sample without introducing an error, the humidity of the transfer media will have to be adjusted to equal that in the product containers. Since there will be a frequency distribution of moisture values in a given batch of lyophilized product, one would also expect a frequency distribution of relative humidity values in the product containers. Performing the transfer of the samples in a properly controlled humidity environment poses an arduous task.

A second problem regarding the use of a pooled sample is the number of samples and the frequency distribution of the moisture values. If there are fewer than 30 samples in the pool, then the moisture value obtained (the average value of the samples in the pool) may vary substantially from pool to pool. Even if the pool contains a statistically significant number of samples and there is consistency in the observed moisture in the various pools, one still does not know the nature of the distribution. For a broad frequency distribution of moisture values, a high probability could exist that there are vials in the batch that either greatly exceed the upper moisture specification for the product or have unacceptably low moisture values. In either case, the potency of the active constituent in these containers could presently or at some later time prior to the expiration date be lower than USP specifications.

It has been proposed that the sample temperature during the gravimetric determination of residual moisture be increased to 60°C, while the P_c is still maintained at 1 Torr. The basic difficulty with using such an elevated temperature is that it may result in the dissociation of hydrates or cause the denaturation of proteins. Consequently, using an elevated temperature may prevent one from distinguishing between residual moisture and bound water.

Thermogravimetric Analysis

In thermogravimetric analysis, the sample is heated at a scan rate of 10°C/min from 30°C to 180°C while under a nitrogen purge. The weight loss of the material is plotted as a function of temperature. The percent weight loss obtained with this method is considered to be an accurate estimate of the moisture content of the lyophilized product [15].

The thermogravimetric instrument consists of a balance encased in a dry glove box. The humidity in the box is maintained at a low level with P_2O_5 and is monitored by a hydrometer. Because of the presence of such a low-humidity atmosphere, one must take special precautions to prevent the operation of the balance from being affected by electrostatic charges. This involves coating the quartz tube that surrounds the sample pan and the Pyrex® tube that surrounds the counterweights with an electrically conductive coating (e.g., gold film). Electrical ground wires are attached to the coated surfaces to drain off any charge buildup and to maintain these components at ground potential throughout the analysis. The residual moisture and bound water values of the sample are calculated from the loss-in-weight profile of the thermogram. Thermogravimetric measurements can make accurate determinations of moisture in samples weighing only 2 mg. Although this analytical method would not require pooling of the sample for most lyophilized products, the problems associated with the differences between the relative humidity in the glove box and that in the container, as discussed above, still lead to some uncertainty about the accuracy of the moisture measurement.

In using this method, one would have to ensure that the weight loss was not a result of the decomposition of one or more of the constituents in the cake, which would result in a false high moisture content. For example, if the formulation contains a material such as sodium bicarbonate ($NaHCO_3$), there will be a loss in weight at about 50°C as a result of losing CO_2 [15]. By monitoring the emission of the gases from the thermogravimetric instrument, one can determine if the loss in weight is associated with a loss in water (m/e = 18) or in CO_2 (m/e = 44) resulting from the thermal decomposition of the sample.

Karl Fischer Method

When iodine is placed in solution, the following reaction occurs:

$$I_2 + 2e^- \rightleftharpoons 2I^- \tag{6}$$

which leads to the formation of a reddish-brown solution. In the above reaction, the free iodine is oxidized to form an iodide ion [41]. A solution that contains an iodide salt will be colorless. This absence of color is important because when the solution turns reddish-brown, the color change will represent the end point to the titration of the moisture content in the solution (i.e., when free iodine [I_2] is present in the solution).

When I_2 is added to a solution containing a substance that has a lower oxidation potential, e.g., sulfurous acid (H_2SO_3), the following reaction will result:

$$SO_3^{2-} + I_2 + CH_3OH + H_2O \rightarrow SO_4^{2-} + 2H_3O^+ + 2I^- + CH_3HSO \tag{7}$$

In 1935, Karl Fischer developed a reagent that contained SO_2, I_2, and pyridine (an amine). When a sample (containing moisture) dissolved in methanol was added to this reagent, the following reaction occurred:

$$SO_2 + I_2 + CH_3OH + H_2O \rightarrow 2HI + CH_3HSO_4 \qquad (8)$$

and the quantity of I_2 that is oxidized will be quantitatively related to the amount of water present in the solution. It is important to realize that the amine (pyridine), although it does not enter into the reaction, is necessary for the analysis, and its removal from the reagent could have a serious effect on the accuracy of the moisture analysis.

From equation 8, it can be seen that the end point to the titration occurs when the reddish-brown color appears, indicating an excess of I_2 and an absence of H_2O in the solution. Since the Karl Fischer reagent is added to the sample dissolved in methanol, overtitration of the end point can occur if the sample contains a small amount of moisture. The same volume of methanol is also titrated with the Karl Fisher reagent to serve as a blank. The difference between the amount of reagent used to reach the endpoint with the sample and with the methanol blank will represent the quantity of water in the sample [38].

In using the Karl Fischer method, one must be aware that pyridine is a toxic material and that its presence can be readily detected by its obnoxious odor. Until recently [43], attempts to replace the pyridine have not met with much success, the principal difficulty being that the resulting reagent systems were not stable over a sufficient time. Diethanolamine has been successfully used in place of the pyridine [43]. In addition, new pyridine-free reagents have been found to have fewer measurement errors in samples having low moisture values. As with the pyridine, diethanolamine cannot interact with any of the constituents in the formulation without affecting the accuracy of the moisture determination.

One cannot use the Karl Fischer method if

- the sample does not adequately dissolve in the Karl Fischer reagent, methanol, or other solvent system, such as another amine or

- the sample moisture does not adequately extract into the solvents [38].

Coulometric Karl Fischer Analysis

The endpoint for the Karl Fischer titration can also be determined by measuring the electrical conductance of the solution. As the Karl Fischer reagent reacts with water, there will be a reduction in the conductivity of the solution. The endpoint is determined when the conductivity of the methanol and sample solution approaches zero.

For lyophilized products having moisture values ranging from 1% to 5%, a minimum sample size of approximately 40 mg is required. For samples having masses less than 20 mg, pooling the samples becomes necessary. Once again pooling the samples provides a measure of the average moisture content, and one should be concerned about any significant change in the observed average moisture content when the moisture in additional pooled lots is determined. Some instruments have increased sensitivity and can detect as little as 10 μg of water. For a 1 mg sample, this would represent a moisture value of 10%.

Comparative Study of Gravimetric, Thermogravimetric and Karl Fischer Moisture Analyses

A comparative study of gravimetric, thermogravimetric, and Karl Fischer methods of moisture analysis of six types of vaccines lyophilized in a buffered gelatin-sorbitol system and a control sample of sodium tartrate dihydrate ($Na_2C_4H_4O_6 \cdot 2H_2O$) was conducted by May et al. [38]. The following is a brief summary and commentary regarding this study.

Gravimetric Measurements. Gravimetric measurements were performed on a sample that consisted of a pool of 5 to 12 specimens. The reported moisture results (percentage residual moisture ± standard deviation [s.d.]) were based on an average of 12 to 13 samples. It should be noted that May et al. indicated neither the actual temperature of the sample nor the chamber pressure used to conduct the analyses. In addition, the reader should be aware that 12 to 13 samples do not represent a statistically significant sample pool.

Thermogravimetric Measurements. The thermogravimetric measurements were based on the results of a single vial. The reported results (percentage residual moisture ± s.d.) were based on an average of 9 moisture determinations.

Karl Fischer Measurements. The Karl Fischer measurements were based on the results of the contents of one vial; there was no indication, however, as to the number of samples that were used in reporting the results (percentage residual moisture ± s.d.).

Comparison Vaccine Results. The Karl Fischer and the thermogravimetric methods gave comparable results for the percentage residual moisture content in the vaccines; however, the gravimetric method gave consistently lower moisture values.

Sodium Tartrate Dihydrate Reference. Only the gravimetric and thermogravimetric methods were used in the determination of the moisture in the $Na_2C_4H_4O_6 \cdot 2H_2O$ reference. Note that the Karl Fischer method could not be used in determining the water of hydration of $Na_2C_4H_4O_6 \cdot 2H_2O$ because of its insolubility in alcohol [11].

The results of the gravimetric method indicated the loss of about one water of hydration, while the thermogravimetric method indicated the percentage residual moisture equivalent to two waters of hydration. One should also note that the temperature range of the thermogravimetric method was from 30°C to 180°C. Dissociation of the $Na_2C_4H_4O_6 \cdot 2H_2O$ (i.e., loss of the two waters of hydration), occurs at 150°C [11]. Based on this it comes as no surprise that the thermogravimetric method was able to determine the waters of hydration for $Na_2C_4H_4O_6 \cdot 2H_2O$.

Conclusion. These results indicate that using $Na_2C_4H_4O_6 \cdot 2H_2O$ as a standard for the gravimetric determination of residual moisture in a sample is not suitable for the prescribed temperature range of 20°C to 30°C. The rationale for such a statement is that the gravimetric method should only remove the residual moisture from the sample, but the above results for the gravimetric determination indicate that the dihydrate was dissociated to a monohydrate. Perhaps the temperature range used in the gravimetric measurements is too broad and should be limited to 20°C ± 2°C to prevent the dissociation of $Na_2C_4H_4O_6 \cdot 2H_2O$. If dissociation of the standard still occurs at this lower temperature limit, then it is suggested that another hydrate reference substance that does not undergo dissociation during a gravimetric moisture determination be selected.

The moisture determinations, especially those for the gravimetric determination of the moisture in vaccines, had standard deviations for the mean values that were relatively high. This would suggest that with such large standard deviations, the frequency distribution of the moisture values is relatively broad, and for a determination consisting of a pool of 5 to 12 specimens, there was considerable variation from determination to determination. For example, the average moisture for the live rubella and mumps virus vaccines, was 0.41% with a standard deviation of 0.26%. The relatively large deviations in the average moisture values could be attributed to the transfer of the specimens to the weighing container. Note that since the $Na_2C_4H_4O_6 \cdot 2H_2O$ is relatively stable at ambient temperature and not affected by the humidity in the transfer system, the average moisture content determined by the gravimetric method was 7.91% with a standard deviation of only 0.02%. The Karl Fischer and thermogravimetric determinations were performed on individual specimens and, in spite of the lower number of determinations, the standard deviations for these measurements indicated a more narrow frequency distribution than was found for the moisture content in the vaccines.

The importance of the above work is, in my opinion, not in the reported moisture values for the various vaccines, but in the apparent negative impact that the process of pooling the specimens and the effect that the transfer environment had in creating a relatively large deviation in the average moisture value for a given vaccine.

Water Pressure Determination

Beckett [40] described a method for determining the moisture content in a lyophilized material that involves the measurement of the partial pressure of water. In principle, the method involves trapping the moisture desorbed from a lyophilized sample onto a cold surface (cold trap). After the desorption process is completed, the cold trap is isolated from the rest of the system and then warmed to ambient temperature. The pressure increase in the isolated portion of the apparatus is then related to the quantity of water desorbed from the sample. The following describes the apparatus, the measurement procedure, and a commentary.

Apparatus. The apparatus is illustrated in Figure 10.4 and consists of the following:

- *Test Chamber.* The test chamber is designed to contain the sample. The test chamber is connected to the system by means of a vacuum seal.

- *Cold Trap.* A cold trap is used for trapping the water vapor and has a volume of V_t. The trap can be chilled by using a mixture consisting of either dry ice and acetone or dry ice and methanol contained in a Dewar flask. Such refrigeration mixtures can provide trap temperatures $\leq -50°C$.

- *Vacuum Pump.* The vacuum pump is the mechanical pumping system for evacuating the test chamber and the cold trap.

- *Vacuum Valve #1.* This valve is used to isolate the mechanical vacuum pump from the cold trap.

Figure 10.4. The Apparatus for Determining the Moisture in a Lyophilized Sample by Measurement of the Increase in Pressure of a Closed System

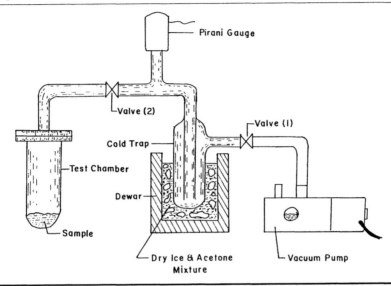

- *Vacuum Valve #2.* This valve is used to isolate the test chamber from the cold trap.

- *Pressure Gauge.* A Pirani or McLeod gauge is used to determine the pressure in the system.

Analytical Method. The moisture is determined by first placing a sample of a known mass (m) into a test chamber. The test chamber is connected via a vacuum seal to the test system, and valve #2 is in a closed position. By opening valve #1, the cold trap portion of the apparatus (at ambient temperature) is evacuated to less than 100 mTorr (P_t) by the mechanical vacuum pump. A McLeod or Pirani gauge is used to determine the pressure in the system. The cold trap is then cooled by a mixture of dry-ice and methanol or dry ice and acetone to a temperature T_t. Valve #2 is then opened slowly, and the test chamber containing the system is evacuated to a pressure approaching P_t (i.e., P_t' or less than 100 mTorr). Upon desorption of moisture from the sample, valves #1 and #2 are closed, and the trap also warms to ambient temperature. The moisture content in the trap is allowed to warm to ambient temperature, and the moisture content in the product is determined from the increase in the trap pressure P_2.

Moisture Determination. When the trap temperature is increased from T_t to ambient temperature (T_0), then barring any leaks in the system, the quantity of moisture in the trap can be determined using the ideal gas law. The ideal gas law can be used provided that the pressure increase resulting from the desorption of the trapped water vapor from the sample does not exceed 17.5 Torr (i.e., the vapor pressure of water at

20°C). Should the partial pressure of water vapor increase to 17.5 Torr, the partial pressure vapor in the cold trap would remain at 17.5 Torr in spite of the presence of a liquid phase of water in the cold trap. Because of the relatively low quantities of moisture associated with lyophilized products, it would be highly unlikely that the trap pressure would reach 17.5 Torr as a result of warming the trap to ambient temperature.

Assuming that the change in trap pressure ΔP is a result of the presence of the trapped water vapor and that $P_2 >> P_t'$, then the mass of water (m_w) in the sample can be determined from

$$m_w = \frac{9.8 \times 10^{-4} P_2}{V_t} \quad \text{g} \tag{9}$$

where P_2 is expressed in Torr, the volume of the cold trap (V_t) is expressed in L, and the ambient cold trap temperature is 293K. Equation 9 assumes that the volume of water vapor present when the trap is refrigerated approaches 0.

Commentary. Assuming that $P_t' \approx P_t$, then, in the absence of any ice in the system or the presence of any other condensed phase, the increase in pressure in the trap as a result of warming the system to ambient temperature is obtained from the ideal gas law:

$$\Delta P = \frac{\sum \left(n_{g,i}' R[T_o - T_t']\right)}{V_t} \tag{10}$$

where $\sum n_{g,i}'$ is the sum of the ith number of moles of gas present at T_t, which is equal to the number of moles of gas present at an ambient temperature T_o. Since V_t can be considered to be constant, ΔP is directly dependent on the increase in the trap temperature $(T_o - T_t')$.

Consider the increase in pressure in a cold trap with a V_t of 100 cm³ when the trap temperature (T_t') is increased from 223.1K (–50°C) to T_o = 293.1K (20°C) and there is no condensed phase present. For a trap pressure of 50 mTorr having a gas composition of 28.4 mTorr of water vapor and 21.6 mTorr of nitrogen, the pressure increase (ΔP) resulting from the increase in temperature will be 15 mTorr. The value of ΔP would be 15 mTorr even when the partial pressure of water vapor approached 0. The error in the moisture measurement induced by the temperature increase would represent 3.6 x 10⁻⁶ g of water, or an error of 0.36% for a 1 mg sample or 0.04% for a 10 mg sample. Therefore, for a low volume trap $(V_t \leq 100$ cm³) and $P_t' \approx P_t$, correcting the pressure by assuming only nitrogen is present would not cause a serious error in determining the moisture content provided that the mass of the sample were ≥ 10 mg.

For a 10 mg sample having a residual moisture content of 1%, and assuming that there is a 100% transfer of moisture from the sample to the trap, the increase in the trap pressure, correcting for the expansion of the nitrogen, would be 257 mTorr. Such a pressure increase would be easily determined with the Pirani pressure gauge.

As with any analytical method, when determining the moisture content of the water in a lyophilized product by using the above method, one should be concerned about the following:

Apparatus. The construction of a trap system having a total volume of only 100 cm³ could present difficulties. For example, because of the impedance of the trap, ascertaining when the pressures in the trap and test chamber pressures are equal may be difficult. Using a larger trap volume would eliminate this problem but would have the effect, as seen in equation 10 of reducing the sensitivity of the apparatus.

When operating at pressures lower than 100 mTorr, hydrocarbon vapors may backstream from the pumping system and deposit in the cold trap, resulting in a false high-pressure reading when the isolated trap is warmed.

Vacuum leaks in the system could result in moisture entering into the system from the environment and cause false high moisture values that would be directly proportional to the time that the trap is refrigerated. In addition, such leaks could cause the trap pressure to drift during the moisture determination and cause the determination of the final pressure to be uncertain.

As will be discussed in Chapter 11, the Pirani gauge will indicate false high-pressure readings in the presence of water vapor. Such a high-pressure reading would indicate false high moisture values. The actual gauge error will be a function of the partial pressures of the water vapor and the other gases present. For accurate measurement with the Pirani gauge, one would have to connect a residual gas analysis (RGA) system to the trap system. The value of m_w from equation 9 would now be based on the change in the partial pressure of water vapor in the trap.

Transfer of the weighed sample to the test chamber could result, depending the humidity conditions at the time, in either high or low moisture values.

Vacuum Pumping Effect. As long as there is ice present in the trap and the trap temperature is maintained at –50°C, the vapor pressure of water vapor in the trap, will remain at approximately 28 mTorr. In the presence of gas flow through the trap, water molecules can be swept from the trap into the vacuum system. Should the pressure in the trap become significantly lower than 28 mTorr then any ice in the trap (perhaps stemming from moisture from the sample) would start sublimating and enter the pumping system. These conditions would result in false low moisture values.

High Vacuum Moisture Determination

Two recurring problems that appear in the previously described methods for moisture analysis are the necessity of exposing the sample to an environment having a different relative humidity and the necessity of pooling the samples to attain a sample large enough to meet the sensitivity requirements of the analytical method. Ideally, the analytical method should be designed to accurately determine the total moisture (i.e., that moisture adsorbed on the surface of the cake and walls of the container) and the gas phase. The analysis should be performed on individual samples in order to establish the frequency distribution that is so important for establishing the probability that the moisture content, for both the high and low specifications, in a batch of lyophilized product is well beyond normal chance. The following is an apparatus that was designed, constructed, and tested to meet the above requirements for determining the moisture content of a lyophilized product.

Apparatus. The apparatus still requires development and is not commercially available. The general description of the apparatus is illustrated by Figure 10.5 [44]. Since all of the components of the apparatus have been previously described [44], only the key components will be described in this text. It should be noted that all of the components of the apparatus are fabricated from 316 stainless steel, and most are connected with copper flat ring seals.

The basic components of the apparatus, as illustrated by Figure 10.5, are the sample section (60), the gas trap section (62), and the vacuum pumping system (64). The entire section outlined by the dotted lines represents a heating system. The normal operating temperature of the system is about 70°C, however, the heating system can provide higher temperatures to bake out the system in order to remove any undesirable adsorbed gases on the surfaces. The following briefly describes each of these basic components.

Sample Section. The sample holder (1) is a specially designed fixture that performs two important functions: (1) provide an energy conductivity path to the sample container and (2) provide a means for retaining and removing the closure. The latter apparatus has been designed and constructed, but the full details are not illustrated in Figure 10.5.

The sample holder is placed in a detachable vacuum chamber (2). It is important that there be good thermal contact between the sample holder and the vacuum chamber. An iron-constantan thermocouple is used to monitor the temperature of the sample holder throughout the moisture determination of the sample. The sample temperature is controlled by chilling the vacuum chamber by means of a Dewar flask (5) containing liquid nitrogen or heating the vacuum chamber by means of the resistance heater (4). The sample holder is attached to a cross (9) that provides the thermocouple feedthrough and the mechanical motion for retaining and removing the closure to the sample. The cross (9) is connected to a high vacuum valve (10),

Figure 10.5. A High-Vacuum Moisture Determination Apparatus as Shown in U.S. Patent 5,016,468

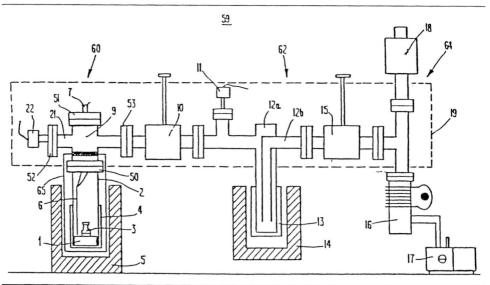

which permits the chamber to be isolated from the gas trap section (62) of the apparatus.

Gas Trap Section. The gas trap is a single unit containing three flanges. Two of the flanges are used to connect to the sample section and the vacuum pumping system through vacuum valves 10 and 15, respectively. The third flange of the gas trap is connected to a vacuum gauge (11) (capacitance manometer). The trap can be refrigerated by means of liquid nitrogen contained in the Dewar flask (14) or heated by means of the electrical heater (13).

Vacuum Section. The vacuum section consists of an air-cooled, high-vacuum oil diffusion pump (16) that is backed by a mechanical vacuum pump. The high vacuum valve (18) allows for backfilling the system to 1 atm.

Moisture Determination Method. The following is a brief description of the method used to determine the total moisture content in a lyophilized sample.

Establishing the Blank. The first step in determining the moisture content of a sample is the bakeout of the entire system represented by the dotted lines in Figure 10.5, the sample vacuum chamber (2) and the trap system to a temperature > 125°C while the system is being evacuated by the vacuum section. The heating section is then maintained at approximately 70°C while the vacuum chamber is returned to ambient temperature. The cold trap is then refrigerated with liquid nitrogen for no less than 1 hr. Valves 10 and 15 are then closed, and the trap is heated to 70°C. The pressure of the trap system is recorded as a function of time. The blank will provide the rate at which gases can condense leaks into the system during the moisture determination.

Preparing the Sample. A plastic glove bag (65) is attached to flange (50) of the vacuum chamber (2) like that shown in Figure 10.6. Note that such an attachment must be leak-proof. The sample vial (with a clean outer surface), tools, copper gasket and a balloon filled with helium are placed in the bag. Because of the weight of air displaced by the balloon is greater than the mass of the balloon, the balloon will rise to the top of the bag. One inlet of the bag is attached to a source of helium gas, and the other is clamped shut.

   The next step is to purge the glove bag with helium in order to reduce the relative humidity to zero. The glove bag is inflated with helium, and the pressure is released by discharging the excess pressure into a fumehood or to an outside vent. The purging is repeated until the helium-filled balloon rests on the bottom of the bag. When the balloon is resting on the bottom of the bag, nearly all of the air has been displaced from the bag. The bag is then maintained under a positive helium pressure.

   With valve 10 in a closed position, the vacuum chamber is removed. The crimp seal is removed from the sample vial and the vial placed in the sample holder. The mechanism for maintaining the container-closure seal and removing the closure is connected to the vial. A new, clean, copper gasket is installed, and the vacuum chamber is sealed to the sample section.

Cooling the Sample. The sample temperature is reduced by filling the Dewar flask with liquid nitrogen. For a sample temperature approaching –150°C, the vapor pressure of water in the vial will be about 1×10^{-9} Torr or 1×10^{-6} mTorr. With the

Figure 10.6. A Sample Preparation Apparatus for Moisture Determination as Described in U.S. Patent 5,016,468

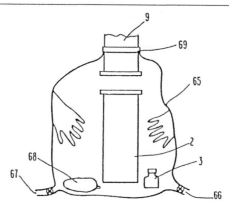

diffusion pump at ambient temperature, valve 10 is opened (see Figure 10.5). With the mechanical vacuum pump (17) operating, valve 15 is opened, and the pressure in the chamber reduced to less than 50 mTorr. The diffusion pump is turned on, and the entire system is evacuated to < 1 × 10^{-6} Torr.

The closure is removed from the vial, and the pressure in the chamber is again evacuated to < 1 × 10^{-6} Torr. At this point, the gases removed from the container will be noncondensable (nitrogen, oxygen). Most inert gases will also be removed, while some gases such as CO_2 will be retained by the sample along with the water vapor.

Desorbing the Water Vapor. The first step is to refrigerate the cold trap using liquid nitrogen. The temperature of the sample is then increased by means of the sample heater to 25°C ± 2°C. When the pressure in the cold trap is again < 1 × 10^{-6} Torr, valves 10 and 15 are closed. The cold trap is then heated to 70°C, and the pressure is recorded as a function of time.

Determining the Moisture Content. The mass of water (m_w) desorbed from the vial is determined from the following expression

$$m_w = \frac{9.8 \times 10^{-4}(P_v - P_b)}{V_t} \quad \text{g} \tag{11}$$

where P_v is the maximum pressure in the trap from the desorption of the gases from the sample and P_b is the maximum pressure obtained for the blank of the system.

Calibration of the Moisture Analyzer. The moisture analyzer was calibrated by using the inherent leak rate from the atmosphere into trap system. In performing this test, the cold trap was chilled for various time periods. The trapped moisture was then determined by increasing the trap temperature to 70°C. Figure 10.7 shows the measured water content as a function of a moisture leak rate of 2.08 × 10^{-11} g/sec.

Figure 10.7. A Plot of the Measured Moisture Content as a Function of the Moisture Content Determined from a Leak Rate of 2.08 $\times$ 10^{-11} g/sec

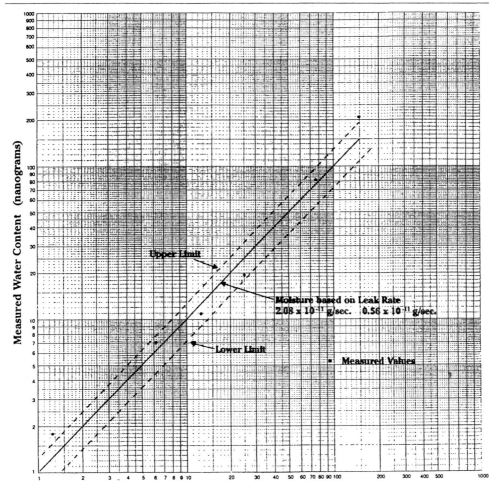

Water content (nanograms) (Leak Rate 2.08 x 10^{-11} g/sec)

Conclusion. While the above method offers the advantage of measuring the total moisture content in a container, there are still some pitfalls to overcome. One major problem, in spite of heating the system to 70°C to limit adsorption of water vapor on the walls of the trap system, appears to indicate a discontinuity in the measurements that may be related to adsorption of water vapor; there is also concern regarding the accuracy for the higher range of moisture values. In addition, the sample preparation and measurement technique is time-consuming, and determining the frequency distribution of the moisture values representing a given batch of lyophilized product would require a substantial amount of time and effort.

As of this writing, it is my opinion that the field of lyophilization awaits a rapid, accurate, and reliable method for determining the total moisture in a product.

RECONSTITUTION

The term *reconstitution* is defined by the *Oxford English Dictionary* [45] as the dehydration of a material to its original state by the addition of a suitable liquid. It is of interest to note that the dictionary states that the word (reconstitution) first appeared in the *National Food Journal II,* page 595: *The clauses . . . prohibiting the addition of milk or coloring matter or water, the reconstituted milk . . . will remain in force.* In 1945, in the publication the *ABC of Cookery* (Min. of Food), XVII page 59, the following statement was made: *If dried eggs are used, they may be added dry with other ingredients and the water needed for reconstitution added with the mixing liquid* [45]. This last statement is quite applicable to a lyophilized formulation that consists of a dry mixture containing an active constituent and other constituents that upon the addition of a diluent result in a system designed for a specific purpose.

This section on reconstitution of a lyophilized product will first consider the diluent and its potential impact on the constituents of the original formulation. The rate of reconstitution will be considered not only as a means of determining when the formulation may be safely used but also for the insight it can provide into the physical nature of the cake and into how such physical properties are related to the formulation and the lyophilization process. The last section will address the important issue of particulate matter in the reconstituted product.

The Diluent

For the most part, the common diluent used in reconstituting a formulation is water. The diluent is selected during the development of the formulation and it is my hope that this section will encourage researchers to give more consideration to selection of the diluent during the development stages of the formulation. The thermal properties of a formulation may be more *lyophilization-friendly* if an electrolyte, such as sodium chloride, is contained in the diluent rather than in the formulation.

The properties of the diluent can be an important consideration when reconstituting a lyophilized product.

pH of the Diluent

If the formulation does not contain a pH modifier (see Chapter 2), and the solubility of the active constituent is pH dependent, then special care must be given so that the user is instructed on the proper preparation of the diluent. Unless water having a pH that can range from as high 7.0 and as low as 5.0 is acceptable, then a diluent having the proper pH value should be supplied with the lyophilized formulation.

Temperature

For the most part, the reconstitution of pharmaceutical or biotechnology products is performed using a diluent, usually water, at ambient temperature (20°C to 25°C).

Lyophilized products generally cannot tolerate the warm or even boiling water that often is used to rehydrate precooked food products [25]. In preparing this text, I found that except for food products, the literature is silent on the effects that the temperature of the diluent can have on the reconstituted formulation. Perhaps the silence is because temperature is not an important consideration for pharmaceutical or biotechnology products or that little emphasis has been given to the topic.

Particulate Matter

The diluent used for reconstitution of the lyophilized formulation must be itself free of particulate matter if the final product is to meet USP specifications. The topic of particulate matter will be considered later in this chapter, but, for the moment, the reconstitution of a cake requires a diluent that meets the basic USP specifications.

Diluent Formulation

As previously stated, the nature of the diluent should be given careful consideration during the early stages of the development of the formulation. As was previously exhibited, the addition of NaCl to make the formulation isotonic had a major impact on Tc. Without the NaCl (0.9% wt/v) in the formulation, Tc was -20°C; however, making the formulation isotonic lowered the Tc to < -50°C. Such a decrease in Tc could make lyophilization of the formulation nearly impossible for some freeze-drying equipment, greatly increase the duration of the lyophilization process, affect the cosmetic properties of the cake, and increase the reconstitution time. Unless the active constituent requires an isotonic solution to achieve the necessary stability in the liquid phase prior to the start of the lyophilization process, it would make sense from both a quality and an economic perspective, to specify an isotonic solution as a diluent rather than as a constituent in the initial formulation. Sometimes the rush or pressure to complete the development of a formulation results in overlooking factors that could later have a major impact on the cost of manufacturing and the quality of the final product.

Reconstitution Time

The time required for reconstituting a lyophilized product has been found to range from what appears to be instantaneous to in excess of 1 hr. This section will consider the impact that the physical nature of the cake, the formulation and the lyophilization process can play in determining the time required for reconstitution.

Surface Area and Physical Properties

The surface area of a lyophilized product has a major impact on its reconstitution properties. As a general rule, an increase in the surface area per unit volume of the cake will tend to decrease the reconstitution time. As the surface area is reduced because of collapse or meltback, the time required to reconstitute the product may be greatly extended. In some instances, the rate at which a lyophilized cake reconstitutes is considered to be an indication of the quality of the product [25]. However, the reader should be cautioned that the effect of surface area on the time required for complete reconstitution is very much dependent on the constituents in the formulation. For example, I have seen products that showed complete meltback yet had very rapid reconstitution and products showed only slight signs of collapse yet required excessively long reconstitution times.

Methods for Reconstitution

The following are some suggested methods for reconstituting lyophilized pharmaceutical, biotechnology, tissue, and food products.

Pharmaceutical, Biological, and Biotechnology Lyophilized Products. The reconstitution of these products is accomplished by inserting a sterile needle through the closure and injecting, via a syringe, the correct volume of the diluent. If a relatively large volume of the diluent with respect to the total volume of the container (e.g., ≥ 50%) is required, then the container could become pressurized if the vial was stoppered at or near 1 atm. Pressure buildup in the vial could make the addition of the correct quantity of diluent a difficult task. The pressure buildup can be prevented by the insertion of a clean sterile vent syringe needle, equipped with a 0.2 μm filter, through the closure prior to adding the diluent. Care should be taken that the tip of the vent needle does not extend too far into the vial, thus permitting the flow of the reconstituted formulation into the vent needle where it could come into contact with the work area or healthcare personnel. The vent needle should be removed immediately after the addition of the diluent.

This discussion should not come as any great surprise to anyone working in the healthcare industry. The important points are the volume of the vial, the relative diluent volume required, and the pressure at which the secondary drying was terminated. This information should be taken into consideration during the preparation of instructions for safely reconstituting the product.

Some products may require some gentle swirling of the diluent to accelerate the reconstitution process. Shaking of the vials should be avoided to prevent foaming and possible denaturing of the protein constituents. However, if the reconstitution time is excessive (e.g., > 3 min), then the user may run out of patience or become frustrated and resort to shaking the vial to accelerate the reconstitution process. In spite of written instructions to the contrary, it is highly unlikely that the average user will gently swirl a mixture of solid cake and diluent for an excessively long time. If a product requires more than 5 min to reconstitute, then steps should be taken to decrease the reconstitution time rather than depending on the patience of the user.

Tissue. Reconstitution is generally performed by soaking the lyophilized tissue in a physiological saline or balanced salt solution. The rehydration of arteries, skin, and cornea may require only a few minutes, while bone could take up to several hours [25].

Foodstuffs. The recommended method for rehydrating dried food products is to place the product on the surface of the water. Placing of the product on the water will allow the water to penetrate the product from the bottom and permit trapped gas to escape upward. Foodstuffs also have the advantage that they can be reconstituted with hot or even boiling water [46].

Relationship of Lyophilization Process and Reconstitution

The following is a brief discussion as to the effect the various lyophilization processes can have on the reconstitution process.

Freezing Process. The freezing process can affect the reconstitution process by the nature of the frozen matrix. For a homogeneous cake structure like that shown in Figure 1.1a, reconstitution could be rapid and complete in a relatively short period of time. In the case of the formation of a heterogeneous cake, as illustrated in Figure 1.3b, the formation of the crust could also cause aggregation of the protein and result in an extension of the reconstitution process. In the case of blood plasma, that the denaturing of lipoprotein occurs during the freezing process has long been suspected [24].

Primary Drying Process. The primary drying process can have the greatest impact on the reconstitution of the product. Collapse of the cake, as illustrated by Figure 10.2, can certainly prolong the reconstitution process, especially for serum-based products. Also *puffing*, as illustrated by Figure 10.1f, can cause aggregation of proteins that would prolong or even prevent complete reconstitution of the product. However, it is the production of meltback during the primary drying process that can have the most serious effect on reconstitution properties. The meltback of a formulation containing protein compounds can result in a material where the interaction between molecules is such that an almost insoluble compound is formed. For such a product, reconstitution could be even a matter of days, and it is possible that the potency of the resulting solution would not meet USP specifications.

Secondary Drying. The complete removal of moisture from a formulation containing proteins in extending the reconstitution time [26]. Excessive aggregation of the protein can lead to the formation of an unacceptable opalescent solution upon reconstitution [47].

Effect of Storage

During storage, blood plasma can form a brown color as a result of the formation of glucose amides. Depending on the quantity of the plasma involved in the browning reaction, the time required for reconstituting the plasma can be prolonged, or the solubility of the plasma is limited. Lyophilized plasma that has been stored for years but did not show signs of the browning reaction was found to have good reconstitution properties. The browning reaction can be inhibited by a reduction of the residual moisture content of the plasma and by lower storage temperatures [26].

Specifications for Particular Matter

The particulate matter in reconstituted lyophilized product is an important consideration. When such particulate matter is present in the reconstituted product in sufficient quantities, it can render the final product unacceptable, even if all of the other properties of the formulation meet USP specifications. The following will be a brief overview of the sources of and the requirements and means for determining particulate matter in a reconstituted formulation.

USP Limits for Particular Matter for Injectable Solutions

The following are the USP limits for particulate matter.

- *Large Volume:* The USP limits for solutions > 100 mL is fifty 10 μm or larger particles per mL [48].

- *Small Volume.* The USP limits for solutions ≤ 100 mL is a total of 10,000 particles not exceeding 10 μm [48].

Sources of Subvisible Particles

Extrinsic Particles. Extrinsic subvisible particles can enter the lyophilized formulation from various external sources, such as the production atmosphere, the manufacturing equipment (e.g., the freeze-dryer), a filter, the filling devices, and the container and closure system [48].

Intrinsic Particles. Intrinsic subvisible particles are particulate matter that stems directly from the formulation or the lyophilization process. Such particles can be generated as a result of freezing a supersaturated solution, generated during storage of the product, or can stem from the selection of a surfactant (e.g., pluronic F-69) [48].

Although there is no supporting evidence at this time, I would not be surprised to find that intrinsic subvisible particles can stem directly from the lyophilization process (i.e., during the freezing process when there is a large change in the concentrations of the various excipients and the formation of a crust or glaze on the upper surface of the product). These latter layers may appear to reconstitute, but a portion of them could be denatured to such an extent that they will remain insoluble. Puffing and partial meltback during the primary drying process could all be sources of such particulate matter. Exothermic reactions occurring during secondary drying and the production of a hazy solution [1] could represent process-related sources of these intrinsic particles. There is little doubt that there is a need for additional investigations into the role that the lyophilization process plays in the generation of intrinsic subvisible particles.

Measurement of Particulate Matter

The following is a general description of a method for determining particulate matter in a reconstituted formulation.

Reconstitution. The formulation is first reconstituted using a particle-free diluent. A particle-free diluent is defined as a solution in which there all particles are less 10 μm [48]. It is important the reconstitution be conducted in accordance with the manufacturer's instructions.

Instrumentation. The determination of the particulate matter in a reconstituted formulation requires the use of a HIAC instrument (i.e., an instrument that determines the particulate matter using the light obstruction principle). A given volume of the reconstituted solution passes by a window that is illuminated by a parallel beam of light. The flow of the liquid is controlled so that particles will pass by the window singly. A reduction in the intensity of the light reaching the photomultiplier tube results from the light scattered by the particle and is recorded as a signal. The amplitude of this signal will be proportional to the area and, hence, the diameter of the particle [48].

Additional Tests of the Reconstituted Formulation

The reconstitution of a lyophilized formulation should result in a solution that mimics the original formulation. Besides the obvious determination of the concentration of the active constituent or potency, other tests should be performed to ensure that the composition and concentration of the other constituents in the formulation are within the established range of the original formulation as outlined in Chapter 2. In order to avoid being repetitious, the following is merely an outline of the various characteristics of the formulation that were suggested in Chapter 2. For more details regarding these characteristics, the reader is referred to Chapter 2.

Colligative Properties

A significant change in the colligative properties would be an indication that the original formulation has undergone significant chemical changes as a result of the lyophilization process. For example, a difference in the freezing point depression of water between the original and the reconstituted formulation would be an indication of a major change in the composition of the formulation. If the original formulation contained sucrose, and there is a decrease in the depression of the freezing point of water with respect to the original formulation (i.e., the freezing point of water occurs at a lower temperature), then it is possible that during the freezing process there was a lowering of the *p*H that caused the sucrose to dissociate and form glucose and fructose. Testing the solution for the presence of glucose would confirm that such a reaction had occurred. An increase in the freezing point depression of water (i.e., the freezing point depression of water occurs at a higher temperature) would indicate some interaction between the constituents, and tests should be conducted to determine the concentration of each constituent in the reconstituted formulation.

Concentration Properties

While several test methods were suggested in Chapter 2, a simple check of the refractive index of the reconstituted formulation would be useful in determining if there has been a change in the concentration and composition of the formulation as a result of the lyophilization process. While there may not have been a change in the colligative properties, a change in concentration and composition of the constituents may have occurred. A measure of the refractive index would confirm that there was no significant change in the concentration and composition of the original formulation as a result of the lyophilization process.

PRODUCT STABILITY

Introduction

The principal reason for lyophilizing a formulation or substance, as pointed out in Chapter 1, is to greatly enhance its stability. Long-term stability is achieved by slowing down the kinetic clock, i.e., the mechanism that controls biological growth or governs the chemical reactions, so that the useful life of a system may be extended from only hours or days to months or years. The slowing of the kinetic clock is generally achieved by reducing the residual moisture content of a material or its storage temperature. Our concern in this section will not be limited to *what makes the kinetic clock tick* but will include means by which the *ticking* of the clock can be determined.

Real-Time Stability

Stability can be defined as that time span required for a lyophilized substance to undergo a change in composition or a loss of potency, e.g., a change in the configuration of protein. For example, it may require 4 years under well-defined storage conditions to reduce the potency of an active constituent by 10%. Assuming that the loss in potency is linear with respect to time, it would require about 5 months before there would be a 1% loss in potency. Because a 1% change in potency may be difficult to determine with a high degree of certainty, more than 1 year of storage may be required before the product stability can be estimated to be 4 years. For this reason, the determination of the *real-time* stability of a lyophilized product can be a long and often arduous task.

Accelerated Determination of Stability

The speed of the kinetic clock can be increased or accelerated by increasing the temperature at which the product is stored. By this means, the time required to estimate the stability of a substance can be reduced from a year or more to a matter of months. The use of *accelerated* studies can provide a relatively rapid estimate of the actual *real-time* stability of the material [49].

This section will consider the key product properties that affect the stability of a substance, provide a general review of statistics and the use of a control chart, evaluate *real-time* stability measurements, and determine stability by *accelerated* studies.

Key Product Properties Affecting Stability

The following will be a general review of key properties that can affect stability or the *ticking of the product's kinetic clock*.

Structure

Chapter 2 stressed the importance of obtaining the active constituent of the lyophilized product in a crystalline state rather than an amorphous state. The reason is that crystalline materials have a higher stability than materials with an amorphous structure [15].

Moisture

The relationship between the stability of a formulation and the formulation's moisture content was introduced in an earlier section of this chapter by citing the work done by Flosdorf [27]. Flosdorf and others [13,26,50] showed that the principle effect of excess moisture was to reduce the stability of a lyophilized product.

Residual moisture in the cake was found to contribute to undesirable reactions between the constituents of a formulation. For example, Lee and Hannan [28] showed that excessive residual moisture was responsible for a reaction (browning) involving the amino group of casein and the reducing group of glucose. When the residual moisture content was lowered, the interaction between the casein and glucose was negligible.

The general rule is that less moisture means better quality and/or stability of lyophilized products. But such empirical rules always have exceptions and the following are examples of higher residual moisture improving the quality or stability of various lyophilized products. In one case, the performance of the product appears to be contrary to the initial findings regarding the observed effects of residual moisture. For lyophilized cyclophosphamide, the addition of moisture to cakes containing mannitol or sodium carbonate (Na_2CO_3) increases stability [15]. Excess moisture has been found to inhibit the oxidation of lipids in foods. The role that water plays in inhibiting oxidation is very complex, but, in general, the increase in the moisture content forms a monolayer that tends to limit the adsorption of oxygen on the surface. Complete removal of moisture from live cells has been found to be incompatible with the preservation of life [26]. Live bacteria require sugars to buffer their moisture content, which must be maintained between 1% and 3% [25].

The percentage of mobile spermatozoa (obtained from Holstein bulls) in freeze-dried samples was unaffected at residual moisture values of 30%. When the residual moisture was decreased from 30% to 2%, the percentage of mobile spermatozoa decreased to zero, i.e., no mobile spermatozoa were detected after the reconstitution of product having moisture values less than 2%. However, using a sample of spermatozoa having moisture values approaching 0% did result in the pregnancy of a cow and the birth of a calf [51].

Based on the above, the reader should realize that when it comes to residual moisture and stability, there is no common law. The relationship between residual moisture and the stability of a lyophilized formulation must be made on an individual basis. In addition, it is best to document both the upper and lower residual moisture limits.

Container System

The container-closure system, which will be considered in greater detail in Chapter 12, can be a key factor in determining stability. For example, storage of BCG (bacillus Calmette-Guérin) vaccine in vials rather than ampoules was found to be unacceptable because the container-closure system could not be maintained at the 20 mTorr necessary for product stability. Packaging of this particular vaccine was thus limited to ampoules [50].

The nature of the atmosphere in the container-closure system is a factor that can sometimes affect the stability of a lyophilized product. BCG vaccine stored under a dry nitrogen environment had consistently poor stability [50]. However, the results of studies of live bovine brucellosis vaccine, conducted at 37°C for 14 days and at 4°C for 14, 20, and 70 days, indicated that the storage environment was not a factor. No significant difference in stability of the latter vaccine was observed when stored under vacuum, nitrogen, or argon [33]. Once again, the factors affecting the stability are best considered as product specific.

Review of Statistical Methods and Control Charts

In preparing a lyophilized product like that shown in Figures 1.3a and 1.3b, it should be realized that not all of the lyophilized cakes will have exactly the same properties. The variation in the cakes may not only be cosmetic but also in the residual moisture content and in the potency of the active constituent. But stability is a property that is assigned equally to all of the cakes in a batch. Since stability is considered as a group or batch property, stability measurements must reflect a group or batch measurement.

The following will be a brief review of the use of statistics to characterize a batch or batches of a lyophilized product. (Those readers who wish to examine the use of statistics in further depth are referred to the many published texts on this subject.) The batches of lyophilized product will be characterized in terms of a frequency distribution that will be defined by an *arithmetic mean* and a *standard deviation*. The frequency distributions will be shown to vary within an overall general frequency distribution that is characterized by a control chart. The control chart will define the reliability of the stability measurements.

Frequency Distribution

When n number of lyophilized cakes are individually examined for residual moisture or for the potency of the active constituent, the residual moisture (%wt) or potency of the cakes varies over a range of values. A frequency distribution can be represented by a bar graph, as illustrated by Figure 10.8, of the ratio N_i/N_o as a function of an increment x_i. N_i represents the number of measured values (residual moisture or potency) that fell within the cell defined by $x_i \pm \Delta x_i$, and N_o is the total number of individual measurements. To be statistically significant, the frequency distribution must be based on an individual sample $(N_o) \geq 25$ [52]. The number of samples (N_o) used to simulate Figure 10.8 was 100.

Arithmetic Mean

The arithmetic mean $(\bar{x})$ is determined from

$$\bar{x} = \frac{x_1 + x_2 + x_3 + \ldots x_n}{N_o} \tag{12}$$

where x_n is the nth value of samples. The value of the arithmetic mean, represented by the dashed line, for the data points of the frequency distribution shown in Figure 10.8 was 1.02%. Figure 10.8 shows that the mean represents a percent moisture value around which the rest of the moisture values are distributed.

Figure 10.8. A Bar Chart Showing a Simulated Frequency Distribution for 100 Individual Measurements of the Residual Moisture in a Lyophilized Product

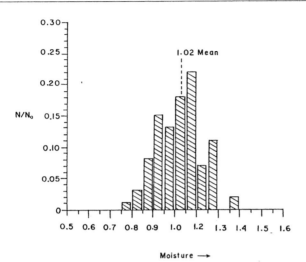

Standard Deviation

The arithmetic mean is useful in providing some information regarding the nature of the frequency distribution but does not provide information as to how the data are actually distributed about the mean. Figure 10.9 shows two frequency distributions having the same mean value. This figure shows, however, that the data for distribution A are dispersed quite differently about the $\bar{x}$ than are the data for distribution B (i.e., x values of distribution A extend beyond the values normally found in the distribution B. For this reason, it is necessary to characterize a frequency distribution by identifying both the mean ($\bar{x}$) and how those data are dispersed about the mean.

The dispersion of the data about the mean value can be represented in terms of the standard deviation (σ). The standard deviation is defined as

$$\sigma = \sqrt{\frac{\Sigma x_i^2}{N_o} - \bar{x}^2} \tag{13}$$

where N_o represents the number of values in the distribution.

From equation 13, σ for the frequency distribution shown by Figure 10.8 was determined to be 0.010%. The frequency distribution is defined by combining $\bar{x}$ and σ to obtain

$$\bar{x} \pm \sigma$$

Examination of the data for frequency distributions A and B shown in Figure 10.9 would indicate that $\sigma_A > \sigma_B$. A knowledge of $\bar{x}$ and σ provides a basis for comparing frequency distributions.

Figure 10.9. Frequency Distributions

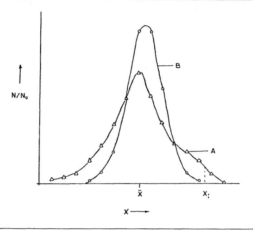

Two frequency distributions based on a sample size of 500 and having the same mean value x but different standard deviations. The A distribution is represented by the line Δ-Δ-Δ, while the line O-O-O is used to represent the B frequency distribution.

From a selection of samples from the same batch of lyophilized product, the value of $\bar{x}$ and σ will vary from sample lot to sample lot as a result of sample size and the sample selection process. It is important for the reader to understand that in order for the sample lot to be representative of the frequency distribution of the batch, the number of units in the batch must be large enough so that the sampling process does not alter the original frequency distribution of the batch. For that reason, a particular property of a lyophilized product can be represented by a distribution of sample distributions. How these sample distributions vary among themselves can be represented by a control chart.

An example of a control chart is illustrated in Figure 10.10. The control chart shows that sampling of the batch produced a distribution of arithmetic means and standard deviations. The arithmetic means were found to be distributed around a grand mean value ($\bar{\bar{x}}$) and the standard deviations about a mean standard deviation ($\bar{\sigma}$). The distribution of $\bar{x}$ is defined between the mean limits of $\bar{x}_1$ and $\bar{x}_2$, while the standard deviations are within the limits of $\bar{\sigma}_1$ and $\bar{\sigma}_2$. Without going into details (the reader can obtain more detail information from reference [52]) the use of samples sizes less than 25 (e.g., < 10) would effectively increase the control chart limits for $\bar{x}$ and $\bar{\sigma}$. Such an increase in the limits could have a major effect on the results of the stability measurements.

The grand mean ($\bar{\bar{x}}$) can be obtained from equation 12, where $\bar{x}_1$ is substituted for x_1 and N_0' represents the number of distributions measured rather than the number of samples in a given frequency distribution. The upper and lower limits for $\bar{x}$ and $\bar{\sigma}$ can be obtained from

Figure 10.10. Control Charts for a Product Parameter of a Lyophilized Formulation

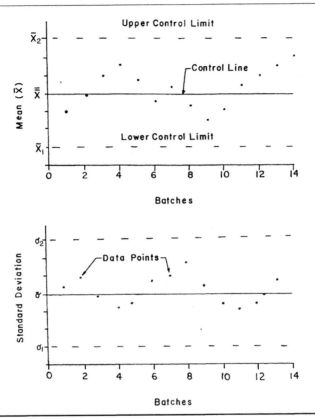

$$\overline{\overline{x}} \pm \frac{3\overline{\sigma}}{\sqrt{N_o'}}$$

and

$$\overline{\sigma} \pm \frac{3\overline{\sigma}}{\sqrt{2N_o'}}$$

The importance of control limits on the control chart is that they (the limits) represent the range of distributions that would still be representative of the original distribution. As long as the $\overline{x}$ and σ remain within the limits, it would have to be said that there was no significant change in the observed property of the product. It is for this reason that it is so important that the batch size be sufficiently large so that its (the batch's) frequency distribution is not altered as a result of the sampling process.

For a significant change to occur in a product property, such as potency, the mean value would have to fall outside the control limits for $\bar{\bar{x}}$. A significant change in $\bar{x}$ would not necessarily result in an associated change in σ, e.g., a value of σ that is outside of the control limits for the standard deviation.

Real-Time Stability Studies

The objective of real-time studies of a lyophilized formulation is to determine actual product stability under defined storage conditions of temperature and relative humidity. In performing these studies, samples are removed from the storage environment and analyzed. Examples of real-time stabilities of lyophilized biological products are listed in Table 10.1. For each of the products listed in Table 10.1, the biological activity was generally reported at discrete time intervals, not as a plot of the change in biological activity as a function of time. In the absence of such a plot, it is not possible to gain an understanding as to the function that is governing the loss in biological activity.

Because of the rather limited amount of information that has been published regarding real-time stability studies, the remainder of this section will consider real-time stability in terms of two key properties of a hypothetical lyophilized product: residual moisture and activity or potency of the product.

Moisture Content

The role that moisture plays in the stability of a lyophilized product has already been considered and found to be product dependent. It is important, therefore, to show during real-time studies that the residual moisture content remains within the specified limits for the product. If during storage the residual moisture value falls outside the specified moisture limits, then the observed stability of the product will not be representative of the original lyophilized product, and the stability studies should be terminated.

Consider the key values for a control chart that, as shown in Table 10.2, are based on the moisture values obtained from a hypothetical batch of a lyophilized product stored at 20°C ± 2°C. The control chart values are based on 10 sample lots, taken immediately after completion of the lyophilization process, in which a lot consisted of 25 samples. The grand mean value of the moisture ($\bar{\bar{M}}$) was 1.00% and the upper and lower limits were 1.30% and 0.70%, respectively. This control chart will serve as a basis for determining significant changes in the residual moisture content of the product during the real-time stability test. The established upper moisture limit for this product is 5% by weight. At moisture values greater than 4%, the hypothetical active constituent was found to lose potency at a relatively rapid rate.

With the assumption that $\bar{\sigma}$ (0.49%) for the measurements remains unchanged throughout the entire analysis, the moisture values and the quantity of the active constituent present in the product at various times during the real-time stability studies are listed in Table 10.3. Each moisture and active constituent value is the mean of 25 individual samples. Based on these data, the expiration date of the product was set for 36 months.

A plot of the percent residual moisture as a function of time is shown by Figure 10.11. Despite the fact that the data points in Figure 10.11 represent the mean

Table 10.1. Results of Real-Time Studies of the Stability of Various Lyophilized Formulations

Product	Temperature (°C)	Stability	Reference
Blood plasma	ambient	5–6 months	[26]
Blood plasma	4	9 months	[26]
Blood plasma	< 0	5 yr	[26]
Vaccines			
influenza	−20	12–36 months	[53]
influenza	2 to 8	24 months	[53]
measles	−20	12–36 months	[53]
measles	2 to 8	24 months	[53]
rubella	−20	12–36 months	[53]
rubella	2 to 8	24 months	[53]
marek (HVT)[1]	−20	12–36 months	[53]
marek (HVT)	2 to 8	24 months	[53]
flow pox	−20	12–36 months	[53]
flow pox	2 to 8	24 months	[53]
Viruses			
ectromelia	4	1,295 days	[20]
equine encephalitis	4	555 days	[20]
herpes	4	1,329 days	[20]
rabies	4	1,394 days	[20]
chickenpox	unspecified	70 months	[20]
influenza	unspecified	5 years	[20]

1. HVT is the freeze-dried herpes virus of turkeys [53].

Table 10.2. The Key Values for a Control Chart for the Residual Moisture in a Hypothetical Lyophilized Formulation While Stored at 20°C ± 2°C Based on 10 Sample Lots, Each of Which Consisted of 25 Individual Samples

Control Chart Values	Percentage Residual Water
Grand Mean ($\bar{\bar{M}}$)	1.00
Upper Control Limits	1.30
Lower Control Limits	0.70
Mean Standard Deviation ($\bar{\sigma}$)	0.49
Upper Control Limits	0.69
Lower Control Limits	0.29

Table 10.3. Example of a Hypothetical Increase in the Mean Percentage of Residual Moisture and the Decrease in the Mean Active Constituent of a Lyophilized Formulation While Stored at 20°C ± 2°C Based on a Sample Size of 25 Individual Samples

Time (months)	Mean Percent Free Water	Mean Active Constituent (μg)
0	1.00	3.100
3	1.22	3.095
6	1.13	3.093
9	1.37	3.085
12	1.28	3.074
15	1.41	3.062
18	1.38	3.051
21	1.52	3.030
24	1.65	2.995
30	1.57	2.986
33	1.66	2.933
36	1.63	2.910
39	1.84	2.885
42	1.76	2.856
45	1.94	2.772
48	1.89	2.732

moisture values of 25 individual samples, this plot shows, as a result of a relatively large value for σ, considerable scatter in the data. From the slope of the line, the mean residual moisture content in the product is increasing at a rate of $6 \times 10^{-4}\%/$ day or $1.8 \times 10^{-2}\%/$month.

The question that now arises is, from the results listed in Table 10.3 and plotted in Figure 10.11 and for a given batch, *What will be the odds against 1 that at or before the expiration date the moisture content of one of the products in the batch will equal or exceed 4% by weight?* For an initial batch size of 50,000, the odds against 1 can be obtained by determining the ratio

$$\frac{|\bar{x} - x_i|}{\bar{\sigma}}$$

where x_i is the ith value of $\bar{x}$ in the frequency distribution x at a given storage time and $\bar{\sigma}$ represents the mean value of σ given by the data shown in Table 10.3. Ignoring the criteria in Table 10.4 that the sample size must be 30 or more, the odds against 1 that a product vial will have a moisture value $\geq 5\%$ before the 36-month

Figure 10.11. A Plot of the Percent Residual Moisture ($\overline{M}$) Content as a Function of Time

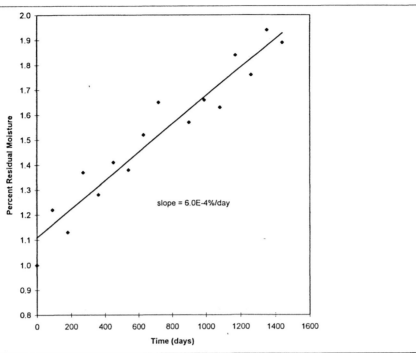

expiration date is listed in Table 10.5. Table 10.5 shows that the odds against 1 that there will exist 1 product vial with moisture values $\geq 4\%$ would still approach 1 in a 1,000,000 at the expiration date. However, when the storage time exceeds 48 months, the odds against 1 is of the order of 10,000; thus, there is a finite probability that the batch may contain product whose moisture content would $\geq 4\%$.

Product Stability

For determining the stability of a hypothetical active constituent for a batch of lyophilized product, the control chart values listed in Table 10.6 provide a basis for determining when there has been a significant change in the quantity or potency of the active constituent in the formulation. From this table, the grand mean value of the active constituent was given as 3.100 μg while the lower control limit was determined to be 3.004 μg. Table 10.7 shows that, for the first 21 months of storage, the observed quantity of active constituent was within the limits of the control chart. Because the observed values of the active constituent were within the control limits, there was no significant change in the quantity of the active constituent. A plot of the active constituent as a function of time is shown in Figure 10.12. This plot shows that there was no significant change in the quantity or potency of the active constituent until the 1,000-day mark, or 33 months. At times greater than 1,000 days, there is a

Table 10.4. The Odds Against 1 That Dispersions Will Occur from a Mean $\bar{x}$ Based on a Sample Size of 30 or more [54]

$|x - x_i|/\sigma$	Odds Against 1	$|x - x_i|/\sigma$	Odds Against 1
0.67	1.00	2.60	106.3
0.70	1.07	2.80	194.7
0.80	1.36	3.00	369.3
0.90	1.72	3.20	726.7
1.00	2.15	3.40	1,483
1.20	3.35	3.60	3,142
1.40	5.19	3.80	6,915
1.80	12.92	4.00	1.57×10^4
2.00	20.98	5.00	1.74×10^6
2.20	34.96	6.00	5.00×10^8
2.40	59.99	7.00	3.90×10^{11}

Table 10.5. The Value of $|\bar{x} - x_i|/\sigma$ and Mean Percent Moisture at Various Time Intervals During Storage Where σ Is the $\bar{\sigma}$ Given in Table 10.2

Time (months)	Mean Percent Free Water	$|\bar{x} - x_i|/\sigma$
0	1.00	6.12
3	1.22	5.67
6	1.13	5.86
9	1.37	5.37
12	1.28	5.55
15	1.41	5.29
18	1.38	5.35
21	1.52	5.06
24	1.65	4.80
30	1.57	4.96
33	1.66	4.78
36	1.63	4.84
39	1.84	4.41
42	1.76	4.57
45	1.94	4.20
48	1.89	4.31

Table 10.6. Key Values for the Control Chart for the Active Ingredient in a Hypothetical Lyophilized Product That Is Stored at 20°C and Is Based on 10 Sample Lots Where the Sample Size in Each Lot Was 25

Control Chart Values	Active Constituent (µg)
Grand Mean $\bar{\bar{A}}$	3.100
Control Limits	
Upper	3.106
Lower	3.004
Mean Standard Deviation ($\bar{\sigma}$)	0.010
Control Limits	
Upper	0.014
Lower	0.006

rather steep decrease in the quantity of the active constituent, which would be consistent with the effects of an increase in the residual moisture.

Table 10.7, like Table 10.5, provides an insight as to stability of product having a 36 month expiration date. This table, with respect to Table 10.4, shows that it is highly improbable that there will be product present in which the active constituent or potency is less than 2.79 µg. However, after 42 months, there is a sharp increase in the probability, as a result the decrease in the odds against 1, of finding a sample where the quantity of the active constituent is less than 90% of the original potency.

All of this shows the value of establishing a control chart for important product properties such as the residual moisture content and the quantity or potency of the active constituent. Although it is gratifying to have this supporting data regarding the properties of the lyophilized product, such assurance or confidence does not come without a price. For the initial batch size of 50,000 units, the determination of real-time stability required 1,800 samples, or 3.6% of the batch. For smaller batch sizes, the required number of samples might represent 18% of the total batch. While I recognize the dilemma that such a statistical approach presents, one must also realize that as the sample size decreases, so does the confidence that the product does not contain samples whose properties are not within the defined specifications.

Accelerated Studies

The discussion of real-time studies has shown that the need to wait such long periods to establish the stability of a lyophilized product represents a real impediment to the development of a lyophilized formulation. In fact, as stability improves, its measurement becomes more costly and time-consuming. This section will be concerned with means by which one can *accelerate* the kinetic clock. While such accelerated studies require less time and fewer samples, the results are often treated as an indicator, not as an adequate substitution for real-time studies. This section will not

Figure 10.12. A Plot of $(a - x)$ as a Function of Time

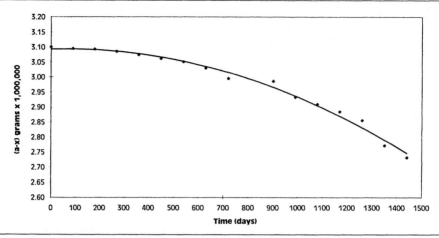

"a" is the initial quantity of the active ingredient and "x" is the quantity of active ingredient present after some given time.

Table 10.7. The Value of $|\bar{A} - A_i|/\sigma$ and the Mean Quantity of the Active Constituent (A) at Various Time Intervals During Storage Where σ Is the $\bar{\sigma}$ Given in Table 10.6

Time (months)	Mean Active Constituent (μg)	$|\bar{A} - A_i|/\sigma$
0	3.100	31.0
3	3.095	30.7
6	3.093	30.2
9	3.085	29.3
12	3.074	28.3
15	3.062	26.8
18	3.051	25.2
21	3.03	23.3
24	2.995	21.1
30	2.986	18.4
33	2.933	15.5
36	2.91	12.3
39	2.885	8.9
42	2.856	5.3
45	2.772	1.5
48	2.732	2.5

only describe the means for conducting meaningful stability measurements but also provide guidelines for when using this method to determine the stability of a product is not applicable.

Arrhenius Rate Equation

It is well known that most chemical reactions, such as the denaturation of an active constituent, e.g., a change in configuration or the unfolding of a protein molecule, will proceed at a higher rate with an increase in temperature. As a result of the efforts of van't Hoff and Arrhenius during the latter part of the 19th century, a general relationship was developed between the specific reaction rate (k_a), the energy (E_a) for activating the reaction, and the temperature (T) of the system [55]. The relationship between these terms, now generally known as the Arrhenius rate equation, is expressed as

$$k_a = B \exp^{\frac{-E_a}{R_o T}} \tag{14}$$

where R_o is the gas constant expressed in energy/mole-K and T is expressed in K. The preexponential term B is generally referred to as the frequency factor [55]. Any further discussion of the nature of B is considered to be beyond the scope of this text. Those readers wishing to learn more about the frequency factor B are referred to [55] or any other suitable text on chemical kinetics.

Description of the Method

The following is a general description of the steps used to determine the stability of an active constituent at a given storage temperature by means of the Arrhenius rate equation 14.

Determination of k_a at a Given Elevated Temperature T_r. In using the Arrhenius rate equation, one determines the values of k_a at a series of elevated temperatures. The elevated temperatures selected for use with heat-liable materials such as biological, biotechnology, and pharmaceutical products generally range from 35°C to 70°C and are strongly dependent on the nature of the active constituent. For example, the temperature range for biological constituents could be considerably lower than that used for determining the stability of a pharmaceutical.

Calculation of k_a' at a Storage Temperature $T_{r,s}$. The value of k_a' at the storage temperature $T_{r,s}$ is determined by the extrapolation of the plot of $\ln k_a$ as a function of $1/T_r$. From k_a', the time necessary to reduce the quantity or potency of the active constituent to its lowest limit can be determined. The reader should be made aware at this point that it is a lack of knowledge of the equation for k_a that often results in a discrepancy between the results of accelerated and real-time stability studies.

Activation Energy (E_a) and Stability. From the slope of the plot, one can obtain a value for E_a. The stability of the product will be directly related to the magnitude of E_a (i.e., higher E_a results in greater stability). For values of $E_a > 20$ kcal/mole, the active constituent would be quite stable and require temperatures considerably higher than ambient in order to achieve a significant value for k_a. For example, the N-N bond is 225 kcal/mole and is stable at ambient temperature. Temperatures in excess of 1000°C would be required for dissociation. Activation energies less than 20 kcal/mole and approaching 5 to 6 kcal/mole would be sufficiently low that k_a would have a significant value even at ambient temperatures. Knowledge of E_a can provide some insight as to the mechanism involved in the denaturation of the active constituent.

General Reaction Expression

The general expression for a chemical reaction at a given temperature T can be expressed as

$$\frac{dx}{dt} = k_a(a - x)^{n_r} \tag{15}$$

where a represents the original quantity or potency of the active constituent present in the product at $t = 0$; x is the amount of active constituent that was denatured or lost potency at a time (t) while $(a - x)$ is the amount of active constituent remaining, and n_r designates the order of the reaction. The n_r is defined from the best fit with the experimental data, and there may not be any *connection between the form of the stoichiometric reaction and kinetic order* [56]. It should be noted that n_r need not be an integer but can have values such as 3/2.

Equation 15 is used to determine the value of k_a at a given T_r and satisfy the first requirement for determining the stability of an active constituent at a given storage temperature by means of the Arrhenius rate equation. However, in order to determine k_a, it is necessary to first determine the value of n_r. The reader should note that investigators using accelerated stability studies often either assumed or ignored the value of n_r. Without first establishing the n_r of the reaction, little credence can be given to the results of stability data obtained by the Arrhenius rate equation. The value of n_r, for a given T_r is obtained from the slope of the plot of log (dx/dt) as a function of the log $(a - x)$. If the function proves to be nonlinear, then the reaction(s) involved in the loss of concentration is complex, and n_r is ill-defined. If this should occur, then the use of the Arrhenius rate equation is not applicable, and one must rely solely on real-time stability measurements.

The results of the accelerated studies of a hypothetical active (A_x) constituent at a temperature of 60°C is shown in Table 10.8. Note that concentration of $(a - x)$ and x are expressed in units of moles/L and time in sec. The values of $(a - x)$ are the mean values of the concentration of 25 individual samples. A plot of the log (dx/dt) as a function of the log $(a - x)$ for the data shown in Table 10.8 is shown by Figure 10.13. As seen from the plot, the slope of the line drawn through the points signifies that the reaction is first order.

Once the order of the reaction is established, then one can use the integrated form of equation 15 to determine the value of k_a for a given T_r. The integrated form for some common orders of reaction are as follows:

Table 10.8. Data Generated from the Accelerated Studies of a Hypothetical Active Constituent (A_x) at 60°C, Where the Values of $a - x$ Represent the Mean Value of 25 Individual Samples

Time (sec)	$(a - x)$ moles/L	x moles/L
0	0.04269	0
604,800	0.04200	0.00069
1,209,600	0.04132	0.00137
1,814,400	0.04065	0.00203
2,419,200	0.04000	0.00269
3,024,000	0.03935	0.00335
3,628,800	0.03872	0.00397
4,233,600	0.03809	0.00459
4,838,400	0.03748	0.00521
5,443,200	0.03688	0.00582
6,048,000	0.03628	0.00641
6,652,800	0.03569	0.00699
7,257,600	0.03511	0.00757
7,862,400	0.03455	0.00814

$$k_r = \frac{x}{t} \quad \text{conc/time (zero order)}$$

$$k_r = \frac{1}{t} \ln \frac{a}{a - x} \quad \text{1/time (first order)}$$

$$k_r = \frac{1}{t} \ln \frac{a}{a(a - x)} \quad \text{1/conc time (second order)}$$

It is of interest to note that the units for a first-order reaction is 1/time, while the units of a second-order reaction are 1/(concentration · time). The units for an ith-order reaction will be 1/(time · concentration$^{(nr-1)}$) where the concentration is expressed in moles/L. This points to a very important difference between real-time and accelerated stability studies: the former can be expressed in terms of quantity (mass, potency, etc.), but the latter requires that the quantities used be expressed in terms of concentration (moles/L).

The value of k_a at 60°C for the hypothetical constituent (A_x) listed in Table 10.8 was calculated, using $(a - x)$ equal to 0.04200 moles/L and the integrated expression

Figure 10.13. A Plot of log (*dx/dt*) as a Function of log (*a – x*) for the Accelerated Studies of a Hypothetical Active Constituent (*A_x*) at +60°C Obtained from Table 10.8

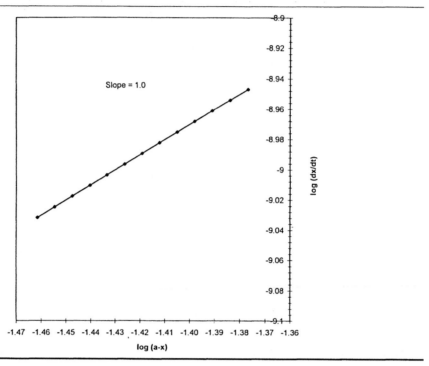

for the first-order reaction, to be 2.7×10^{-8}/sec. As an exercise, the reader may wish to use other values of (*a – x*) and their associated times in Table 10.8 to determine the value of k_a using the integrated expression for the first-order reaction. The fact that k_a will remain constant is another means for establishing the order of the reaction.

To satisfy the second requirement for determining the stability at a given temperature $T_{r,s}$, the values of k_a at the T_r can be determined only from the proper integrated form of equation 15. Once the k_a have been determined over a range of elevated temperatures (e.g., 30°C to 60°C), then from equation 14, one can obtain the value of k_a' at a lower storage temperature by extrapolation of a plot of ln k_a as a function of $1/T_r$. The stability (i.e., the time required for the product to reach a given concentration at $T_{r,s}$) is obtained from inserting the value of k_a' into the appropriate integrated form for the order of the reaction and solving for *t*.

The k_a values for the hypothetical active constituent A_x at temperatures of 30°C, 40°C, and 50°C were determined and found to be 7.46×10^{-9}/sec, 1.18×10^{-8}/sec, and 1.81×10^{-8}/sec, respectively. A plot of ln k_a as a function of $1/T$ (K) is shown by Figure 10.14. The general function for the Arrhenius rate equation was determined to be

Figure 10.14. An Arrhenius plot of ln k_a as a Function of 1/T(K) × 1,000 for Temperatures Ranging from 30°C to 60°C for the Hypothetical Active Constituent (A_x).

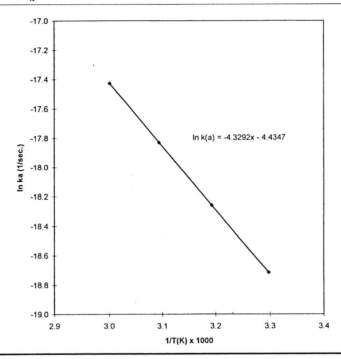

$$\ln k_a = -4.3292x - 4.4347 \tag{16}$$

where from the slope ($-4.3292 \times 1,000 \times R$), E_a is determined to be 8,603 cal and B is obtained from the intercept as 0.01185 per sec. From equation 15, the k_a values at 20°C and 4°C were determined to be 4.59×10^{-9}/sec and 1.96×10^{-9}/sec. From the integrated form for the first-order reaction, the extrapolated time for the active constituent A_x to reach 90% of its initial concentration was found to be 8.8 months at 20°C and 20.8 months at 4°C.

In using accelerated studies to determine the stability of a lyophilized product, one must be careful to determine the order of the reaction for T_r. If there is a significant change in the order of the reaction with temperature, then the reactions involving the loss in concentration are complex, and one cannot ascertain the stability of the product using the Arrhenius accelerated method. Attempting to do so is a common error that perhaps explains the reason for the lack of agreement between the stability ascertained by the accelerated method and those determined from real-time studies of the product at the storage temperature.

A second area of concern is that there will be a frequency distribution of concentration values of a product at any given temperature. Without knowledge of such a frequency distribution, the observed change in concentration at a given temperature may not be significant. To determine if the observed concentration values of the accelerated method are significant, one would have to establish control charts as described in a previous section of this chapter.

Results of Arrhenius Accelerated Studies

The following are some stability studies conducted using the Arrhenius rate expression. I have used these studies to emphasis the possible pitfalls one can encounter in using accelerated studies to determine the stability of an active constituent.

Case #1. Accelerated studies were performed on measles vaccine at testing temperatures of 20 to 25°C (2 to 3 months), 37°C for 2 to 3 weeks, and 45°C for 2 to 3 days [57]. The accelerated studies were performed from a linear regression analysis for each of the temperatures (20 to 25°C, 37°C, and 45°C) from the log of the potency as a function of time. The slope of the line for each of the temperatures is referred to as the thermal rate of degradation.

   The $\log_{10}$ of the degradation rate was then plotted as a function of the $1/T$, were T is expressed in K. The plot was then extrapolated to a temperature of 5°C. The predicted loss in potency at 5°C from the slope was $-0.019 \log_{10}$ per month or $0.23 \log_{10}$ per year. Based on actual stability data at 5°C, the potency was found to be 0.24 to 0.23 $\log_{10}$ over 12 months when tested every 3 months.

   It should be noted that Cowderly et al. [57] did not indicate which of the integrated forms (for a given order of reaction) they used to determine the stability at the extrapolated storage temperature of 5°C. In addition, the Arrhenius plot was performed using $\log_{10}$ rather than $\log_e$ as required by equation 14.

Case #2. First-order rate constants for freeze-dried hGH (%/month) as a function of time and average moisture content were determined for temperatures of 40°C and 25°C. For a given moisture content, the rate constant at 25°C is lower than that for 40°C [3].

Comment. The significance of these results is in question because the authors indicate the reaction to be first order, but the units for the rate constant were expressed in %/month, not 1/time. In addition, the authors offer no information to indicate if the reactions were first order in nature. While these authors went to great lengths to describe their experimental techniques, the use of uncommon and undefined measurement parameters such as *time average percent* distracts from the significance of the results.

   These cases are just two of other instances in the literature where either accelerated stability studies have been improperly used or the terminology is not consistent with accepted convention. I hope that future accelerated stability studies will be conducted with greater care and diligence so we may better understand the role that water plays in the stability of various compounds.

SYMBOLS

a	original quantity or potency of active constituent present in the product at $t = 0$
A	active constituent
A_s	surface area of the cake
A_x	hypothetical active constituent used in accelerated studies
B	preexponential term generally referred to as the frequency factor
C_o	heat-transfer coefficient of the bottom of the container
ΔP	change in trap pressure
E_a	energy for activating a chemical reaction
k_a	specific reaction rate for a reaction involving the active constituent (units depend on order of reaction)
k_a'	specific reaction rate at the product storage temperature
k_d	dissociation constant
m_w	mass of water in the sample
n	the number of individual sample examined
n_r	the order of the reaction
N_i	the number of measured values (residual moisture or potency) that fell within the cell defined by $x_i \pm \Delta x_i$
N_o	the total number of individual measurements
N_o'	the number of distributions
$\Sigma n'_{g,i}$	sum of the number of moles of gas present at T_t
P	pressure of water vapor above the cake
P_b	maximum pressure obtained for the blank of the system (high vacuum system)
P_o'	the saturation pressure for water vapor
P_t	initial pressure of the cold trap portion of the apparatus, at ambient temperature
P_t'	final cold trap pressure
P_v	maximum pressure in the trap from the desorption of the gases from the sample (high vacuum system)
P_2	increase in the cold trap pressure
R	total radius
R_m	mean radius of the frequency distribution of the pores

R_o gas constant expressed in energy/mole-K

R_p average pore radius

R_1 radius of the adsorbed gas layer in the pore

R_2 radius of the meniscus of the liquid adsorbate

t' the thickness of the adsorbed layer

T temperature of a system expressed in °C or K

T_a transition temperature or incipient melting that is used to replace the term $T_g(w_g)$.

Tc collapse temperature of a frozen matrix

T_g' glass transition temperature which is equivalent to Tc

T_o ambient temperature

T_p product temperature during the lyophilization process

T_r elevated temperature expressed in K

$T_{r,s}$ storage temperature for a compound x

T_s shelf-surface temperature

T_t cold trap temperature when cooled by a mixture of dry-ice and methanol or dry ice and acetone

V_a volume of gas adsorbed on the surface

V_L volume of the liquid in the pores of a matrix

V_m molar volume of the liquid adsorbate

V_t volume of the cold trap

x amount of active constituent that was denatured or the loss in potency at a time t

x_i the ith value of a measurement of a property x

REFERENCES

1.   T. A. Jennings and H. Duan, *PDA J. Pharm. Sci. & Tech.*, 49: 272 (1995).

2.   F. Franks, in International Symposium on Biological Product Freeze-Drying and Formulation, Bethesda, USA (1990), *Developments in Biological Standardization*, 74: S. Karger, Basel, 1991, p. 9.

3.   M. J. Pikal, K. Dellerman, and M. L. Roy, loc. cit., p. 21.

4.   R. H. M. Hatley, loc. cit., p. 105.

5.   J. A. Bashir and K. E. Avis, *Bull. Parenter Drug Assoc.*, 27: 62 (1973).

6.   P. DeLuca and L. Lachman, *J. Pharm Sci.*, 54: 617 (1965).

7.   R. K. Garrell and R. E. King, *J. Parenter. Sci. and Tech.*, 36: 2 (1982).

8.   P. P. DeLuca, in International Symposium on Freeze-Drying of Biological Products, *Developments in Biological Standardization*, 36: S. Karger, New York, 1977, p. 41.

9.   J. M. Flink, *Freeze Drying and Advanced Food Technology* (S. A. Goldblith, L. Rey, and W. W. Rothmayer, eds.), Academic Press, New York, 1975, p. 309.

10. D. Simatos and G. Blond, *Freeze Drying and Advanced Food Technology* (S. A. Goldblith, L. Rey, and W. W. Rothmayer, eds.), Academic Press, New York, 1975, p. 401.

11. *Handbook of Chemistry and Physics* (R. C. Weast, ed.), 65th ed., CRC Press, Inc., Boca Raton, Florida, pp. 1984–1985.

12. J. Flink, *Freeze Drying and Advanced Food Technology* (S. A. Goldblith, L. Rey, and W. W. Rothmayer, eds.), Academic Press, New York, 1975, p 143.

13. J. E. Shields and S. Lowell, *American Laboratory*, November, p. 81 (1984).

14. L .H. Jensen, in International Symposium on Biological Product Freeze-Drying and Formulation, Bethesda, USA (1990), *Developments in Biological Standardization*, 74: S. Karger, Basel, 1991, p. 53.

15. T. R. Kovalcik and J. K. Guillory, *J. Parenter. Sci. & Tech.*, 42: 29 (1988).

16. T. A. Jennings, *J. Parenter. Drug Assoc.*, 34: 109 (1980).

17. N. A. Williams, Y. Lee, G. P. Polli, and T. A. Jennings, *J. Parenter. Sci. & Tech.*, 40 136 (1986).

18. N. A. Williams and T. Dean, *J. Parenter. Sci. & Tech.*, 45: 94 (1991).

19. A. T. P. Skrabanja, A. L. J. De Meer, R. A. De Ruiter, and P. J. M. Van Den Oetelaar, *PDA Journal of Pharm. Sci & Tech.*, 48: 311 (1994).

20. D. Grief and W. Rightsel, *Aspects Théoriques et Industrials de la Lyophilzation* (L. Rey, ed.), Herman, Paris, 1964, p. 369.

21. T. Nei and S. Fujkawa, in International Symposium on Freeze-Drying of Biological Products, *Developments in Biological Standardization*, 36: S. Karger, New York, 1977, p. 243.

22. T. Haby, A. Thakur, D. Nowotnik, Y. W. Chan, K. Linder, and S. Varia, *PDA Journal of Pharm. Sci & Tech.*, 51: 68 (1994).

23. P. J. Dawson and D. J. Hockley, in International Symposium on Biological Product Freeze-Drying and Formulation, Bethesda, USA (1990), *Developments in Biological Standardization*, 74: S. Karger, Basel, 1991, p. 185.

24. E. Maltini, *Freeze Drying and Advanced Food Technology* (S. A. Goldblith, L. Rey, and W. W. Rothmayer, eds.), Academic Press, New York, 1975, p. 121.

25. L. Rey, *Aspects Théoriques et Industrials de la Lyophilzation* (L. Rey, ed.), Herman, Paris, 1964, p. 23.

26. G. J. Rosenberg, loc. cit. p. 335.

27. E. W. Flosdorf, in *Freeze Drying*, Reinhold Publishing Co., New York (1960).

28. C. H. Lee and R. S. Hannan, *Nature* 165: 438 (1950).

29. J. E. Jewell, R. Workman, and I. D. Zeleznick, in International Symposium on Freeze-Drying of Biological Products, *Developments in Biological Standardization*, 36: S. Karger, New York, 1977, p. 181.

30. H. T. Meryman, in *Cryobiology* (H. T. Meryman, ed.), Academic Press, New York, 1996, p. 1.

31.   L. Rey, *Proc. Roy. Soc. Lond. B.*, 191: 9 (1975).

32.   R. L. Alexander et al., U.S. Patent 4,537,883.

33.   L. Valette, C. Stellmann, P. Précausta, Ph. Desmettre, and M. Le Pemp, in International Symposium on Freeze-Drying of Biological Products, *Developments in Biological Standardization*, 36: S. Karger, New York, 1977, p. 313.

34.   M. J. Pikal and S. Shah, in International Symposium on Biological Product Freeze-Drying and Formulation, Bethesda, USA (1990), *Developments in Biological Standardization*, 74: S. Karger, Basel, 1991, p. 165.

35.   J. P. Earl, P. S. Bennett, K. A. Larson, and R. Shaw, loc. cit., p. 203.

36.   E. B. Sekigmann, Jr., in *Freeze Drying and Advanced Food Technology* (S. A. Goldblith, L. Rey, and W. W. Rothmayer, eds.), Academic Press, New York, 1975, p 175.

37.   L. F. Shackell, *Am. J. Physiol.*, 24: 325 (1909).

38.   J. May, R. M. Wheeler, N. Etz, and A. Del Grosso, in International Symposium on Biological Product Freeze-Drying and Formulation, Bethesda, USA (1990), *Developments in Biological Standardization*, 74: S. Karger, Basel, 1991, p. 153.

39.   J. R. Pemberton, in International Symposium on Freeze-Drying of Biological Products, *Developments in Biological Standardization*, 36: S. Karger, New York, 1977, p. 191.

40.   L. G. Beckett, in *Biological Applications of Freezing and Drying* (R. J. C. Harris, ed.), Academic Press Inc., New York, 1945, p. 614.

41.   I. M. Kolthoff and E. B. Sandell, *Textbook of Quantitative Inorganic Analysis*, The MacMillian Company, New York, p. 614.

42.   K. Fischer, *Angew Chem.*, 48: 394 (1935).

43.   E. Scholz, *American Laboratory,* August, 1981, p.89.

44.   T. A. Jennings, United States Patent 5,016,468.

45.   *The Oxford English Dictionary*, 2nd ed. (prepared by J. A. Simpson and E.S.C. Weiner), Clarendon Press, Oxford, 13: 357 (1989).

46.   U. Hackenberg, *Aspects Théoriques et Industrials de la Lyophilzation* (L. Rey, ed.), Herman, Paris, 1964, p. 519.

47.   R. A. Buffi and R. L. Garnick, in International Symposium on Biological Product Freeze-Drying and Formulation, Bethesda, USA (1990), *Developments in Biological Standardization*, 74: S. Karger, Basel, 1991, p. 181.

48.   P. K. Gupta, E Porembski, and N. A. Williams, *J. Pharm Sci. & Tech.*, 48: 30 (1994).

49.   E. R. Garrett and R. F. Carper, *J. Am. Pharm. Assoc.*, 64: 515 (1955).

50.   P. W. Muggleton, *Aspects Théoriques et Industrials de la Lyophilzation* (L. Rey, ed.), Herman, Paris, 1964, p. 407.

51.   E. V. Larson and E. F. Graham, in International Symposium on Freeze-Drying of Biological Products, *Developments in Biological Standardization*, 36: S. Karger, New York, 1977, p. 343.

52.   *Manual on Presentation of Data and Control Chart Analysis*, 6th ed., prepared by Committee E-11 on Quality and Statistics, ASTM Manual Series MNL 7, ASTM, Philadelphia, Penn., 1992.

53.   J. Peetermans, G. Colinet, A. Bouillet, E. D'Hondt, and J. Stephenne, in International Symposium on Freeze-Drying of Biological Products, *Developments in Biological Standardization*, 36: S. Karger, New York, 1977, p. 291.

54. *Handbook of Chemistry and Physics* (C. D. Hogman, ed.), 31st ed., Chemical Rubber Publishing Co., Cleveland, Ohio, 1949, p. 227.

55. K. J. Laidler, *Chemical Kinetics*, McGraw-Hill Book Company, Inc., New York, 1950.

56. W. J. Moore, *Physical Chemistry*, 3rd ed., Prentice-Hall, Inc., Englewood Cliffs, N.J., 1964.

57. S. Cowdery, M. Frey, S. Orlowski, and A. Gray, in International Symposium on Freeze-Drying of Biological Products, *Developments in Biological Standardization*, 36: S. Karger, New York, 1977, p. 297.

11

Vacuum Technology

INTRODUCTION

The Need to Understand Vacuum Technology

The previous chapters have showed that the primary and secondary drying processes occur at subatmospheric pressure, i.e., generally at pressures lower than the vapor pressure of water at the triple point (4.58 Torr). In Chapter 8, an important relationship was established between the pressure in the chamber (P_c), the product temperature (T_p), and the shelf-surface temperature (T_s). While an understanding of those relationships proves helpful in defining the necessary parameters for the primary drying process, the rate of sublimation was shown to be directly related to the rate of heat transfer to the product. If the pressure in the chamber was too high, the heat-transfer rate would approach zero, and, for all practical purposes, there would be no sublimation of ice crystals. We understood that the condenser temperature (T_c) had to be sufficiently low to condense water vapor and that an increase in T_c could result in an increase in the P_c, which, for a given T_s, would cause an increase in T_p. However, for drying to take place, the pressure in the container must exceed the pressure in the chamber. Thus, for the drying process, one should understand not only the flow of gases from a given container but also the nature of the flow of gases from the drying chamber to the condenser.

For the secondary drying process, which was the subject of Chapter 9, the final moisture content of the product was shown to be dependent on T_p and the partial pressure of water vapor above the surface of the cake. It was shown that by using a residual gas mass spectrometer, one could ascertain the partial pressure of the gases in the chamber. In order to determine the partial pressure of water vapor, one must have an accurate measurement of P_c.

In order to understand the nature of gas flow during the drying process and various means for determining the total and partial pressures in the chamber, one must have a fundamental knowledge of the principles of vacuum technology.

Objectives of the Chapter

I wish to make it clear that the following is not a thesis on the principles of vacuum technology. For a more comprehensive understanding of this topic, the reader is referred to the publication by Dushman [1]. The objective of this chapter is to provide the reader with a general overview of the principal concepts of vacuum technology that are applicable to the lyophilization process. The following is a brief introduction of the topics that will be considered in this chapter:

- *Kinetic Theory of Gases:* This section will consider the key properties of gases. The gas properties described in this section will provide the basis for understanding the remaining topics in this chapter.

- *Pressure and Its Measurement:* Since *pressure* plays such an important role in the lyophilization process, the reader will be provided with a definition of the term pressure and its various units of measurement which contribute their own confusion to communications regarding the lyophilization process. Various types of pressure gauges that have been or are currently used to measure subatmospheric pressures will be examined. It will be shown that although a gauge may indicate a pressure value, one needs to be aware of the actual parameter being measured. This will be of particular interest when, in Chapter 13, we consider the impact that the vacuum freeze-dryer has on the lyophilization process.

- *Mean Free Path:* The mean distance between gas collisions. An understanding of the *mean free path* is an important prerequisite for understanding the nature of gas flow that may stem from a container-closure system or the gas dynamics of a freeze-drying system.

- *Gas Flow:* An understanding of the nature of *gas flow* will enable the reader to make prudent decisions when selecting the size and design of an elastomer closure or the necessary configuration of a vacuum freeze-dryer based on the pressure needed for the lyophilization of a given formulation.

- *Vapor Throughput of a System:* This topic will prove useful when examining the container-closure system (Chapter 12) or a vacuum freeze-dryer (Chapter 13).

- *Real Leaks:* This type of leak stems from penetrations in a system. The application of real leaks will be addressed when the integrity of a container-closure system or a vacuum freeze-dryer is discussed. This section will only consider the nature of such leaks. The means for their detection is system dependent and, for that reason, will be deferred to specific applications in the next two chapters.

- *Virtual Leaks:* This type of leak stems from a source of gas from within a system. As a result, this type of leak can appear to be a real leak but prove to be more harmful to the lyophilized product. In Chapters 12 and 13, it will be shown how virtual leaks in a container-closure system can impact the quality of a product during storage or can be a source of contamination from the freeze-dryer.

Thus, it is the intent of this chapter to provide the necessary background in vacuum technology for future discussions of the container-closure system and the vacuum freeze-dryer.

KINETIC THEORY OF GASES

Since the transport of gases involves not only the primary and secondary drying process but also the container-closure system and the operation of the freeze-dryer, it is essential that one have a basic understanding of the nature of gases. Because gases by their very nature are in constant motion, their properties have been generally defined as what is known as the *Kinetic Theory of Gases*.

Postulates

The Kinetic Theory of Gases is based on two major postulates [1]:

1. Matter is composed of small particles; i.e., atoms or molecules, and these particles can be grouped according to chemical composition, configuration, and mass. Atoms of helium and argon can be easily distinguished from one another by mass alone, because the latter is 10 times greater than the former. In the case of molecules, the mass of the molecule can no longer be a distinguishing criterion. The chemical formula for dimethyl ether is $(CH_3)_2O$ and ethyl alcohol is C_2H_5OH. Both of these molecules have the same number of carbon, hydrogen, and oxygen atoms and, therefore, the same mass; however, the structures of the two molecules are quite different. Thus for molecules, one must also take into account their chemical structure.

2. Gas molecules are in constant motion, and the motion is related to the temperature of the gas. For atomic gases, such as helium and argon, the temperature of the gas will be directly related to their kinetic energy ($1/2\ mv^2$, where m is the mass of the atom and v is its velocity [distance per unit time]). The total energy of molecules such as ethyl ether or ethyl alcohol, will be defined not only by their kinetic energy but also by the rotational and vibrational energy associated with the chemical bonds. It is assumed that when in the gas state the kinetic or translational energy of the molecule far exceeds the combined rotational and vibrational energies.

Gas Pressure

The second postulate indicated that the translational or kinetic energy of gas particles is directly related to the temperature of the gas. In fact, a measure of the translational energy of the gas will define its temperature. When confined to a closed cubic box with each side having the dimension *s*, the gas particles will strike the walls of the chamber. If we assume that there is no interaction between the walls of the chamber and the particles, such collisions with the surface can be considered perfectly elastic in nature. When a gas particle strikes a wall of the vessel, it will impart momentum (*mv*). If the collisions are perfectly elastic, the change in momentum of the particle upon striking the wall will be $2m|v|$. The term $|v|$ denotes the absolute velocity of the particle. For N particles striking a surface area per unit time, the force per unit area exerted by the particles will be defined as

$$P = 2Nmv \tag{1}$$

where P is the pressure. For those readers who desire a more rigorous derivation of pressure, I suggest references [1,2]. The reason that I defined pressure in the above terms will become apparent when the units and measurement of pressure are considered.

Units of Pressure

Any discipline will suffer because of poorly defined terms or parameters, and lyophilization is no exception. A term that would certainly be high on any list of terms that requires standardization would be the units of pressure. Depending on the age, manufacturer, and location of the freeze-drying equipment, the pressure of the vacuum gauge could be expressed in any of the units listed in Table 11.1. This table not only lists the various units but also provides the reader with means for converting one set of units to another. Conversion of units from one form to another is not difficult. First select the row in which the current units are expressed; then select the column that contains the desired units and multiply the present pressure value by the factor shown in the appropriate column. It can be seen from Table 11.1 that the most common pressure units used today are based on a measurement of the height of column of mercury or water, not on a force per unit area.

I know of two instances where the pressure units were expressed in percent vacuum. Logic would dictate that, for such a gauge, zero vacuum would be 1 atm; however, in the systems just mentioned, such logic did not prevail, and zero vacuum was the lowest pressure in the chamber. It is gratifying that this unit of pressure in a freeze-dryer has not become widely adopted. I have and will continue to encourage those technical societies whose membership consist of individuals interested in the lyophilization process to adopt a standard unit of pressure. It is further hoped that the pressure unit adopted will be the Pascal (Pa) (Newtons/m^2), because it represents a true force per unit area rather than the height of a column of liquid.

Partial and Vapor Pressures

The pressure in a drying chamber may be maintained at a given value throughout an entire lyophilization process. In most instances, the predominant gas in the lyophilizer during the primary drying process is water vapor. As was shown in Chapter 9, a major change in the composition of the gases can occur at the completion of the secondary drying, when the primary gas component (assuming a nitrogen gas bleed is used to maintain P_c) in the chamber is nitrogen. Since P_c remains constant, the total pressure in the chamber can be expressed as the sum of the partial pressures:

$$P_c = P_{N_2} + P_{O_2} + P_{H_2O} + \ldots + P_i \tag{2}$$

Table 11.1. Pressure Units Used for Vacuum Gauges and Their Conversion Factors (read across or down to convert one unit to another)

	bar (760 mm @ 0°C)	Pa (Newtons/m²)	Torr (1 mm Hg)	mTorr (.001 mm Hg)	mbar (.75 mm Hg)	μbar (.000075 mm Hg)	micron (0.001 mm Hg)	inches Hg
bar	1	1E+5	750.1	7.5E+5	1,000	1E+6	7.5E+5	29.61
Pa (Newtons/m²)	0.010	1	7.5E-3	7.5	10	1E4	7.5	2.97E-4
Torr	1.33E-3	133.3	1	1,000	1.333	1.3E+3	1000	3.94E-2
mTorr	1.33E-6	0.133	1E-3	1	1.33E-3	1.33	1	3.94E-5
mbar	0.001	100	0.750	750	1	1000	750	2.96E-2
μbar	1E-6	0.10	7.5E-4	0.750	1E-3	1	0.750	2.96E-5
micron	1.33E-6	7.5	1E-3	1	1.33E-3	1.33	1	2.94E-2
inches Hg	3.37E-2	3.38E3	25.3	2.5E+4	33.7	3.37E+4	2.5E+4	1

1 atm = 1.013 bar = 760 Torr

m = milli

μ = micro

Hg = mercury

E±x = 10±x

where P_i indicates the partial pressure of the *i*th gas species. If the *i*th gas behaves in an ideal fashion, then its equation of state can be expressed as

$$P_i V = n_i RT \tag{3}$$

where V is the volume of the system, n_i in the number of moles of the *i*th gas in the chamber, R is the gas constant, and T is the temperature expressed in K.

Consider the unique collapsible drying chamber illustrated in Figure 11.1. I invented this chamber to control pressure in the chamber without having to resort to the use of a gas bleed system. Since the chamber is assumed to be completely free of any leaks, once the chamber has reached its operating pressure, the valve between the drying chamber and the vacuum system can be closed. P_c is then controlled merely by varying the volume of the chamber rather than having to bleed a gas into the chamber. The uniqueness of this drying chamber can be demonstrated by the results shown in Table 11.2.

Table 11.2 shows that the partial pressure of nitrogen obeys the ideal gas law shown by equation 3; however, the partial pressure of water vapor remained

---

Figure 11.1. A Collapsible Freeze-Drying Chamber with the Condenser Temperature Maintained at –40°C

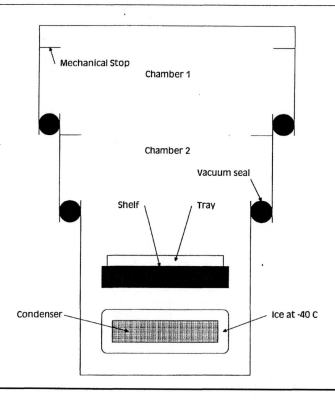

Table 11.2. The Change in the Partial Pressures of Nitrogen and Water Vapor with a Change in Volume

Chamber Volume	P (nitrogen)	P (water vapor)	Total Pressure
V_o	100	100	200
$V_o/2$	200	100	300
$V_o/3$	300	100	400
$V_o/4$	400	100	500
$V_o/5$	500	100	600
$V_o/6$	600	100	700
$V_o/7$	700	100	800
$V_o/8$	800	100	900
$V_o/9$	900	100	1,000
$V_o/10$	1,000	100	1,100

Pressure units in mTorr

constant at about 100 mTorr. While both gases were partial pressures in the chamber, the partial pressure of water vapor was governed by the phase diagram of water vapor shown in Figure 3.1. Since water vapor has a partial pressure and a vapor pressure, the decrease in the volume of the chamber only served to condense more ice onto the condenser surfaces. This invention demonstrates that every vapor pressure is a partial pressure, but not every partial pressure is a vapor pressure.

Mean Free Path

The pressure of a system was defined by equation 1 in terms of the force exerted by gas-wall elastic collisions, but gas particles can also undergo gas-gas collisions. It will be shown later in this chapter and subsequent chapters that it is important to know if the number of gas-wall collisions >> the number of gas-gas collisions or if the number of gas-gas collisions >> the number of gas-wall collisions. The following will describe a means for determining the gas-gas collisions from a knowledge of the *mean free path* (M.F.P.)—the mean distance that a molecule will travel between gas-gas collisions.

In an effort to simplify the description of the M.F.P., I have chosen to disregard the first postulate of the Kinetic Theory of Gases by assuming that, for an instant, only one of the molecules in a closed system is in motion. The remainder of the gas molecules will be in a fixed position so that the density of the gas will remain constant. The reader should realize that under these conditions, the gas-wall collisions will be zero as will the pressure in the system. One can now describe, with the above conditions in mind and the aid of Figure 11.2, the M.F.P of a gas.

Figure 11.2 shows a system in which one gas particle having a diameter *d* passes through the system in which all other gas particles are in a fixed position. For the sake of simplicity, we shall make another assumption: all of the other gas particles have the same diameter *d*. The one gas particle that is free to move will have a velocity (*v*) and sweep out a volume per unit time (V_t) defined by

$$V_t = \pi d^2 v \quad \text{volume/unit time} \tag{4}$$

From a knowledge of V_t, the number of *gas-gas collisions* (Z) per unit time made by the one free gas particle can be determined from equation 5 which stems from a knowledge of the temperature of the gas (T_g) and the original system pressure (P_c) and use of ideal gas equation 3:

$$Z = \frac{d^2 \pi v P_c}{kT_g} \quad \text{collisions/unit time} \tag{5}$$

where *k* is the Boltzmann constant.

The M.F.P. or mean distance between gas-gas collisions can be can now be defined as

$$\text{M.F.P.} = \frac{v}{Z} = \frac{kT}{\pi d^2 P} \quad \text{distance/collision} \tag{6}$$

Equation 6 shows the M.F.P. is directly related to the gas temperature and inversely related to the square of the diameter of the particle and the pressure in the system.

Figure 11.2. Gas Collisions

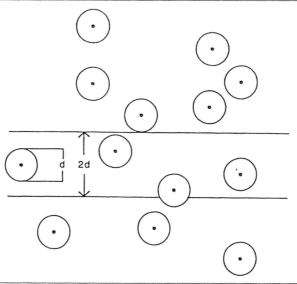

Gas collisions of a gas particle in a system in which all other gas particles are in a fixed position.

Table 11.3 shows some calculated M.F.P.s for some gases at a temperature of 25°C for pressures of 1 atm and 1 mTorr. For 1 atm, the M.F.P. is very small (i.e., gas molecules can only travel a relatively short distance between collisions). But at 1 mTorr, the smaller diameter He atom can, in a pure He environment, travel 14.7 cm between collisions, while the larger diameter CO_2 will, in a CO_2 environment, travel only 3.3 cm between collisions. At a pressure of 100 mTorr, a typical pressure used during a lyophilization process, the M.F.P. for He will be 0.14 cm, while for CO_2 it will be 0.033 cm.

Gas Flow

Gas flow at subatmospheric pressures will fall within three general categories: *viscous, transition* and *molecular*. The category for the gas flow will depend on the Knudsen number (K.N.) [1]. The K.N. for gas flow is defined as

$$\text{K.N.} = \frac{\text{M.F.P.}}{r} \quad \frac{1}{\text{collisions}} \tag{7}$$

where r is the radius of the system. When K.N. is < 0.01, i.e., when $r \gg$ M.F.P., the nature of the gas flow will be considered *viscous* in nature (i.e., the number of gas-gas collisions >> the number of gas-wall collisions). When the M.F.P. > r, then K.N. > 1, and the flow will be *molecular* in nature. Molecular gas flow will occur when the number of gas-wall collisions far exceed the number of gas-gas collisions. Gas flow in the *transition* range will occur when 1 > K.N. > 0.01. The following is a brief description of gas flow and will be followed with expressions for gas flow in the viscous and molecular flow categories [1].

General Gas Flow Expression

The nature of gas flow is analogous to the passage of an electric current through a resistance circuit where the latter is defined by Ohm's law. The electrical current (I) through a resistance circuit (R') is expressed as

$$I = \frac{V'}{R'} \quad \text{amperes} \tag{8}$$

Table 11.3. The Effect That Particle Diameter and Pressure Have on the M.F.P. at a System Temperature of 25°C

Gas	Diameter (Å)	M.F.P. (cm) (1 atm)	M.F.P. (cm) (1 mTorr)
He	2.18	1.93×10^{-5}	14.7
O_2	3.64	6.97×10^{-6}	5.4
Ar	3.67	6.97×10^{-6}	5.3
CO_2	4.65	4.40×10^{-6}	3.3

where V' is the electrical potential difference (voltage) across the resistor. Since the conductance (C) of an electrical circuit is the reciprocal of the resistance R', equation 8 can be shown as

$$I = CV' \quad \text{amperes} \tag{9}$$

If Q is substituted for I and the terms F and $(P_1 - P_2)$ are substituted for the terms C and V', respectively, one obtains

$$Q = F(P_1 - P_2) \quad \frac{\text{pressure volume}}{\text{time}} \tag{10}$$

where Q is gas flow expressed in units of (pressure volume) per unit time or, since pressure x volume represents mass, Q can also be expressed as mass per unit time. F is the *conductance* of the system and will be dependent on K.N., while $(P_1 - P_2)$ represents the pressure difference or driving force. As the value of $P_1 \rightarrow P_2$, Q will approach 0.

Viscous Flow

When K.N. < 0.01, then the gas flow is defined as *viscous* and the number of gas-gas collisions will far exceed the number of gas-wall collisions. The conductance (F_v) of a system in viscous flow through a straight tube having a radius (r) is defined from the Poiseuille equation as

$$F_v = \frac{\pi r^4}{8 l_g \eta} P_a \quad \text{L/sec} \tag{11}$$

where l_g is the length of the *gas conduction path*, η is the coefficient of viscosity of the gas at a temperature (T), and P_a is the arithmetic mean pressure [1]. Substituting F_v for F in equation 10 and setting the pressure difference across a conduction path as $(P_1 - P_2)$, the gas flow Q_v for viscous flow becomes

$$Q_v = \frac{\pi r^4}{8 l_g \eta} P_a (P_1 - P_2) \quad \frac{\text{mTorr liters}}{\text{sec}} \tag{12}$$

I have found that the Poiseuille equation 12 is not applicable for determining the gas flow in a freeze-dryer. A modified version of equation 12 that was found to be applicable has the form

$$Q_v = \frac{7.5 \pi r^4 S}{l_g (\text{M.F.P.})} \sqrt{\frac{T}{M}} P_a (P_1 - P_2) \quad \frac{\text{mTorr liters}}{\text{sec}} \tag{13}$$

and is expressed in terms of the M.F.P. rather than η and contains a term S that is defined as a gas flow impedance factor resulting from the hardware in the freeze-dryer chamber (e.g., the shelves of the dryer). Equation 13 shows that Q_v is pressure dependent, i.e., as the pressure (P_a) increases, the M.F.P. will decrease, which will result in an increase in Q_v.

Molecular Flow

For *molecular flow* (i.e., when K.N. > 1), the molecular gas conductance (F_m) is expressed as

$$F_m = \frac{30.48\pi r^3}{l_g}\sqrt{\frac{T}{M}} \quad \text{L/sec} \tag{14}$$

and the molecular gas flow (Q_m) is expressed as

$$Q_m = \frac{30.48\pi r^3}{l_g}\sqrt{\frac{T}{M}}(P_1 - P_2) \quad \frac{\text{mTorr liters}}{\text{sec}} \tag{15}$$

Equation 15 shows Q_m is independent of the M.F.P. and that Q_m can be increased only by an increase in ($P_1 - P_2$).

Effect of Bends on Gas Conduction

The effect of bends in the length of the conductance path (l_g) can have a serious effect on gas flow. This effect of bends on the conductance of a tube will be angle (β) dependent by a factor B(r), where the value of B(r) will vary from 0 for a $\beta \to 0°$ to 10 for $\beta \to 90°$. By increasing the number of bends in the conductance tubing, one is effectively increasing the length of l_g. The effect that bends can have on the gas flow of a system will be addressed again in the section of this chapter that discusses real leaks and in later chapters.

Vapor Throughput

The *vapor throughput* (Q_t) of a vacuum system is defined as the mass per unit time of a gas at a given pressure. Consider the vacuum system illustrated by Figure 11.3. This figure shows two chambers that are interconnected by a tube that has a radius r and a length l_g. If we assume that the gas flow will be viscous, then the gas flow in the tube (Q_v) between the two chambers will be defined by equation 13. With m_i representing the number of grams of ice in the chamber A, the time required to transport all of the water from chamber A at a pressure P_1 to chamber B at a pressure P_2 is defined by the *throughput time* (t).

From equation 3, the mass of water vapor (m_w) in chamber A at a temperature (T) can be determined from the chamber volume (V) and the pressure P (expressed in mTorr) as

$$\alpha PV = 6.24 \times 10^4\, m_w T \quad \text{mTorr liters} \tag{16}$$

where α is the *fractional partial pressure* of water vapor in the chamber,

$$\alpha = \frac{P_w}{P} \tag{17}$$

and P_w is the partial pressure of water vapor in the chamber. The limits for α will be from 0 to 1. When α approaches 1, the volume of water vapor in the chamber will approach V, and when α approaches 0, the volume of the water vapor will also approach 0.

The *throughput time* (*t*) required when K.N. is < 0.01 (or the gas flow is viscous) to transport m_w grams of ice from chamber A to chamber B is defined as

$$t = \frac{\alpha PV}{Q_v} \quad \text{sec} \tag{18}$$

where *PV* is defined by the ideal gas law for a given gas temperature. From a knowledge *t* and the mass of ice in chamber A, one can ascertain Q_t as

$$Q_t = \frac{m_i}{t \,(\text{time})} \quad \frac{\text{mass of water vapor}}{\text{unit time}} \tag{19}$$

Assuming that the volume of chamber A (see Figure 11.3) is 50 L, the conduit between chambers A and B has a l_g of 50 cm and a radius (*r*) of 4 cm. The impedance factor (*S*) for the system is 0.1. Since the diameter of water vapor is similar to that of CO_2, the assumed M.F.P. of water vapor at 1 mTorr will be 3.3 cm. With this information and equations 13 and 16 through 19, one can determine Q_t and the energy that will be required to transport 1,000 g of ice from chamber A to B. Table 11.4 lists the time (*t*), Q_t, and necessary refrigeration capacity [H(1)] to transport 1 L of water between the two chambers for various pressures when $\alpha = 1$ and $S = 0.1$. The entire system is considered to be adiabatic, and the temperature of the water vapor approaches the temperature at the sublimating surface. The nature of the gas flow from chamber A to chamber B is considered, for this example, to be viscous.

As the water vapor pressure is increased, the time required to transport 1 L of water from chamber A to chamber B decreases. This decrease results in an increase in the Q_t of the system. However, in terms of energy, based on a heat of sublimation of 670 cal/g, the horsepower [H(1)] necessary to condense the vapor also increases with pressure. Although only 1 min will be required to transport 1 L of water vapor at a pressure of 500 mTorr, the required capacity of the condenser refrigeration

Figure 11.3. Gas Flow Via a Tube

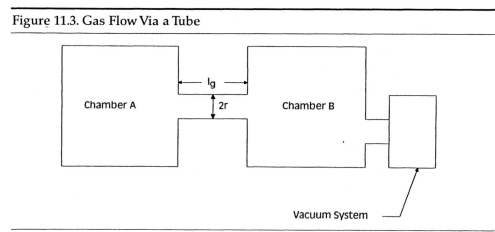

A vacuum system in which gases are flowing from chamber A to chamber B via a tube having a radius *r* and a length l_g.

Table 11.4. The Time (t), Throughput (Q_t), and Energy Required to Transport 1 kg of Water from Chamber A to Chamber B at Various Pressures when $\alpha = 1$ and $S = 0.1$

P_1	P_2	T_g (°C)	M.F.P.	Q_v	t (min)	Q_t	H(1)
*50	28	−45	0.066	5.58E5	1415	0.707	0.044
100	40	−40	0.033	5.52E6	146	6.84	0.42
200	65	−33	0.0165	4.77E7	17	57.4	3.56
250	78	−31	0.0132	9.45E7	9	112.8	6.99
300	90	−29	0.011	1.65E8	5	195.6	12.1
400	115	−26	0.0083	3.97E8	2	494.5	28.8
500	140	−24	0.0066	7.82E8	1	907.8	62.8

*Results for 50 mTorr are in agreement with published data [3].

Heat of sublimation of ice taken at 670 calories.

M.F.P. in cm

Q_t in g of water per min

H(1) energy in horsepower (hp)

$y\text{E}\pm z = y \times 10\pm z$

system will have to be 62.8 hp. If the capacity of the refrigeration system is 5 hp, then increasing the water vapor pressure to 250 mTorr will cause an increase in P_2. and a reduction in Q_v. Thus, there will be a limit to the gas conductance of a freeze-dryer based either on its configuration (e.g., the piping between the chambers or the impedance factor) or the capacity of the refrigeration system.

Differences in the gas conductance between dryers can impact that transfer of a lyophilization process that is based on a chamber pressure that is composed mainly of water vapor. A system having a high Q_t will operate at a lower chamber pressure and product temperature than a system which has a lower value of Q_t. Consequently, the drying process will be freeze-dryer dependent. Such a difference in process can be avoided by operating the dryers at a common elevated pressure by bleeding in a noncondensable gas, such as nitrogen, into the chambers. At a higher common pressure, the Q_v of the systems will be similar. Because the product temperatures will also be similar, the partial pressure of water vapor or α term will also be similar. This will result in a reduction in Q_t for both systems; however, the Q_t for each system and the drying processes will be similar. Thus the chamber pressure, for a given T_s, serves not only to control the process but permits the drying process to be independent of the configuration of the freeze-dryer.

Gas Thermal Conductivity

The last property of gases to be considered is the *thermal conductivity* of gases. The effect of the thermal conductivity of gases was shown to be an important consideration during the primary drying process (see Chapter 8). It will be shown later in this chapter that the thermal conductivity of gases can also play an important role in determining P_c during the lyophilization process.

In the simplified definition of gas pressure, it was assumed that the gas collisions with the walls of a chamber are elastic in nature (i.e., the particles impact the surface and immediately rebound without imparting any energy to the surface and there is no change in energy of the gas particle). In reality, the gas particles will spend some finite time on the surface and can leave the surface with an increase or decrease in energy. The energy transferred to a surface as a result of the thermal conductivity (q_g) of gases can be expressed as

$$q_g = 4/3A\sqrt{\frac{273.2}{T_1}}(T_2 - T_1)\sum \Lambda_i P_i \quad \text{energy/unit time} \tag{20}$$

were A is the area of the surface on which the gas particles impinge, T_1 is temperature of the surface and T_2 is the temperature of second surface where $T_1 \neq T_2$. Λ_i represents the thermal conductivity of the gas and P_i the partial pressure of the *ith* gas particle that strikes the surface. The thermal conductivities of some common gases in the drying chamber during a lyophilization process were listed in Table 8.1. The significance of the thermal conductivity of gases in determining the pressure of a system will be considered in the next section of this chapter.

PRESSURE GAUGES

While previous chapters stressed the need for measuring and controlling P_c during the drying processes, this section will consider the various means that have been used or are currently being used for determining P_c. It is important for the reader to understand that, unlike the T_p or T_s measurements that can involve more than one sensor, the pressure in the drying chamber is often determined by a single pressure transducer.

U Tube

The pressure gauge used by Shackell [4] during his studies of the drying process that would become known as lyophilization was a simple device known as a U tube. The U tube gauge, like that shown in Figure 11.4, is a device whose configuration is implied by its name. As shown in Figure 11.4, one end of the gauge is attached to the drying chamber, while the other end of the gauge is open to the atmosphere. The liquid contained in the tube is mercury. The lowest point on the gauge is indicated by a dotted line. In a steady state, i.e., in the absence of pressure fluctuations, the force due to the height of the column of mercury on the left-hand side of the dotted line will be equal and opposite to the force exerted by the column of mercury on the right-hand side. The equality can be represented by

$$P_v + \rho g h_v = P_o + \rho g h_o \qquad (21)$$

where ρ is the density of mercury, g is the acceleration due to gravity, h_v is the height of the mercury on the vacuum side of the tube having a pressure of P_v, and h_o is the height of the mercury on the atmospheric pressure (P_o) side of the gauge. Solving equation 21 for P_v, one obtains

$$P_v = P_o + \rho g(h_o - h_v) \qquad (22)$$

If P_o is 1 atm (760 Torr) and P_v is 100 mTorr, then the absolute difference $|h_o - h_v|$ would be 759.8 mm. If the pressure in the chamber was increased to 200 mTorr, the absolute difference $|h_o - h_v|$ would be 759.7 mm. While this gauge would be useful at pressures $\geq$ 1 Torr, the above has shown that the gauge does not have sufficient precision to be useful at pressures generally used during the lyophilization process.

In using this gauge, one could enter the correct value for g; however, variations in P_o could introduce serious errors in the determination of P_v. Under a high atmospheric pressure of 32 in. of Hg or 812.8 mm Hg (see Table 11.1), which was recorded in Siberia [5], the value of $|h_o - h_v|$ for a P_c of 100 mTorr would be 812.6 mm Hg, signifying that the pressure measurement was in error. If the pressure measurement were made during a typhoon, when the atmospheric pressure was 25.9 in. or 657.9 mm Hg [5], the reading for $|h_o - h_v|$ would be 657.5 mm Hg, and the gauge would indicate a false high pressure of 102 mm Hg.

Figure 11.4. A U Tube Pressure Gauge Containing Mercury

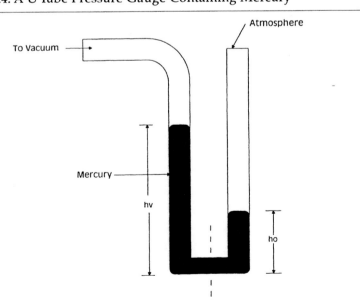

The lowest point on the gauge is indicated by the dotted line.

With all mercury gauges operating at ambient temperature, the vapor pressure of mercury is about 1 mTorr, and there is always the possibility of mercury vapors diffusing into the drying chamber. In addition, since the mercury levels must be visible for making measurements, breakage of the glass tube or the accidental spilling of the mercury would constitute a health hazard to the operator.

McLeod Gauge

Fundamental Principles of the McLeod Gauge

The McLeod gauge, like that illustrated in Figure 11.5, also determines the low pressure in the drying chamber using liquid mercury. While the operation of the U tube was not affected by the presence of water vapor in the drying chamber, the McLeod gauge determines P_c by compressing a sample volume of the gas from the chamber. If water vapor is present in sufficient quantities, which could be true during primary drying and a portion of secondary drying, compression of the gases may result in a partial pressure of water vapor reaching the equilibrium vapor pressure of water at ambient temperature. As in the case of the collapsible freeze-drying chamber (see Figure 11.1), further compression of the water vapor will not result in an increase in pressure in the gauge but in the condensation of water and a false low pressure reading. Thus, in order to obtain a correct pressure measurement with a McLeod gauge, the gases in the gauge must behave in an ideal fashion,

---

Figure 11.5. A McLeod Gauge

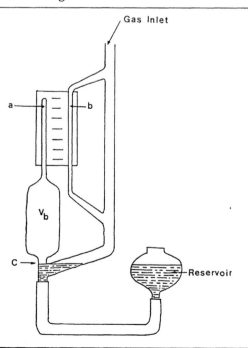

$$P_v V_g = P'V'$$ (23)

where V_g is the sample volume of the gas obtained from the drying chamber, P' is the pressure of the compressed chamber gases and V' is the volume of the compressed gases. Rearranging equation 23 results in

$$P_v = \frac{P'V'}{V_g}$$ (24)

and

$$P' = P_v + \rho g h_d$$ (25)

where ρ is the density of the mercury and h_d is the difference in the height of the mercury levels in tubes "a" and "b" shown in Figure 11.5. The volume V' can be expressed as

$$V' = V_b + b_o h_d$$ (26)

where V_b is a known volume, b_o is the volume per unit length of the capillary tube "a," and h_d is the difference in the height of the mercury in capillary tubes "a" and "b." Substituting equations 25 and 26 into equation 24, one obtains

$$P_v = K h_d^2$$ (27)

where K is referred to as the gauge constant.

The above reveals two fundamental principles regarding the McLeod gauge. The first is that the McLeod gauge is one of the few vacuum gauges, that has a relatively high degree of accuracy and that can be constructed from first principles. The second, from equation 27, is that the observed pressure will be a function of the square of the difference in the mercury levels.

Operation of the McLeod Gauge

The McLeod gauge is connected to the drying chamber in a similar fashion to that shown for the U tube (see Figure 11.4). Unlike the U tube that provides a constant measurement of the pressure in the chamber, the McLeod gauge determines P_c from a gas sample. In order to ensure that water vapor does not enter the gauge, a desiccant is placed between the gauge and the chamber to remove any water vapor. Because of the presence of mercury vapors, I prefer to use a liquid nitrogen trap between the gauge and the vacuum chamber. The liquid nitrogen trap prevents water vapor from entering into the gauge and mercury vapors from entering the vacuum system.

When P_c is being measured, the mercury reservoir (see Figure 11.5) is lowered so the mercury level falls below point C on the gauge. The gauge is allowed to remain in this position for a period of time to allow the partial pressures of the noncondensable gases such as N_2, O_2, and H_2 to become equal to P_c. For measuring the pressure in a system where the major gas component is water vapor, one must provide sufficient time for the pressures to reach a steady state. This is determined by varying the time interval that the mercury level is allowed to remain below point C of the gauge (see Figure 11.5). The correct time period is ascertained when an extension of the time period does not alter the pressure reading.

The mercury level is then raised so that mercury flows into the volume (V_b) of the gauge and into the capillary tube "b" and the side arm of the gauge. The height of the mercury in capillary "b" is increased until the mercury is level with the inside top of capillary "a." The top of the capillary tube is usually designated by a solid line across the gauge scale. With the mercury level in capillary tube "b" at the indicator line of the gauge, the pressure is read directly from the level of the mercury in capillary tube "a."

The main role of the McLeod gauge is to calibrate the other gauges that are used to determine P_c during the lyophilization process. Because making a pressure reading with a McLeod gauge is a manual operation and creates the possibility of mercury vapors entering the vacuum chamber, the McLeod gauge is not considered a suitable instrument for measuring P_c during a lyophilization process.

Thermal Conductivity Gauges

Fundamental Principles of a Thermal Conductivity Gauge

The basic principle of the thermal conductivity gauge is the thermal properties of gases given by equation 20. The thermal conductivity (Λ_i) of the gases can be equated to P_c if one could measure the energy removed from a hot surface (T_2) and transported to a surface at a lower temperature (T_1). If the mass of the hot surface is low, then gas molecules striking the surface will not only remove energy from the surface but also alter its temperature. If the distance between surfaces is less than the M.F.P., then the gases will not suffer any thermal loss in transferring energy between the two surfaces as a result gas-gas collisions. With the lower temperature surface acting as a heat sink for a given temperature range, the change in temperature of the low mass hot surface is related to P_c.

Currently, there are two types of thermal conductivity gauges being used to monitor P_c during a lyophilization process: the thermocouple and Pirani gauges. The following will be a brief discussion of the operation of each of these gauges.

Operation of the Thermocouple Gauge

The general construction of the thermocouple gauge is illustrated by Figure 11.6a. This figure shows that the main components of the gauge are a power supply, a filament, an iron-constantan thermocouple, a current-regulating system and a meter. The filament consists of a thin platinum wire having a diameter of about 0.01 mm [1]. Spot welded to the center of the filament is a thermocouple junction that is either copper or iron constantan with a wire diameter of 0.01 mm. The power supply and current-regulating circuit provide sufficient electrical current (30 to 50 mA) to the filament to raise its temperature to between 100°C and 200°C. The gauge is designed so that the distance between the filament and the walls of the gauge ensures that the number of gas-wall collisions will be more frequent than the number of gas-gas collisions. The current leads for the filament and the leads for the thermocouple wires are connected to electrically insulated and vacuum seals.

The general configuration of a thermocouple gauge tube is illustrated by Figure 11.6b. The reader should note that the tube has a metal housing and the connection to the vacuum system is made using a pipe-threaded connection. The reason for the pipe-threaded connection are for convenient mounting and to provide a good heat sink for the outer walls of the gauge. Since P_c is determined by the transfer of heat between the filament and the walls of the tube housing, the temperature of the walls will be an important factor in determining the accuracy of the gauge.

At pressures approaching 1 mTorr, there is relatively little heat transfer between the hot filament and the walls of the tube, and the temperature of the filament will be at a maximum. For pressures ≥ 2 Torr, the transfer of heat between the filament and the walls will be at its maximum, and the temperature of the filament will be at its minimum. A further increase in pressure does not result in a significant change in filament temperature. With a nitrogen atmosphere, the temperature

Figure 11.6. A Thermocouple Gauge

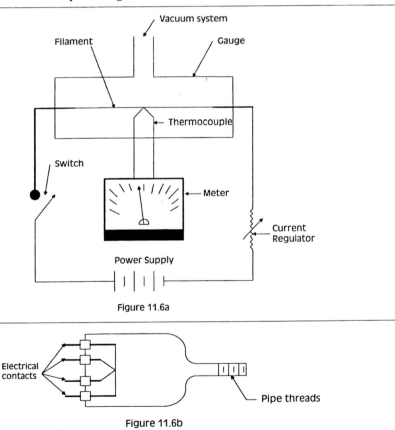

Figure 11.6a

Figure 11.6b

Figure 11.6a illustrates the key components and electrical circuit for the gauge. Figure 11.6b shows the configuration of the gauge tube.

output of the thermocouple gauge is calibrated with respect to the pressure of a system determined by a McLeod gauge. The temperature scale is then equated to a pressure scale relative to nitrogen. The resulting pressure scale is nonlinear (e.g., at the low pressure range, a relatively large change in the filament temperature results from a small change in pressure and at the higher pressure end of the scale, a small change in filament temperature results from a large incremental change in pressure). The accuracy of the thermocouple gauge is about 2% f.s.d. (full scale deflection of the gauge). For a gauge having a range of 0 to 2 Torr, the accuracy would be of the order of ±40 mTorr. This implies that there is a fair amount of uncertainty with regard to pressure reading at the low pressure end of the scale. The importance of these statements is that one must always be cognizance that the pressure readings of a thermocouple gauge reflects the temperature of a filament, not an actual pressure reading (i.e., force per unit area). In addition, a factor that is often overlooked is the outside of the gauge that not only serves as a protective housing but is also a key element in the operation of the gauge.

Although the thermocouple gauge is of a rugged construction and can provide continuous readings of a vacuum system, there are a number of aspects of the gauge that one should be aware of while using it.

Effect of the Thermal Conductivity of the Gases. As seen in equation 20, the heat transfer resulting from the thermal conductivity of the gases is not only related to the pressure and temperature differential between a hot and cold surface but also on the individual thermal conductivity of the gases. Since it is a general practice to calibrate the thermocouple gauge in a dry nitrogen atmosphere, the fact is that water vapor could be the predominant gas species in the chamber during the primary drying and a portion of the secondary drying processes (see Figures 8.10 and 8.12). Because the thermal conductivity of water vapor is greater than that of nitrogen (see Table 8.1), the thermocouple gauge will indicate a false high pressure. If P_c is 100 mTorr and consists mainly of water vapor, the pressure indicated by the gauge will be about 150 mTorr. If the thermocouple gauge is used to maintain P_c at 100 mTorr, the true P_c will be only about 65 mTorr. However, at the end of the secondary drying process when the composition of the gases in the chamber is mainly nitrogen, the thermocouple gauge will indicate the correct pressure of 100 mTorr. The difficulty with using a thermocouple gauge, especially for monitoring and controlling P_c, is that if one is uncertain of the composition of the gases in the chamber, then one is also uncertain of P_c.

Effect of the Temperature of the Tube Housing. As illustrated in Figure 11.6b, the housing of the tube is equipped with pipe threads. The gauge is constructed so that it can be connected to a good heat sink. A heat sink is necessary because energy is being transported from the hot filament to the tube housing. In the absence of an adequate heat sink, there can be an increase in the temperature of the tube housing. An increase in the housing temperature will result in an increase in the filament temperature. Such an increase in filament temperature will result in false low pressure readings. It is the practice of some manufacturers of freeze-drying equipment to use a section of elastomer hose to connect to the gauge. If there is inadequate air circulation, the result could be an increase in the temperature of the tube housing and an inaccurate pressure measurement.

If the tube is connected to a heat sink whose temperature is significantly lower than ambient temperature, this would tend to increase the thermal conductivity of the gases. This increase would lower the filament temperature and result in false high pressure reading.

Thermocouple gauges are used to measure the pressure not only in the drying chamber but also in the condenser chamber and in the foreline that leads to the vacuum pump. If the filament of the gauge becomes coated with oil or decomposed organic materials, the resulting film may serve as an insulating layer. In the presence of an insulating layer, the temperature of the filament will tend to be higher, and the output of the gauge indicates a false low pressure reading.

Operation of a Pirani Gauge

The general circuit diagram for a Pirani gauge is shown in Figure 11.7. This circuit diagram shows two filaments, a sensor and reference (ref.). These filaments are typically fabricated from thin gauge platinum wire. The sensor filament is exposed to the vacuum system, while the reference filament is housed in a sealed vacuum chamber at a pressure < 0.1 mTorr. The filaments are connected to form part of a

Figure 11.7. The Components and Circuitry for a Pirani Vacuum Gauge

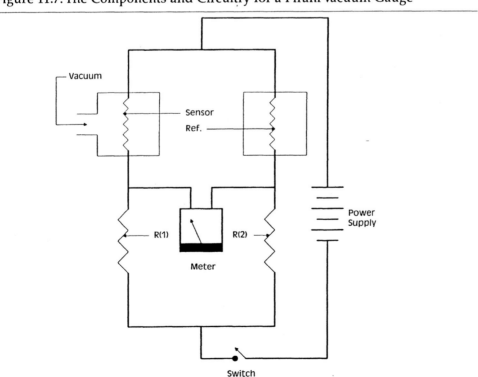

bridge circuit. Instead of measuring the temperature of the filament, the Pirani gauge compares the resistance of the filaments. The resistance of the platinum filament will be a function with temperature (i.e., an increase in temperature results in an increase in electrical resistance). As the pressure in the vacuum chamber approaches 0.01 mTorr, the resistance of the sensor filament [R(sensor)] will equal the resistance [R(ref)] of the reference filament. Because $R_1 = R_2$, the output voltage of the bridge circuit will approach 0 when R(sensor) = R(ref). As the pressure surrounding the sensor filament is increased, the thermal conductivity of the gases will cause a decrease in the temperature and an associated decrease in the resistance of R(sensor). Under these conditions, R(ref) > R(sensor) and imbalance of the bridge circuit results in an output voltage. Since R(ref), R_1, and R_2 will remain constant, the output voltage will be related to a change in R(sensor). The magnitude of the output voltage from the bridge is related to the pressure in the vacuum chamber.

Although the Pirani gauge has a higher accuracy and a greater operating pressure range than does the thermocouple gauge, it is still plagued with the same measurement errors as the thermocouple gauge. If calibrated with dry nitrogen, the Pirani gauge will indicate a false high pressure in the presence of water vapor. Any insulating contamination on the surface of the sensor filament will result in false low temperature readings.

It is hoped that, when looking at the results of pressure measurements by a thermocouple or Pirani gauge, the reader will bear in mind that the readings are a measure not of a force per unit area but a filament temperature.

Capacitance Manometer Gauge

The capacitance manometer gauge, as illustrated in Figure 11.8, determines the pressure in a vacuum chamber from the deflection of a diaphragm. In essence, the pressure readings are a result of the force per unit area on the diaphragm [6].

Construction of the Capacitance Manometer Gauge

Figure 11.8a shows the general construction of the pressure sensor. The gauge consists of two chambers and two electrically insulated plates. One of the chambers is open to the vacuum system, while the second chamber is sealed and serves as a reference chamber. The reference chamber is evacuated and maintained at low pressures through the use of *getters*, chemically active substances such as barium [1] that react chemically with most gases that leak into the chamber. With the use of a getter, it is possible to maintain the reference P_c at < 0.01 mTorr for a number of years.

The electrically insulated plates in each of the chambers represent one part of an electrical capacitor whose capacitance varies inversely with the distance between the plate and the flexible diaphragm (i.e., a decrease in distance results in an increase in capacitance). Because the capacitance measurements are performed in a vacuum, where the dielectric constant approaches 1, the observed change in the capacitance of the circuit can be attributed entirely to the distance between the sensing plates and the flexible diaphragm, not to the dielectric properties of the gas. From Figure 11.8a, the capacitance plate associated with the sensing chamber will tend to decrease as P_c increases. The capacitance gauge shown in Figure 11.8a is not necessarily representative of the actual gauge construction. In the actual construction of

Figure 11.8. Cutaway Illustration of the Construction of a Capacitance Manometer Gauge and Its Electrical Circuitry

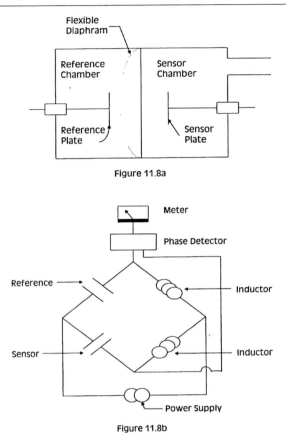

Figure 11.8a

Figure 11.8b

Figure 11.8a shows a simplified construction of the gauge, while Figure 11.8b shows the general circuit diagram.

the gauge head, both capacitance plates can be located in the reference chamber, and the sensing chamber can be virtually empty. I chose the gauge configuration shown in Figure 11.8a to simplify the explanation of the construction and operation of the gauge.

Operation of the Capacitance Manometer Gauge

The electrical circuit used to determine the pressure in the sensing chamber is shown by Figure 11.8b. This figure reveals that the sensor and reference capacitors are contained in two arms of an a.c. bridge circuit. Each arm of the bridge circuit consists of a capacitor (sensor or reference) and an inductor. The output of the bridge is sensed by a phase detector. The degree that the current and voltage from the output of the bridge circuit are out of phase is related to deflection of the diaphragm and the pressure in the sensing chamber.

The degree that the current and voltage are out of phase is illustrated by Figure 11.9. In this figure, it can be seen that the current leads the voltage and that the circuit will have a capacitive reactance. The degree that the two parameters are out of phase is shown by the angle Φ. There will be a linear relationship between Φ and P_c. As Φ approaches 0, the circuit becomes resistive and the pressure in the sensor will approach 0.

Zeroing of the Gauge

Because of the linear relationship between the output voltage and pressure (e.g., 1 mV = 1 mTorr), it is important to periodically zero the gauge to ensure that the pressure-voltage function passes through 0. This is accomplished by placing the sensor in a vacuum system that is capable of reaching < 0.01 mTorr [6]. At this pressure, the pressure in the sensor chamber will approach that of the reference chamber. After about 4 hr at this pressure, which allows the diaphragm to relax, the circuit in Figure 11.8b is adjusted so that the gauge reads 0. When properly zeroed, the accuracy of the capacitance manometer gauge is $\leq \pm 2\%$ of the pressure reading.

The chief advantage of the capacitance manometer gauge is that the pressure readings are independent of the composition of the gas. These gauges are sensitive to high temperatures and excessive pressures, and it is recommended that the user check with the gauge manufacturer prior to sterilizing the gauge with steam.

Residual Gas Analysis (RGA)

I have found the residual gas analysis (RGA) instrument to be a very useful analytical instrument during the lyophilization process [6,7]. Other investigators have also

---

Figure 11.9. The Determination of Angle Φ from Current and Voltage Wave Forms

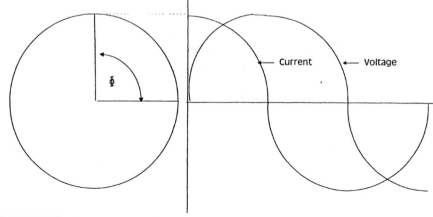

found this instrument to be useful in their studies of the lyophilization process [8,9]. The usefulness of the RGA instrument in determining the partial pressure of water vapor in a drying chamber during the final stages of the secondary drying process has already been demonstrated (see Chapter 9). The following briefly describes the overall instrument, an RGA quadrupole mass spectrometer, ion formation and fragmentation patterns, and the procedure for determining the partial pressure of a gas.

Overall Instrument

The general components and configuration of a differentially pumped RGA system used in determining the partial pressures of various gas in a freeze-dryer are illustrated by Figure 11.10; the RGA system is similar in design to one reported previously [7]. *Differentially pumped* means that the mass spectrometer is contained in a separately pumped system. The reason for a separate pumping system is that the mass spectrometer requires pressures ≤ 0.01 mTorr, and pressures in the freeze-dryer are generally in excess of 20 mTorr. The following briefly describes the major components of an RGA system.

Figure 11.10. A Differentially Pumped Mass Spectrometer Used in Determining the Partial Pressure of Gases in a Freeze-Drying Chamber

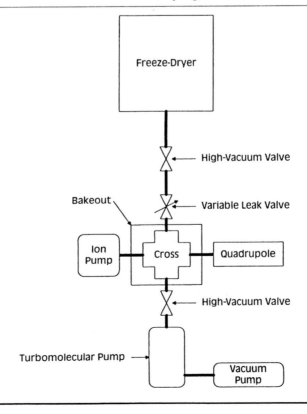

Stainless Steel Cross. The cross both supports the other components and provides a low-volume test chamber for determining mass spectrum of the residual gases. The other components of the system are attached to the cross using high-vacuum copper gaskets.

High-Vacuum Metal Valves. The two high-vacuum metal valves have metal-metal rather than metal-elastomer seal. It is essential that the RGA system be constructed without or with limited use of elastomer seals. Such seals could serve as a secondary source of absorbed gases. Desorption of such gases from an elastomer seal could interfere with the accuracy of the partial pressures or the detection of hydrocarbon vapors from the freeze-dryer. One of the valves will serve to isolate the RGA system from the freeze-drying chamber, while the second valve is between the stainless steel cross and the vacuum pumping system.

Variable Leak Valve. Figure 11.10 shows that a variable leak valve is positioned between the RGA system and the high-vacuum valve to the drying chamber. With this high-vacuum valve in a fully open position, the multiturn variable leak valve serves to balance the flow of gases into the chamber to the pumping speed of the turbomolecular pump to maintain a given operating pressure in the system.

Quadrupole Mass Spectrometer. The quadrupole detector head of the mass spectrometer is also connected to the stainless steel cross.

Pumping System. The RGA system that I used contains two high-vacuum ($< 1 \times 10^{-6}$ Torr) pumping systems and a backup pump. The high-vacuum pumps (ion pump and a turbomolecular pump) are connected to the stainless steel cross. The backup vacuum pump is an oil-mechanical pump and is connected to the output of the turbomolecular pump. The turbomolecular pump is essentially a very high-speed multifan pump. The fans in the pump rotate at about 100,000 rpm. At such speeds, gases that strike the specially shaped blades are given a downward momentum vector toward the backup mechanical vacuum pump. The gases compressed by the turbomolecular pump are at pressures < 1 atm and must be further compressed by the mechanical pump before they are discharged to the environment. The turbomolecular pump will pump all residual gases stemming from the freeze-dryer. In Chapter 12, this pump will be shown to be quite useful in pumping inert gases such as helium.

The ion pump has no mechanical moving parts and once in operation does not require a backup pump. The pump has crossed magnetic and electric fields. Because of the configuration of these fields, electrons move towards the anode but take a circular path that enhances the likelihood that they will ionize gases in the pump. The path of the positive gas ions is not affected by the magnetic field, and, the positive ions will impact the titanium cathode. The impact of the high-energy gas ions ($\approx 6,000$ V) with the titanium cathode surface results in the *sputtering* of titanium atoms from the cathode surface. Sputtering is a process by which material leaves a surface as a result of the impact of high energy ions. In addition to sputtering the titanium surface, the gas ions are removed by being embedded in the titanium metal. The sputtered titanium deposits on other surfaces in the pump. Since titanium is chemically active, it will react with all atmospheric gases except inert gases.

The reacted gases are then buried by a fresh layer of titanium and removed from the system.

Generally, the turbomolecular pump will be used during RGA, and the ion pump will be used to maintain the system in a ready state at pressures of the order of 1×10^{-7} Torr. With the use of the ion pump, the turbomolecular pump can be placed in a standby mode (i.e., operating at reduced speed) which extends the life of the pump.

Bakeout. A means to periodically bakeout the stainless steel cross and its attached components that comprise the RGA system should be provided. Prior to bakeout, it is advisable to remove the electronic head of the quadrupole mass spectrometer, turn off the ion pump, and disconnect the high voltage lead to the ion pump. Pumping of the system during the bakeout will be performed by the turbomolecular pump.

The bakeout of the system at temperatures $\geq 150°C$ will accelerate the desorption of residual gases from the walls of the system and allow the system to achieve lower operating pressures. It is imperative that during the bakeout the temperature of the system be increased at a slow rate, e.g., $1°C/min$, until the maximum bakeout temperature is reached. The system is then maintained at the maximum temperature for ≥ 8 hr. The system is then allowed to cool at the same rate that it was heated. The slow heating and cooling rates of the bakeout procedure are important to allow even expansion and contraction of the metal components. Rapid heating or cooling could result in distortion of the bolts of the flanges and increase the chance of creating a leak path via a loose flange.

Quadrupole Mass Spectrometer

Because of its compact size, the quadrupole mass spectrometer is ideally suited for use in a differentially pumped RGA system. A general illustration of the key components of the quadrupole mass spectrometer is shown by Figure 11.11. When the system is in operation, the filament reaches temperatures that are high enough for the emission of electrons ($\approx 2000°C$). The grid plate is maintained at an electrical potential of about 150 V with respect to the filament. In traveling to the grid, the electrons acquire sufficient energy to fragment and ionize the gas molecules that are in the analyzer. The ionized gases are further accelerated by a set of plates to form an ion beam. The ion beam then passes though a region surrounded by four precisely machined and positioned metal poles. By applying electric potentials on the poles, only ions of a given mass/charge ratio (m/e)—where m is the mass of the ion and e is the charge of an electron—will pass through the slit in the ion plate and be collected by the detector (generally a Faraday cup). The other ions in the beam are deflected and are captured by the ion plate. The ions that pass through the slit are neutralized by electrons from ground passing through a low impedance meter and indicate the ion current for a given m/e. As the potential applied to the poles is varied, the m/e of the ion beam that can pass through the slit is varied. A measure of the ion current as a function of the m/e generates a mass spectrum of the gases in the analyzer.

Figure 11.11. The Key Components of a Quadruple Mass Spectrometer

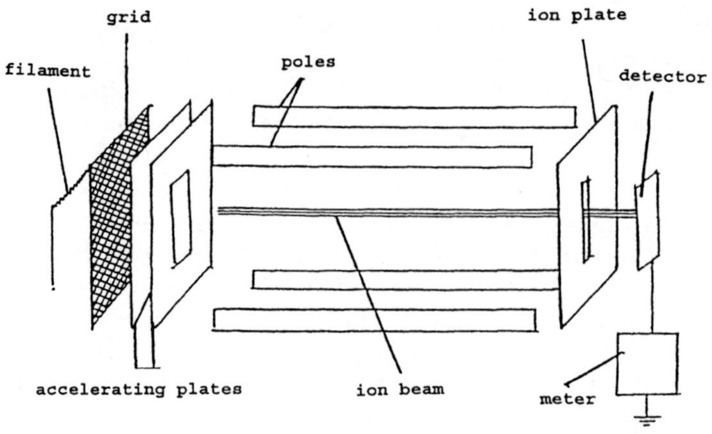

Ion Formation and Fragmentation Patterns

When accelerated electrons collide with the molecules in a gas, the following reactions can occur.

$$Ar + e^- \rightarrow Ar^+ + 2e^- \tag{28}$$

$$Ar^+ + e^- \rightarrow Ar^{++} + 2e^- \tag{29}$$

$$C_xH_yO_z + e^- \rightarrow C_{x-a}H_{y-b}O_{z-c}{}^+ + C_{x-d}H_{y-e}O_{z-f} + 2e^- \tag{30}$$

where Ar represents an atomic gas species and $C_xH_yO_z$ a hydrocarbon molecule. The following is a brief discussion of these reactions.

In reaction 28, the impact of the electron results in the ionization of the atom. For reactions 28 through 30, the results of a mass spectrum are given in terms of m/e. If in reaction 28 the Ar represents an argon atom, then the ion current would appear at m/e equal to 40. In reaction 29, the argon atom is doubly ionized and the ion current will appear at m/e equal to 20. Since neon also has an atomic weight of 20, a singly charged Ne^+ will also appear at m/e equal to 20. Both nitrogen and carbon monoxide have molecular weights of 28, and the RGA system does not have sufficient resolution to resolve the nitrogen and carbon monoxide peaks.

In reaction 30, the impact of the electrons on the hydrocarbon $C_xH_yO_z$ results in the fragmentation of the molecule. Some of the ion fragments will be in greater abundance than other fragments. The result is that the hydrocarbon molecule displays a fragmentation pattern consisting of m/e ratios with a defined relative ion current.

A fragmentation pattern for acetone is shown in Table 11.5. In determining the fragmentation pattern of a substance, the m/e value with the highest current is assigned a relative intensity of 100. The ion currents for the remainder of the other

Table 11.5. The Fragmentation Pattern for Acetone

m/e	Relative Intensity	Possible Ion Species
12	3–5	C^+
13	5–7	CH^+
14	22–23	CH_2^+
15	100	CH_3^+
16	23–27	CH_4^+
26	7–9	$C_2H_4^+$
29	3–4	$C_2H_5^+$
43	49–53	$C_2H_3O^+$
58	5	$C_3H_6O^+$

peaks associated with the substance are then assigned values relative to the major peak. In Table 11.5, the most intense peak had a *m/e* of 15 and was most likely a CH_3^+ ion. The intensities of the other ion currents are then assigned values relative to the ion current of the CH_3^+ ion. By using this means for assigning the relative intensities for a mass spectrum, the pattern obtained will be independent of the partial pressure of the gas species in the system.

Procedure for Determining the Partial Pressure of a Gas

The following is the recommended procedure for determining the partial pressure of a gas using a differentially pumped, residual gas mass spectrometer.

Preparation of the Instrument. The mass spectrometer should be clean, baked, and at a pressure $< 1 \times 10^{-7}$ Torr. The filament should be turned on and allowed to outgas.

Test mass spectrums of the system should appear like that shown in Figure 11.12. This spectrum shows peaks at *m/e* of 14 (N^+), 16 (O^+), 17 (OH^+), 18 (H_2O^+), 20 (A^{++} or Ne^+), 28 (N_2^+), and 40 (A^+) and no other peaks with *m/e* values like those in Table 11.5 to indicate the presence of hydrocarbons in the analyzer system. If such peaks are observed, it is recommended that a bakeout of the system be performed. In the absence of a bakeout, it may be difficult to ascertain the presence of any hydrocarbons in the freeze-dryer.

The electronic components of the mass analyzer should be allowed to warm up for no less than 1 hour to allow the various circuit components to stabilize.

Blank Mass Spectrum of the RGA. After successful completion of the preparation, a blank mass spectrum of the residual gases in the analyzer system is recorded.

Figure 11.12. Mass Spectrum of a Clean Residual Gas Analysis System

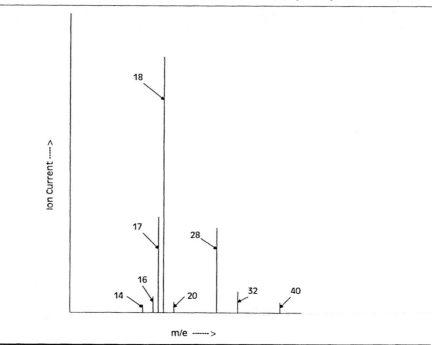

Mass Spectrum of the Freeze-Dryer. The high-vacuum isolation valve to the freeze-dryer is opened. The connection between the mass spectrometer and the freeze-dryer is allowed to equilibrate to the composition of the gases in the freeze-dryer.

The mass spectrometer is set to read total pressure, and the variable leak valve is adjusted to permit the P_c to be maintained at approximately 1×10^{-6} Torr while the system is being pumped by the turbomolecular pump.

The mass spectrum of the system is taken, and the chamber pressure is recorded. The leak valve is closed and the RGA system is allowed to pumpdown to lower than 1×10^{-7} Torr. If the mass spectrum is the last one to be taken, the isolation valve to the freeze-dryer is placed in a closed position.

Interpretation of the Data. The mass spectrum is first corrected for the residual gases in the mass spectrometer system by subtracting the ion current of a given m/e of the blank reading from the ion current of the same m/e obtained from the mass spectrum of the gases in the freeze-dryer.

The corrected ion currents are then adjusted according to the sensitivity for the gas species. The sensitivity for gases such as nitrogen, oxygen, and water vapor is about 1 and needs no correction.

The total ion current is determined from the sum of the corrected and adjusted ion currents. The fraction of the ion current for the m/e for a given gas species will represent the fraction of the chamber pressure, or a partial pressure of the gases in the drying chamber.

Determining the partial pressure of a hydrocarbon that displays a fragmentation pattern would require knowledge of the sensitivity of the compound based on the corrected ion current of the major ion species.

The usefulness of the mass spectrometer will be further demonstrated in the next two chapters.

VACUUM LEAKS

There is no vacuum system that does not leak. The differences are in the nature of the leak and the leak rate [11–13]. This section will consider *real* and *virtual* leaks. The discussion will be on the evaluation and detection of such leaks. Chapters 12 and 13 will consider leaks in the container-closure system and the freeze-dryer.

Real Leaks

Real leaks stem from leak paths that permit atmospheric gases to enter the vacuum system. Such leak paths can be further characterized, as illustrated in Figure 11.13, as being either straight or serpentine. The following discusses the nature of the gas flow through the two leak paths.

Straight Leak Path

A *straight leak path,* as illustrated by "a" in Figure 11.13, is one in which the nature of the gas flow (molecular, transition or viscous) is governed by the K.N. number. Since the M.F.P. of gases like nitrogen at 1 atm is of the order of 7×10^{-6} cm, the radius of the leak path would have to be $\geq 7 \times 10^{-4}$ cm. For a leak path (l_g) of 0.5 cm, the gas flow (Q_v) into a vacuum chamber is determined by equation 12.

The rate of pressure rise (ROR) [12] for an isolated chamber can be expressed as

$$\text{ROR} = \frac{Q_v}{V_e} \quad \frac{\text{mTorr}}{\text{sec}} \tag{31}$$

Figure 11.13. Examples of Two Types of Leak Paths: (a) Straight and (b) Serpentine

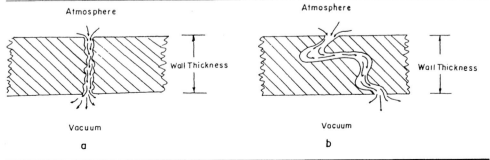

where V_e is defined as the *effective volume* (i.e., the total inside volume of the vacuum chamber less the volume occupied by other components in the chamber [12]). Based on the above leak path dimension and pressure assumption, Table 11.6 shows the ROR for various values of V_e, where the ROR is expressed in mTorr/hr. Table 11.6 clearly shows that for a given leak path the ROR of the system will be inversely dependent on the effective volume. While the leak is quite apparent for systems having a low V_e, the presence of the leak becomes obscure at larger values.

Serpentine Leak Path

An illustration of a *serpentine leak path* is shown by "b" in Figure 11.13. In this figure, there are a number of bends in the leak path. If the total length of the leak path remains constant, the effective length ($l_{g,e}$) of the leak path will be increased, and gas conductance or the ROR of the chamber will decrease.

Consider that the l_g for the serpentine leak path will be the same as that for the straight leak path (i.e., 0.5 cm). The effect that the number of 90° bends will have on the gas conductance and the *ROR* for a chamber having a V_e of 10 L is shown in Table 11.7. This table shows that as the number of bends in the leak path increases, there is an increase in $l_{g,e}$ and a decrease in Q_v and ROR for the 10 L system. The

Table 11.6. The Rate of Pressure Rise of Isolated Chambers at Ambient Temperature and Having Various Effective Volume (V_e) Values, Resulting from a Real Leak Path with a Radius of 5×10^{-5} cm and a Length (l_g) of 0.5 cm and Having a Gas Flow of 4,320 (mTorr liters)/hr, Where the M.F.P. of the Gases is 7.6×10^{-6} cm and the Nature of the Gas Flow Is Viscous (based on equation 13 where $S = 1$)

ROR (mTorr/hr)	Effective Volume (V_e) (L)
4,320	1
2,160	2
1,080	4
540	8
270	16
135	32
68	64
34	128
17	256
8	512
4	1,024
2	2,048

Table 11.7. The Rate of Pressure Rise and Gas Flow (Q_v) for an Isolated Chamber having an Effective Volume (V_e) of 10 L at Ambient Temperature Resulting from a Real Leak Path with a Serpentine Configuration Defined by a Radius of 5×10^{-5} cm and a Total Length (l_g) of 0.5 cm but at Various Effective Lengths ($l_{g,e}$), Depending on the Number of 90° Bends Where the M.F.P. of the Gases is 7.6×10^{-6} cm and the Nature of the Gas Flow Is Viscous (based on equation 13 where $S = 1$)

Number of Bends	$l_{g,e}$ (cm)	Q_v (mTorr Liters)/min	ROR (mTorr/min)
0	0.5000	72	7
5	0.5025	72	7
10	0.5050	72	7
50	0.5250	69	7
100	0.5500	66	7
250	0.6250	58	6
500	0.7500	48	5
1,000	1.0000	36	4

reader should understand that the actual l_g for the data listed in Table 11.7 remained at 0.5 cm and that the increase in $l_{g,e}$ was a result of the number of 90° bends in the leak path.

Molecular Flow Leak Paths

For a system having a low ROR, where a single leak path would be $<< 1 \times 10^{-4}$ cm., the gas flow can often be a result of a large number of leak paths whose radius is $< 7 \times 10^{-6}$ cm. Since K.N. approaches 1, the gas flow through the leak path will be molecular in nature. If leak paths are considered to be straight and have a l_g equal to 0.5 cm, then from equations 15 and 32, the ROR of a chamber having a $V_e = 10$ L, as a function of the number of leak paths, is shown by Figure 11.14. A comparison of the ROR for a 10 L chamber for a single leak path under viscous flow in Table 11.7 with the ROR resulting from leak paths under molecular flow shows that it would require 1 million leak paths of molecular flow to equal the ROR of 1 leak path under viscous flow. This suggests that for most vacuum systems the number of leaks having molecular flow will far exceed those with viscous flow. Leak detection methods will be dependent on the nature of the system; however, the above results indicate that real leaks having viscous flow characteristics will be easier to identify than those with molecular flow.

Figure 11.14. A Log-Log Plot of the Number of Leak Paths

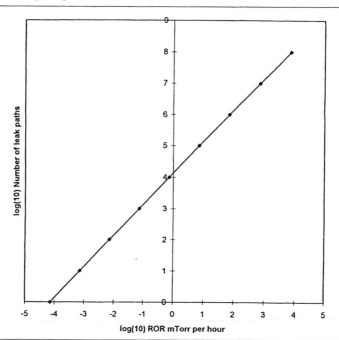

The number of leak paths is shown as a function of the ROR for a 10 L system for leaks in the molecular flow region.

Virtual Leaks

Virtual leaks in a vacuum system stem from a source of gas within the system [12]. Some internal sources of virtual leaks are outgassing of elastomer materials [8], adsorbed gases on surfaces, vapor pressure of fluids, rough (unpolished) surfaces and defective welds. Of all of the forms of virtual leaks, defects in welds will be the hardest to detect and locate. The determination and location of virtual leaks will be considered in greater detail in the next two chapters.

SUMMARY

This chapter considered the basic principles of vacuum technology as it relates to the lyophilization process. From the Kinetic Theory of Gases, the concept of pressure was established to be a force per unit area. Various units of pressure were listed, and a conversion table was provided. Only one pressure unit, Pa, expressed the pressure in terms of force per unit area (i.e., Newtons/m^2). In addition, since gas flow plays a major role in the drying (especially the primary drying) process, the various forms of gas flow, i.e., molecular and viscous, were considered. It was shown that the system can become conductance limited in the absence of a bleed gas. The effect of bends in the tubing was shown to impede the gas flow by making the tube appear to be longer than its actual length. The quantity of gas transported from one chamber to another per unit time, for a given pressure was defined as Q_t or the throughput.

Various forms of vacuum gauges were considered. Some of the gauges, like the U tube and the McLeod gauge, determined the pressure of the gas from first principles, but the units are expressed in terms of a length rather than a force per unit area. Other gauges, like the thermocouple and Pirani gauges, determined the pressure from the temperature of a hot filament. One must exercise some caution in using these gauges because the output will be dependent on the thermal conductivity of the gases. The temperature variation of the outer wall of the latter gauges can result in either false high or false low pressure readings. The capacitance manometer gauge uses the deflection of a diaphragm to cause a change in phase in a bridge circuit. The pressure in the system is related to the change in phase of the output of a bridge circuit and is not directly related to the capacitance of the gas, as the name of the gauge may imply. The last pressure gauge considered was the RGA system. In determining the partial pressure of water vapor in a chamber, it is essential that one not only correct for the gases in the analyzer but also have accurate total pressure readings for the system.

The characteristics of real and virtual leaks were introduced. The means for detecting and locating leaks will be considered as they relate to the topics of the last two chapters.

SYMBOLS

a.c.	alternating electrical current
A	area of the surface
α	fractional partial pressure of water vapor
b_o	volume per unit length of the capillary tube
C	electrical conductivity
d	diameter of a gas particle
e	charge on an electron
η	coefficient of viscosity of gases
f.s.d.	full scale deflection of a gauge
F	gas conductance of a system
F_m	molecular gas conductance
F_v	viscous gas conductance
g	acceleration due to gravity
h_d	difference in the height of the mercury levels
h_o	height of the mercury on the atmospheric pressure side of the U tube

h_v	height of the mercury on the vacuum side of the U tube
H(1)	energy in horsepower
I	electrical current in amperes
k	Boltzmann constant
K	McLeod gauge constant
l_g	length of the gas conduction path
$l_{g,e}$	effective length of a serpentine leak path
Λ_i	thermal conductivity of the ith gas particle that strike the surface
K.N.	Knudsen number
m_i	grams of ice
m	mass of an atom or molecule
M	gram molecular weight
M.F.P.	mean free path of gases
n	number of moles of gas
P	pressure of a closed container
Pa	pressure in Pascals
P_a	average pressure across a conduction path
P_c	chamber pressure
P_o	atmospheric pressure side of the U tube
P_v	pressure on the vacuum side of the U tube
P_w	partial pressure of water in mTorr
$(P_1 - P_2)$	pressure difference or driving force
Q	gas flow expressed pressure volume per unit time
Q_m	molecular gas flow
Q_v	viscous gas flow
Q_t	mass per unit time for a given pressure
r	radius of the system
rpm	revolutions per minute
R(ref)	reference platinum filament for a Pirani gauge
R(sensor)	sensing platinum filament for a Pirani gauge
S	gas impedance factor
ROR	rate of pressure rise in an isolated chamber

R_1	balancing resistor for a Pirani bridge circuit
R_2	balancing resistor for a Pirani bridge circuit
ρ	density of mercury
t	time (in sec)
T_c	condenser temperature
T_p	product temperature
T_s	shelf-surface temperature
v	velocity of an atom or molecule
V	volume of a container
V'	voltage across a resistor element
V_b	a known volume in the McLeod gauge
V_e	inside volume of the vacuum chamber less the volume occupied by other components in the system
V_t	volume per unit time

REFERENCES

1. S. Dushman, *Scientific Foundations of Vacuum Technique* (J. M. Lafferty, ed.), 2nd ed., John Wiley & Sons, Inc., New York 1962.

2. W. J. Moore, *Physical Chemistry*, 3rd ed., Prentice Hall, Englewood Cliffs, N.J. (1964).

3. T. A. Jennings, *J. Parenter. Sci. and Tech.*, 40: 95 (1986).

4. L. F. Shackell, *Am J. Physiol.*, 24: 325 (1909).

5. *The New Encyclopaedia Britannica*, 15th ed., 1988, vol. 1, p. 676.

6. J. G. Armstrong, *J. Parenter. Drug Assoc.*, 34: 473 (1980).

7. T. A. Jennings, loc. cit., p. 62.

8. K. S. Leebron and T. A. Jennings, *J. Parenter. Sci. and Tech.*, 35: 100 (1981).

9. H. Willemer, International Symposium on Biological Product Freeze-Drying and Formulation, Bethesda, USA, 1990 in *Develop. Biol. Standard* (Karger, Basel, 1991), 74: 123 (1991).

10. S. L. Nail and W. Johnson, loc. cit., p. 137.

11. T. A. Jennings, *J. Parenter. Sci. and Tech.*, 36: 151 (1982).

12. T. A. Jennings, ibid., 44: 22 (1990).

13. K. Kinnarney, in *Good Pharmaceutical Freeze-Drying Practices* (P. Cameron, ed.), Interpharm Press, Inc., Buffalo Grove, Ill. (1997).

12

Container-Closure System

INTRODUCTION

This chapter will address the role that the container-closure system plays in the lyophilization process and the various effects that the individual components and the combined system can have on the quality of the lyophilized product. My approach here is to first familiarize the reader with the general properties of both the closure and the container and how these properties can relate to the lyophilization process. In some instances, additional information not related to lyophilization is included. The objective for including such additional information is to make the reader aware of what components are suitable and why other components are not generally suitable for the process or product. When combined with the addition of a crimp seal, the container and closure form a packaging system that serves two important functions: (1) to protect the lyophilized product from any harmful effects of the environment and (2) to offer a convenient means for reconstituting the lyophilized formulation and delivering a sterile product.

The first topic will be the closure and the general requirements for its performance. It is important that the reader be aware of the various chemical compositions of the closure and the effects that these properties may have on the lyophilization process and on the quality and stability of the final product. The physical properties of the closure will be shown to impact the drying process and the effectiveness of the seal that the closure forms with the container.

The reader will then be given a brief tour of glass vials and ampoules. In Chapter 8, we considered the impact that the heat-transfer properties of the bottom of the container had on the lyophilization process and found that, from knowing the frequency distribution of the C_o values of a lot of vials, one is able to develop a lyophilization process in which the probability of generating process defects in a lyophilized batch becomes well beyond normal chance. In this chapter, we will focus our attention on the top of the vial were the container-closure seal is formed.

The next to last section of this chapter will deal with the formation of the container-closure seal or the glass seal in the case of the ampoule. We shall consider not only the forces that are involved in the sealing mechanism but also the various sealing points. It is considered important for the reader to have a general understanding of the mechanics involved in the formation of a container-closure seal.

The last segment of this chapter will describe the various means for testing the effectiveness of the seal. Each test method will be described, along with its inherent advantages and disadvantages. The objective here is make the reader aware that it may be prudent to perform more than one test to evaluate the effectiveness of a seal.

It is my hope that upon completion of this chapter, the reader will have a deeper understanding of and a greater appreciation for the complexity of the container-closure seal. Perhaps the seal will no longer appear merely as a closure attached to a container via the crimping of an aluminum device.

THE CLOSURE

As stated above, a primary function of the closure is to become an integral part of the packaging system that will protect the lyophilized product from chemical and biological contamination during storage, shipping, and handling. The second function of the closure is to offer a convenient means for reconstituting the lyophilized product while still maintaining the product in a sterile environment. Finally, the closure provides a means to remove all or a portion of the reconstituted formulation without compromising the efficacy of the remaining product [1–3].

A criterion of the closure that is routinely used but may be overlooked by those examining the properties of a container-closure system is that the closure should reseal itself after each insertion of a hypodermic needle. The insertion of the needle should not produce coring [4] nor should its removal allow gases, microorganisms, or particulate matter to enter the container [3]. In spite of the fact that a number of pharmaceutical companies produce lyophilized products in a multidosage form, the technical literature is generally silent about the effects that multiple penetrations can have on the closure's ability to maintain an effective chemical and biological barrier.

The following sections will consider various closure systems based on their chemical composition and the means for their manufacture. The key property of the closure, its ability to maintain an effective seal over the shelf life of the lyophilized product, will be considered with respect to its impact on the quality and stability of the final product.

Closure Composition and Formulation

The closure is, in some aspects, similar in nature to a lyophilized formulation. Both systems generally consist of a number of constituents that; when processed, produce a final product with a defined configuration. For the lyophilized product, it is a highly porous system that when subjected to a diluent quickly reconstitutes to the original formulation. For a cured or vulcanized closure, the processing produces an elastomer material having a defined configuration and physical properties that will mate with the upper flange surface of a glass container to form an effective seal. The balance of this section is devoted to the constituents that generally compose the closure formulation.

Polymer Component

The basic polymer component is, as in the lyophilized formulation, the key or active constituent and will determine the functionality of the final closure. The polymer component will determine key properties such as gas permeation and sorption, coring, and resealing [3]. The following is a list of some of the polymers that are used and the general application of the final closure.

- *Natural Rubber.* Natural rubber (cis-1,4-polyisoprene) [3,5] and closures prepared from this polymer are generally recommended for sealing liquid-fill products that have *p*H values ranging from 4.0 to 7.5 [6]. It has been my experience that natural rubber closures are not the generally preferred closure for use with a lyophilized product because of their sorption of water vapor, especially when the closure is steam sterilized at 15 psig and 121°C.

- *Butyl Rubber.* The basic polymer ingredient in butyl rubber is poly(isobutylene-co-isoprene) [3,5]. Closures fabricated from halogenated butyl polymers such as chloroprene [5] have been found to be especially suitable for use with lyophilized product because they offer protection from oxygen and water vapor permeation [6].

- *Silicone Rubber.* Closures prepared from the compound polydimethylsioxane [3,5] have been generally found suitable mostly for solutions have high *p*H values [6].

- *Nitrile Butadiene.* Nitrile butadiene polymers have applications where closures are required to seal containers containing oil-based products [6].

- *Coated Closures.* Closures are sometimes coated with polypropylene or Teflon® in order to reduce particulate matter and to prevent leaching of other constituents of the closure formulation from entering into the pharmaceutical solution. At this time, such coated closures are not commonly used in conjunction with lyophilized products [5].

Reinforcing Agent

Closure formulations generally include reinforcing agents in order to obtain the durometer hardness or stiffness. These properties, which will be discussed in greater detail in the next section, provide the closure with resistance to the compression set [3] (to be defined later) so that the container-closure seal will maintain its effectiveness over the shelf life of the product.

Other Constituents

There are other constituents that are added to the closure formulation to serve special functions during and after the curing process. Pigment obviously serves only as a coloring agent; constituents known as activators, curing agents, or accelerators are added to expedite the cross-linking of the basic polymer ingredient to generate polymerization. Antioxidants are often added to the closure to prevent deterioration from strong oxidizing agents such as ozone [3].

Formation of the Closure

A lengthy discussion of the details of the actual molding process that is used to fabricate the closures is considered to be beyond the scope of this text. What is important for the reader to understand is that the various closure formulations described above are subject to a controlled amount of heat for a specified period of time. It is during this time period that curing or vulcanization takes place to create cross-linking between the polymer molecules. It is these cross-links that are formed between the polymer molecules that provide the matrix that will not only govern the shape of the closure but also influence its physical characteristics [5]. Because of the extensive cross-linking that takes place during the curing process, the closure can be, in a sense, viewed as one giant molecule.

Physical Properties

The physical properties of the closure are just as important as the composition of the formulation. For example, the closure configuration must be designed in such a manner that, when positioned on the container for lyophilization, it will offer little impedance to the transport of water vapor during the drying process (this topic will be considered later in this section). It is important that the closure be designed such that vibrations, which may occur during the transport of the vial from the filling line to the freeze-dryer, do not result in a change in position of the closure with respect to the sealing surface of the container. Such changes in the position of the closure on the container may impede gas flow or cause dislodging of the closure from the container during the stoppering operation. For properly seated closures, the tops of the closures should appear as a level surface. The appearance of an uneven surface of closure tops, when viewed in a tray prior to loading onto a dryer shelf, is an indication of misplacement of the closures resulting from the configuration of the closure or improper placement of the closure upon completion of the filling operation.

The vapors from the product flow from the closure as a result of vents that are formed in the lower portion of the closure (see Figure 12.1). Upon completion of the drying process, the stoppering mechanism of the freeze-dryer depresses the closure into the container such that the primary sealing surface of the closure (see Figure 12.1) is contacting the flange surface of the container (see Figure 1.1e). With the closure fully seated on the rim of the container, the seal should be sufficient to provide adequate protection for the lyophilized product until the vial is crimped sealed.

The physical properties of the elastomer material used to form the closure will be dependent on the nature of the basic polymer and on the cross-linking that occurred during the curing process. Since the cross-linking process occurs, even at ambient temperature, during the first three to six months after the initial curing, there will also be a change in the physical properties of the closure [5]. The following is a brief description of the key physical properties of the closure.

Modulus

The *modulus* of a closure is the measure of stress versus strain of the elastomer material [5]. The modulus for elastomer materials is measured in terms of the amount of stress, determined by the force per unit length (e.g., lb/in.), that is necessary to stretch the material a given percentage [5].

Figure 12.1. The General Features of an Elastomer Closure Used in the Lyophilization Process

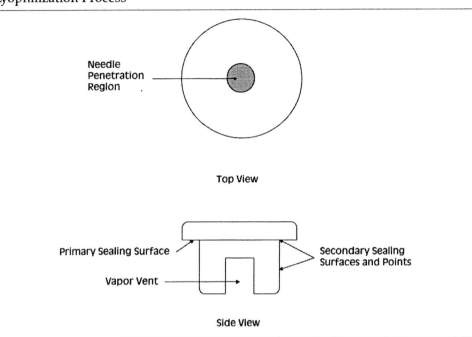

Compression Set

The *compression set* of an elastomer material is defined as the resistance of the substance to permanent deformation. This property of the elastomer material is measured in terms of the percentage of the material that fails to return to its original dimensions after being subject to a given compression load and temperature for a defined period of time. An elastomer material having a low compression set (< 2%) means that upon release of the compression load, the elastomer material returned to a relatively high percentage of its original dimensions [5]. Elastomer materials exhibiting a high compression set would represent a poor choice for the seal requiring a crimp seal because the resulting compression set would reduce the force exerted by the closure on the rim of the glass container.

Durometer Hardness

The mechanical resistance of an elastomer material to indentation is referred to as the *durometer hardness*. This property of the material is a function of its modulus and visoelastic behavior [5]. The hardness of the material is ranked on a scale from 0 to 100. A value of 0 denotes no resistance to indentation; a value of 100 means that the material offers complete resistance to indentation. The durometer hardness of an elastomer formulation can be modified by the addition of reinforcing agents or coating with polypropylene or Teflon® [5]. The durometer hardness for natural rubber is about 40; the value of polypropylene is on the order of 80. It is important that the hardness value be low enough to allow the closure to conform to any surface imperfections associated with the sealing surface of the glass container. In addition, as the hardness of the elastomer material is increased, so is the tendency for coring to occur upon the insertion of a syringe needle [5]. Coring will be addressed in a later segment of this chapter.

Glass Transition Temperature

The *glass transition temperature* of a closure is that temperature at which the material loses it elastic properties and becomes hard and brittle [3]. Most of us have witnessed a demonstration in which a rubber ball is first bounced to demonstrate its elasticity and then shattered after being immersed in a bath of liquid nitrogen. The glass transition of an elastomer material is dependent on the nature of the polymer; butyl polymers will have glass transition temperatures from -70°C to -80°C, while the polymer butadiene-acrylonitrile has glass transition temperature ranges between -25°C to -30°C [3].

Vapor Transport Properties

Since an important aspect of the closure is to provide a seal with the container to protect the lyophilized product from the surrounding environment, its physical configuration must also take into account the transport of water vapor from the container to the drying chamber. The configuration of the closure should not impede the flow of vapors from the container. Any impediment to the flow of vapors from the container will prolong the drying process. The following is a discussion of the gas flow under near molecular and viscous conditions from a 20 mm closure.

   A common configuration of a vapor vent for a 20 mm closure is illustrated in Figure 12.1. The typical vent for the latter closure has a width of 5 mm and height of about 6 mm. When positioned on the container for the lyophilization process, the vent dimensions for the transport of water vapor would be about 5 mm × 3 mm or an area of 15 mm². With the molecular diameter of the water molecule at 4.7×10^{-8} cm and the mean free path at 1 mTorr being 3.4 cm [7], the nature of the gas flow from the closure, based on the Knudsen Number (K.N.) (see Chapter 11) will be molecular at pressures less than 10 mTorr and viscous in nature only at vial pressures equal or greater than 1,300 mTorr. At a pressure of 50 mTorr, the K.N. will be 0.272 cm, which would signify that the nature of the gas flow would be more molecular than viscous (i.e., K.N. is < 0.01). With the assumption that the gas flow is molecular, the gas flow (Q_m) from the closure at a vial pressure (P_1) of 50 mTorr and temperature of 227K into a chamber at a pressure (P_2) of 20 mTorr can be approximated from the following expression:

$$Q_m = 3.64 KA \sqrt{\frac{T}{M}} (P_1 - P_2) \quad \frac{\text{mTorr liters}}{\text{sec}} \tag{1}$$

where $K = 1$ when the radius (a) and length (l) of the vent are equal and A is the cross-sectional area of the vent [8]. The value of Q_m is approximately 116 mTorr liters/sec or 1.48×10^{-4} g/sec of water vapor. Under the above flow conditions, the time required to sublimate 1 g of water would be about 1 hr. The latter assumes equal gas flow from both vents of the closure.

Under viscous flow conditions, when the pressure in the vial is about 1,300 mTorr and the chamber pressure is much lower than the vial pressure, the gas flow (Q_v) can be approximated from [8]

$$Q_v = 37a^2 P_1 \quad \frac{\text{mTorr liters}}{\text{sec}} \qquad (2)$$

where P_1 is the pressure in the vial at a temperature of about 259K and is greater than the chamber pressure. The Q_v for the closure vent would be 3,006 mTorr liters/sec or 3.33×10^{-3} g/sec. If the gas flow is equal for both the closure vents, the time required to sublimate 1 g of water would be 2.5 min.

From the above discussion, it appears that, at least for the 20 mm closure illustrated in Figure 12.1, the gas flow through the closure, depending on the chamber pressure, does not represent a limiting factor for the rate of the primary drying process. Pikal et al. [9] determined the sublimation rate of ice through 20 mm and 13 mm closures and found that the 13 mm closure had a resistance to gas flow 5 times higher than that of the 20 mm closure. I believe that the results reported by Pikal et al. [9] to mean that, for a given container pressure, the gas flow from a 20 mm closure will be 5 times higher than that of a 13 mm closure. It is my opinion that in describing the above system, it is more meaningful to express the results in terms of gas flow (mass per unit time) rather than in an obscure term like resistance to flow.

Diffusion and Permeation

The base polymer used in the closure formulation will govern not only the closure's application and physical properties, as described above, but also its interaction with gases. Since one of the primary functions of the closure is to protect the lyophilized product from the environment, we must be concerned about the transport of vapors through the elastomer material, which may be harmful to the quality or stability of the product. However, before considering the permeation of a gas through a closure, it is first important to understand the diffusion process.

Diffusion

When a gas of a given species interacts with the surface of an elastomer material at a given temperature, some of the adsorbed gas molecules will remain on the surface long enough to be absorbed. Some of the absorbed gas will proceed further into the elastomer materials, while other gas molecules may again reach the elastomer surface and desorb back into the gas phase. When the number of gas molecules that are desorbed from the surface equals the number that are absorbed, a state of equilibrium exists between the gas molecules in the gas phase and the absorbed phase [10]. Under equilibrium conditions, the concentration of the absorbed gas will be uniform throughout the elastomer material. The gas moves throughout the elastomer material by *diffusion*.

The diffusion of a gas in a substance is a result of concentration gradients in the elastomer material. As long as these concentration gradients exist, gas molecules will tend to migrate from the region of high concentration to that of a lower concentration. It may be helpful to the reader to understand that the diffusion process is analogous to gas flow. Gas molecules in a chamber are, according to the Kinetic

Theory of Gases (see Chapter 11), in constant motion; however, gas flow will occur only when a pressure gradient exists. The absorbed gas molecules in an elastomer material will also be in constant motion, but there will be no gas diffusion unless there is also a concentration gradient. For example, if the pressure of the gas above the surface of the elastomer were to be changed (e.g., increased), then the concentration of the gas at the surface will change. An increase in the surface concentration of the gas produces a concentration gradient of the gas in the elastomer material, and the diffusion process will commence. The mass (m) of gas that will diffuse through an elastomer material, having a cross-section A, from a layer d_1 having a concentration C_1 to a region d_2 that has a concentration C_2 in a time (t) can be expressed as

$$m_d = \Delta A \frac{(C_1 - C_2)t}{d_1 - d_2} \quad \text{mass} \tag{3}$$

where Δ is the *diffusion coefficient* and is dependent on the temperature and the chemical and physical properties of both the gas and the elastomer material.

Permeation

Gas permeation is a process in which gas is transported from one side of a material such as a closure to the other. The permeation of a gas through an elastomer material at a temperature T, a thickness (d), and a cross-sectional area A involves a number of processes [10]. At a pressure P_1, the gas in contact with side d_1 of the elastomer is adsorbed on the surface. A portion of this adsorbed gas is then absorbed, as a result of a *solubility coefficient* (S), at the gas-surface interface to create a region in the closure with a gas concentration C_1. With the formation of the concentration gradient of the gas, gas will diffuse through the elastomer material to side d_2, where the concentration of the absorbed gas at the gas-surface interface is C_2. If the rate at which the gas is desorbed from surface d_2 is greater than the number of gas molecules that strike the surface from pressure P_2 of the gas at the gas-surface interface, then the result is the transport of water vapor through the closure.

The fundamental expression for the permeation (q) of a gas through an elastomer material, in terms of the diffusion coefficient (Δ) and the solubility coefficient (S), is given as

$$q = \frac{\Delta S A(P_1 - P_2)t}{d} \quad \frac{\text{atm cm}^3}{\text{cm}^2} \left(\frac{\text{mass}}{\text{unit area}} \right) \tag{4}$$

where $C_1 > C_2$ and $P_1 > P_2$.

In determining the permeability for a given gas through an elastomer material maintained at a given temperature, it is difficult to ascertain the terms Δ and S for a given time (t). As a result, these three terms are combined into a single term k_p, which is known as the *permeation constant* and has the units of sccm/(cm^2 atm sec) where the term *sccm* means that the units for the volume of the gas are given in cm^3 at standard temperature and pressure (STP) [8]. By substituting k_p into equation 4, one obtains an expression for the rate of permeation (Q_p) of a gas through an elastomer material:

$$Q_p = \frac{k_p(T)A(P_1 - P_2)}{d} \quad \frac{\text{atm cm}^3}{\text{cm}^2 \text{sec}} \tag{5}$$

where $k_p(T)$ represents the permeation constant for a given gas and elastomer material at a temperature T [1]. In examining equation 5, the reader should make special note of the following:

- *Effect of the Closure Formulation.* For a given range of temperatures, the permeability of gases through a closure will be dependent on the composition of the closure [1]. This statement is borne out by Table 12.1, which lists the permeation constants (k_p) for various materials at temperatures ranging from 20°C to 30°C [11]. This list shows that butyl has a lower permeability for water vapor than all of the other materials with the exception of viton and Teflon®.

- *Temperature Dependence of k_p.* For a given gas and closure composition (see Table 12.1), the value of k_p will increase with temperature. The increase in k_p is a direct result of the increase in the solubility of the gas and the diffusion coefficient of the gas. In essence, the concentration of the gas in the closure is increased along with the mobility of the gas molecules with respect to any concentration gradient.

- *Effect of the Pressure Differential $(P_1 - P_2)$.* The permeation of a gas through a closure will be directly related to the pressure differential $(P_1 - P_2)$. For a given temperature, the permeation of the gas will continue as long as the pressure differential exists and will cease only when $P_1 = P_2$.

- *Selection of the Closure Formulation.* From the above, it should be apparent that one must exercise some caution in selecting the composition of a

Table 12.1. Permeation Constants $(k_p \times 10^8)$ for Helium, Oxygen, and Water Vapor for Various Elastomer Materials at Temperatures Ranging from 20°C to 30°C (sccm sec^{-1} cm^{-2} atm^{-1}) [11]

Elastomer	Helium	Oxygen	Water Vapor
Viton	9–16	1.0–1.1	40
Buna N	5.2–6	0.7–6.0	760
Buna S	1.8	1.3	1800
Neoprene	10–11	3–4	1400
Butyl	5.2–8	1.0–1.3	30–150
Polyethane	—	1.1–3.6	260–9,500
Proply	—	20	—
Silicone	—	76–460	8,000
Kel-F	—	0.2–0.7	—
Polymide	1.9	0.1	—
Teflon®	—	0.04	27

[Source: Reprinted with permission from R. N. Peacock, *Journal of Vacuum Science and Technology*, Vol. 17, copyright 1980. American Vacuum Society.]

closure material; for example, if the permeability of oxygen is the major concern, then one should (based on Table 12.1) select closures whose main polymer is Buna N, Buna S, Neoprene, or butyl. However, the selection is limited to butyl only if the permeation of water vapor is a key issue. It would be prudent to check with the closure manufacturer regarding the permeation constant $[k_p(T)]$ for gases that may affect the quality or the stability of the lyophilized product under various storage conditions.

Outgassing

Outgassing is the release rather than the absorption of gases from a material [12]. The release of the gas will occur when there is an absence of equilibrium between the elastomer surface and the partial pressure of the gas species at the gas-elastomer interface. As long as the number of gas molecules leaving the elastomer exceeds the number being absorbed, gas will leave the elastomer material, and concentration of the gas in the closure will decrease. The rate of outgassing (Q_o), at any one instant, can be expressed as

$$Q_o = \frac{k_p(T)A(P_s - P_1)}{d_x} \quad \frac{\text{atm cm}^3}{\text{cm}^2\text{sec}} \tag{6}$$

where P_s is the pressure of the gas species exerted by the closure material at the gas-elastomer interface and d_x is the distance from the gas-elastomer interface where the concentration of the gas species is C_x. The value of C_x will be greater than the gas concentration at C_1 or C_2. As the value of C_x approaches C_1 and C_2, $P_s \rightarrow P_1$ and $Q_o = 0$. Since k_p is temperature dependent, the value of Q_o, for a given system pressure P_1, will vary inversely with time. As the temperature is elevated, the initial outgassing rate will increase, and the time required for P_1 to approach P_{1s} will decrease. An important consequence of equation 6 is that gas permeation of a given gas species cannot take place in the presence of outgassing. Only after Q_o approaches 0 can the permeation of the gas species commence.

All elastomer closures, depending on the environmental conditions (temperature and partial pressures of the gas species), will outgas. The composition of gases involved in the outgassing process are water vapor, carbon monoxide or nitrogen, and carbon dioxide [1]. The remainder of this section considers the following topics: (a) distinguishing between the effects of outgassing and gas permeation on lyophilized products, (b) outgassing of water vapor, (c) methods of determining the outgassing of water vapor from closures, (d) means to determine the frequency distribution for the outgassing of water vapor from closures, and (e) the outgassing of other gas species from closures.

Outgassing or Gas Permeation

When there is an increase in the residual moisture content of a lyophilized product, it is sometimes difficult to separate the water that entered the product as a result of outgassing and permeation [1]. One reason for the difficulty is that during testing the closure is often exposed to an elevated temperature and high relative humidity. The elevated temperature favors both outgassing and permeation of a given gas species. Since both processes will be dependent on the partial pressure of the gas species in the container, an initial increase followed by a decrease in the partial pressure would be consistent with both mechanisms of the water vapor transport for an elastomer material. An increase in the partial pressure of a gas in the container, resulting from permeation or outgassing, can be distinguished during storage of the product at a given temperature by using different atmospheric environments. For example, if the gas species were water, some sealed containers would be stored in a controlled relative humidity environment, while others would be stored in a dry environment. An increase in the moisture content of a lyophilized product over a period of time in the dry environment would have to stem entirely from the outgassing of water vapor from the closure. Deducting the moisture increase obtained from outgassing from the increase obtained from those vials stored at a higher relative humidity is a measure of the moisture associated with permeation of the closure.

Outgassing of Water Vapor

Closures for injectable lyophilized products must be sterilized. One current method of sterilization is to autoclave the closures at 121°C and 15 psig for some specified period of time. In order to achieve sterilization, the temperature of all of the closures must reach 121°C in a wet steam environment. Under these conditions, the moisture content in the closure will reach a state of equilibrium with the steam. Upon cooling of the closures, the permeability constant for water vapor [$k_{p,w}(T)$] will be lowered before the closures can achieve the equilibrium moisture concentrations that were present prior to the sterilization process. The result is that the sterilized closures contain a relatively high concentration of water vapor.

Various methods have been used to dry the sterilized closures prior to their use in the lyophilization process. One method was to vacuum dry siliconized closures for 1 hr immediately after the autoclave sterilization at 122°C to 125°C (closures sterilized and dried by this means were identified as <<SV1>>) [13]), while other closures were dried under vacuum, without any additional heating, for 8 hr [13]. As equation 6 showed Q_o to be dependent on the pressure of the drying chamber and $k_{p,w}(T)$, it is unfortunate that Pikal and Shah [13] did not specify the actual pressures and temperatures used in the drying process. Without knowing the temperatures and pressures, other investigators cannot duplicate their results. It has been my experience that the moisture content in the closures can be reduced by placing the closures in a dry-heat gas circulating oven at temperatures ranging from 100°C to 110°C for no less than 4 hr.

Closures that have been steam sterilized can be dried in plastic bags containing a filter that will allow the passage of steam but acts as a barrier to microorganisms. The plastic portion of the bag presents a special problem because the closures that are adjacent to the plastic portion of the bag will take longer to dry than those in contact with the filter; consequently, the closures in the bag may contain a relatively wide range of moisture values.

Determining the Outgassing of Water Vapor

The following describes two methods that have been used to determine the rate of outgassing of moisture from closures. The first method determines the presence of closure outgassing by measuring the increase in moisture of a substance over a period of time. A measure of the rate of increase in the partial pressure of water over a period of time with a residual gas analysis (RGA) system is the second means for determining the average rate of outgassing from closures. Based on the results of the first and second methods, I will propose a third method for ascertaining the frequency distribution of moisture in the closures.

Increase in Moisture of Moxalactam Disodium. Pikal and Shah, using the above described <<SV1>> method for sterilizing and drying the closures [13], determined the outgassing of butyl closures by measuring the moisture content in vials containing 250 mg and 1,000 mg of moxalactam disodium at 25°C. The authors report that during the first 6 months, there was an increase in the moisture content of the moxalactam disodium, as shown in Figure 12.2. The results in Figure 12.2 show that after 6 months the moisture content in the moxalactam disodium remained relatively constant, and it was concluded that the water vapor in the closure and that in the container had reached a state of equilibrium. Based on these results, the authors concluded that the moisture must have stemmed from outgassing of the closures and did not result from moisture transmission *through the stopper* [13].

With the desire to benefit those who wish to determine the rate of outgassing of water using the above method, I would like to make the following comments regarding these results.

Figure 12.2. The Increase in Moisture of Moxalactam Disodium During Storage at 25°C

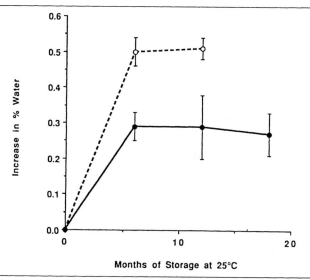

Vials are 20 mm finish vials of 10 mL volume. Rubber stock #1816 butyl from West Co. Stopper treatment was <<SV1>>. Data represent the mean of 6 lots (250 mg dose) or 8 lots (1,000 mg dose). The mean initial water content was 0.68%. Key O = 250 mg; ● = 1,000 mg dose [13]. [Source: *Developments in Biological Standardization*, Vol. 74, Page 166, Figure 1. Reprinted with permission.]

While Pikal and Shah [13] may be correct that moisture was a result of out-gassing and not due to permeation of water vapor through the closure, the fact the moisture content became constant after a period of time is, however, not conclusive proof that the permeation of water vapor through the stopper was not the source of the increased moisture in the moxalactam disodium. Both equations 5 and 6 will give constant moisture values of residual moisture in a material when equilibrium is established between the moisture in the closure and that in the product container. One factor that does support their conclusion [13] that the moisture stems from out-gassing is the relatively low $k_{p,w}(T)$ values (see Table 12.1) for the butyl closure for-mulation. However, with such a low $k_{p,w}(T)$ value for butyl closures, one would *not* expect such a major increase in the moisture content during the first six months. These results would have been more conclusive had Pikal and Shah provided data taken under conditions favorable and not favorable for water vapor permeation (i.e., under low and high relative humidity conditions).

A second point that I would like to make is that by connecting the initial mois-ture value with the first moisture value taken after six months, Pikal and Shah [13] are implying that the increase in moisture content is a linear function of time during this period. Since the moisture values remain relatively constant for storage times greater than six months, one is uncertain just when equilibrium between the mois-ture in the closure and the moxalactam disodium was established. Additional mois-ture data taken during the first six months would have confirmed either that the increase in moisture was linear or that it was nonlinear with respect to time. This in-formation is important because it would reveal how quickly equilibrium was estab-lished when the source of the water results from outgassing.

Outgassing Using Residual Gas Analysis. For a number of years, I determined the rate of outgassing of closures at a given temperature by measuring the partial pres-sure of water vapor over a period of time. The measurements are made at subat-mospheric pressures when no equilibrium exists between the closures and the partial pressure of water vapor in the chamber. Under these test conditions, any ef-fect resulting from the permeation of water vapor is eliminated (see equation 5) be-cause $P_1 = P_2$ and, therefore, $Q_p = 0$. A general descriptive outline of the procedure follows.

Outgassing of the Freeze-Drying System. A clean and dry FTS Systems freeze-dryer having a total effective volume (total system volume less the volume of any hard-ware including two trays) of about 60 L and at a given shelf temperature is evacu-ated to less than 100 mTorr, with the condenser at ambient temperature. Dry nitrogen gas is then bled into the system while the system is still being evacuated, so that the output of an MKS capacitance manometer gauge (supplied by MKS In-struments, Inc., Andover, Mass.) in the chamber is maintained at 2,000 mTorr. The *purging* or *conditioning* of the dryer is continued for a period of no less than 12 hr.

The gas bleed is turned off, and the pressure in the freeze-dryer chamber is evacuated to less than 100 mTorr. Using the technique for determining residual gases described in Chapter 11, a mass spectrum of the system and the pressure of the dryer are recorded. The evacuation of the dryer is then terminated.

Mass spectrums of the freeze-dryer at various chamber pressures are recorded periodically over a period of no less than 4 hr. An example of the results of a typical mass spectrum of the clean, dry, empty chamber is shown in Figure 12.3. This figure shows the partial pressures of water vapor, nitrogen, and *other* gases in the chamber as a function of time.

Outgassing of the Freeze-Dryer and the Closures. A number of closures (250 to 1,000) are placed in the trays, and the freeze-dryer is again evacuated to 100 mTorr, and the system is purged with dry nitrogen gas for a period of no less than 12 hr with the condenser at ambient temperature.

   The purging of the freeze-dryer containing the closures is terminated, and the pressure in the chamber is evacuated to < 100 mTorr. The total chamber pressure and the partial pressures of water vapor, nitrogen, and *other* gases (e.g., oxygen and argon) are recorded. The partial pressures of water vapor, nitrogen, and *other* gases in a freeze-dryer containing 324 closures is shown in Figure 12.4.

Determination of the Average Outgassing Rate of Water Vapor of Closures. The following is a brief outline of the steps that are taken to determine the average rate of outgassing of closures at a given temperature.

   The total outgassing rate of water vapor $Q_{w,t}$ from N number of closures at a temperature T is determined from the following expression:

$$Q_{w,t} = \frac{ROR_{w,c} - ROR_{w,b}}{V_e} \quad \frac{mTorr\ liters}{hr} \tag{7}$$

where $ROR_{w,c}$ is the rate of pressure rise for the partial pressure of water vapor with N number of closures at a temperature T in the system having an effective volume

---

Figure 12.3. The Partial Pressure of Water Vapor in an Empty Freeze-Dryer

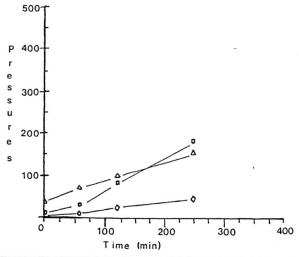

The partial pressure of water vapor (denoted by □), nitrogen (denoted by △), and "other" gases (denoted by ◇) in a clean, empty dryer at ambient temperature, as a function of time.

Figure 12.4. The Partial Pressure of Water Vapor in a Freeze-Dryer Containing
325 Closures

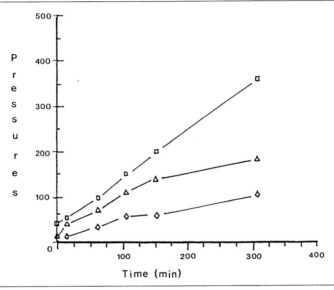

The partial pressure of water vapor (denoted by □), nitrogen (denoted by △), and
"other" gases (denoted by ◇) in a freeze-dryer containing 325 closures at ambient temperature, as a function of time.

of V_e and $ROR_{w,b}$ is the rate of pressure rise for a clean, dry, empty, vacuum system
at a temperature (T) having the same V_e. It should be noted that, because of the relatively large V_e of the system, the value of V_e was not corrected for the volume occupied by the closures. For those vacuum systems where the volume of the closures
with respect to V_e is significant, then the corrected effective volume V_e' (in L) will be
expressed as

$$V_e' = V_e - V_c \tag{8}$$

where V_c is the volume occupied by the closures.

The average outgassing rate of the closures ($Q_{w,avg}$) is defined as

$$Q_{w,avg} = \frac{Q_{w,t}}{5N} \quad \frac{mTorr\ liters}{hr} \tag{9}$$

where the integer 5 stems from the assumption that 20% of the surface of the closure
is exposed to the contents of the vial.

From the ideal gas law, the average outgassing rate of the closures can be
expressed in terms of mass of water per unit time ($Q_{m,avg}$) as

$$Q_{m,avg} = 9.0 \times 10^{-7} Q_{w,avg} \quad g/hr \tag{10}$$

The time (in hr) required to increase the moisture content of 1 mg of lyophilized material by 1% wt/wt, assuming that the *sticking coefficient* (S_c) approaches 1, is obtained from

$$t = \frac{S_c \times 10^{-5}}{9.0 \times 10^{-7} Q_{w,avg}} \tag{11}$$

where S_c is defined as the ratio of the number molecules of a given species that remain on the surface to the number of molecules that strike the surface in a given time period [8].

When the above method for determining the outgassing of closures is used, the results can vary depending on the sterilization process and the temperature of the closure. For closures steam autoclaved in a bag, the average time required to add 1% moisture to 1 mg of material ($S_c \to 1$) at 4°C to 8°C could require as long as 20 years. Increasing the temperature of the closure to between 20°C and 25°C tends to reduce the time to between 18 months and 2 years. However, at temperatures near 40°C, the outgassing could decrease the time to about 10 hr. These results indicate that the rate of outgassing increases dramatically with temperature, and the results of stability testing at elevated temperatures ($\geq 40°C$) could result in erroneous results (e.g., the testing of the product could be performed at moisture values that exceed product specifications). Stability studies, under the latter conditions, would not be examining the stability of the lyophilized product but the effect that the combination of temperature and moisture from the closure would have on the product.

One would still have to exercise caution even if the results of the outgassing at 40°C indicated that the residual moisture in the product would not exceed product specifications. The reason for such caution stems from the results representing the average outgassing rate of the closures but not providing information about the nature of the frequency distribution for the outgassing from the closures. If the frequency distribution were sharp and rather well defined, then the probability that a closure would outgas at a rate that would cause the moisture in the product to exceed the specified limits would be beyond normal chance. However, if the frequency distribution was broad, which may occur when sterilization is performed in a bag, then there could be closures present that would cause the residual moisture in the product to exceed product specifications at elevated temperatures. If the sample size that is set aside for stability testing is small, then the presence of such closures may not become apparent during stability testing but become apparent when qualification batches of the product are made to satisfy regulatory requirements. The next section will consider the means for determining the frequency distribution for the outgassing rate for elastomer closures.

Frequency Distribution for the Outgassing of Water Vapor

It is my opinion that it is very important to know the frequency distribution of the outgassing rate of the closures. From such information, one could determine the probability of increasing the residual moisture content of a product to its upper limit during normal storage or during accelerated stability studies of the lyophilized product. Such information would enable a manufacturer to adjust the process for drying the closures so that the probability that a closure will have unacceptable outgassing properties would be very small. It has already been shown that the previous method for determining the average outgassing rate of the closures is unsuitable for determining such a frequency distribution.

The method described by Pikal and Shah [13] offers a possible means for establishing the frequency distribution for the outgassing of water vapor from a batch of closures. The reported moisture data was the mean value of 6 to 8 lots. As has been repeatedly stressed throughout this text, a minimum of 25 to 30 measurements of a property, such as the outgassing of water vapor from a closure, is required for the results to be considered statistically significant. However, it would appear that the frequency distribution of the outgassing of the closures could be determined by the method described by Pikal and Shah [13] if the lot size for the measurement were increased and the following test criteria are established: (1) The testing must be done under conditions that eliminate the possibility of the transport of water vapor into the vial by the permeation of water vapor through the closure or from real leaks in the container-closure seal. (2) The hygroscopic nature of the moxalactam disodium is sufficient that $S_c \rightarrow 1$ (i.e., the partial pressure of water vapor in the headspace of the vial would approach 0).

Outgassing of Other Gas Species from Closures

It is possible that the outgassing from the closure may involve gases other than water vapor [1,14]. Pikal and Lang [14] showed that closures could be the source of sulfur and hydrocarbon vapors. Such vapors from the closures were shown to be the source of haze or turbidity in the reconstituted lyophilized product. Although Pikal and Lang were able to demonstrate that the haze or turbidity stemmed from the closures, they were unable to determine the presence of such gases in the dryer.

Leebron and Jennings [1] were able to detect such vapors in the gas phase by means of a RGA using a quadrupole mass spectrometer (see Figure 11.11). The apparatus used in this analysis is shown in Figure 12.5. The apparatus consists of a glass vacuum chamber into which a tube containing a number of elastomer closures is placed. The pressure in the chamber was determined by means of a thermocouple gauge, and the system was evacuated by a mechanical pump. A liquid nitrogen trap was positioned between the vacuum pump and the apparatus to prevent any harmful contamination of the pump from any vapors from the closures. The trap also served to prevent the backstreaming of any hydrocarbon vapors from the pump from entering the test apparatus and interfering with the test results. A connection to the differentially pumped mass spectrometer was made by means of a variable leak valve. The variable leak valve was used to control the leak rate into the mass spectrometer so as to maintain its pressure at 1×10^{-6} Torr. The leak rate for the clean, dry test apparatus shown in Figure 12.5 was about 1 mTorr/min [1].

Before placing the closures into the test apparatus, the clean, dry apparatus was first evacuated, and a mass spectrum of the residual gases was recorded. The results of the mass spectrum of the empty apparatus are shown in Figure 12.6. The mass spectrum shows that the main constituents of the gases in the apparatus were water vapor (m/e [mass to charge ratio] = 18), nitrogen (m/e = 28), oxygen (m/e = 32), and argon (m/e = 40). At the highest sensitivity, the mass spectrum did indicate the presence of carbon dioxide (m/e = 44). There were no significant peaks having m/e > 45 to indicate the presence of any significant amounts of hydrocarbon vapors in the test apparatus [1]. The presence of any additional peaks with m/e > 45, when

Figure 12.5. The Apparatus Used for Determining the Composition of Gases
Outgassing from a Number of Closures

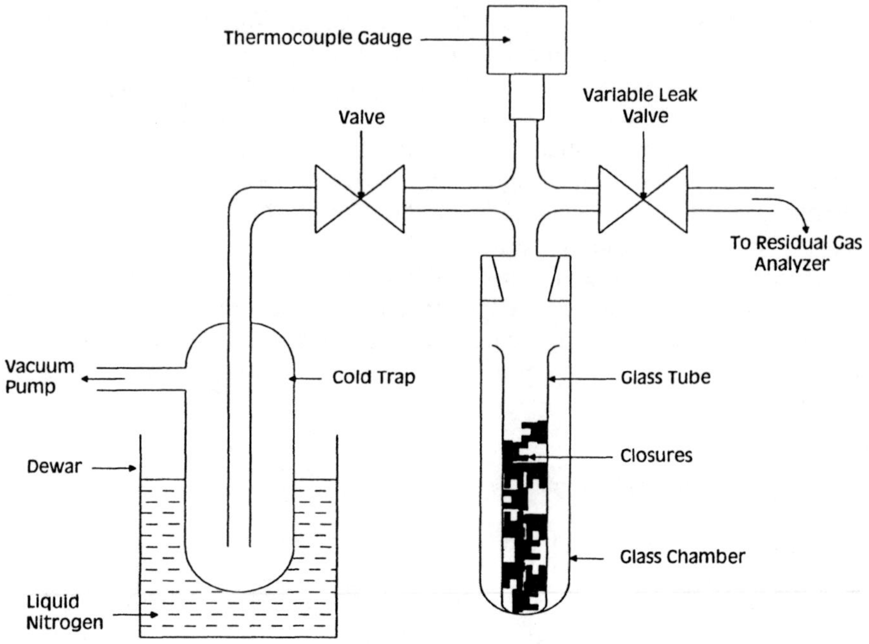

the closures are placed in the apparatus, must be attributed to the outgassing of the closures.

Eleven closures were placed in the apparatus (see Figure 12.5), and the pressure was reduced to 15 mTorr. The closures and the test apparatus were at ambient temperature. After 1 hr at a pressure of about 15 mTorr, a mass spectrum was taken of the gases in the apparatus. The mass spectrum, shown in Figure 12.7, indicated the presence of two additional groups of peaks at m/e = 54 to 57 and m/e = 68 to 70. Because these peaks were not observed in the clean, dry, empty chamber, they were associated with outgassing of the closures. A pressure rise test indicated a ROR of 5 mTorr/min and would indicate that there was some outgassing of the closures.

The closures were allowed to remain overnight in the apparatus without evacuation by the vacuum pump. The closures were removed, the chamber was evacuated to a pressure of 20 mTorr, and a mass spectrum, as shown in Figure 12.8, was recorded at a sensitivity of 10^{-10}. This mass spectrum of the gases in the apparatus showed a significant increase in the intensity of the groups of peaks at m/e = 54 to 57 and m/e = 68 to 70. By employing frequent bakeouts (increasing the temperature to > 125°C) of the mass spectrometer system, the spectrum of the analyzer was kept free of any contamination of the vapors from the closures. Thus, the persistence of these latter groups of peaks would indicate that they were a result of desorbed gases from the walls of the apparatus. Efforts to clean the apparatus with common solvent

Figure 12.6. Mass Spectrum of Gases from an Empty Test Apparatus

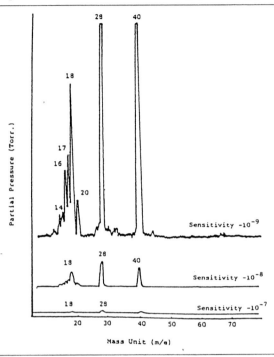

Test apparatus pressure 20 mTorr, mass analyzing system pressure 1×10^{-6} Torr, and sensitivities of 10^{-7}, 10^{-8}, and 10^{-9}. (Note that units of the ordinate axis of the plot should be in A, not Torr.) [Source: *PDA Journal of Pharmaceutical Sciences*, Vol. 35, No. 3, May/June 1981, Page 101, Figure 2. Reprinted with permission.]

proved futile. The glass portion of the apparatus could be cleaned only when subjected to hot concentrated nitric acid (HNO_3) that contained a few drops of hydrofluoric acid (HF). The thermocouple gauge (see Figure 12.5) could not be cleaned and had to be replaced.

The outgassing of a tenacious transparent film from the closures was demonstrated by placing 480 vials in a Stokes 24 ft^2 dryer (supplied by Stokes Vacuum, Inc., Philadelphia, Penn.). Each of the vials contained an elastomer closure from the same batch of closures used in the test chamber shown in Figure 12.5. The shelves of the dryer were maintained at 22°C ± 2°C, and the chamber pressure was lowered to 10 mTorr. The closures were maintained in the dryer under these conditions for 24 hr and then removed. Vials that were removed from the drying chamber did not exhibit a meniscus when distilled water was added to the vial. Vials that were not exposed to the closures at low pressures did exhibit a meniscus with distilled water. The lack of a meniscus indicates that there was no adhesion between the water and the glass. It was concluded that the absence of a meniscus is an indication of the presence of an intermediate film between the glass surface and the water [1]. The reader should

Figure 12.7. Mass Spectrum of the Gases from the Empty Test Chamber

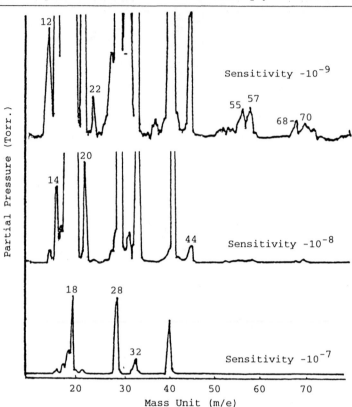

Test apparatus pressure 15 mTorr, mass analyzer system pressure 1×10^{-6} Torr, and sensitivity of 10^{-9}. (Note that units of the ordinate axis of the plot should be in A, not Torr.) [Source: *PDA Journal of Pharmaceutical Sciences*, Vol. 35, No. 3, May/June 1981, Page 102, Figure 4. Reprinted with permission.]

note that the lack of a meniscus is a simple but effective means for determining the presence of a nonpolar film on a surface. One will also observe the lack of a meniscus with water when silicone oil is present on a glass surface.

The above has shown that determining the outgassing properties of closures is an important aspect of the lyophilization process. Knowing the outgassing properties can provide assurances that moisture values in the product will not increase beyond specifications during storage of the product. In addition, quality control testing of the outgassing properties of incoming closures can avoid the loss of batches due to haze or turbidity.

Figure 12.8. Mass Spectrum of the Gases from the Empty Test Chamber

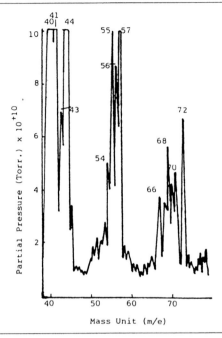

Test apparatus pressure 20 mTorr, mass analyzer system pressure 1×10^{-6} Torr, and sensitivity of 10^{-10}. (Note that units of the ordinate axis of the plot should be in A, not Torr.) [Source: *PDA Journal of Pharmaceutical Sciences*, Vol. 35, No. 3, May/June 1981, Page 102, Figure 5. Reprinted with permission.]

Coring

Coring is the production of closure fragments as a result of the insertion of a needle. Such fragments may be located in the tube of the needle; the injection of the diluent into the closure can also lead to the addition of closure fragments to the product. These fragments of the closure would represent a foreign particulate matter in the reconstituted product and, thereby, act as a source of contamination. Preston [4] expressed concern that if coring is indeed considered a life-threatening situation, then the acceptable level for coring should be zero. He further states that *A generally acceptable pass/fail limit in the pharmaceutical industry is a 10% coring rate and no one seems to understand why it is considered to be appropriate.* The following is a brief discussion of the conditions under which coring can occur and a means for testing the coring tendency of a closure.

Conditions for the Coring of Closures

The potential of coring increases as the temperature of the closure reaches its glass transition temperature [3], which was considered earlier in this chapter. It was shown that the temperature at which the closure loses its elastic properties is strongly dependent on the composition of elastomer formulation. For butyl polymer-based closures with glass transition temperatures ranging from –70°C to –80°C, coring as a result of the glass transition temperature should not be a major concern. However, a product stored in a freezer and having a closure whose polymer is butadiene-acrylonitrile with a glass transition temperature of –25°C to –30°C [3] should be allowed to warm to ambient temperature before inserting the needle.

Multidosage products also represent a potential coring problem. Preston [4] found that the tendency for coring increased with the number of punctures. It seems logical that the potential for coring as a result of multiple insertions of the needle would be, for a given elastomer composition, greater for a 13 mm closure than for 20 mm closure.

I have found that closures sealed at low pressures (< 500 mTorr) tend to form, especially for the 20 mm closure, an indentation in the middle of the closure where the needle will be inserted. The stress on the closure will be removed once the pressure in the vial is increased as a result of reconstituting the product. What is not known is the effect that the stressed elastomer closure will have on the tendency of the closure to produce fragments.

The literature is rather silent on the topic of the coring of closures. Perhaps it is time for a study to be conducted in order to provide guidelines for the manufacturer of the lyophilized product and instructions for the user regarding the proper means for inserting the needle.

Core Testing of the Closure

A means for testing closures for coring is described in PDA Bulletin No. 4. In this method, the needle is inserted into the chuck of a drill press, and the needle is inserted into the closure. With the drill press, the needle is always inserted perpendicular to the surface of the closure. It would be of interest to learn how the results of this method compare with those needle insertions that are performed by hand. The basis for establishing such a comparison is that I have yet to witness a doctor or nurse in an office or hospital setting, either in real life or as dramatized on television, inserting a needle attached to syringe by means of a drill press.

Siliconization of Closures

Siliconization of the closures is performed to prevent clumping of the closures in high-speed filling equipment. The process involves coating the closures with a medical-grade silicone fluid in an appropriate solvent [2]. The adhesion of the silicone fluid to the closure is a physical phenomenon whereby detachment of the silicone fluid from the stopper during product processing can become a source of product contamination. Using the meniscus of distilled water as a means for detecting nonpolar surface films on a glass surface, I have noted on a number of products that the silicone fluid will tend to creep down the walls of the vials. The presence of a silicone layer can be detected by slowly filling the vial with water. At the point where there is no longer any meniscus, there is silicone film on the glass surface.

CONTAINER

The container is certainly an important component of the lyophilization packaging system. At first glance, the function of the container is merely to provide a vessel for the formulation and a surface that will interface with the closure to form an effective seal. But the container plays a far more important role in the lyophilization process. It is the intent of this section of the chapter to examine the various roles that the container plays and its impact on the lyophilization process and the possible effects on the quality and stability of the lyophilized product.

Although some containers used to produce lyophilization are fabricated out of molded plastic, the vast majority of the containers are fabricated from silicone oxide glass. The following discussion will address issues concerning glass vials rather than those related to vials fabricated from polymers.

Historical

The origins of man-made glass go back some 5,000 years. In a fashion similar to that of the lyophilization process, glass making started as an art form, and the composition and method for producing glass were carefully guarded secrets. It was not until the beginning of the 20th century, when optics became an important part of our technical culture, that science and technology were introduced into the glass manufacturing process. As our understanding of the nature of glass has changed, so has its definition. In 1950, the definition was that in a *glassy state, there are uncrystallized materials*, which has given way to *glass is a frozen-in supercooled liquid* [15]. The reader should realize how closely the above definitions for silicone glass resemble our present understanding of the nature to the interstitial region of a frozen formulation; for the lyophilization process, depending on the nature of the formulation, both definitions would appear to be applicable. Unlike the lyophilization process, the composition of silicone glass can be carefully controlled and reproduced. This is unfortunately not necessarily true for the composition of the interstitial region of a frozen matrix, where it appears that nature can produce a number of systems with varying physical and chemical characteristics.

Dissolution and Leaching of Glass

Dissolution

When one thinks of the *dissolution of glass*, the first reaction that comes to mind is that with HF:

$$SiO_2 + 6HF \rightarrow H_2[SiF_6] + 2H_2O \tag{12}$$

in which the complex $H_2[SiF_6]$ is soluble is water [15]. One, however, tends not to think of the solubility of glass in water. Although glass is only slightly soluble, water will react with a glass surface in the following manner:

$$\equiv Si\text{-}O\text{-}Si\equiv + H_2O \rightarrow \equiv Si\text{-}OH + HO\text{-}Si\equiv \tag{13}$$

The solubility of glass increases at higher pH values.

Glass is also dissolved in some organic compounds, especially with those compounds that form complexes that contain hydroxy groups (OH). Compounds commonly found in lyophilized formulations such as citrate, gluconate, and tartrate compounds and some sugars can attack a glass surface [15]. The reader should be aware that such glass complexes will most likely be incorporated in the interstitial region as a constituent of the formulation. The presence of such glass compounds in the interstitial region may or may not affect the thermal properties of the formulation. In order to prevent or limit such contamination, the initial surface layer should be removed prior to the filling operation. Methods of cleaning the vial will be considered later in this section.

Leaching

While dissolution results in the removal of glass from a surface, *leaching* is an ion exchange process in which hydronium ions (H_3O^+) replace alkaline ions such as Na^+ and K^+ in the glass structure. When water reacts with a glass surface, as represented by reaction 13, the first reaction to occur is the leaching of the alkaline ions from the surface. Since these ions are replaced by H_3O^+ ions, the result is that the water will tend to become more alkaline, which leads to the dissolution of the glass [15].

Glass Types

In 1978, glass was classified into five different types based on its ability to resist corrosion or dissolution.

Type I	Glass in this category has a very thin (hydration) surface layer that is subject to the effects of leaching and dissolution. These glasses typically contain the compound B_2O_3 and are generally referred to as borosilicate glass [5,15].
Type II	This type of glass has, below the hydration surface layer, a second layer that contains a reduced quantity of alkali ions, which serves to protect the glass from leaching and to limit corrosion [15]. This type of glass is often referred to as dealkalized soda lime glass [5].
Type III	This glass is the typical soda lime glass and is characterized by having a number of layers that are subject to leaching and dissolution [15].
Type IV	This glass is formed after dissolution, but leaching of the alkali ions can still occur at the surface [15].
Type V	The dissolution of this form of glass is. so rapid that there is no leaching [15].

Cleaning of the Vials

Each of the above types of glass has, at least initially, a surface layer that can interact with the liquid formulation. In order to remove the active layer, one should pretreat the vials or wash the vials in such a manner as to remove the active layer. One possible means is to invert the vial and rinse the vial with a jet of hot ($\geq 80°C$)

Water for Injection. In this manner, the silicates from the active surface will be dissolved and flow out of the vial. The time required for the rinse will depend on the temperature of the water and the type of glass. It is doubtful that one could ever passivate surfaces for glass types III through V; for that reason, these types of glass products would not be good candidates for a lyophilized product.

The Physical Properties of Vials

The physical properties of a vial not only impact the drying process, as shown in Chapter 7, but also influence the effectiveness of the final crimp seal to provide adequate protection against the environment. The following considers the effects of the physical properties of two types of glass vials and ampoules on the lyophilization process and on the quality of the final product.

Effect of Vial Configuration

It was shown in Chapter 7 that the fill height can have a major impact on the nature of the frozen matrix. The fill height for a given fill volume is sometimes dictated by the configuration of the vial. Figure 12.9 illustrates this point by plotting the relative fill height as a function of A_p, where A_p is equal to the ratio of vial height to vial diameter. This plot shows that when $A_p = 1$, the relative fill height is 0.6 and near 1 when $A_p = 2$. However, when A_p became ≥ 3, the relative fill volume increases rapidly with each incremental change in A_p. The significance of this can be seen at an $A_p = 3.75$ when the fill height is 3.14. With this vial configuration, the given fill volume occupies more than 90% of the volume of the vial and represents the upper limit to which A_p can be increased for the given fill volume. At this fill height, freezing would most likely produce a heterogeneous cake, and the time required to complete the primary and secondary drying processes would most likely be significantly prolonged. The reader should understand that the values of A_p and fill height will change with the selected fill volume but the nature of the plot shown in Figure 12.9 will remain the same.

Molded and Tubing Vials

Vials used in the preparation of lyophilized products can be fabricated by two different processes: molding and tubing. This section will consider the construction, advantages, disadvantages, and impact that each type of vial can have on the lyophilization process and the quality of the final product.

Effect of the Configuration of the Base of the Vial. Illustrations of a molded vial and a tubing vial are shown in Figure 12.10. An examination of each vial shows that the base is concave with respect to the surface. The percentage of the cross-sectional area (A) that is in actual contact with the surface is quite small, usually 2% to 5%. It is this small contact area that will, depending on the chamber pressure, supply the necessary refrigeration to freeze the formulation and the energy required for the sublimation process (primary drying) and the desorption of the water vapor from the cake (secondary drying). The variation in the value of A from vial to vial will govern the nature of frequency distribution for the heat-transfer rate coefficient C_o

Figure 12.9. A Plot of Relative Fill Height as a Function of A_p (ratio of vial height to vial diameter) for a Constant Fill Volume

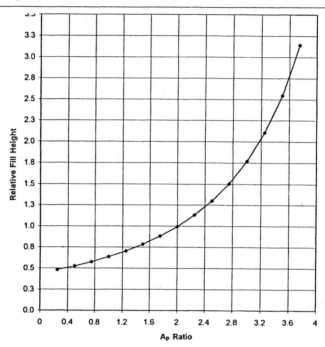

Figure 12.10. Molded and Tubing Vials Used in the Manufacture of Lyophilized Products

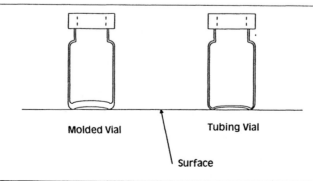

[16]. The temptation to increase the heat-transfer rate by increasing the surface contact area may generate more problems than it solves. Since the description of a glass is a supercooled liquid, attempts to increase the contact area may result only in broadening the C_o frequency distribution and lead to an increase in the number of defective cakes. Partial meltback in cakes may result from vials having a low C_o value or small contact area; complete meltback of cakes may result from vials having a high contact area and a high C_o value.

To qualitatively determine if there is a problem with variation in the contact area of the vial, one needs only to take a number of vials and drop them so they remain upright on a flat horizontal surface. Vials that immediately come to rest on the surface are assumed to make contact with the surface along the entire contact perimeter of the vial. Vials that vibrate for some period of time do so because the vial is making contact at a few points along the contact perimeter; for that reason, the vial will tend to rock back and forth between these points.

Molded vials are generally preferred over tubing vials for vial volumes $\geq$ 50 mL. In order to improve the movement (reduce friction) of these vials along a surface, the bottom surface of such vials often contains knobs or ridges molded along the contact perimeter. While such molded features can reduce the friction between vial and the surface, they can also decrease the C_o of the vials. The effects of reducing C_o have been previously described.

Configuration of the Sealing Surface. A marked difference in the finish of the upper sealing surface between molded and tubing vials was pointed out by Morton and Lordi [17]. The basic difference between the vials is shown in Figure 12.11. These authors use ink blots of the sealing surface to ascertain the contact area with a flat surface. As seen by the single ink-blot area and illustrated by a cross-sectional view of the top flange, the upper flange of the molded vial provided a uniform contact area. The ink blot of the tubing vial showed two distinct rings, which indicates the presence of a small dip in the surface profile for the tubing vial. These results will be of concern when discussing the container-closure seal.

Figure 12.11. Differences in Vial Finish Between Molded and Tubing Vials

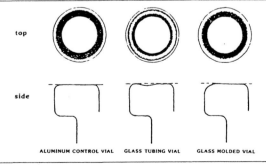

[Source: *PDA Journal of Pharmaceutical Sciences,* Vol. 42, No. 1, Jan./Feb. 1988, Page 28, Figure 11. Reprinted with permission.]

Glass Stress. The breaking of vials that contain a mannitol solution has already been discussed in a previous chapter [18,19]. The breakage of vials with a mannitol solution was found to be associated with a recrystallization phenomenon that commenced at a temperature of about –27°C. For the glass vial to break, the forces resulting in the recrystallization process had to generate mechanical stresses in the glass. If the glass was already stressed, perhaps the stress resulted during the cooling portion of the vial manufacturing process. These stresses in a vial can be observed as interference colors resulting from changes in the refractive index of the glass resulting from stressed regions in the glass. A simple method for viewing such stresses in the glass is to examine the glass under polarized light. The stress will appear as a band of colors in the region of the stress. For large-volume glass-molded vials, stress lines were often observed starting from and normal to the bottom of the vials. Placing such vials, particularly with low fill volumes, onto cold shelves could cause significant temperature gradients across the stress lines and increase the probability of vial breakage.

Glass Ampoules

Glass ampoules, as illustrated by Figure 12.12, are often used as a packaging system for lyophilized products. These containers are generally fabricated out of type I glass [20] and are similar in construction to tubing vials. One notable exception is the replacement of the sealing flange of the tubing vial with a long stem that is used to

---

Figure 12.12. The General Feature of a Glass Ampoule Used in the Lyophilization Process

---

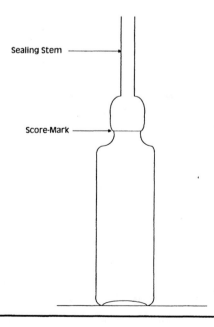

Sealing Stem

Score-Mark

form a glass seal. The vial also contains a score-mark, as seen in Figure 12.12, which provides a means to break the seal and reconstitute the lyophilized product. This feature of the ampoule has always been of concern because of the possible introduction of chemical and biological contamination into the product. In addition, breaking the glass seal will always present the possibility of introducing glass fragments into the container.

The second feature on which the ampoule differs from the tubing vial is that the sealing flange of the latter is replaced by a sealing stem. The sealing stem provides a means for the formation of a glass seal. The nature of the glass seal for ampoules will be considered in the next segment of this chapter. The formation of a glass seal provides the ampoule with a distinct advantage over the serum vial in that it does not require an elastomer closure. The ampoule does not have to contend with the problems that may be associated with the possible contamination of the product as a result of gas permeation or outgassing.

CONTAINER–CLOSURE SYSTEM

The principal objective of the container-closure system is to provide chemical, microbial, and possibly radiation protection—the latter is accomplished by vials containing amber—throughout the shelf life of the lyophilized product when subjected to various process, storage, and handling conditions [17]. The container-closure system consists of three main components: the closure, the container, and the crimp seal. The chemical and physical nature of the closure and container were considered previously because of their direct effect on the lyophilized product.

A detailed discussion of the crimp seal is beyond the scope of this text because the crimp seal does not have direct access to the lyophilized product. While there are several designs of crimp seals (e.g., single piece, multiple piece, and Flip-Off), the principal material used to create the necessary sealing force between the container and the closure is aluminum [5]. The main concern regarding the crimp seal is that the wall thickness of the aluminum is within tolerances to ensure the necessary force is applied to the container-closure system [17].

The concern that will be addressed in the remaining portion of the chapter will be if the seal is adequate to protect the lyophilized product from chemical and microbial contamination [5,17,21,22,23]. In addition, this section will briefly discuss the glass seal of the ampoule [20,24].

Description of the Sealing System

The reader will be provided with a description of what takes place during the crimping operation, which is followed by a discussion of the nature of the container-closure and the ampoule seals.

The Crimping Operation

The crimping operation is a two-step procedure and is applicable only to the container-closure system. The first step is that the crimping instrument must exert a compression force on the closure. If such a compression force is insufficient, then the closure may not make full contact with the sealing surface flange of the vial [5,17,25]. However, when the applied compression force is too high, the closure may be forced into the neck of the vial and be deformed as a concave surface. Since the deformation, and its accompanying stress to the elastomer material, will occur in the region designated for the penetration of the needle, the overcompression of the closure can lead to an increased likelihood of coring and its associated contamination of the lyophilized product [25].

The second force that is applied to the container-closure system is a horizontal force. The function of this second force is to produce a mechanical seal that will maintain the compressive force between the container and the closure [5]. The force is maintained by bending the aluminum sheet so that it contacts the underside of the glass flange of the vial.

The reader should be aware that, once the crimping operation is complete, there will be a gradual change in the compression forces that were exerted during the crimping operation. This decrease in the compression forces can result from *creep* in the aluminum crimp seal. The effect of *creep* of the aluminum seal is generally terminated after about 25 days but has been known to change slowly over a period of a year from the time of the initial crimping operation [25]. The loss in compression can also be attributed to the compression set of the elastomer material, which was described earlier in this chapter. If the elastomer closure has a relatively high compression set, then upon release of the compression force by *creep* associated with the aluminum crimp seal, the elastomer would not expand to maintain the same force between the crimp seal and the flange of the vial. As a result, the compression force on the closure would now be solely dependent on the *creep* properties of the crimp seal. With an elastomer closure fabricated from an elastomer material having a low compression set, the closure will expand and the *creep* of the crimp seal will have less impact on the exerted compression forces.

Container-Closure Seal

The seal of the container-closure system, as shown by Figure 12.13, occurs at three locations [5]. The primary sealing surface is formed by the closure along the surface

Figure 12.13. Sealing Points for the Container-Closure System

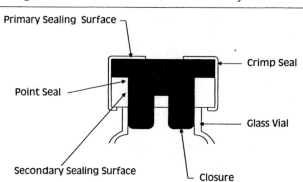

of the glass flange, and the secondary seal occurs between the vertical walls of the vial and the closure. A third seal, as seen in Figure 12.13, is formed between the closure and the vial at the corner defined by the vial flange and the vertical walls [5,25]. Attempts have been made to form the primary seal along the vertical walls of the vial and the closure by making the dimensions of the closure slightly larger in diameter than the neck of the vial. This force fit between the container and closure can result in considerable frictional resistance to the stopping process. In addition, the closures may not fully seat with the vial flange, and the frictional resistance may have a major effect on the compression force necessary to form the crimp seal. Incompatibility between the diameter of the throat of the vial and the closure can cause a convex distortion of the top of the closure. This would add to the *creep* of the aluminum crimp seal, and the stressed elastomer would increase the possibility of coring.

Measurement of Compressive Forces

As described above, the container-closure seal is formed by compressive forces exerted on the container and the closure by the formation of the crimp seal. Since it has been shown that compressive forces on the container-closure seal will change over a period of time [25], as a result of the *creep* of the aluminum, it is of interest to measure such forces. The following describes means for determining such compressive forces.

Twist Method

One of the earlier qualitative methods used to determine the effectiveness of a crimped container-closure seal was to twist the crimp seal. The reliability of such a test method is in question. Connor [26] points out this method is not really measuring the compressive forces of the seal but the frictional forces between the interfaces of the closure and the crimp cap, the closure and the glass, and the crimp seal and the glass.

Seal Force Monitor

The seal force monitor (supplied by West Co., Phoenixville, Penn.) measures the total force necessary to form the crimp seal of a container-closure system and determines the integrity of the seal [26]. The instrument determines the effectiveness of the seal by comparing it with a standard force profile of the applied force exerted to the crimp seal as a function of time. For a given container-closure system, the upper and lower limits of the force profile must first be established. If the force profile for the crimping of the container-closure system falls within these limits, then the seal is considered to be acceptable. Based on the limits of the force profile, vials may be rejected because of defective seals stemming from missing stoppers or crimp seals or from defects in the glass vial flange.

West Seal Tester

The West Seal Tester model WG-005 (supplied by West Co., Phoenixville, Penn.) determines the compression forces by exerting a downward force (σ_1), determined by a pressure gauge (0 to 50 lb) shown by Figure 12.14, to a small area on the top

Figure 12.14. A West Seal Force Tester

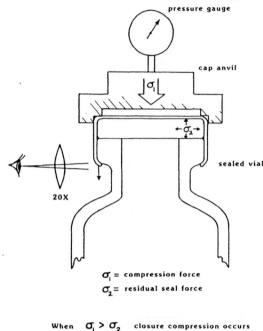

pressure gauge

cap anvil

σ_1

σ_2

sealed vial

20X

σ_1 = compression force

σ_2 = residual seal force

When $\sigma_1 > \sigma_2$ closure compression occurs

and seal skirt moves in relation to vial

[Source: *PDA Journal of Pharmaceutical Sciences*, Vol. 42, No. 1, Jan./Feb. 1988, Page 25, Figure 4. Reprinted with permission.]

circumference of the seal [17]. The principle on which the instrument functions is that when the downward thrust (σ_1) is greater than the compression force of the seal (σ_2), there will be further compression of the seal. The force at which σ_1 just exceeds σ_2 is detected, as seen in Figure 12.14, by movement of the lower edge of the crimp seal with respect to the flange of the vial as determined by an optical device having a 20× magnification. Movements of the crimp seal on the order of 0.0001 in. and measurements of the compression force are recorded to the nearest 0.5 lb. Compression measurements are taken at three different locations on the crimp seal, and the data are reported as the lowest, highest, and average compression. As with any instrument that relies on the judgment of an operator, there will be some variability regarding the initial movement of the crimp seal [17].

Constant Rate of Strain Testing

The Instron Constant Rate of Strain Testing Machine model 11113 (supplied by the Instron Corp., Canton, Mass.), as illustrated by Figure 12.15, determines the compression forces of the crimped container-closure system by applying a force on the

Figure 12.15. An Instron Residual Seal Force Tester

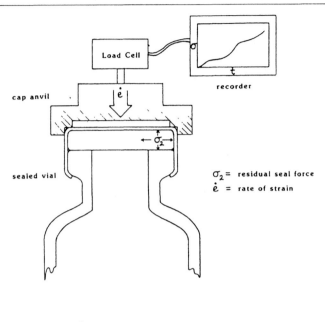

When $\dot{e}$ results in $\sigma > \sigma_2$, σ v. time curve

follows closure stress deformation response

[Source: *PDA Journal of Pharmaceutical Sciences*, Vol. 42, No. 1, Jan./Feb. 1988, Page 25, Figure 3. Reprinted with permission.]

top of the seal to deform the seal at a given rate, i.e., de/dt cm/sec [17]. A typical value of de/dt would be 0.05 cm/sec. As shown by Figure 12.15, the force applied to a crimped container-closure system and the change in distance is transferred to a 50 kg load cell. The output of the load cell is then recorded as the compressive strain necessary to deform the aluminum seal. When the applied de/dt no longer results in deforming the aluminum seal, an additional force will be necessary to compress the closure. From the above results, one obtains a measure of the percentage compression of the closure (%C_p) [17].

The %C_p is defined as

$$\%C_p = \frac{1}{\text{CFT}}[(h_v + \text{CFT} + S_t) - S_h] \times 100 \tag{14}$$

where CFT indicates the elastomer closure flange thickness, h_v is the vial height, S_t is the measured seal thickness, and S_h is the sealed package height.

On-Line Testing

Joyce and Lorenze [27] report an on-line instrument for measuring seal force with the use of a Westcapper (West Co., Phoenixville, Penn.), a load cell (Andreuss-Peskin, Natick, Mass.), an IBM computer, and LABTECH® Notebook software. This system provides the operator with quantitative measurements during the adjustment of the capping machine. In addition, the instrument proves useful in troubleshooting should there be a malfunction of the machine because the system provides benchmark reference data.

CONTAINER–CLOSURE LEAKS

It was stated at the beginning of this chapter that the main function of the container-closure system was to protect the lyophilized product from chemical or microbial contamination from the environment. For a lyophilized product to become contaminated, there must be some path for the gas or the organism to enter the vial.

Consider an initial container-closure seal that is formed at a chamber pressure < 500 mTorr. An increase in the vial pressure may stem from any of the following:

- A defective seal between the container and the closure

- The presence of lyophilized material in the container-closure seal

- Inadequate seating of the closure because of insufficient force exerted by the stoppering mechanism of the dryer

- The outside diameter of the closure exceeds the inside diameter of the throat of the vial, which results in breaking of the seal during the crimping operation [22]

- Gas permeability or outgassing of the closure

The defective seal between the container and the closure may be a result of one or more imperfections in the primary sealing surface of the container or the closure (see Figure 12.13). The presence of lyophilized material in the container-closure seal stems from a formulation containing an insufficient solid content to maintain a self-supporting matrix structure during the drying process. The seal that will be most affected by cake deposits will be in the secondary sealing area located in the throat of the container. Inadequate seating of the closures during the stoppering operation may not always be associated with lack of sufficient force exerted by the stoppering mechanism but may be a result of uneven shelves or trays. I have experienced a number of instances when the throat diameter of the vial was not large enough to accommodate the closure, with the result that a seal could not be formed. Such instances may have stemmed from a misunderstanding that the primary seal (see Figure 12.13) occurs between the flange of the vial and the closure, not as a force fit between the neck of the vial and the closure. While the above mechanisms for an increase in vial pressure would be defined as real leaks, the separation of the pressure increase in the vial resulting from gas permeability or outgassing may be just as difficult to distinguish as between a real and virtual leak (see discussion of real and virtual leaks in Chapter 11). All of the above examples represent leak paths for a container-closure system. While it is recognized that all container-closure systems leak, it is the relative leakage that is of concern and whether or not the leakage falls within the limits of acceptable standards. The remainder of this section will be a discussion of the various methods used to determine the integrity of the container-closure seal.

Bubble Test

The *bubble test* is one in which the container-closure system is pressurized and then submerged into a clear liquid. The presence of bubbles in the liquid would be an indication of a leak path in the seal. This test would generally not be applicable to lyophilized products because such container-closure systems are generally sealed at pressures ≤ 1 atm [5]. However, because of the product's sensitivity to heat and its relatively low mass, these products are often transported by air. At an altitude of about 30,600 ft or 15 km, the pressure in an unpressurized cargo compartment is about 82 Torr [8]. If the container-closure systems are sealed at 600 Torr, then the vials would become pressurized at 10 psig. Such an internal pressure in the container-closure system would generate an internal stress on the seal and cause the aluminum seal to be stressed such that it would decrease the compression force on the container-closure seal. Should the seal permit the release of gas from the container, then the original seal would be compromised and no longer have the same sealing properties (compression) that were established during the crimping operation. The reader should understand that the generation of the above 10 psig, as a result of an altitude of 15,000 km, is considered to be a worst-case scenario. The key point being made is that, when there is a significant decrease in environmental pressure, the container-closure seal will experience force opposite to the compression force used to form the crimp seal. It is for this reason that I suggest that a bubble test be conducted on lyophilized products even when they are sealed at pressures near 1 atm.

Standard Method

The standard method for performing the bubble test on a container-closure system is to first pressurize the container-closure system (e.g., 3 psig) [28]. The pressurized system is then immersed in water containing a surfactant (0.5% polysorbate) using an apparatus like that shown in Figure 12.16 [28]. The formation of the seal was performed by hand crimping an aluminum seal. The aluminum seal was then pierced to form a small hole (0.5 mm diameter) on the side of the seal, as shown in Figure 12.16, to direct the release of any gas from the seal into a viewing path. The viewing path consisted of a black background and a high intensity light source positioned at a 45° angle to the hole in the aluminum seal. The presence of any bubbles was optically (3× magnification) determined. If no bubbles appeared after 15 min, the seal was considered adequate.

 With the bubble test, Morton et al. [28] were able to establish a relationship between the leakage and $\%C_p$. The results showed that, for known defective metal-stoppered vials, the average cutoff for the formation of bubbles was found to occur at a $\%C_p$ of 11.5% ± 1.4%. These results indicate the importance of setting limits for the $\%C_p$ for a given container-closure system.

Figure 12.16. Apparatus for the Bubble Test

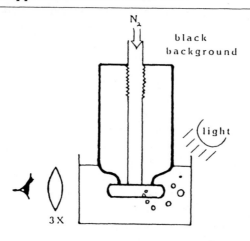

[Source: *PDA Journal*, Vol. 43, No. 3, May/June 1989, Page 105, Figure 1. Reprinted with permission.]

Vacuum Bubble Test

While it has been pointed out that the bubble test may not be suitable for lyophilized products because they are stoppered at pressures lower than 1 atm [5], it has been shown that conditions could arise when the internal pressure in the vial could exceed that of its environment. The issue that needs to be addressed is if the presence of such low pressures could cause a change in $\%C_p$ sufficient to break the seal and allow gas to escape from the vial or permit gas to flow from a real leak associated with the primary sealing surface (see Figure 12.13).

The apparatus used for this test method is similar to that shown in Figure 12.16, except that no external source of gas was used to pressurize the vial. Instead of water containing 0.5% polysorbate surfactant, the liquid was Dow Corning® 704 diffusion pump fluid that has a vapor pressure << 1 mTorr. All containers were sealed with a Flip-Off crimp seal. A 1/16 in. diameter hole was drilled in the aluminum skirt of one of the crimp seals, like that shown in Figure 12.13. The other vials tested did not contain the hole. Sufficient silicone oil was added to the glass container so as to completely cover the crimp seal. The container was then placed in a FTS Systems laboratory freeze-dryer. The chamber door was then closed, and the system was evacuated with the condenser at ambient temperature. The chamber pressure as a function of time was recorded.

In order to remove the trapped gas between the crimp seal and the container-closure, the system was evacuated to about 1 Torr and maintained at this pressure for about 5 min. The chamber was then backfilled with air to 1 atm. The system was again evacuated, and the pressure at which a steady stream of gas was observed leaving the vial was noted.

The following are the results of a preliminary study of this form of the bubble test. I am fully aware that additional testing is required before any decisive conclusions can be made regarding the merits of this test method. All of the vials tested were stoppered at 600 Torr or 1 atm and then hand crimped sealed. The vials that were stoppered at 600 Torr were crimp sealed more than 1 year prior to the test. The vials stoppered at 1 atm were crimp sealed just prior to the test. The bubbles streaming from the crimp seal were observed through the door of the dryer. All vials exhibited bubbling during the initial evacuation process. The gas release from the vial with the 1/16 in. diameter hole was similar to those without the hole. After the second evacuation, only one of the crimped seals (the vial with the hole) exhibited a steady stream of gas from a given location at the shoulder of the crimp seal when the chamber pressure was < 10 Torr. The rate of the stream of gas appeared to increase as the chamber pressure decreased; however, its location with respect to the hole remained unchanged. Other smaller streams of gas were observed as the pressure approached 1 Torr. It is of interest to note that this technique not only indicated the presence of a real leak in the seal but also its location. Such information would be useful if the cause of the leak is found to be associated with the settings of a crimping machine.

The reader should understand that the above results merely demonstrate a technique by which leaks in lyophilized vials can be observed by means of a vacuum bubble test. Without a more detailed study of this technique, the present test should be considered only qualitative in nature. Perhaps from a knowledge of the rate of the bubble emission and the pressures in the chamber and vial, one may be able to ascertain the actual leak rate.

Pressure Rise

Some lyophilized products, upon completion of the secondary drying process, are stoppered while at the secondary drying pressure. Such pressures are generally ≤ 100 mTorr. An increase in the pressure of a crimped container-closure system can result in a virtual leak such as outgassing of the closure [5,12], permeation of gases such as oxygen and water vapor through the closure [12], or result from real leak paths. Such leaks may increase the pressure in the vials from an initial 50 mTorr to 2–100 Torr over some time period [12]. The following is a brief description of the means that Crist [12] used to ascertain an increase in vial pressure upon storage.

The test vials used by Crist [12] were 10 mL and 50 mL amber vials. Some of the vials contained product and were designated as 10P and 50P, while other vials were empty and were identified as 10E and 50E. The 20 mm elastomer closures were gray butyl and were supplied by the West Company (West S-87J-1888). Prior to the test, the closures were washed, siliconized, autoclaved, and dried. At the completion of secondary drying, all of the vials were stoppered at a chamber pressure of 18.3 mTorr and were removed from the freeze-dryer.

After crimp sealing, the pressure in the vials was determined by means of a capacitance manometer gauge (MKS Baratron® type 102A) (supplied by MKS Instruments, Inc., Andover, Mass.) to which was attached a chromatography needle. The direct pressure measurement (P_1) was found equal to

$$P_1 = \frac{P_v V_v + P_g V_g}{V_v + V_g} \tag{15}$$

where P_v is the vial pressure and P_g is actual atmospheric pressure (740 Torr) not corrected to sea level, while V_v is the headspace of the vial (the total volume of the vial less the volume of the lyophilized product and closure, i.e., 13.6 mL,) and V_g is the internal volume of the MKS gauge (3.9 mL). From equation 15, the actual internal pressure of the vial is determined as follows:

$$P_v = \frac{P_1(V_v + V_g) - P_g V_g}{V_v} \tag{16}$$

With this technique, Crist [12] was able to show that after a few days of storage, the vial pressure increased to about 2 Torr and, after a month, the pressure increased to between 5 and 10 Torr. It should be noted that the testing was performed in an environment in which the partial pressure of water vapor was about 10 Torr. The ROR in empty vials was found to be greater than in vials containing product. The reason for the greater increase in pressure for the empty vials is attributed to the adsorption of water vapor by the lyophilized product. In the presence of a significant real leak, Crist [12] found that the pressure in the vials increased after just a few days to 20 Torr. Crist concluded that, because the vial pressure became constant with time, and since Pikal and Shah [13] found that the moisture in their lyophilized product remained constant with time, the constant pressure in the vial resulted from a state of equilibrium between the water vapor in the vial and that in the closure. Crist further concluded that the source of water vapor was a result of the outgassing of the closure rather than permeation of water vapor through the closure.

The description of the paper by Crist [12] was included in this text because it was felt this method for determining the increase in the headspace of a sealed vial as a function of time was a unique and useful technique. However, the reader is cautioned regarding the following:

The increase in the vial pressure may result from gases from the closure. The results reported by Pikal and Shah [13] showed that the moisture content in lyophilized moxalactam disodium first increased and then remained stable over a period of time. Crist [12] offers only conjecture that the increased pressure in the vial is primarily a result of the increase in the partial pressure of water vapor. It is most unfortunate that, in spite of showing that *the presence of (hydrophilic) product in the vial causes a lower vial pressure*, such a statement was not further substantiated with residual moisture data.

Given the assumption that water vapor is the source of the increase in pressure in the vial, the fact that the pressure attains a steady state does not in itself (see equations 5 and 6) rule out the possibility that the permeation of water contributed to the increase in the pressure in the vial. For example, if the partial pressure of water in the vials reached a maximum value of 10 Torr and the partial pressure of water vapor in the environment was also 10 Torr, then the permeation rate (equation 5) and the outgassing rate (equation 6) would result in a constant pressure value in the vial. If the objective of the work by Crist [12] was to verify the results of Pikal and Shah [13] by showing that the source of the water vapor stems from outgassing rather than permeation through the closure, then the vials should have been tested under two separate environments. In one of the environments, the partial pressure of water should be maintained at 10 Torr; in the second environment, the partial pressure of water vapor should approach 0. If the results indicate a similar pressure rise for vials stored under both environmental conditions, then one could state with greater certainty that the source of the increase in water vapor in the vial was a result of out-gassing and not associated with the permeation of water through the closure.

Color Dye Test

The color dye test is perhaps one of the most widely used methods to detect the presence of a real leak in a container-closure system [5,23,29]. In this test, the container-closure seal is immersed in a dye solution and then checked visually or by instruments for presence of the dye in the container. Typical colors for the dye are red and blue, while materials exhibiting ultraviolet (UV) fluorescence have also been used an indicators [5]. Although this type of testing has the advantage of simplicity, its chief disadvantage is its slow response time [5].

Helium Leak Testing

Helium leak testing of a system is perhaps one of the most widely used and sensitive methods (leak rates as low as 1×10^{-5} sccs [29]) used for determining the integrity of a packaging system. It is my understanding that some beer manufacturers add a small quality of helium to the can just prior to sealing. If helium is later detected during inspection, the can is rejected because the leak could also result in a loss in carbonation, and the beer could become *flat*. It often puzzled this nonbeer-drinking author why individuals started to talk *funny* after consuming the beer from a number of cans. Could the speech problem be associated with an overdose of helium?

In determining the real leak rate of a container-closure system, the test method may involve the detection of helium entering or leaving a container [5]. It is my opinion that it is far easier to detect helium leaving a packaging system than to detect helium entering it. The following will first describe a study conducted by Kitsch et al. [23] intended to determine the viability of the helium leak detection method as a means of assessing the integrity of a container-closure system and then a method I have used to assess the helium leak rate from uncrimped and crimped container-closure systems.

Verification of the Helium Leak Test for a Container-Closure

In conducting the helium leak test, Kitsch et al. [23] used a Varian helium leak detection system. This system uses a 180° magnetic field mass spectrometer that is permanently set for the He$^+$ (i.e., $m/e = 4$). In addition, the authors constructed test units like that shown in Figure 12.17. Each of the test units was equipped with a micropipette. Five different inside micropipette diameters where used: 0.5 μm, 1 μm, 2 μm, 5 μm, and 10 μm. The diameters of the pipettes were based on the results of a bubble test, not on dimensional measurements. The pipettes were attached to the wall of the test vial with epoxy, which also served as a gas seal. The length of the pipette varied from 1 to 2 cm. A butyl rubber closure and aluminum crimp seal composed the main seal for the container. Helium was added to the test unit. The test unit was then placed in a vacuum system, and the leak rate of helium was determined as a function of the diameter of the pipette. Tests were conducted using two different test methods for filling the test units with helium.

Figure 12.17. Schematic Description of a Modified 10 mL Glass Tubing Vial for Evaluation of Mass Spectrometer Helium Leak Rate Measurements

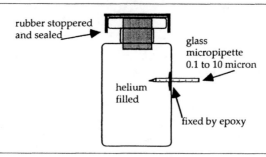

[Source: *PDA Journal*, Vol. 51, No. 5, Sept./Oct. 1997, Page 188, Figure 1. Reprinted with permission.]

Test Method A. In this method, the vials were placed in a plastic glove bag, and the glove bag was filled with helium. The test units were then stoppered and crimped sealed. The authors referred to the leak rates from these test units as being *absolute*. The test unit was then placed in a test chamber that was connected to the helium leak detector. Test measurements commenced when the pressure in the test chamber was < 2 mbar. When the pressure was < 0.2 mbar, the leak rate of the helium was displayed. Leak rate measurements were displayed at 10, 15, and 20 sec intervals. If the readings were identical, then the reading was taken as the leak rate; otherwise, three readings in 5 sec intervals were taken every minute until three identical readings were obtained. The time to reach stable leak rates varied from 5 to 100 sec, depending on the size of the leak diameter. The larger diameter tubes required a longer time to achieve stability because of helium desorption from the walls [23].

The mean leak rates determined for the 0.5, 1, 2, 5, and 10 μm micropipettes were 1×10^{-6}, 4×10^{-5}, 3×10^{-3}, 7×10^{-3}, 1×10^{-2}, and 2×10^{-2} sccm/sec, respectively. The relationship between the leak rate (L_r) and the diameter of the pipette (d_p) was found to be related to the following function:

$$\log L_r = B \times d_p^2 \quad \text{sccm/sec} \tag{17}$$

where B is a constant and must have the units of speed [i.e., length/unit time (sccs)].

Filling Method B. For this method, the test units were stoppered and crimped sealed at ambient temperature and pressure. Helium was added to the test unit by placing it in a pressurized vessel. The authors [23] referred to this method of filling the containers as *charging* or *bombing*. The pressure vessel was then sealed, and the system pressurized to 2 or 3 atm with helium gas. The units were allowed to remain at these pressures for 2 to 4 hr. The primary purpose of the charging procedure was to maximize the concentration of the helium gas in the container.

Results and Discussion. Kirsh et al. [23] reported reasonable agreements between the leak rates obtained from using either method of filling the containers with helium. The importance of the above statement is that the charging method provides a means by which helium can be introduced into crimped sealed containers and the leak rates determined. The charging procedure also permits a means for nondestructive testing of the leak rate and allows for tracking the effectiveness of a given container-closure seal over an extended period of time. The authors recognize that permeation of the helium in the closure could be a limiting factor in determining the leak rate; however, once the closures become saturated with helium, then outgassing of the helium should provide a reproducible baseline. In addition, the effects of the absorption of helium by the glass surface during the charging process can be eliminated by exposing the vials to an ambient room environment for about 12 hr prior to the leak testing.

Commentary. One important aspect of the above method is that the testing can be performed on individual container-closure systems. Because of this feature, one can establish a frequency distribution of the leak rates for a lot of container-closure systems that were crimped sealed under a given set of compression settings [12,17,27].

It is recognized that the above represents the first of what I hope to be a series of papers concerning this technique for determining real leaks in a container-closure system. I look forward to additional publications on this method, where the authors will provide additional information or rationale regarding the following:

The above method determines the total leak rate for a given container-closure system. One important question that must be addressed is, "How does one distinguish between a container-closure system that contains a large number of small leaks that would pose no threat regarding the transport of microorganisms and a system in which a single, major leak path could permit microbial contamination?" In each case, the leak detection system would sense the same leak rate.

The nature and location of the gauge used for the pressure measurements during the leak test is an important consideration. A thermal conductivity gauge would, in the presence of helium, tend to indicate false high pressure readings. Is the pressure gauge located in the test chamber or at the entry port to the leak detection system?

The test chamber was connected to the foreline of the helium leak detector (i.e., between the mechanical pump and the outlet of the diffusion pump). Helium gas must flow counter to the flow of the gases in the diffusion pump in order to reach the mass spectrometer that is located at the inlet of the diffusion pump. Some of the helium gas must be pumped by the mechanical pump that is used to back up the diffusion pump. Future papers should explain how the leak detector was calibrated so that the leak rates could be read directly from the output of the mass spectrometer.

Another key issue that needs to be addressed is the rationale or basis for determining the point that represents a real leak of the container-closure system. Kirsch et al. [23] assessed that the leak rate is determined when the observed leak rate was stable for three consecutive readings. It is important that future publications provide the rationale that supports such a measurement criterion.

In view of the above work and the success that other industries have had in detecting leakage in their packaging systems by detecting (sniffing) helium gas, it would seem logical that the industry would backfill the container-closure systems with a mixture of nitrogen and helium at pressures < 1 atm. This would certainly eliminate any need to *charge* or *bomb* container-closure systems with helium. Such *charging* or *bombing* of the container-closure systems could not be used for human consumption because of the concern of microbial contamination during the testing procedure. However, for a container-closure system having a helium-nitrogen atmosphere, the fact that the system contains helium is sufficient to detect any leak path, especially, if the testing for helium is performed at a pressure < 1 atm.

Determination of the Leak Rate for an Uncrimped Closure

Upon completion of the drying process, the vials are stoppered and removed from dryer. It will be some finite time before the container-closure seal will be formed with the application of a crimp seal. The question that needs to be answered is, *For a given container-closure system, how effective is the seal of the uncrimped closures and the vials in preventing contamination of the lyophilized product from the environment?*

The pressure in the vials used in the above helium leak detection method was ≥ 1 atm [23]. An attempt to use this method for determining the helium leak rate for uncrimped vials would result, because of pressure in the vial, in displacement of the closure during the evacuation of the test chamber. The following describes a method I have used for a number of years to ascertain the average helium leak rate of uncrimped container-closure systems.

Sample Preparation. Based on the size of the vials, two trays of vials with the closure in position for lyophilization are loaded into a two-shelf FTS freeze-dryer. Unless the test is to be conducted at ambient temperature, the vials and closures are first allowed to equilibrate to the shelf temperature of the system.

The condenser is then chilled to < –80°C, and the chamber evacuated with the shutoff valve, the gas solenoid valve, and the variable leak valve to the helium gas supply in their fully open position. All 3 valves are located near the inlet port to the drying chamber. When the chamber pressure, determined with a capacitance manometer gauge, reaches < 50 mTorr, the shutoff valve is closed, and the line to the helium tank is pressurized to 5 psig. The shutoff valve is opened, and the chamber and the helium transfer line are once again evacuated to < 50 mTorr. The above procedure is repeated no less than 5 times. Each evacuation and pressure increase is recorded by a pen recorder. The main objective of this first step is to purge the helium transfer line of any atmospheric gases. The reader should note that the mechanical pump is operating throughout the entire purge operation.

After the last purge of the helium transfer line, the shutoff valve is closed, and the transfer line is pressurized with helium and maintained at 5 psig. The variable leak valve is opened and adjusted such that the leak rate of helium is sufficient to increase the pressure in the chamber to 2 Torr while the system is being evacuated with the mechanical pump. After 2 min at 2 Torr, the solenoid valve is closed, and the system is allowed to evacuate to < 100 mTorr. This purge of the chamber and vials with helium is repeated at least 5 times. After the last purge, the chamber is again filled with helium gas to a pressure of 2 Torr and the vials stoppered.

The chamber is backfilled with dry nitrogen, and the stoppered vials are removed from the dryer and inspected. Any improperly seated vials are removed from the trays, and the number of vials removed is recorded in a laboratory notebook. The properly seated vials are placed back into the drying chamber. The nitrogen transfer line and the chamber of the freeze-dryer are purged with dry nitrogen using the same method described for helium gas.

The filament to the RGA system is outgassed, and a blank mass spectrum of the gases in the chamber is recorded. The freeze-dryer chamber pressure is reduced to less than 100 mTorr, and the mass spectrum of the chamber is recorded as described in Chapter 11. The chamber and condenser are isolated from the pumping system and sets of blank and chamber mass spectrums are recorded in 0.5 hr intervals over a period of 5 hr.

The corrected rate of pressure rise for helium (ROR_{He}) for Y number of uncrimped containers is determined from knowing the partial pressures of other gases in a chamber having an effective volume (V_e) and the increase in the chamber pressure over a period of time. The leak rate for Y uncrimped stoppers is Q_u and is defined as

$$Q_u = ROR_{He} \times V_e \quad \frac{\text{mTorr liters}}{\text{sec}} \tag{18}$$

or in terms of the test conditions and the ideal gas law

$$Q_u = \frac{0.23 \times ROR_{He} \times V_e}{T_s} \quad \text{g He/hr} \tag{19}$$

while the average leak rate of helium per vial ($Q_{a,u}$) is expressed as

$$Q_{a,u} = \frac{Q_u}{Y} \quad \text{g He/hr} \tag{20}$$

and, correcting for the difference in the diameters of the helium and water molecule by the factor 0.457, one obtains an expression for the leak rate in terms of water vapor ($Q_{a,c}$) as

$$Q_{a,u} = 0.45 \times \frac{Q_u}{Y} \quad \text{g H}_2\text{O/hr} \tag{21}$$

The time required to increase the moisture content of 1 mg of material 1% at a given partial pressure of water vapor P_w, assuming a sticking coefficient (S_c) of 1, can be obtained from equation 20 as

$$t_{w,u} = \frac{1 \times 10^{-6} Y P_w}{S_c Q_{a,u}} \quad \text{hr/\%(H}_2\text{O)} \tag{22}$$

where $t_{w,u}$ represents the time expressed in hours.

I apologize to the reader for not presenting actual test data obtained from this test method for determining the leak rate for uncrimped container-closure systems, but existing confidentiality agreements prevent this. However, from typical leak rates of helium, the average time to increase the moisture content of just 1 mg of lyophilized material by 1% was a matter of several days. In addition, the reader should be aware that the assumption of S_c having a value of 1 represents a worst-case scenario. The reader should also note that the results of this testing are in terms of the average leak rate; however, without knowing the nature of the frequency distribution, one cannot assess the probability of the moisture increase in a lyophilized cake, causing it to exceed the moisture specification prior to forming the crimp seal.

Leak Rate for a Crimped Container-Closure System

The following is a procedure for determining the average leak rate of a crimp-sealed container-closure system. The preparation of the equipment was the same as that used for the leak testing of the uncrimped container-closure system. The difference in the test procedures was that, after completion of the purging of the system with helium, the system was typically backfilled to a helium pressure of 17 Torr, which would represent a 100% relative humidity of water vapor at 20°C.

Substituting Q_c for Q_u and $Q_{a,c}$ for $Q_{a,u}$ into equations 18 through 21, inclusive, where Q_c is the helium leak rate for the crimped container-closure vials and $Q_{a,c}$ is the average leak rate expressed in terms of water vapor, one obtains an expression for the time necessary to increase the moisture content for a crimped container-closure system, at a given relative humidity and temperature (T) as

$$t_{w,c} = \frac{2.8 \times 10^{-7} Y P_w}{S_c Q_{a,c}} \quad \text{hr}/\%(\text{H}_2\text{O}) \tag{23}$$

where $t_{w,c}$ represents the time required to increase 1 mg of water 1% for a crimped container-closure system and Y is the number of crimped vials.

Once again, I am not free to provide actual test data because of existing confidentially agreements; however, it was not unusual to determine $t_{w,c}$ as being greater than 5 years. One important difference between the above two methods for using helium to determine the leak rate for uncrimped and crimped container-closure systems and that described by Kitsch et al. [23] is that the outgassing of helium from the closure or the glass is not considered to be a factor in measuring ROR_{He}. However, the above method only provides a measure of the average leak rate for a crimped container-closure system. As in the case for the uncrimped system, the above method does not provide one with a means for assessing the probability of the moisture increasing in a lyophilized cake, thus causing it to exceed residual moisture specifications under normal shelf storage conditions. In order to determine such a frequency distribution, one would have to measure the individual leak rate for at least 30 uncrimped or crimped vials.

Glow Discharge Method

For lyophilized products stoppered at the pressure used during the secondary drying process, generally ≤ 100 mTorr, the pressure in these vials is sufficiently low to support an electrical glow discharge. Glow discharges are generated when the gas in the vial is subjected to a high electric field. This section will consider the nature of the glow discharge, its use in detecting vials containing substantial leak rates, and the effects that the glow discharge may have on the quality of the product.

Nature of the Glow Discharge

First of all, a glow discharge is a gaseous system that consists of a nonuniform distribution of electrical charges resulting from an applied electric field. In the case of a dc applied electric field and at pressures < 1 Torr [7], a glass tube with two platinum electrodes at either end (see Figure 12.18) will, when sufficient voltage is applied, produce a system that will emit light. Light will be emitted from a region

Figure 12.18. Glow Discharge Tube

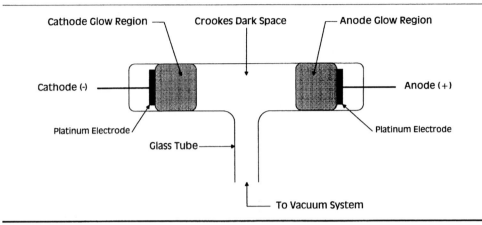

surrounding both the cathode (negatively charged electrode) and the anode (positively charged electrode) and are appropriately referred to as the cathode and anode glow. Separating the two glow regions is a region without any glow that is referred to as the Crookes dark space. At the cathode, the glow is a result of the emission (loss in energy) when an atom in an excited state goes to its ground state. At the anode region, electrons have acquired sufficient kinetic energy to cause both ionization and excitation of the gas. The Crookes dark space is where the electrons generated in the cathode region have not achieved sufficient energy to ionize or excite the gas molecules in the system. Thus, the electric fields at the electrodes are higher than that in the Crookes dark space. If the dc applied voltage is replaced with an alternating electric field, such as that generated by a Tesla coil, then the glow discharge can be maintained throughout the tube. Since free electrons are required to cause the ionization and excitation of the gas, what is the source of these initial electrons?

Wu and Crist [22] propose that the glow discharge results from the dissociation of the gases as a result of the high electric fields. If the pressure in the system is < 1 Torr, then the electrons from the initial ionization are accelerated to sufficient energies to cause additional ionization or excitation of the gas in the system. The first ionization potentials for the atoms of some common gases found in the freeze-drying chamber are listed in Table 12.2 [7,30].

In order to ionize a gas such as helium, which has an atomic diameter of 2.2×10^{-8} cm, the required electric field, based on the first ionization potential of 24.587 eV, would have to be on the order of 1,117 million V/cm. Such field strengths could not be generated (hopefully) by a handheld Tesla coil, let alone generated by any known Van der Graaff generator [30]. The voltage output of a Van der Graaff generator is limited to tens of millions of volts, which would be insufficient if the distance between the spheres is only 1 cm. If there is no means available to ionize a gas by means of an electric field, then the question still remains, *What is the mechanism for generating the glow discharge?*

Table 12.2. Molecular Diameter, First Ionization Potentials, and the Visible Color from the Strongest Emission Lines for Some Common Gas Atoms

Gas and (Molecular Diameter × 10⁸ cm) [7]	First Atomic Ionization Potential (V) [30]	Visible Color from the Strongest Emission Lines in Vacuum
Hydrogen (2.75)	13.598	Red [30]
Helium (2.2)	24.587	Yellow [30]
Neon (2.6)	21.564	Red (735) [7,30]
Argon (3.67)	15.759	Violet to Orange [7]
Nitrogen (3.75)	14.534	Blue to Orange [7]
Oxygen (3.64)	13.618	Green to Red [30]

The answer is that an *ionizing agent* must be present [31]. The source of such ionizing agents can be cosmic rays, radioactivity, photoelectric emission, X rays, or collisions between molecules. For a glass container, such as a serum vial, one would not anticipate photoelectric emission as being a candidate as an ionizing agent; however, glass may contain a certain small quality of potassium. The natural abundance of the potassium isotope $_{19}K^{40}$ is 0.01%, and the isotope has a half-life of 1.28×10^9 years. When this isotope decays, it can emit an β^- (electron) particle that has an energy of > 1 million V, which is sufficient energy to ionize gas in a chamber [30]. The principal point is that there are numerous means for producing electrons or momentarily ionizing the gas. These free electrons are present in the system, and the electric fields generated from the Tesla coil are sufficient to accelerate these few electrons and generate a glow discharge.

The primary use of the glow discharge is to determine qualitatively if the container-closure system is below a given pressure limit. If, for a given applied voltage, no glow discharge is formed, then one would assume that there was an increase in the pressure of the container resulting from either a real or virtual leak and that any electrons in the system cannot acquire sufficient energy to generate the glow discharge. The entire vial is then said to be in the Crookes dark space.

Electro-Technic VC-105 Automatic Vacuum Tester

Wu and Crist [22] report that an instrument, known as an Electro-Technic VC-105 Automatic Vacuum Tester, is designed to evaluate quantitatively the pressure in the vial by measuring the current generated by a glow discharge. The key components of the instrument are illustrated by Figure 12.19, and the instrument functions in the following manner. The vial is passed by a rod that is maintained at some predetermined high voltage (40 kV to 200 kV). The presence of a glow discharge in the vial will cause a flow of current in the conducting plate that is in contact with the vial. The current from the glow discharge is sensed by a logic circuit, and the vial is accepted. Should there be no current (no glow discharge) when the high voltage is

Figure 12.19. Key Components for the Automatic Testing of a Container-Closure System Using a Glow Discharge

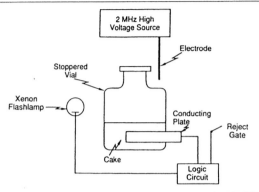

[Source: *PDA Journal,* Vol. 43, No. 4, July/Aug. 1989, Page 180, Figure 1. Reprinted with permission.]

applied, the reason may be that there is an insufficient source of electrons to initialize the glow discharge. The absence of a glow, because of an insufficient initial source of electrons, could result in an indication of a false high pressure in the vial. To prevent false high-pressure rejection of vials, a high intensity xenon strobe light is flashed if, after a given time period, no glow discharge is detected by the logic circuit. The high intensity xenon light causes photoionization of the gases and, if the pressure in the vial is low enough, a glow discharge will result and the vial will be accepted. If no glow discharge is generated, then the pressure in the vial must exceed the upper pressure limit, and the vial is rejected. By varying the applied voltage to the rod, the pass or reject pressure can be varied from 50 mTorr to 300 mTorr.

Effect of the Glow Discharge on the Properties of the Lyophilized Cake

The formation of the glow discharge produces a variety of gas species. For example, the following are some of the gas species that can be formed in a glow discharge containing oxygen:

$$O_2 + e^- \rightarrow O^* + O + e^- \quad (24)$$

$$O_2 + e^- \rightarrow O^+ + O + 2e^- \quad (25)$$

$$O^+ + e^- \rightarrow O^{++} + 2e^- \quad (26)$$

$$O_2 + e^- \rightarrow O_2^- \quad (27)$$

$$O^* \rightarrow O + h\upsilon \quad (28)$$

$$O^{++} + 2e^- \rightarrow O + 2h\upsilon \quad (29)$$

While the above list of possible reactions in a gas discharge is by no means complete, it does show the formation of chemically active nascent oxygen (O), positive (O^+)

and negative ions (O_2^-), and oxygen atoms in an excited state (O^*). The term *hv* (*h* is Planck's constant and *v* is the frequency of the photon) denotes the energy of a photon that can be released upon the neutralization of a ion or the photon energy released when the electron in an atom drops from an excited state to a ground state. If one includes similar reactions for gases such as nitrogen, carbon dioxide, and water vapor, the generation of glow discharge can produce a wide variety of chemical active species and photons that can interact with the lyophilized product.

When the product Activase® was subjected to a series of 12 glow discharge tests that had a duration 0.6 sec, the only noticeable effect on the product was an increase in particulate matter from 10 μm to 25 μm. However, when the glow discharge was operated for 7 sec, there was a change in the Activase®, and it no longer met USP XXI specifications [22].

In using high voltage, one can produce O_3 (ozone) that could prove harmful to operating personnel. Wu and Crist [22] eliminated the ozone hazard by passing the gas from the testing chamber over a manganese dioxide catalyst, where the ozone was converted to oxygen (O_2).

Commentary

Wu and Crist [22] reported that as a result of outgassing from the closure, the pressure in the vial increased from an initial pressure of 18.3 mTorr to 2 Torr in less than a day after stoppering. Since a glow discharge cannot be sustained at pressures ≥ 2 Torr, one is left with the uncertainty if the glow discharge method of evacuating the pressure in a vial can be used for a container-closure system in which the closures were previously steam sterilized.

Microbial Transport Tests

Although gas leakage into a container-closure system, from either a real or virtual leak source, can affect the quality and stability of the lyophilized product, the transport of microorganisms across the container-closure seal represents a far more potentially dangerous source of product contamination. In this section, two methods for determining the effectiveness of the container-closure seal in preventing the transport of microorganisms into a lyophilized product will be described.

Microbial Ingress Test Method

The microbial ingress test method challenges the effectiveness of the container-closure seal to prevent the transport of microorganisms into the vials. The vials are prepared by filling with a given amount of sterile growth media such as soybean casein digest or tripticase broth. All vials are stoppered under a vacuum and then crimp sealed. The test vials are then exposed to various environments that contain a microorganism such as *Enterobacter coli* [5]. Such testing environments would include the following:

- Exposure of the vials to an aerosolized microbial suspension environment for a given period of time [5].

- Typical conditions in which the lyophilized product will be stored. The test vials are periodically checked for microbial contamination [5].

- Some of the test vials are immersed in microbial media that contains 1×10^8 CFU/mL.

Of all of the ingress microbial test methods, this test is considered to offer the most severe challenge to the effectiveness of the container-closure seal [5].

Guazzo [29] has expressed concerns regarding the microbial ingress testing of the container-closure seal. He has described such tests as . . . *insensitive, costly, time consuming, and frequently bearing little resemblance to real-life storage challenge conditions.* Guazzo [29] also states that there does not appear to be any established standard method for conducting such tests, and the number of procedures are about as varied as the number of manufacturers.

Microbial Regress Test Method

In the regress test method, a solution of microorganisms is added to the test vials. In the work described here, the microorganism was *Pseudomonas aeruginosa*, which had a concentration of $\geq 3 \times 10^8$ CFU/mL [28]. Note that the test container-closure systems reported here were sealed using a screw cap rather than a crimp seal. It is my opinion that the following test procedure is applicable only for container-closure systems using crimped seals. The following is a brief description of the test procedure.

The sealed vials containing the solution of microorganisms are inverted into a sterile saline solution so that the container-closure seal is completely immersed. The vial is then pressurized to 3 psig for a period of 15 min. The vial is then removed, and the saline solution is passed through a 0.2 μm filter. The filter is placed on a sterile plate containing Tryptic Soy Agar (TSA), and the system is incubated for 72 hr at temperatures between 30°C and 35°C. The microbial concentration in the saline solution is then determined by the plate count method [28].

Four tests were conducted. Each test consisted of three known defective container-closure systems that contained untreated butyl closures and nine nondefective systems that contained coated closures. One of the nondefective container-closure systems served as a control and did not contain the solution of microorganisms during the testing [28].

The results of the tests showed that for the defective vials, microbial transport did not occur when the %C_p became ≥ 12.1. The results for the nondefective container-closure systems were somewhat ambiguous, and a %C_p ranging from 13.1 to 16.2 was required to prevent the transport of the microorganism from the vial to the sterile saline solution [28]. What is important here is that the above method offers a direct means for assessing the effectiveness of the container-closure seal to ensure against the transport of microorganisms. It would be of interest to know of the above test method with respect to a crimped container-closure system.

Miscellaneous Tests

The following are some additional testing techniques that have been reported but not widely used in the industry. They are worth mentioning because they may prove useful for some lyophilized products.

Vacuum/Pressure Decay

In vacuum/pressure decay method, the container-closure system is placed in a chamber that is then either evacuated or pressurized. After a short period of time, to allow for a change in pressure, the pressure in the test chamber is measured. For crimped container-closure systems, the test method can be automated; testing, depending on the method and nature of the packaging system, can be performed as high as 300 systems/min with leak rates as a low as 10^{-3} sccs [29].

Carbon Dioxide Tracer Gas

Helium gas can be replaced with carbon dioxide. For this method, the container-closure systems are first filled with carbon dioxide and then crimp sealed. The crimp-sealed vials are then placed in a vacuum system, and the chamber is evacuated. The increase in the partial pressure of carbon dioxide in the chamber is detected with an infrared (IR) detector. This test method has been shown not to have the sensitivity of helium, and leak rates are limited to about 10^{-2} sccs [29].

Oxygen as a Tracer Gas

For those products that are sensitive to oxygen, such as reducing agents, the vial contains a coulometric detector [5]. The test vials are first flushed with nitrogen and then exposed to oxygen gas. Any oxygen gas that enters the vial is detected by the coulometric detector. Rather than measuring the leak rate into the container-closure system, this technique determined the effectiveness of preventing oxygen from entering into the vial [29].

AMPOULE SEAL

The ampoule seal is formed by closing off the stem of the ampoule (see Figure 12.12) with a hot flame. For borosilicate glasses (type I glass), the temperatures required to make the seal are between 700°C and 1,100°C [20]. The ampoule seal is made by a procedure that is sometimes referred to as *tip sealing* [24]. In this method, the open tip of the ampoule is placed in a flame and melted until a seal is formed. In the second sealing method, referred to as *draw sealing*, the glass stem is melted and drawn down until a seal is formed. The short length of glass that was separated during the drawing process is then discarded.

MacKenzie et al. [24] showed that the use of *draw sealing* produces an effective ampoule seal. The effectiveness of the seal was demonstrated by first filling the ampoules with a partial pressure of argon. The ampoules were then sealed. An ampoule containing the argon was then placed in a high vacuum system equipped with a mass spectrometer. The leak rate of the argon from the ampoule was determined, and then the ampoule was broken to release all of the argon gas in the vial. The largest leak rate for drawn seals was about 5×10^{-12} Torr liters/sec and would correspond to a leak path of 1×10^{-5} cm. It was also shown that drawn seals do not contain channels (real leak paths) like those formed when using the *tip sealing* method [24].

Leak Testing of Ampoules

The leak testing of ampoules presents its own set of unique problems in that it is necessary to ascertain not only if the ampoule leaks but also the source of the leak.

Water Bath

Although the conventional method described above often proves inadequate for determining the leak paths in ampoules, one simple method is a water bath. If the bath is operated at temperatures greater than ambient, beads of water are sometimes observed on the inside of the glass, indicating the possibility of a leak [20].

Dark Field Microscopy

With the use of dark field microscopy, it is possible to observe small capillary channels that did not completely close during the sealing process. These capillary channels were not only observed at the tip, where the final seal was made, but also on the bottom of the vial that was generated during the fabrication of the ampoule. If such leaks are observed, they can be sealed with the use of neoprene dissolved in toluene. A neoprene coating has been found to withstand temperatures as low as -70°C and to be stable for about 12 months [20].

SYMBOLS

a	radius of the vent of the closure in cm
A	cross-sectional area
CFT	elastomer closure flange thickness
C_1	concentration of an absorbed gas species at side of the closure defined by d_1
C_2	concentration of an absorbed gas species at side of the closure defined by d_2
de/dt	stress on the top of the seal at a given force per unit time
d_p	diameter of the pipette expressed in cm
d_1	side of the elastomer material that interfaces with the pressure of a given gas species P_1
d_2	side of the elastomer material that interfaces with the pressure of a given gas species P_2
Δ	diffusion coefficient for a given gas species and elastomer material at a temperature T
h_v	vial height
hv	energy of a photon (h is Planck's constant and v is the frequency of the photon)
$k_p(T)$	permeation constant for a given gas species and elastomer formulation at a temperature T

K.N.	Knudsen Number (see Chapter 11 for details)
l	length of the vapor path of the closure in cm
L_r	leak rate expressed in (std cc)sec
m/e	mass to charge ratio
N	number of closures in a vacuum chamber
P_1	pressure in the vial in mTorr
P_2	pressure in the drying chamber
%C_p	percentage compression of the closure for a crimped container-closure system
$Q_{a,c}$	average leak rate expressed in terms of water vapor for a crimped container-closure system
$Q_{a,u}$	average leak rate expressed in terms of water vapor for an uncrimped container-closure system
Q_c	helium leak rate for the crimped container-closure vials
Q_m	gas flow from the closure under molecular flow conditions
$Q_{m,avg}$	average rate of outgassing of water vapor expressed in terms of mass of water per unit time
Q_u	helium leak rate for uncrimped container-closure vials
Q_v	gas flow from the closure under viscous flow conditions
$Q_{w,t}$	total outgassing rate of water vapor from N number of closures at a temperature T
ROR	rate of pressure rise for a system
ROR_{He}	corrected rate of pressure rise for helium
$ROR_{w,b}$	rate of pressure rise for a clean, dry empty vacuum system at a temperature (T) in the chamber having an effective volume of V_e
$ROR_{w,c}$	rate of pressure rise for the partial pressure of water vapor with N number of closures at a temperature (T) in the chamber having an effective volume of V_e
sccs	leak rate expressed in standard cm^3/sec and is approximately 0.1 Pa m^3/sec (SI units) [29]
sccm	standard cubic centimeters
S	solubility constant for a given gas species and elastomer material at a temperature T
S_h	sealed package height
S_c	ratio of the number molecules of a given species that remains on the surface to the number of molecules that strike the surface in a given time period

S_t seal thickness

$t_{w,u}$ time, expressed in hr, necessary to increase the moisture content of 1 mg of lyophilized material by 1% for an uncrimped container-closure system

$t_{w,c}$ time, expressed in hr, necessary to increase the moisture content of 1 mg of lyophilized material by 1% for a crimped container-closure system

σ_1 a downward force exerted by the West Seal Tester (supplied by West Co., Phoenixville, Penn.) to a small area on the top circumference of the seal

σ_2 compression force of the seal exerted on the container-closure system by the crimp seal

t time

T temperature expressed in °C or K

V_e effective volume of a vacuum system (total system volume – volume of hardware)

V_e' corrected effective volume of the vacuum system

V_c volume occupied by the closures

Y number of uncrimped or crimped container-closure systems

REFERENCES

1. K. S. Leebron and T. A. Jennings, *J. Parenter. Sci. & Tech.*, 35: 100 (1981).

2. L. C. Li, J. Parasrampuria, and Y. Titan, *J. Parenter. Sci. & Tech.*, 47: 270 (1993).

3. G. H. Hopkins, International Symposium on freeze-drying of biological product, Washington, D.C., 1976, in *Develop Biol. Standard.* (Karger, Basel), 36: p. 139 (1977).

4. W. A. Preston, *J. Parenter. Sci. & Tech.*, 40: 207 (1986).

5. D. K. Morton, *J. Parenter. Sci. & Tech.*, 41: 145 (1987).

6. West Company information on elastomers (1/24/98).

7. J. Yardwood, *High Vacuum Technique*, John Wiley & Sons, Inc., New York, 1955.

8. S. Dushman, *Scientific Foundations of Vacuum Techniques*, 2nd ed. (J. M. Lafferty, ed), John Wiley & Sons, Inc., New York, 1962.

9. M. J. Pikal, M. L. Roy, and S. Shah, *J. Pharm. Sci.*, 73 1224 (1984).

10. G. J. van Amerongen, *Rubber Chemistry and Technology*, 37: 1065 (1964).

11. R. N. Peacock, *J. Vac. Sci. Technol.*, 17: 330 (1980).

12. B. Crist, *J. Parenter. Sci. & Tech.*, 48: 189 (1994).

13. M. J. Pikal and S. Shah, International Symposium on Biological Product Freeze-Drying and Formulation, Bethesda, USA, 1990, in *Develop Biol. Standard.* (Karger, Basel), 74: p. 165 (1991).

14. M. J. Pikal and J. E. Lang, *J. Parenteral Drug Assoc.*, 34: 62 (1980).

15. H. Scholze, *Glass—Nature, Structure and Properties* (translated by M. J. Lakin), Springer-Verlag, New York, 4th ed., 1990.

16.   T. A. Jennings and H. Duan, *PDA J. Pharm. Sci. & Tech.* 49: 274 (1995).

17.   D. K. Morton and N. G. Lordi, *J. Parenter. Sci. & Tech.*, 42: 23 (1988).

18.   T. A. Jennings, *J. Parenter. Drug Assoc.*, 34: 67 (1980).

19.   N. A. Williams, Y. Lee, G. P. Poli, and T. A. Jennings, *J. Parenter. Sci. & Tech.*, 40: 135 (1986).

20.   H. Melton, D. Greiff, and T. W. G. Rowe, International Symposium on freeze-drying of biological product, Washington, D.C., 1976, in *Develop Biol. Standard.* (Karger, Basel), 36: p. 145 (1977).

21.   W. R. Frieben, R. J. Folck, and A. Devisser, *J. Parenter. Sci. & Tech.*, 36: 112 (1982).

22.   V. Wu and B. Crist, *J. Parenter. Sci. & Tech.*, 43: 179 (1989).

23.   L. E. Kirsch, L. Nguyen, and G. S. Moeckly, *PDA J. Parent. Sci. & Tech.*, 51: 187 (1997).

24.   A. P. MacKenzie, D. G. E. Welkie, M. Lagally, M. Pace, and F. I. Elliott, International Symposium on freeze-drying of biological product, Washington, D.C., 1976, in *Develop Biol. Standard.* (Karger, Basel), 36: p. 151 (1977).

25.   J. S. Amneus, *J. Parenter. Drug Assoc.*, 32: 67 (1978).

26.   J. T. Connor, *J. Parenter. Sci. & Tech.*, 37: 14 (1983).

27.   M. A. Joyce and J. W. Lorenz, *J. Parenter. Sci. & Tech.*, 44: 54 (1990).

28.   D. K. Morton, N. G. Lordi, L. H. Troutman, and T. J. Ambrosio, *J. Parenter. Sci. & Tech.*, 43: 104 (1989).

29.   D. M. Guazzo, *PDA J. Pharm. Sci. & Tech.*, 35: 378 (1996).

30.   *CRC Handbook of Chemistry and Physics*, 65th ed. (R. C. West, ed.), CRC Press, Inc., Boca Raton, Florida, 1984.

31.   F. W. Sears and M. W. Zemansky, *College Physics* (1957), 2nd ed., Addison-Wesley Publishing Company, Reading, Mass.

13

Effect of Vacuum Freeze-Dryers— Present and Future

INTRODUCTION

The main focus of this chapter will not be on the design and construction of the vacuum freeze-dryer, although the advantages and disadvantages of batch and continuous dryers will be considered in the last section, but on the impact that the freeze-dryer can have on the lyophilization process, the final product, and the container-closure system. For those readers who desire a more in-depth description of the actual design and construction of the dryer, references [1–6] are recommended. As a basis for our discussions, refer to the diagram of the freeze-dryer in Figure 1.2.

IMPACT ON THE LYOPHILIZATION PROCESS

A principal function of the freeze-dryer is to provide the necessary operating parameters to conduct the lyophilization process. In providing these operating parameters, the question that will be addressed is, *Does the freeze-dryer play a passive or an active role in conducting the lyophilization process?*

A passive role would be that the dryer provides a set of operating parameters, such as chamber pressure (P_c) and shelf-surface temperature (T_s), that is independent of the design, construction, and instrumentation of the dryer. In other words, the passive freeze-dryer merely serves as a conduit for the defined lyophilization process and does not, in itself, affect the lyophilization process. In this instance, the dryer should provide a final product that is both uniform in appearance and performance [5].

For a dryer to play an active role in the lyophilization process would imply that the design, construction, and instrumentation of the dryer will, in some way, affect the nature of the lyophilization process. By affecting the operating parameters by some means, a freeze-dryer playing an active role will, in some manner, place its own impact (*marking*) on the lyophilization process. This *marking* of the process is often revealed when a process is transferred from one freeze-dryer to another [5]. The reader may expect that there may be some marking when transferring the process from dryers supplied by different manufacturers; however, the author must warn the reader that marking is also possible between dryers of the same model and manufacturer. This is not to imply that freeze-dryer manufacturers are not diligent in their efforts to produce a quality product, but freeze-dryers are highly complex pieces of equipment that require the integration of components from numerous sources in order to produce a system that will function over temperature ranges of > 121°C to < -60°C and pressures from > 15 psig to ≤ 20 mTorr.

Impact on the Lyophilized Product

A second and equally important function of the freeze-dryer is to provide a safe environment for the final product. The quality of the product should not be compromised by any source of contamination stemming directly or indirectly from the freeze-dryer. An example of a direct source of contamination would be a leak in the heat-transfer system that would permit the vapors from the heat-transfer fluid to be adsorbed by the product. An indirect source of contamination would be exemplified by the presence of a high partial pressure of oxygen in the system as a result of a significant real leak in the dryer. The oxygen could then react with a reducing agent, e.g., stannous chloride ($SnCl_2$), and affect the performance of the final product. While the oxygen did not come directly from the dryer, its presence was a result of a defect in the vacuum integrity of the dryer.

Effect on the Container-Closure System

Throughout our discussion of the freeze-dryer, we must be cognizant of how a freeze-dryer can impact the container-closure system with regard to the latter's role during the lyophilization process or in its protection of the final product from the environment once it is removed from the dryer. The dryer can affect the role of the container during the lyophilization process by providing the container with an additional source of energy. For example, the additional source of energy could, if sufficient, result in an increase in the product temperature (T_p). An increase in T_p may only lead to a higher rate of drying but, if excessive, result in mobile water being present in the interstitial region of the matrix, which could cause collapse or even meltback of the product.

The effectiveness of the container-closure system may also be compromised by the freeze-dryer. Should the shelves of the dryer not be flat, uniform, and parallel with respect to one another, then there may be vials where the closure is not fully seated. As seen in the previous chapter, the primary seal between the container and the closure occurs on the flange of the vial. If during the stoppering process there is insufficient force exerted on the stoppers because of the configuration of the shelves, then those vials not fully seated will not be protected from the environment prior to completion of the crimp sealing operation. The dilemma that one faces here is that the stoppered vials will require an additional inspection step prior to performing the crimping operation. It would be far more prudent to have shelves where it can be verified that all of the closures are fully seated.

From the above discussion, it should be apparent that it is the intent of this chapter to make the reader aware of those areas of the freeze-dryer that can have an impact on the lyophilization process, the final product, and the container-closure system. I have made an effort to be rigorous in examining the impact of the freeze-dryer but must advise the reader that, because of the freeze-dryer's complexity and its continued evolution with improved technology, the reader should be conscious of potential areas where ill-advised innovation would be counterproductive.

Safety Considerations

The freeze-dryer should not only provide a means for producing a lyophilized product but also ensure the safety of both the operating personnel and the facility [5]. For example, because of the wide range in temperatures that the heat-transfer fluid may undergo during the course of a manufacturing process (i.e., from -50°C to >121°C), most freeze-dryers are equipped with an expansion tank. The level of the heat-transfer fluid in the expansion tank will increase during the sterilization process (> 121°C) and decrease during the freezing process (-50°C). The level of the excess heat-transfer fluid in the expansion tank is often indicated by a glass tube that is positioned along side of the tank. In many instances, the tubing is unprotected from accidental breakage with a metal shield. Since the expansion tank is often positioned in the vicinity of the electric motors that drive the compressors and vacuum pump(s), an accidental breakage of such a tube could release a large volume (gal) of heat-transfer fluid that could not only represent a potential fire hazard but also harm operating personnel if the fumes of such fluids are inhaled. It is surprising to find facilities containing one or more freeze-dryers containing such large volumes of flammable material and equipped with only standard water sprinkler systems. For a substance like silicone oil, activation of the sprinkler system may actually cause the fire to spread rather than be extinguished.

One final word regarding the manufacturers of freeze-dryers. I know of no freeze-dryer manufacturer who does not make an attempt to supply the industry with dryers of the highest quality. The basic problem, in the opinion of this author, is that the manufacturers of this equipment do not use a freeze-dryer to manufacture a pharmaceutical, biological, biotechnology, or diagnostic product. Because of this, the equipment manufacturers are not often aware that their equipment may have an adverse effect on the lyophilization process, the lyophilized product, and the performance of the container-closure system.

THE DRYING CHAMBER

The various features of the drying chamber and their potential impact on the lyophilization process and the final product will be considered here.

Chamber Design or Configuration

The ideal shape of a freeze-dryer chamber, from the viewpoint of constructing a vacuum chamber, would be spherical in nature. Because it is spherical, the pressure is uniformly distributed around the entire sphere. A sphere configuration would ensure complete drainage after cleaning, sterilizing, or defrosting a condenser surface. However, such a configuration would not lend itself well when it is equipped with shelves, a stoppering mechanism, and a door.

In order to make the most efficient use of space, the drying chamber is generally rectangular in configuration. Because of the immense pressure exerted on the walls of the chamber during steam sterilization and the drying process, the thickness (mass) of the walls and door must be increased. In order to withstand such forces, the ideal chamber should be equipped with reinforcement ribbing on the outside and inside of the chamber. Placement of reinforcement ribbing on the inside of the chamber would serve to protect the chamber from implosion [7], but it would also reduce the effective volume of the shelves that can be accommodated by the chamber. Placement of reinforcement ribbing on the outside chamber serves to protect both personnel and the facility from a rupture of the chamber during steam sterilization.

In general, the reinforcement ribbing is placed on the outside of the chamber. While such a placement may not protect the chamber against an implosion, should such an occurrence take place, there will be a sudden increase in P_c, but there would be little danger to the operating personnel or the facilities. A rupture of the chamber containing wet steam at 121°C and 15 psig would, however, pose a serious threat to personnel. For example, the release of a 3 in. diameter flange weighing less than 1 lb would have the impact of more than 100 lb. Thus, a poorly designed chamber can pose a serious threat to operating personnel plus a loss in the use of the equipment.

By the addition of more steel to withstand the above extreme pressures, the mass of the dryer is also increased. This increase in the mass of the walls of the drying chamber can impact the lyophilization process. The mass of the walls acts as a heat sink. When the shelves are being chilled to conduct the freezing process, there will be a loss in refrigeration resulting from radiant energy heat transfer between the walls and the shelves (Stefan-Boltzmann Law, see Chapter 8) and gas convection currents resulting from gases rising from the walls of the chamber and falling as a result of contact with the shelves. In order to achieve the necessary T_s for the shelves, one must also refrigerate the walls.

I have demonstrated the effect that the walls of the dryer have on limiting T_s during the freezing process by lining a small production dryer with aluminum foil. Without the foil being present, the lowest T_s attainable was about -45°C. With the aluminum foil lining in place, T_s values lower than -55°C were obtained. The emissivity of the aluminum foil is lower than that of the stainless steel that was used to construct the walls and door of the dryer. Thus, the presence of aluminum foil on the walls of the chamber served to reduce the radiant energy heat transfer between the walls and the shelves. Because of the low mass of the thin foil, the energy transfer between the chamber walls and the shelves resulting from gas convection was reduced. By reducing heat transfer between the walls of the chamber and the shelves, a substantiality lower T_s was obtained.

The lower T_s obtained with the aluminum foil made it possible to lyophilize a formulation with a relatively low collapse temperature (Tc). This would not have been possible had the inside walls of the freeze-dryer not been lined with aluminum foil.

Construction

The following discusses the impact that the construction of the dryer chamber can have on the lyophilization process and the final lyophilized product.

Materials

While seldom used today, at one time, many freeze-dryers were constructed with cold-rolled steel. Cold-rolled steel is less expensive than stainless steel, and it is easier to machine and form weld joints. But because the chamber will be subject to cleaning or sterilization that involves water, the inside walls of the drying chamber, door, and shelves are generally subjected to a chemical treatment that deposits several layers of a phenolic resin material that is then cured by baking at elevated temperatures [4]. The baking of the resin serves not only to cure the resin but to remove any volatile compounds that may remain in the resin.

When using a freeze-dryer constructed with resin-coated cold-rolled steel, one must exercise some caution. First of all, it should be established that the resin is not a source of vapors that could affect the quality of the final product. I was involved in the lyophilization of a product in a resin-coated freeze-dryer. It was observed that during the pumpdown of a drying chamber, with a chilled external condenser, the time required to reach 100 mTorr was within the performance specifications of the dryer. However, when the dryer chamber was isolated from the condenser chamber (see Figure 1.2), there was a sharp increase in the chamber pressure. It was later established that the rate of pressure rise (ROR) of the chamber was not a result of a real leak but from a virtual leak in the chamber. The ROR of the virtual leak was found to decrease when the shelves were heated to about 60°C and the chamber was subjected to a gas purge with dry nitrogen for more than 8 hr. After numerous purges of dry nitrogen with a T_s of 60°C, the ROR of the chamber when isolated from the condenser chamber was reduced to about 1 mTorr/min.

Had the ROR been a result of a real leak, then heating the shelves to 60°C and purging the chamber with dry nitrogen would not have any effect on the initial leak rate. If the leak rate was a result of a leak in the heat-transfer fluid system to the shelves or from the shelves themselves, the ROR would not have been reduced by heating the shelves. Therefore, it was concluded that the source of the virtual leak was a result of the release of volatile compounds from the resin coating. Tests on the product lyophilized in the dryer with the virtual leak from the resin did indicate the presence of foreign hydrocarbons in the product or the loss in activity of the active constituent. Therefore, in using a freeze-dryer constructed with resin-coated cold-rolled steel, it would be prudent to ensure that the resin does not represent a source of undefined vapors prior to using the system to produce a lyophilized product.

Current freeze-dryer chambers are fabricated with stainless steel. For dryers that do not require steam sterilization, 304 stainless steel should be adequate. For those dryers requiring steam sterilization, 316L stainless steel is recommended [35]. Because 316L stainless steel contains molybdenum, it is more expensive than 304 stainless steel. However, it does offer protection from chloride stress corrosion at the grain boundaries that are formed when making a weld joint. The welding of 304 stainless steel alters the composition of the stainless steel and a grain boundary is formed about 1 to 2 in. from the weld. The boundary represents that region where altered stainless steel interfaces with the original stainless steel. It is along this grain boundary that chloride stemming from such sources as the atmosphere, the insulation covering the chamber, or even the product itself will react, especially during steam sterilization when there is both heat and moisture present for some period of time, with the metal. The end result of chloride stress corrosion is the formation of a real leak path in the chamber. Such a corrosion process occurs slowly over a period of time (2–4 years), and the presence of real leaks from chloride stress corrosion may not be noticeable at first but will gradually increase with time. In addition, as will be described later in this chapter, the presence of such leaks can be difficult to detect and costly to repair.

Finish

While some prefer a *satin* or 240 grit finish [6] over a *mirror* or electropolished surface on the inside walls of the drying chamber, one cannot agree with the rationale that a satin finish is preferred over a mirror finish because the latter is more difficult to clean and marks easier. A smooth surface is easier to clean because there is less surface area and locations where particles can become lodged, and water can drain easier from the surface. Because the freeze-dryer should provide a safe environment for the product during the lyophilization process, a mirror finish on the walls of the chamber is preferred over a satin finish.

In addition, a highly polished (*mirror*) surface will have a lower emissivity than that of the satin finish and, as discussed earlier in this chapter, the reduction of the emissivity will reduce the radiant energy heat transfer between the walls of the chamber and the shelves. While the reduction of the radiant energy heat transfer will certainly be beneficial during the freezing process, the reduction of the radiant energy from the walls of the dryer will be of particular importance during the primary and secondary drying processes. With a major reduction in the radiant energy heat transfer from the walls, those vials or trays having a line of sight with the walls will only receive energy from the shelves and, therefore, will not dry faster as a result of an external source of radiant energy [8]. The mirror finish will not, however, prevent the heat transfer resulting from gas conduction that can occur during the freezing process.

Insulation

The nature of the insulation that covers the outside walls of the chamber is an important consideration that is often overlooked. One reason for the lack of interest in the insulation of the chamber is that it is, for the most part, covered with a thin aluminum or stainless steel sheet. But the insulation is an important consideration, because it has been shown that heat transfer between the walls of the chamber and the shelves can greatly affect not only the T_s but also any vials having a line of sight to the chamber walls. The effect that the chamber walls can have on the lyophilization process and quality of the final product is based on experience rather than conjecture.

While studying the heat-transfer properties of vials at various low pressures [9], it was found that calorimetric measurements were being affected by radiant energy from the walls and door of the dryer. In order to eliminate such stray thermal radiation, the entire test region on the shelf had to be shielded with aluminum foil.

It was only after carefully shielding the region containing the test vials could one achieve confidence in the measurements.

The most typical insulation materials used for the drying chamber are rockwool, Ameflex™, and polyurethane. What makes a good material for insulation is a low thermal conductivity of heat. For example, the thermal conductivity of rockwool is about 1×10^{-3} (cal cm^2)/(sec cm^3 °C). This means that 1 cm^3 of rockwool will conduct 1 cal/sec for an insulation thickness of 1 cm when there is a temperature differential of 1°C across the insulator. If the chamber is cooled by the shelves to a temperature of -40°C and the ambient room temperature is 20°C, then the rate at which energy will be transferred to the chamber will be 60 mcal/sec or 216 W-hr or 0.29 hp-hr. For a surface area of 1 m^2, the rate of energy transfer would be 2,160 kW-hr or 2,894 hp-hr. Increasing the thickness of the rockwool to 10 cm (3.4 in.) would reduce the heat transfer to 216 kW-hr. While using insulating material over the surface of the chamber will reduce the thermal conductivity from the environment to the chamber, it will not prevent the radiant energy heat transfer. Placement of a second metal sheet, having a low emissivity, between the drying chamber and the two layers of insulating material will serve to reduce the radiant energy heat transfer between the outer metal cover of the dryer and the outer surface of the chamber.

In evaluating the insulation of a dryer, one should be aware of how the insulating material is maintained in place. For example, does the insulation have a permanent configuration, such as a sheet, or it is made up of material packed between two barriers? In the latter case, one must be concerned that over a period of time the material may tend to settle and become packed. Should the former condition occur, then there could be a region, perhaps near the top of the dryer, where there would be insufficient thermal insulation.

It is hoped that, as a result of the above discussion, those individuals designing or purchasing a freeze-dryer will pay more attention to the composition and configuration of the insulation used in the construction of a dryer chamber.

Leaks

It was pointed out in Chapter 11 that all vacuum systems leak, and a freeze-dryer chamber is no exception. With repeated calibration checks of gauges and lyophilization processes, seals and connections become worn and, with time, the potential for leak paths is enhanced [2]. Therefore, our emphasis in this section will be concerned with a significant increase in the leak rate in the drying chamber. Another intent of this section is not only to make the reader cognizant that such leaks will develop with time but, more importantly, the impact that such leaks can have on the lyophilization process and the quality of the product. While it is important to recognize a major increase in the leak rate of the drying chamber, it is paramount that the reader be aware of the various means that can be used not only to locate the source of the leak but also in sealing it.

Real Leaks

While the nature of real leaks in a vacuum system was described in Chapter 11, this section will be mainly concerned with the impact that a significant increase of such leaks in the drying chamber can have on the lyophilization process and the quality of the final product [10,11].

Gas Composition. One of the effects of a real leak is to affect the composition of the gases in the drying chamber during both the primary and secondary drying processes [11]. When a major real leak occurs, there will be an increase in the partial pressure of oxygen. Should the formulation contain, as one of its constituents, a reducing agent (e.g., $SnCl_2$), the exposure of such a constituent to oxygen, especially during the secondary drying process when the product is near ambient temperature, could result in the formation of Sn^{+4} and a reduction in the activity of the final product [12].

The problem may be further complicated at the storage temperature of the final product, when the specific reaction rate between oxygen and $SnCl_2$ may require some time before the final product is not within its specifications. One could be misled in believing that the oxygen resulted from the container-closure system when, in fact, the oxygen was present in the vial prior to the stoppering operation. Even backfilling the drying chamber with nitrogen prior to stoppering may not be effective in reducing the oxygen gas that would be present in the headspace of the vial because the oxygen already in the vial would have to leave the vial while there is viscous flow of nitrogen into the vial.

Microorganism Contamination. The possibility of microorganism contamination of the drying chamber increases when there is a major real leak in the drying chamber. The possibility of microorganism contamination is heightened if only the door of the drying chamber interfaces with an aseptic area, and the remainder of the surface area of the drying chamber is exposed to an environment that has a higher microbial density, e.g., ≥ 35 CFU/ft^3 [10]. Although there have not been any published incidents of microorganism contamination as a result of a real leak path, the absence of such published information, especially in an industry that is closely regulated by a government agency, should not be taken as proof that such contamination is not possible or has never occurred.

The average effective diameter of an airborne microorganism is about 12 μm. If the density of the colony forming organisms (CFU) in the drying chamber is found to exceed 1 CFU/10 ft^3, then the lyophilization process can no longer be considered as performed under aseptic conditions [10]. The nature of real leak paths was illustrated in Figure 11.13. It is estimated that in order for a microorganism to enter the drying chamber, the average diameter of the leak path must be of the order of 1×10^{-3} cm. Since the mean free path of nitrogen and oxygen at 1 atm is about 7.5×10^{-6} cm, the nature of the gas flow through the latter leak path, assuming the worst case, will be viscous. If one further assumes that the density of the microorganisms that flow through the leak path is the same as that of the surrounding atmosphere, then knowing the volume of gas that will flow will also be an indication of the number of microorganisms that may have entered the drying chamber.

Based on the above assumptions, the rate (R_m) at which microorganisms can enter a drying chamber can be expressed as

$$R_m = Q_v \times CFU_d \quad \frac{CFU}{min} \tag{1}$$

where Q_v is the viscous gas flow through the leak path expressed in terms of atm cfm, while CFU_d is the density of the microorganism in the surrounding environment.

Since it has been shown that there is a relationship between Q_v and the ROR resulting from a real leak path in the drying chamber, one can obtain

$$\text{ROR} = \frac{R_m}{V_e} = D_m \quad \frac{\text{CFU}}{\text{ft}^3 \text{ day}} \tag{2}$$

where V_e is the effective volume of the chamber and D_m is the microorganism density per unit time [10].

The importance of equation 2 is that D_m will be inversely dependent on the effective volume of the drying chamber (V_e) and directly proportional to R_m, which will be related to the D_m in the surrounding environment.

Figure 13.1 shows a family of log-log plots of the ROR as a function of D_m for various values of CFU_d (i.e., 1 CFU/ft^3, 5 CFU/ft^3, and 35 CFU/ft^3). The vertical dotted lines shown in Figure 13.1 indicate the upper limits for a dryer to be considered sterile and aseptic. The upper limit for sterility was arbitrarily set at 3×10^{-3} CFU/ft^3, while the upper limit for the environment in the dryer to be considered aseptic was set at 1 $CFU/10$ ft^3 [10].

If one assumes that the V_e of a drying chamber is 1 ft^3 (the effective volume for a development drying chamber), then the maximum ROR for a drying chamber contained in an environment that has a CFU_d of 1 CFU/ft^3, would be 630 mTorr/hr. When the CFU_d of the surrounding environment is increased to 5 CFU/ft^3, then the maximum ROR would have to be reduced to about 12 mTorr/hr and, for an

Figure 13.1. A Family of Log-Log Plots for the Rate of Pressure Rise (ROR) for a Leak Path Length of 1 cm as a Function of the Microorganisms/(day ft^3) for Surrounding Environment of a Given Bioburden

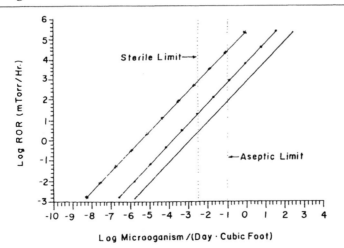

The lines (———), (•–•–•), and (*–*–*) denote a surrounding environment of 35 CFU ft^{-3}, 5 CFU ft^{-3}, and 1 CFU ft^{-3}, respectively. [Source: *Journal of Parenteral Science and Technology*, "A Model for the Effect of Real Leaks . . . Freeze Dryer,", Vol. 44, No. 1, Jan./Feb. 1990, Page 24, Figure 2. Reprinted with permission.]

environment having a CFU_d of 35 CFU/ft^3, the ROR would have to be further reduced to 2 mTorr/hr. It has been my experience that leak rates of only 2 mTorr/hr and 12 mTorr/hr for a freeze-drying chamber having a V_e of just 1 ft^3 would be highly unlikely. Therefore, for relatively small drying chambers, the CFU_d of the surrounding environment would have to be about 1 CFU/ft^3 in order to ensure that the drying chamber would remain sterile for at least 24 hr.

Based on the assumption that Q_v of the drying chamber remains unchanged but the value of V_e is increased to 10 ft^3, then the ROR of the drying chamber contained in environments of 35 CFU/ft^3 and 5 CFU/ft^3 could be increased to 20 mTorr/hr and 120 mTorr/hr, respectively. While leak rates of 120 mTorr/hr would be attainable for a drying chamber having a V_e of 10 ft^3, it would not be likely that such a chamber would have a leak rate of 20 mTorr/hr. In order to maintain sterility over 24 hr, when the surrounding environment is 35 CFU/ft^3 and the leak rate is unchanged, V_e would have to be increased to about 25 ft^3. What is important for the reader to understand from the above discussion is that in order to maintain sterility in a drying chamber, one must take into account (worst case scenario) not only the ROR of the chamber but also its V_e, the magnitude of the CFU/ft^3 in the surrounding environment, and the duration that the drying chamber is maintained at low pressures. Certainly, as the length of the drying process is increased, the likelihood of maintaining sterile conditions in the drying chamber is decreased.

Since the filling and loading of the dryer is often done under aseptic conditions, one would assume that maintaining aseptic conditions throughout the drying process results in a safe environment. This is particularly true since, once the door of the sterile dryer is opened to the aseptic environment, the dryer can no longer be considered sterile.

The second dotted line in Figure 13.1 [10] represents the upper limit of the ROR to maintain a drying chamber under aseptic conditions for an environment having a given CFU_d. Based on a dryer having a V_e of 1 ft^3, the upper leak rate for surrounding environments of 1 CFU/ft^3, 5 CFU/ft^3, and 35 CFU/ft^3 would be 20,000 mTorr/hr, 630 mTorr/hr, and 63 mTorr/hr, respectively. Since a leak rate of 63 mTorr/hr can be easily obtained for most freeze-dryers having a V_e of just 1 ft^3, these results would suggest that maintaining an aseptic condition in the drying chamber for 24 hr is more plausible than maintaining a sterile environment. But, once again, one must not lose sight of the fact that the typical drying process exceeds 24 hr, and so one must exercise some caution, especially when lyophilization is being conducted in a drying chamber that has a relatively small V_e. For a drying process having a duration of 5 days, the maximum *ROR* for a dryer housed in a typical environment of 35 CFU/ft^3, would be limited to just 12 mTorr/hr. As in the discussion for maintaining a sterile environment in a small drying chamber, it would be equally unlikely that a dryer with a V_e of only 1 ft^3 would have a ROR $\leq$ 12 mTorr/hr. The reader should take note that when using a freeze-dryer having small V_e, the ROR, the nature of the surrounding environment, and the duration of the drying process should be given careful consideration in order to ensure (worst case scenario) that the lyophilization of the formulation is performed under aseptic conditions.

Virtual Leaks

While the above has shown that real leaks in the drying chamber can represent a potential source of chemical and microbial contamination of the formulation during the drying process, virtual leaks (see Chapter 11) represent a myriad of potential sources of contaminants for the formulation prior to, during, and after the drying process. The following discussion cannot possibly account for every possible source of a virtual leak, but it will provide the reader with a general overview of the nature and potential sources of such leaks. The impact of substances from such leaks on the quality or activity of a lyophilized product is product dependent.

Virtual leaks are a result of gases entering the drying chamber from an internal source. These leaks could be gases stemming from the closures, the heat-transfer fluid (shelf or condenser), and the backstreaming of oil vapors from the mechanical pump [12,13]. The difficulty with virtual leaks is that they tend to develop over a period of time, and it is difficult to ascertain at what point they can affect the quality of the final product [12].

Comment: If the final product is not tested for traces of pump oil or heat-transfer fluid, how could one know when the presence of such substances represents a real problem? This is not to say that such testing is not performed; however, I do not know of a manufacturer of lyophilized product who routinely checks for the presence of these substances in the final product.

Virtual leaks can stem from the cleaning of vacuum components with solvents. For example, vacuum components are often cleaned with trichloroethylene (TCE). However, relatively inert materials, such as Teflon™, can absorb large amounts of TCE, which may result in the generation of a virtual leak that could require weeks under vacuum to completely remove from the system [14]. In order to avoid such virtual leaks in a freeze-dryer chamber, one should avoid cleaning elastomer seals with oils, chlorinated compounds such as TCE, esters like ether or β,β'-dichloroethyl ether, and ketones such as acetone. These substances can be absorbed into the elastomer seal, especially a gasket that serves as a seal for the door, and thereby represent a potential source of product contamination over an extended period of time.

Other sources of virtual leaks may result from the construction of the dryer. For example, if improper welding techniques are employed, pits in the welded joints may act as a virtual leak and could be a source of contaminants that are introduced when the freeze-dryer is constructed. Solvents or disinfectants used during the manufacture or the initial preparation of the dryer may find their way into such defective weld joints [12]. The time that it may require to completely outgas such a source of contamination could be extensive, e.g., a matter of months or, in an extreme case, even years.

Leak Testing

The following will describe some general guidelines for not only identifying the nature of the source of the leak (real or virtual) but also means for identifying its location and possible steps for eliminating (repairing) the leak path.

Dryer Preparation Leaks and the Nature of the Leak

Before endeavoring to determine if a significant leak is present and the nature of such a leak (real or virtual), one must ensure that the complete dryer is clean and dry [13]. Since water vapor in a drain line could give misleading results, it is best that any water trapped in the drain lines be removed. One way to remove such water is to seal the chamber and then pressurize the system (1–5 psig) with dry nitrogen. The drain lines are then opened, and the gas is discharged into the drain lines for some period of time (for most systems, 5–10 min should be sufficient). I was once asked to determine the poor performance of a freeze-dryer. Preliminary results indicated that the dryer appeared to have a serious virtual leak. A visual inspection of the dryer revealed that there was still condensate (estimated at more than 50 L) from the defrosted condenser coils in the condenser chamber. The problem was that the drain valve was opened, but the air release valve on top of the condenser was not opened to allow all of the water to drain from the condenser. The latter could be avoided by a careful visual inspection of the system and an effective Standard Operating Procedure (SOP) for maintenance of the dryer.

Upon completion of the preparation of the dryer, the presence of leaks in a chamber can be determined from a measure of the ROR. From the plot of ROR as a function of time, one can ascertain the presence and magnitude of real leaks, virtual leaks, and a combination of real and virtual leaks [12]. But before conducting the test, it is important to properly prepare the dryer for the determination of the leak rates and specify the conditions under which the test will be conducted.

With the dryer at ambient temperature (including the condenser) and the data collection system on, the chamber is first evacuated to about 100 mTorr and then the system is purged with a flow of dry nitrogen sufficient to maintained P_c at about 2,000 mTorr. The objective of this conditioning of the dryer is to make its performance independent of its location (i.e., the performance of the dryer will be the same regardless of the nature of the ambient environment). It is sometimes suggested that the shelves be chilled to 0°C to 10°C and the condenser temperature lowered to –55°C [13]. However, these operating conditions would tend to mask any virtual leak that may exist in the heat-transfer or condenser refrigeration system.

The gas purge is then turned off, and the freeze-dryer evacuated to a P_c of 100 mTorr. When the P_c reaches 100 mTorr, the dryer chamber is isolated from the vacuum system in order to conduct the pressure rise test [15]. The ROR of the system is recorded over a period of no less than 1 hr. Typical ROR plots for freeze-dryers are shown in Figure 13.2 [12]. Time zero (0) indicates the time that the dryer chamber was isolated from the vacuum system. Plot A would be typical of a freeze-dryer having only a real leak present, while plot B indicates a typical leak profile with the presence of a small virtual leak and a real leak in the system. The virtual leak could be a result of outgassing from the walls of the dryer and not necessarily represent a source of product contamination. Plot C of Figure 13.2 represents the presence of a major virtual leak [12]. It should be understood by the reader that the pressure increases are not to scale. If a major virtual leak rate like plot C were present in the system, then the pressure rise may be of the order of several Torr, depending on the vapor pressure of the heat-transfer fluid at ambient temperature. Once the equilibrium vapor pressure of the virtual leak is obtained, the pressure will continue to increase as a result of the real leak into the chamber. The reason why the leak rate for plot C appears to reach a plateau is that a leak rate of 30 mTorr/hr will have little impact on the pressure of a chamber that is already very high, e.g., 10 Torr.

A leak rate plot, taken from the data supplied by Waltrick [13], is shown in Figure 13.3. It is assumed, by the manner in which the data were presented, that the leak rate is for the drying and condenser chambers. This plot shows that data can be best represented by a fourth order polynomial expression and is similar to plot B in

Figure 13.2. A Family of Plots of the Rate of Pressure Rise (ROR) of a Dryer as a Function of Time

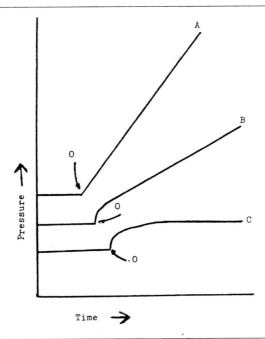

Plot A represents a real leak, plot B represents a combination of virtual and real leaks, and plot C represents a virtual leak rate in the dryer. [Source: *PDA Journal*, Vol. 36, No. 4, July/Aug. 1982, Page 153, Figure 3. Reprinted with permission.]

Figure 13.2. The sharp increase in pressure during the first 4 min of the test is representative of a virtual leak, while the leak rate for times greater than 10 min appears to be somewhat linear in nature and indicative of a real leak.

In determining the leak rate, by measuring the ROR of a system, one should be aware of the effect V_e can have on the results [12]. Note that the relationship between the real leak rate and the product (ROR $\times$ V_e) has already been discussed in Chapter 11. For a system with a V_e of 500 L, the ROR was observed to be 60 mTorr/hr, while for a second freeze-dryer having a V_e of 2,000 L, the ROR would be 15 mTorr/hr. While one dryer would appear to have a lower leak rate, the actual leak rate (Q) in both systems would be the same, i.e., 8 mTorr liters/sec. In addition, one should also take into account the outgassing of stainless steel. It has been shown that stainless steel (304L) may contain up to 100 monolayers of absorbed water vapor. The water is most likely bound in the oxide layer (passivated layer) on the stainless steel. It has also been shown that other gases such as CO_2 may desorb from the stainless steel surface [16].

Figure 13.3. A Leak Plot Obtained from the Data Supplied by Waltrick [13]

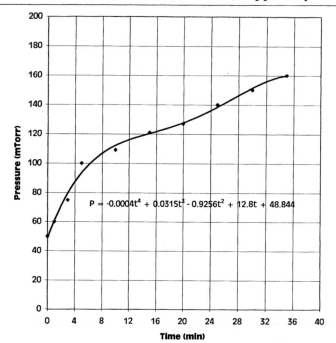

Soap Solution

In the early 1950s, leak testing of the freeze dryer was sometimes performed by pressurizing the chamber and painting any suspecting connection with a soapy solution [7]. This is suitable for locating real leaks in a freeze-dryer chamber, but it is inadequate for detecting virtual leaks. Since the method depends on the formation of soap bubbles, the method is inadequate for detecting small leaks in a system. While this method can be safely used on drying chambers that have been certified as pressure-coded vessels, the reader should exercise extreme care when using this leak detection method with a dryer that has not been certified, e.g., a research dryer. The accuracy of a typical gas pressure regulator may not be adequate for controlling the output pressure between 1 psig and 5 psig. For that reason, the reader should be aware that for a small dryer having a chamber diameter of 2 ft, the total pressure exerted on the door when 5 psig is applied to the system would be 2,262 lb. If the door of the chamber were to suddenly give way, serious injuries could be inflicted on operating personal or there could be substantial damages to the facility. I have found the soap solution test to be quite useful in testing the gas bleed lines to the drying chamber. The failure of a hose connection will only result in a loss of gas and will not represent a physical hazard.

Acetone

Acetone is sometimes used to determine the source of real leaks while the chamber is being evacuated [11,15]. Acetone is a hydrocarbon liquid having a gram molecular weight of 58.08 and a relatively low surface tension of 32.7 dynes/cm at 20°C; water at the same temperature has a surface tension value of 72.5 dynes/cm [17]. Because of its lower surface tension and higher vapor pressure, acetone will flow though small leak passages and cause changes in the P_c. In using acetone for leak testing, one must exercise extreme care not to breathe its vapors or to allow the liquid to come into contact with the eyes or any part of the skin. In addition, because it is highly flammable, one should make certain that there are no open flames present in the working area and that all electrical power has been turned off in the area where the testing is being conducted.

When acetone first comes into contact with a leak path, the leak path in momentarily sealed by the liquid. One would observe, depending on the V_e of the system, a decrease in P_c. When the acetone vaporizes, then there will an increase in P_c. The assumption here is that, at the time of the test, the real leak rate into the chamber is equal to the pumping speed of the vacuum system. The introduction of additional gas into the chamber will result in an increase in pressure. The question that now needs to be addressed is, *How much acetone must enter the system in order to generate a significant pressure response, e.g., 10 mTorr, in order to indicate the location of the leak?*

Figure 13.4 is a plot of the quantity of acetone/min that would be necessary to result in a pressure rise in the chamber of 10 mTorr as a function of the V_e of the chamber. This plot shows that for chambers having a V_e of about 2 ft³ (about the volume of a laboratory freeze-dryer), acetone could be used to detect a leak path that would permit a flow of about 3 mg of acetone/min. But for a production dryer of 100 ft³, the leak path would have to be sizable in order to permit the passage of about 200 mg of acetone/min. Acetone may be used to detect leaks in small systems, but it is useful in production size dryers only to detect very large real leak paths.

Helium Leak Testing

Real leaks in a freeze-dryer can be detected using a residual gas analysis (RGA) system or leak detector and helium as a trace gas [12,18]. While helium is readily available and safe to use, the principal reason for using helium is that it is not commonly found in the atmosphere. Therefore, if helium is sensed by the mass spectrometer, it must have entered via a real leak path.

While helium leak testing is simple in theory, it can sometimes prove frustrating. I took part in determining the location of real leaks in a freeze-dryer. From the large V_e of the dryer and the excessive ROR, there was distinct possibility that one or more very large leak paths were present. In performing the helium testing, one sprays a suspected location with a small stream of helium and then allows a few minutes for the helium to reach the mass spectrometer. After probing near a pressure gauge and waiting a reasonable time, I started to move to another location on the dryer. But in transit, the audio output of the mass spectrometer indicated an usually high helium signal that lasted for several minutes. Returning to the gauge location, I waited until the helium had cleared the system and again probed with a stream of helium around the pressure gauge connection. Once again, when moving away from

Figure 13.4. The Quantity of Acetone/Min Necessary to Increase the Pressure in the Chamber 10 mTorr as a Function of the Effective Volume of the Chamber

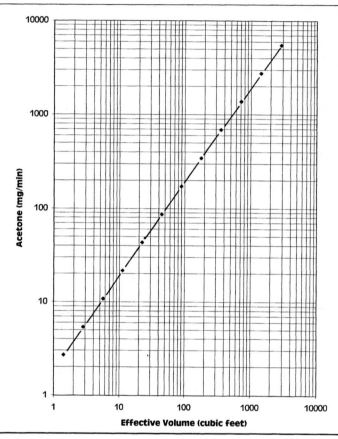

this location, the mass spectrometer indicated the presence of a very large leak path. The long delay time indicated that the leak was not at the gauge connection but that the helium had reached the leak by some other path. When helium was sprayed directly under the insulation of the dryer, the helium reached the mass spectrometer rather quickly. After further investigation, the leak path was located some 5 or 6 ft from the pressure gauge and under the insulation.

The above incident reveals two dilemmas one faces when conducting helium leak testing. The first is the helium may travel to another leak path, and one might falsely identify the location of the leak path. The second dilemma is that leaks may be located in a region that is inaccessible to the helium, in which case the leak path would go undetected. Fowler [19] described a special helium probe that contained a vacuum line. With such a device, it was possible to spray a suspected leak location, and the vacuum line would prevent excess helium from straying to other parts of the system and indicating false signals. If no leaks can be located using the latter helium gas probe, then one must isolate specific regions of the dryer with a sealed plastic cover. The sealed plastic cover is then pressurized with helium. This technique will only indicate that there is a significant leak path in a general area of the dryer. The next step is to locate the leak path using a helium gas probe on all welds or connections in this isolated region.

There are two golden rules in helium leak testing. The first rule is to use only a small stream of helium (i.e., a steady flow of individual helium bubbles is formed when the probe is placed in a container of water). For the second rule, one should commence leak testing from the top to bottom. Because helium is lighter than air, the second rule will reduce the chance of helium contacting a location not previously inspected.

It is one task to locate the real leak path, but stopping or repairing the leaking component is another problem. If the leak is in a flange, it would be best to replace the gasket. However, if the leak is in another location, such as a welded seal, then replacement can be both difficult and costly. In the past, a red resin paint was often applied to the surface of the source of the leak path. The paint would not penetrate the leak path but merely place a polymer cover over the source of the leak.

The use of this resin material to seal leaks led to two basic problems. First, the paint was soon being applied to all suspected leak sources. It would not be long before all connections are painted red and, instead of serving as a leak marker, the painted surface only served to obstruct future leak testing. It was often the case that the application of the resin resulted in a decrease in P_c. In such cases, the paint not only served to locate the leak path but also sealed it. The second problem is that in time the paint would age and crack and once again expose the system to the leak path. The presence of the aged resin only served to hinder the locating of the leak path.

Another material, Loctite 290, has been found to be most useful in repairing small leaks [20]. This material is a liquid with a low viscosity, which allows it to penetrate the leak path. What makes this sealant so interesting is that it will remain liquid at atmospheric pressure, but because it is anaerobic in nature, it will polymerize when exposed to a vacuum. The polymerization of the polymer will occur at ambient temperature. The normal temperature range for such a leak seal is -53°C to 149°C, which would appear to make it useful; however, this product has not been verified for use with steam sterilized freeze-dryers. Outgassing studies conducted on this material showed no detectable outgassing until the temperature exceeded 140°C. In the case of very large leaks, Loctite 271 and 277 should be used because of their higher viscosities.

Mass Spectrometer

Virtual leaks can be detected via a residual gas mass spectrometer [12,18]. The mass spectrometer is useful in identifying the nature of the source of the leak (i.e., if the vapors are from the refrigerant used in the refrigeration system, the heat-transfer fluid, or cleaning and sterilization compounds). However, detection of the nature of the leak would prove useful in locating the general area of the leak and the actual leak path.

A second method of determining the location of a virtual leak is to use a mass spectrometer system equipped with a *sniffer* probe [12]. Instead of spraying the suspected site with helium, one would use a tube connected to the inlet to the vacuum system of a mass spectrometer. The sample tube must have a small diameter so that the ratio of compound to air will be greater than one when the probe is placed in the vicinity of a virtual leak. In addition, the use of a small diameter sniffer probe will also allow for a low pressure in the transfer line from the sniffer probe to the inlet of the mass spectrometer. The latter will increase the sensitivity and reduce the response time in detecting the leak. In this instance, the chamber would be pressurized with helium, and the sniffer probe would determine the source of the leak. Because of the need for a vacuum line rather than a helium gas line, the use of the sniffer probe may be useful for detecting leaks in small freeze-dryers, but it is considered impractical for locating leaks in a large product dryer. The reason being that the sniffer may not be able to reach all parts of the dryer. In addition, the pressurizing of a large freeze-dryer with helium would not only be an added expense but would also be considered a waste of a natural resource.

Refrigerant Leak Detection

For leaks in the refrigeration system, one would have to examine the condenser system using a leak detection system designed for halogen-based or the environmental safe refrigerants such as R134a. For halogen refrigerants, the detection system can be as simple as a butane gas torch to which a hose is connected. Air for the combustion of the butane would enter the burner by means of the hose. If the hose were placed near a source of leaking refrigerant, there would be a noticeable change in the color of the flame. Because of the hazard of an open flame, this detection system is being replaced by an electronic device. An electronic handheld device uses a high voltage field to create a corona discharge at its tip. In the presence of a halogen gas, the corona discharge is disturbed, triggering a visual and audio signal. The sensitivity of the device can be varied such that it can detect a halogen gas leak as small as 1/2 oz per year.

Fluorescence

Virtual leaks stemming from the heat-transfer system are serious because they can directly impact the quality of the lyophilized product. Vapors from the heat-transfer fluid can be identified from the fragmentation pattern obtained from a mass spectrometer. Unfortunately, the mass spectrum only indicates the source of a virtual leak; it cannot identify the leak path. It is recommended that the reader pay special attention to leaks involving silicone oil as the heat-transfer fluid. If the shelf(ves) must be removed in order to seal the leak path, then precautions should be taken to prevent the silicone oil from coming in contact with any metal surface of the dryer because it is very difficult to remove and could serve as a secondary virtual leak source that could cause product contamination (haze) for some time after the original virtual leak was repaired. In order to prevent contamination of the interior surfaces of the dryer, a plastic sheet should cover all exposed metal surfaces.

The location of the virtual leak in the heat-transfer fluid can be determined by adding a small amount of a fluorescent compound such as anthracene to the heat-transfer fluid (60% wt/v ethylene glycol, lexsol, methyl alcohol, methylene chloride, n-propyl alcohol, and silicone oil [21]). In the event of a leak in the heat-transfer fluid, the fluid will pass through the leak path and evaporate, leaving a residue of the fluorescent compound. The latter can be detected by means of an ultraviolet light.

It is recommended that the fluorescent compound be added to the heat-transfer fluid during the construction of the dryer. When the fluorescent compound is always present in the heat-transfer fluid, periodic examination of the dryer with ultraviolet light can detect the early stages of a virtual leak prior to it becoming a source of product contamination. In using this method of leak detection, one should check with the manufacturer of the freeze-drying equipment that the fluorescent compound is compatible with the heat-transfer fluid over the entire range of operating temperatures and that it will not affect the flow through the shelves. In addition, when exposed to ultraviolet light, the fluorescence of pure anthracene will appear violet and could be difficult to see in a dark environment. However, 98% pure anthracene will produce an orange radiation that will much easier to detect. Finally, personnel performing this test must be wearing special glasses to prevent eye exposure from the reflected ultraviolet light.

Door

The basic function of the door of the freeze-dryer chamber is to provide a means by which product can enter and exit the drying chamber. In essence, the door should not in itself affect the operation of the dryer, impact the lyophilization process, or present a potential hazard to operating personnel or the facility.

Design and Construction

The door should be designed such that it can withstand the forces exerted by atmospheric forces or the combined effect of temperature during steam sterilization of the dryer. This section will consider various means whereby the design and construction of the door not only affect the lyophilization process but also represent a potential hazard to operating personnel.

Doors should contain safety interlocks to prevent the accidental opening of the door while the chamber is under steam pressure. While there are some dryers that are equipped with such devices, it is cautioned that a dryer should also be equipped with a flashing warning light to indicate that the dryer is pressurized with steam and that no attempt should be made to open the door. Because the control panel of the dryer is often in a separate room, the door should also be equipped with an electrical interlock system that would prevent opening of the steam valve to the drying chamber when the door is in an open position [6].

For those dryers that do require steam sterilization, the doors are generally fabricated from 316L stainless steel [6] because of its resistance to chloride stress corrosion. As a means for visual observation during the lyophilization process, most freeze-dryers are equipped with viewing ports on the door or on the side of the drying chamber. The value of such viewing ports is to permit operating personnel to determine if the product is completely frozen or if premature stoppering has occurred [5]. It is my opinion that the viewing ports are more for physiological reasons than providing any diagnostic tool for the operator. The argument that the viewing ports will provide the operator with a visual means for determining if the product is frozen is rather weak when all one has to know, based on the thermal properties of the formulation, is the T_p. When lyophilizing a formulation in small-volume vials (e.g., ≤ 5 mL), the portion of the vial containing the formulation is often completely obscured from view, thus making any observation regarding its physical state impossible.

It may be argued that as a result of the viewing port, the operator can ensure that the stoppers are fully seated after the stoppering operation and prior to opening the door. The question is then asked, *What steps does the operator take if it is found that the stoppers are not fully seated in the vials?* If the response is that there are no steps that the operator can take to obtain a fully seated container-closure seal, then what is the purpose of the viewing ports in the door? The windows in the door only serve to further complicate the cleaning and sterilization of the dryer. In order to ensure that the dryer is sterile, it must be verified that the temperature on the surface of the viewing port reaches $\leq 121°C$ for the necessary time period. Repeated sterilizations will often cause the inside surface of the window to become highly discolored and, in time, make visual observations in the dryer very difficult.

The windows on the door are often flush with the outside surface of the door. The placement of the windows in such a fashion would enhance the appearance of the dryer but also creates a recess on the inside of the door that can prove difficult to clean and rinse of any cleaning solution.

From the above discussion, it should be apparent to the reader that any physiological benefit that one may obtain from peering into the drying chamber during a drying process is more than offset by the issues such ports raise with respect to providing a safe environment for the product during the lyophilization process. In preparing this text, I found comparatively little information or studies that consider the impact that the design and construction of the door could have on the lyophilization process.

For dryers that do not require steam sterilization, acrylic plastic doors are often used. These doors offer a clear view of the entire chamber [6]. While providing an almost unobstructive view of the entire dryer during the drying process, they also serve as a source of radiant energy and can affect the uniformity of the T_p [5]. The T_p nearest the door will tend to be higher than the T_p at the back of the chamber. Grieff [8] found the loss in water was greater or the sublimation rate was higher in those vials closest to the door.

Thermal Insulation

It is not unusual to observe sweating or condensation on the door of the chamber during the freezing of the product. The sweating is not a result of some form of work being performed by the door; it is the temperature of the outside of the door being below the dew point (i.e., that temperature where condensation will commence on a surface). For example, a room at $20°C$ and having 40% relative humidity will have a dew point of $4.5°C$ [22]. This would mean that not only was the product being frozen but also the door to the chamber was being refrigerated. The presence of water in an aseptic area is not only highly undesirable but also represents a loss in refrigeration.

I have found that attaching a sheet of aluminum foil over the inside surface of the door was sufficient to greatly reduce and in most cases eliminate the formation of condensation on the outside of an acrylic plastic door. For those who wish to visually observe the inside of the dryer during the lyophilization process, a small window can be cut in the aluminum foil. Such a window has been found not to cause any significant loss in refrigeration or the formation of condensation. Since a thin layer of aluminum foil is sufficient to reduce or eliminate the heat transfer from the refrigerated shelves to the acrylic plastic door, it would appear that the principal heat-transfer mechanism involves Stefan's Law and that heat transfer resulting from gas convection is not a major factor.

The presence of aluminum foil on the inside of the chamber door will not only limit or prevent radiant energy heat transfer between the door and the shelves during the freezing process but will also play a major role in reducing radiant energy heat from the door to the product during the primary during process. By reducing the radiant energy heat from the chamber door, a more uniform T_p will prevail across the shelf, thus reducing the impact that the location of the vial can have on the drying rate of the product.

Sweating and condensation can also occur on solid stainless steel doors. Since this type of door is generally associated with chambers requiring steam sterilization, attaching a thin sheet of aluminum foil to the inside of the door is not a reliable practice. For this type of chamber door, insulation can be achieved by attaching several sheets of stainless steel in series to the inside of the door or by applying a cover of thermal insulation over the outside of the door. In the first instance, more than one sheet of stainless steel has been found necessary to reduce the radiant energy heat transfer from the door because the emissivity of the stainless steel, unless it is highly polished, is greater than that of the aluminum foil. Thus, there will be a difference in temperature between the radiation barriers. The presence of the radiation barrier on the inside of the door poses an additional problem, i.e., cleaning. Because these radiation shields are generally not convenient to detach and clean, a door that contains an outer insulating cover and has a highly polished door inside surface is preferred. With the need to insulate the door of the chamber, it should become apparent to the reader that the presence of viewing ports in the door serves to complicate the thermal insulation of the door.

Gorilla Effect

In steam sterilizing the drying chamber, the door must be sealed for pressurization. If the door is sealed at pressures lower than 1 atm, one could cause what I refer to as the *gorilla effect*. Tightening the door handles used to seal the chamber at pressures lower than 1 atm will result in an internal force being exerted on the door handles by the release of the compression forces that were exerted on the door gasket when the chamber was under vacuum. One must remember that the force exerted on the door gasket by a door of a production dryer, having dimensions of 5 ft by 5 ft, when the chamber pressure is at 100 mTorr, will be of the order of 53,000 lb or 26.5 tons. When such a compression force is released, as the pressure in the chamber is returned to atmospheric pressure, the compression forces on the gasket are now translated to the door handles. The force is sufficient so that the door handles can no longer be turned by hand. Instead of evacuating the chamber again to release the force on the handles, the operating personnel open the door by generating the necessary force with the use of a lever created with a section of metal pipe that one end slips over one of the bars of the handle. After repeated opening of the dryer in this manner, the door handles are changed from their original configuration shown in

Figure 13.5. "Gorilla Effect" on the Door Handles for Securing the Chamber Door

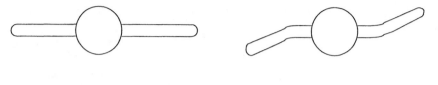

Figure 13.5a   Figure 13.5b

Figure 13.5a represents the handle in its original configuration, and Figure 13.5b shows the result of releasing the door handle in the presence of a force exerted by the chamber door.

Figure 13.5a to that of Figure 13.5b. Since the bending of the door handle bars by hand would require a force beyond human capabilities, the bending of the bars like that shown in Figure 13.5b could only be accomplished by a gorilla. The reader should take note that anytime after the drying chamber is under vacuum, the door handles used to secure the door should be released in order to prevent the occurrence of the gorilla effect.

It might be asked, *Why would the gorilla effect impact the performance of the dryer?* By subjecting the door handles to repeated additional stress conditions, especially during the steam sterilization process when the temperature of the sealing bolts is increased, the bolts will slowly become elongated. After some time, the end of the bolt will contact the end of the threaded bolt hole. When this condition occurs, the bolt may no longer exert sufficient force on the door during the steam sterilization process. A loss in steam pressure is not only a safety hazard but also may prevent the steam pressure from reaching the required value necessary to achieve sterilization of the dryer.

An additional risk of the gorilla effect would be the potential stripping of the bolt or the tapped threads or seizing of the bolt to the tapped threads. In the first instance, it may be a simple matter of just replacing the bolt. However, if the tapped threads are damaged, then this repair would require enlarging the tapped bolt hole. Such a repair could be performed on-site but at the cost of contaminating the clean room and the loss of production of lyophilized product. Seizure of a bolt would prevent opening the door of the dryer and removing the lyophilized product. The defective bolt can be mechanically removed (cutting) but would once again compromise the clean room. One is then faced with either removing the product in a nonacceptable environment, whereupon it should be discarded, or leaving the product in the dryer until the dryer can be repaired and the clean room is once again certified as an aseptic Class 100 environment.

While I have often observed the gorilla effect on freeze-dryer chambers and the resulting elongation of the bolts, I have no knowledge of failures as described in the last two paragraphs. However, it is clear that the gorilla effect does make such scenarios a real possibility. Such scenarios can be easily avoided by merely not tightening the bolts on the dryer door when the chamber is under vacuum.

Shelves

While one of the principal functions of the freeze-dryer is to provide the necessary operating parameters to conduct the lyophilization, the shelves must supply the necessary refrigeration in order for the formulation to acquire a completely frozen state, the energy for the sublimation of the ice during primary drying, and desorption of the water from the cake during the secondary drying process. The remainder of this section will be devoted to how the nature of the shelves can impact the lyophilization process.

Construction

Because of the important role that the shelves play in the lyophilization process, the size of the dryer is generally defined by the area of the shelves. Pilot and small production units have shelf areas ranging from 8 ft^2 (0.74 m^2) to 36 ft^2 (3.3 m^2). Because of their size, these dryers can be constructed as a single unit. Production dryers have shelf areas of 120 ft^2 (11.1 m^2) to 220 ft^2 (20.4 m^2) [2]; however, dryer sizes of 400 ft^2 (37.2 m^2) and 600 ft^2 (55.7 m^2) are not uncommon. One of the difficulties that faces the industry is a lack of standardization on a series of standard shelf sizes. In the absence of such standardization, trays that will fit one dryer may not be effectively used in another dryer having an equivalent shelf surface area.

The shelves of the dryer are constructed in such a manner as to give the heat-transfer fluid a serpentine path through the interior of the shelves. The serpentine path serves to direct the path of the heat-transfer fluid such that it flows through the entire shelf surface area. If the interior of the shelves is not properly constructed and the interior walls of the shelf allow heat-transfer liquid to bypass a segment(s) of the shelf, then there will be a region in the interior of the shelf where there can be a reduction or absence in the flow of the heat-transfer fluid. As a result, a segment of the shelf will tend to be warmer during the freezing process and colder during the drying process. Thus, product in this region may not completely reach a frozen state and not complete the primary drying process, with the end result being defective product.

Since it is not possible to visually inspect the construction of the serpentine path in the interior of a shelf, one needs a means of verifying that there are no regions of the shelf that do not receive adequate heat-transfer fluid flow. One method would be to map every inch of shelf surface over a range of temperatures. This method is not only tedious but also impractical in nature. Another means for determining if there are significant regions of the shelf not receiving adequate heat-transfer fluid flow would be to first evacuate the system to about 100 mTorr. At this pressure, refrigerate the shelves to a low temperature, e.g., –20°C. Allow the shelves to remain at this temperature for at least 1 hr. Backfill the chamber with moist air to atmospheric pressure to form frost or ice on the shelf surfaces. When the chamber has reached 1 atm, open the chamber door and visually inspect the shelves for a uniform coating of frost or ice. A region of a shelf not having a layer of frost or ice would be the first indication of an improperly constructed serpentine flow configuration. Set the temperature of the shelf fluid for 20°C using either the high heat setting or the fastest heating ramp. Observe the melting of the ice on the shelf surface. An isolated region maintaining an ice or frost layer, after the ice on the rest of the shelf surface has melted, would be an indication that there was a significant difference in the flow of the heat-transfer fluid through that portion of the shelf.

Because stoppering of the vials at the end of the lyophilization process will be performed in the dryer, it is essential that the shelves be machined flat and horizontal [2,5]. Shelf flatness values, across any 1 m length of the shelf, has been recommended at ±1 mm to ±0.5 mm [6]. Since for the majority of shelves both sides of the shelf (excluding the top unusable and bottom usable shelves defined in the next section) will be involved in the stoppering operation, it is imperative that shelf surfaces be not only flat but also parallel. If a shelf is not parallel, a gap may form between the shelf and its adjacent shelf, and the closures may not be fully seated by the stoppering mechanism. Since the primary seal for the container-closure system occurs on the flange of the vial, an incomplete seal can result in the product of this vial having a higher moisture content, be subject to oxidation from the atmosphere, or even permit the entry of microorganisms prior to the crimping operation.

While it is important to achieve an uniform T_s, it is more important to have an uniform T_p. Random differences in the T_p may stem from the geometry of the containers or nonuniform loading of the shelves. Thermal radiation from the walls or the door may also affect T_p (e.g., the highest residual moisture was found in those vials located in the middle of the shelf, while lower moisture values were found around the edges of the shelves [8]).

The finish of the shelves is also an important consideration. In the past, shelves were constructed from carbon steel with baked organic coating and, in some cases, aluminum shelves [2]. Currently, shelves tend to be constructed from stainless steel with the preference given to 316L. Although I was not able to obtain any data on the emissivity for a resin-coated shelf at a given temperature, one could assume that the emissivity of the resin-coated surface would be greater than the emissivity of a stainless steel surface having a satin or 240 grit finish [6]. For a lyophilization process requiring a P_c of < 100 mTorr, radiant energy may be a major heat-transfer mechanism. Assuming that the radiant heat-transfer rate for the resin-coated shelves is greater than that for the stainless steel shelves, the drying rate (especially during the primary drying process) would be higher for the resin-coated shelves than for the stainless steel shelves. Further, highly polished shelves provide an easier surface to clean; however, it was also found that a high polish also reduced the energy transfer between the shelves and the vials that stems from radiant energy. A highly polished surface is a good reflector but a poor emitter of radiant energy.

The above examples point out that the emissivity of the surface can be an important consideration in determining the heat transfer to a product during a lyophilization process. As a result, switching a drying process from a dryer whose shelves have a high emissivity to one with a low emissivity could result in an extension of the drying time because of the reduced energy transfer [2]. In order to offset the reduction in radiant energy, one would have to increase the T_s. If the necessary T_s is outside the range of temperatures that have been submitted to a regulatory agency, then one would have to file an amended process and show that the final product meets all specifications.

Unusable and Usable Shelves

It is a common practice to have the first shelf of the dryer designated as an *unusable* shelf, i.e., this shelf is not designed to accept product for lyophilization. *Usable* shelves are shelves onto which product can be loaded for the lyophilization process. The main function of the unusable shelf is to provide a thermal environment during both the freezing and drying process that will be equivalent to the other shelves in the dryer. In the absence of an unusable shelf, as exemplified by Figure 13.6, the drying rate on the shelf will be greater as a result of the additional energy supplied by thermal radiation from the walls of the chamber. Grieff [8] reported that, in the presence of an unusable shelf, the moisture values at the end of the drying process were comparable.

It has been my experience that the T_p on the bottom shelf can also be affected by thermal radiation from the bottom of the chamber. It was observed that the T_p on the bottom shelf took longer to reach the required freezing temperature than product on the other shelves of the dryer. However, during the drying process, the T_p on the bottom shelf appeared to be lower than what was measured on the other shelves. It would seem that during the freezing process, the bottom shelf of the dryer was being warmed by radiation from the bottom of the dryer chamber, while during the drying process the bottom of the chamber acted as a thermal heat sink and caused the temperature of the bottom shelf to be lower than that of the other shelves, thereby reducing the drying rate especially during the primary drying process. Since it was observed that the walls of the chamber affected the temperature of the bottom shelf during the operation of several dryers, it would appear that an unusable bottom shelf would also serve the same function as a top unusable shelf.

Figure 13.6. A Freeze-Dryer Without an Unusable Shelf

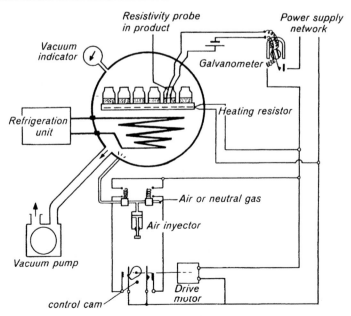

[Source: *Bulletin of the Parenteral Drug Association*, "Advances in Lyophilization Technology," Vol. 31, No. 5, Sept./Oct. 1977, Page 228, Figure 1. Reprinted with permission.]

Spacing

Although not present on all freeze-dryers, most dryers are designed and constructed such that the shelf spacing can be varied by *pairing* the shelves to increase the distance between the shelves [23]. The term *pairing* means that two shelves are positioned such that they constitute a single shelf. While this feature reduces the effective shelf-surface area of the dryer, it does permit the lyophilization of larger volume vials that normally could not be accommodated using normal shelf spacing.

The typical shelf-to-shelf distance for most production dryers is about 4 in. (10 cm) [2]; however, for dryers that have large shelf areas, there can be a substantial increase in the area of the individual shelves. One should exercise caution that the shelf-to-shelf distance is adequate to prevent the formation of a significant pressure differential, especially during the primary drying process, across a given shelf [5]. Such a pressure differential would cause an increase in the T_p and a reduction in the drying rate in those shelf regions having a higher pressure. If the pressure increase is excessive, one may experience collapse, partial meltback, or even complete meltback of the cakes.

Heat Transfer

The shelves of the dryer serve more than just a surface to contain the product either in vial or trays; they also provide the necessary heating and cooling for conducting the lyophilization process. Refrigeration of the shelves is accomplished either by direct expansion of a refrigerant or the passage of a fluid through shelves designed to cause the fluid to follow a serpentine path. The use of direct expansion of a refrigerant offers the advantage of obtaining a lower T_s. The refrigerant in this instance could be generated by a standard electrical refrigeration system using a compressor or liquid gas such nitrogen. Since I could not find any published detailed description of the design and construction of shelves using direct expansion, one can only hypothesize on the basic design, construction, and operation of such a system. The objective here is to make readers cognizant of the nature such a system.

Direct Expansion

To obtain uniform temperature across the shelf and from shelf to shelf, it would be logical that the expansion fluid take a serpentine path like that described in the above. A uniform T_s would also require that the entire serpentine path contain liquid refrigerant and have no significant pressure differential between inlet and out of the shelf. The absence of a significant pressure differential would require a careful design of the thickness of the shelves with respect to the serpentine path of the liquid and gaseous refrigerant. The temperature of the liquid refrigerant would be defined by the phase diagram of the refrigerant. A pressure differential across the shelf would mean a liquid refrigerant temperature differential. It should be pointed out that the liquid refrigerant would be in contact with the bottom surface of the shelf, and the top portion of the shelf must use gas conductivity as a means of transferring energy from the product to the expansion fluid during the freezing process. As the refrigerant fluid temperature is decreased, so too will the vapor pressure of the refrigerant in the shelves. A decrease in the vapor pressure would result in a reduction in the gas thermal conductivity between the liquid phase and the top surface of the shelf. This decrease in gas thermal conductivity could increase the freezing time necessary to obtain a given T_p.

Where the refrigerant is liquid nitrogen, because of its low boiling point (-195.8°C at atmospheric pressure [24]), the cold gas would be used to refrigerate the shelves. A special valve system would have to be devised to improve the heat-transfer efficiency from the cold gas in order to conserve the quantity of liquid nitrogen used during the refrigeration of the shelves. The use of cold nitrogen gas may prove more advantageous than the direct expansion of a mechanical refrigeration system from a standpoint of design and cost of equipment. What is not certain is the impact that the cost of liquid nitrogen will have on operating costs over the life of the equipment, especially for a production dryer.

Heating the shelves presents an additional problem. It may be possible to reverse the refrigeration system and use the hot gas from the compression portion of the refrigeration process to heat the shelves. This may prove undesirable because of the following reasons. By using the compression portion of the refrigeration process, the pressure would tend to be > 100 psi, and construction of a shelf that could withstand such pressures would require very thick metal surfaces. The mass of such shelves would certainly add greatly to the cost of construction and, assuming that the shelf is made of stainless steel, represent a poor heat-transfer medium for both the refrigeration and heating of the shelves. The additional mass of the shelves would also require a more massive support system and a larger hydraulic system for the stoppering system.

For shelves that use direct expansion for refrigeration, heating is accomplished by electrical resistance [23]. The resistance heating would most likely be accomplished using electrical heating elements attached either to the bottom of the shelf or to the bottom of the upper shelf surface. Heat transfer, in the former case, would again rely on the thermal conductivity of the refrigerant gas. Since the heaters in such a system would be connected in a parallel circuit, it may be difficult not only to sense the electrical failure of a given heating element but also to locate and replace the defective unit. A failed heating element would lead to a region having a lower T_s during primary drying. Since the T_s would be lower, the T_p for a given P_c would also be lower (see Chapter 8 for the relationship between T_p, T_s, and P_c). A lower T_p during primary drying would be translated into a lower product drying rate, which could lead to partial or complete meltback of the product in the shelf region with the defective heater. The latter defect would result from incomplete sublimation of the ice prior to the start of the secondary drying process.

Gas Conductance

Where nitrogen gas is used as a heating medium, one may be able to use heated air rather than consume additional liquid nitrogen; however, such a system may prove complex because of the need to control the T_s by introducing cold nitrogen gas. It would appear that one would be forced to heat gas derived from liquid nitrogen in order to reduce the complexity of the system.

Heat-Transfer Fluid

The most common means for refrigerating and heating the shelves of a dryer is by using heat-transfer fluid that is passed through a serpentine path. In this system, the fluid would serve to provide both refrigeration and heating of the shelves [23]. While certainly the most commonly used medium for providing the refrigeration or heating necessary to conduct the lyophilization process, it is not without its share of ambiguity and pitfalls that may mislead the unwary. The following addresses some common areas of concern regarding the design and construction of the heat-transfer system and the impact that the nature of the heat-transfer fluid can have on the lyophilization process.

The heat-transfer fluid should pass through the shelves in parallel rather than in series [6]. When the shelves are connected in series, the temperature of the heat-transfer medium entering the first shelf will be equivalent to that leaving the heat exchanger. If energy is removed from the medium, e.g., by the sublimation of ice, the temperature of the fluid will be lowered. For that reason, the temperature of the fluid entering the second shelf will be lower than that leaving the heat exchanger and the T_s of the second shelf will be lower than that of the first. Because the T_s is lowered, the T_p will also be lowered along with the sublimation rate, the latter assumes that there is a uniform pressure throughout the drying chamber. Thus, the drying rate will be lowered for each subsequent shelf, and one would have to be careful to base completion of the lyophilization process on the shelf that is last in the series or one could experience defective product. If the entire process is based on the drying parameters of the last shelf in the series, then the T_p on the first shelf could become excessive and lead to collapse or meltback of the product.

It has been stated that a thinner and lighter shelf results in a more energy efficient operation of the freeze-dryer. Energy efficiency is further increased by increasing the rate at which the medium passes through the shelf. For a heat-transfer fluid system, this is accomplished by increasing the capacity of the circulating pump that pumps the fluid through the shelves [6]. One must be careful to note that a *thinner shelf* is taken as the distance between the top and bottom plates of the shelf, not the total thickness of the shelf surface. By decreasing the actual thickness of the shelves, one would tend to increase the impedance to liquid flow. To increase the fluid flow through such a thinner shelf, one would need to install a larger capacity pump. The additional work performed by the pump to force the fluid through the shelves would cause an increase in the fluid temperature [6]. In order to compensate for the additional energy input, the dryer would require additional shelf refrigeration. The thinner shelf may permit the drying chamber to accommodate more shelves, but the need for additional circulating pump capacity and offsetting the additional required refrigeration may not significantly increase the overall energy efficiency of the system. Additional circulating pump capacity may be required to offset the change in the viscosity of the fluid. It will be shown later on in this section that the viscosity of the heat-transfer fluid is subject to increase as the temperature is decreased. In selecting a dryer with thinner shelves, one should be careful that the circulating pump can maintain the required flow at low temperatures. With a decease in the flow of the heat-transfer fluid through the shelves, one must be cognizant of the possibility of significant temperature variations across the shelf. Such T_s variations can also lead to variations in the drying rate and the potential for generating defective product. An increase in the amount of product produced in a dryer, while also increasing the amount of defective product, is not a mark of an increase in the efficiency of the dryer.

The presence of air in the shelves can be troublesome. It was stated that air trapped in the system will invariably migrate to the top shelf, *if for no other reason, that this is where it can do the most harm* [6]. If the air does migrate to the upper shelf, which is the unusable shelf, the presence of the air would do no harm because it is the temperature of the bottom of the first shelf that creates the necessary radiation environment for the product on the first usable shelf. Where the trapped air can cause a problem is when it forms a barrier between the shelf surface and the fluid for one or more usable shelves in the dryer. The presence of the gas layer would create an additional energy barrier that would, especially during the primary drying process (see Figure 8.17), result in a significantly lower T_s. Once again, a lower T_s, for a given P_c, would lead to a reduction in the drying rate of the product.

Air in the shelves should not be a problem if the shelves and that of the reservoir tank are first evacuated to a low pressure (≤ 1 Torr), and the heat-transfer fluid is added so that it flows from the lowest point in the system up to the reservoir tank. In this manner, there is little or almost no air to displace while filling the system with the heat-transfer fluid. If air in the shelves is suspected as being the cause for significant shelf-to-shelf temperature variation, then it is recommend that corrective action be taken (i.e., the system should be drained and the entire system leak tested to ascertain how air is entering the heat-transfer system). Any leak is repaired, and the heat-transfer system is backfilled with dry heat-transfer fluid. The heat-transfer fluid is dried by first passing through a desiccant such as silica gel [6].

An ideal heat-transfer fluid would have a low viscosity at low temperatures and be noninflammable, nontoxic, and noncorrosive in nature. In addition, the fluid would be inexpensive. Such a heat-transfer fluid has not been found or developed [2]. It is because heat-transfer fluids can be toxic and inflammable that one should be diligent in making sure that such materials do not enter the freeze-drying system during a lyophilization process. While the presence of a heat-transfer fluid, such as a silicone oil, in a cake will result in a turbid reconstituted solution, the presence of a toxic heat-transfer fluid such as methanol would not be visually detected in the reconstituted product, and one would have to use other analytical methods to detect its presence in perhaps low concentrations.

Powell [2] pointed out an important property of the heat-transfer fluid: its *heat capacity*. The heat capacity of the heat-transfer fluid is an important consideration because it is defined as the quantity of energy required to change the temperature of the substance by 1°C. If the heat capacity of the heat-transfer fluid is high, then as it passes through the shelves, the fluid must undergo a considerable thermal load in order to alter its temperature. With such a heat-transfer fluid, there would be little change in the inlet and outlet temperatures of the fluid even though there has been a considerable exchange of energy between the fluid and the product. For a heat-transfer fluid with a low heat capacity, a small change in energy will result in a change in the temperature of the heat-transfer fluid. The mere work done by the circulation pump could result in a change in the temperature of such a heat-transfer fluid.

The issue of energy transfer in heat-transfer fluids was addressed by Powell [2] by plotting the amount of energy required to pump the fluid as a function of the amount of energy removed from the shelf system. Figure 13.7 is a plot of the ratio of the energy input to pump the fluid to the energy removed by the shelf system as a function of temperature. This plot shows that for silicone oil, the ratio is greater than 0.5 at temperatures < 0°C. This means that, as the temperature of the silicone oil is lowered, the amount of energy necessary to transport the fluid is approaching the amount of energy that can be transferred to the product. The significance of this relationship is that for a constant heat-transfer rate to a product, the amount of energy required to transfer the fluid will have to be increased as the temperature of the fluid is decreased. If the work done on the heat-transfer fluid by the circulating pump remains constant, then the amount of energy transferred to the product will decrease as the fluid temperature is decreased. This decrease in the heat transfer to the product would have the effect of prolonging the freezing process at a given T_p.

It is unfortunate that Powell [2] did not provide further details as to how the values used to construct Figure 13.7 were obtained. While he provides information on the various properties of heat-transfer fluids (boiling point, flash point, hazard rating, cost, etc.), he does not provide any information regarding the heat capacity of the fluids.

In the case of lexsol, the energy input to energy transfer ratio remains near 0.1 for temperatures ranging from 70°C to -60°C. This result would suggest that for a given energy input by the circulating pump, the energy exchange with the product will remain the same over the entire range of temperatures. In spite of its flammability, it would appear from these data that lexsol would be a wise choice for a

Figure 13.7. The Ratio of the Energy Input to Pump the Fluid to the Energy Removed by the Shelf System as a Function of Temperature

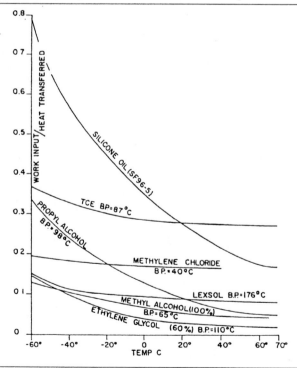

[Source: *Developments in Biological Standardization*, Vol. 36, pp. 117–129, 1976, "Trends in Freeze Drying . . . Materials," Figure 5. Reprinted with permission of the International Association of Biological Standardization.]

heat-transfer fluid unless a $T_s \gg 70°C$ is required to complete the secondary drying process.

Shelf Temperature

Throughout this text, an effort has been made to distinguish been the T_s and the temperature of the shelf fluid. The rationale for such a distinction of terms is that for a given thermal load, the temperature of the shelf fluid necessary to maintain a given T_s will be dependent on the design and construction of the shelves, the heat-transfer fluid refrigeration and heating systems, and the nature of the heat-transfer fluid. Since there are no known standards with regard to the design and construction of a freeze-dryer or are such standards expected in the near future, merely specifying the T_s of a process does not provide a reader with meaningful information regarding the actual T_s that was used in the process. However, if one is provided with the T_p and actual P_c, then one would have a means of determining an *equivalent shelf-fluid temperature*. An equivalent shelf-fluid temperature would be a fluid temperature that would produce a given T_s as defined by the T_p and P_c. The following describes the various means by which the T_s can impact the quality of a lyophilized product.

In order to achieve a reproducible lyophilization process for product contained in vials, it has been shown that it is necessary to understand the frequency distribution of the heat-transfer coefficients (C_o) of the vials [9]. While not stressed but certainly implied, in order to base a lyophilization process on the heat-transfer properties of the vials, one must assume that no significant temperature differential exists across the shelf surface or from shelf to shelf. Such a temperature differential $\geq 5°C$ could offset any reliance one might place in determining the drying process via calorimetric measurements. Temperature variations across a shelf surface are often specified not to exceed 1°C; for some dryers, the variation in the T_s is only 0.5°C [6]. However, the fallacy here is that the temperature distribution is measured when there is no thermal load present and the T_s approaches that of the fluid temperature. However, in the absence of an adequate flow of the heat-transfer fluid resulting from the design of the shelves or the nature of the heat-transfer system, temperature variations across the shelf may well exceed 5°C. In the presence of such T_s variations one would still be faced with variation in the drying rates resulting from a frequency distribution of C_o but further complicated by a nonuniform distribution of T_s.

A major consideration of a freeze dryer is its capacity to reach a specific temperature. For example, temperatures < -40°C may be required to achieve a product with a completely frozen matrix. For a typical dryer, a T_s of -40°C is obtainable, while special refrigeration considerations are often required to achieve temperatures < -40°C. The latter will depend on the capacity of the refrigeration system and the heat-transfer system. In general, an upper T_s requirement is not difficult to achieve. The upper T_s for most dryers is about 60°C, whereas upper temperatures of 80°C are obtainable [2].

The following is an example of the important role that the T_s plays in the lyophilization process and the impact that the design and construction of the dryer has on T_s. Grieff [8] reported that the temperature of the top shelf was lower than the other shelves and for that reason the moisture values were greater for products on the top shelf. (For the effect that T_p and partial pressure of water vapor have on the moisture content of a product, refer to Chapter 9 and the discussion regarding the desorption isotherms.) From the diagram of the apparatus used by Grieff [1] (not illustrated here), there does not appear to be a nonusable shelf present to serve as thermal environment for the top shelf. As a consequence, the temperature of the top shelf, as a result of thermal radiation from the walls of the dryer, should have been warmer rather than colder [8]. It should be noted that the shelves were refrigerated by direct expansion of a refrigerant and were heated by electrical heaters attached to the bottom of the shelves. Grieff further reported that by balancing the refrigeration and heating systems, variations in T_s could be maintained to +0.3°C. (It should be pointed out that I found that the schematic drawing used to describe the refrigeration system for this apparatus a bit ambiguous [1]. The drawing for the refrigeration system [not illustrated here] did not indicate a return line for the refrigeration gas for the top shelf or bottom shelf, and it is not clear if absence of the latter was an omission or represents the actual configuration of the system. If it is not an omission, then the shelves would represent dead-leg for gas flow and the top shelf would be at a higher rather than lower temperature.) Grieff [1] pointed out that the higher moisture cannot be attributed to the method for determining the moisture content or the personnel performing the tests; he attributed the residual moisture to the *position in the array of samples.* He further stated that, in general, each shelf provides a mini-environment and is, in a sense, *in a world of its own.* If each shelf produces its own mini-environment, then the construction and operation of the freeze-dryer is not taking a passive role but an active role in the lyophilization process. It is my opinion that the freeze-dryer must assume a passive role in order for the nature of the product on one shelf to be indistinguishable from the product of another shelf.

SAFE ENVIRONMENT

The second function of the freeze-dryer is to provide a *safe environment* for the product during the lyophilization process. A safe environment is one that is free of any particulate matter, which would include microorganisms, and vapors that may serve as contaminants of the lyophilized product. It was felt that this topic should be considered prior to discussing the impact that other components of the dryer may have on the final quality of the product.

Particulate Matter

A key question that needs to be addressed, especially for the freeze-drying of injectable products in which the formulation is filled in an aseptic Class 100 room, is that if the dryer is sterile, does it provide a Class 100 environment during the initial pumpdown or backfilling of the chamber?

Particulate matter in a freeze-dryer may stem from glass that broke during the lyophilization process. Such glass particles can result during the stoppering of the vials if the vial has not been properly annealed to remove stress lines or is accidently cracked during shipping and handling prior to the filling operation. Any glass chips that may remain in the dryer are often, over a period of time, reduced to a fine powder as a result of loading and stoppering operations. Another source of particulate matter may stem from lyophilized product [25]. Such particulate matter may also stem from broken vials or be a direct result of the drying process. This latter source of particulate matter could also stem from unbroken vials because of a total low solid content (< 2% wt/v) of the formulation. For formulations having a low solid content, the resulting cake may have a poor self-supporting structure; during the drying process, the cake is ruptured, and product passes through the vial or from the tray by means of the vapor flow from the sublimation process. Experience has shown that the rupture of a cake from a low solid content formulation becomes more prevalent as the fill volume is increased.

Personnel in contact with the dryer during loading or unloading operations can also be a source of contamination. In testing a dryer, which will be described later, the presence of lint and hair were found in the dryer. The source of such particulate matter may stem from the operating personnel of the dryer, but it also could have been residuals from the construction of the dryer. I have yet to visit a freeze-dryer manufacturer and witness personnel engaged in the construction of the dryer wearing protective hair-nets and lint-free uniforms. One is puzzled by the fact that such protective covering is commonplace in the construction of a space satellite that will be used in communication systems yet is not deemed necessary in the construction of equipment that will be used in the manufacture of vital healthcare products. The following describes three means for detecting the presence of particulate matter in a dryer.

Swab Test

One means for determining the presence of particulate matter in a dryer is to wipe a surface of the dryer with a damp (alcohol) piece of lintless cloth [25]. The cloth is then stored in a plastic bag, carefully removed, and examined under a low power (20×) optical microscope. In using this technique, one is able to identify both glass and metal fragments. The glass most likely came from broken vials, while the source of the metal fragments most likely was residuals generated during the manufacture of the freeze-dryer. As long as these particles remain on the dryer surface, they will not be a source of product contamination; however, they pose a real threat to the quality of the product should they become airborne as a result of the turbulent flow of gases during either the initial pumpdown of the dryer or the backfilling of the chamber prior to initiating the stoppering operation. This method is useful in detecting if particulate matter is present in the dryer but cannot provide information if such particulate matter can become airborne during the lyophilization process. The following test methods will be useful in determining qualitatively and quantitatively the airborne particulate matter in the dryer.

Filter Test

A second qualitative method known as the *filter test* can be used to detect the presence of airborne particulate matter in the dryer. This method also collects particles on a surface of a paper or cloth filter. A clean piece of dry filter paper or cloth is first placed over a port controlled by a valve from the chamber (e.g., the drain line). The filter is fastened over the port to permit the flow of gas through the filter when the port is opened and the chamber is pressurized. A clean and dry freeze-dryer chamber is evacuated, with the condenser at ambient temperature, to about 100 mTorr.

The chamber is then isolated from the vacuum system and backfilled to 1 atm with air or nitrogen that has been passed through a 0.2 μm filter. The backfilling valve is then closed, and the door to the chamber is sealed for pressurization. By means of the backfilling valve, the chamber is pressurized at 1 to 2 psig, the valve to the port with the filter is opened, and the gas is allowed to flow for about 5 min. The gas backfilling valve is closed, and the filter is removed and examined under a low power (20×) optical microscope. This test is useful in detecting airborne glass and metal particles in the drying chamber. Actual test results showed that the number of particles on the filter surface was often sufficient to turn a white filter cloth gray. In addition, hair and fine metal turnings were sometimes observed. The source of the hair is somewhat in doubt (i.e., it could arise from the plant's operating personnel or during the fabrication of the freeze-dryer). But the metal turnings leave little doubt that they are a result of the construction of the dryer. Small metal turnings on the filter may stem from shedding of the ends of the metal braiding that is used to protect the fluid heat-transfer lines to the shelves.

Counter

The last method for determining particulate matter in a freeze-dryer is more quantitative in nature. This method involves the use of a particle counter [25]. The particle counter is first connected to a port of the dryer. With the freeze-dryer chamber clean and dry, the system is evacuated to a pressure of about 100 mTorr. As with the previous method, the dryer is then backfilled to atmospheric pressure with air or nitrogen that has been passed through a 0.2 μm filter. The chamber door is secured for pressurization, and the P_c is increased to 1 psig. The port to the light-scattering particle counter [26] is opened, and the particulate matter in the gas flowing from the chamber is determined over a period of 1 hr. From the data, the mean particle count per ft^3 can be determined [25].

Vapor Deposits

Vapor deposits may also be a source of product contamination. Such sources of hydrocarbons may stem from a number of sources, e.g., backstreaming from the mechanical pump or from leaks of heat-transfer fluid. In the case of backstreaming of hydrocarbons from the vacuum system, an oil vapor deposit on the shelves may not occur during the actual lyophilization process but may be transported from the condenser system when the dryer is steam sterilized and the temperature of the condenser surfaces reach $\geq$ 121°C at a pressure of 15 psig. However, at the these temperatures, the vapor pressure of the vacuum pump fluid is estimated to be between 1 and 10 mTorr [27]. Therefore, the heat from the sterilization process could not account for the transport of oil vapors throughout the dryer. If the oil vapors stem from the decomposition of pump fluids [27], then the vapor pressure of these by-products of oxidation or thermal decomposition may be considerably higher than the pure oil (e.g., vapor pressure of pump oil at about 20°C by the manufacturers of HyVac® oil for vacuum pumps is 1×10^{-6} Torr). The transport of these lower molecular weight oils may occur even at ambient temperature during the drying process, or during the steam sterilization process. Their vapor pressure at ambient temperature may be sufficient to result in contamination of the dried product prior to the stoppering process. I have not detected hydrocarbon vapors in the drying chamber by means of a mass spectrometer but have observed oil in the condensate from the condenser. The reader should not be complacent regarding the possibility of the contamination of the product by hydrocarbon vapors but take steps to prevent such contamination.

Swab Test

The swab test consists of contacting a given surface area with a known clean material that may or may not be dampened with water or an organic solvent. The resulting swab is stored in a clean container and then analyzed for foreign substances. The presence of vapor deposits on an interior surface of a freeze-dryer can be difficult to detect because it may consist of only a few monolayers. Such a surface layer may be so thin that swabbing the surface may not produce a sufficient amount of material for chemical analysis. In addition, one would need some idea as to the potential sources of surface contamination in order to select the proper analytical method. The question that one must ask is, *Does the fact that the analysis did not indicate the presence of a specific number of substances be considered sufficient proof that the surface is clean?* In using the swab test, being unable to detect the presence of certain compounds on the surface leaves some doubt regarding the actual cleanliness of a surface. In order to confirm that the surface is indeed free of any vapor deposits, one should verify the absence of any vapor deposit with another independent testing method.

Contact Angle

The presence or absence of a foreign film on a surface can be determined by measuring the contact angle that a drop of water has on the surface, such as that shown in Figure 13.8 [25]. For a clean surface, the contact angle α will be quite small because the water will *wet* the surface or there will be adhesion between the water and the surface. In the presence of a surface film, the water will not wet the surface, and the contact angle β will be much greater than α. Since the diameter that the drop makes with the surface will be inversely related to the contact angle, a measure of the contact diameter of the drop will be an indication of the contact area (i.e., as the diameter of the drop increases, the contact angle illustrated by Figure 13.8 will become smaller). The presence of a hydrophilic film will result in the formation of a *bead* of water having a smaller diameter.

In determining if a surface film is present, one should first clean a piece of stainless steel having the same composition and finish as that used in the freeze-dryer chamber. After allowing the surface to air dry, a drop of water is allowed to fall on the surface, and the diameter of the drop is recorded. A similar drop of water is then allowed to fall on a surface of the freeze-dryer. If the diameter of the drop on the metal surface of the freeze-dryer is significantly smaller than that of the clean test surface, then one would conclude that a surface film is present; thus the cleaning procedure of the dryer and the water drop test should be repeated. If it is found that repeatedly cleaning the dryer does not produce a similar diameter water drop as that of the test surface, then one must alter the cleaning procedure so as to remove the vapor film deposit.

If there is a surface deposit present after a cleaning process, then it would be reasonable to assume that the surface deposit must have stemmed from vapors of

Figure 13.8. Using the Contact Angle of a Drop of Water to Verify the Absence or Presence of a Nonwetting Film Deposit

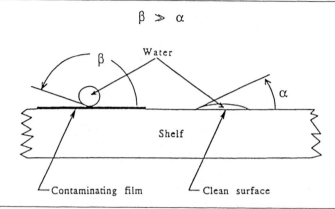

The clean surface is indicated by the low contact angle (α), while the presence of a nonwetting surface is shown to have a larger contact angle β. [Source: *PDA Journal*, Vol. 50, No. 3, May–June 1996, Page 184, Figure 3. Reprinted with permission.]

substances that are not soluble in water. Such substances could come from mechanical pump oil, grease used on a sealing gasket, or vapors from the heat-transfer fluid. The means for removing such surface deposits will be considered in the next section.

Cleaning

In order to be considered clean, a surface must be free of any foreign substances. The surfaces of freeze-dryers can be contaminated from many sources, including operating personnel, the product being lyophilized, and components of the dryer [25]. The following discusses three possible means for cleaning the freeze-dryer. It should be understood that this section will mainly address the various means for cleaning, not validation issues that are beyond the scope of this text.

Manual

Manual cleaning of the dryer is performed by operating personnel using one of two SOPs. The first method requires operating personnel to remove substances from the surfaces of the dryer, in particular the drying chamber, that could represent potential sources of particulate matter or chemical contamination. If the dryer interfaces with an aseptic Class 100 clean room, the cleaning materials must be sterile and lint free. Special care must be taken so that when any particulate matter is removed from the dryer, it is not transferred to the environment of the clean room. The possibility of the latter taking place can be reduced by adding a small amount for Water for Injection (WFI) to dampen but not wet the cloth. Organic solvents should be avoided, as pointed out previously in this chapter, because of the potential absorption of such material in the elastomer seals of the dryer. The outgassing of the solvent during the lyophilization process could pose a source of chemical contamination to the product. The possibility of such product contamination would be greatest during the completion of secondary drying when there is little, if any, flow of gas from the vial or tray and the cake offers a porous adsorption medium with a high surface area. The main function of this cleaning procedure is to remove any potential source of particulate matter from the dryer.

The second cleaning method would be employed during a shutdown period when the clean room and freeze-drying equipment are being serviced. During this period, the drying and condenser chambers can be manually cleaned with a detergent solution. The detergent will not only aid in removing any particulate matter but also remove vapor deposits that may have resulted during repeated lyophilization processes. One such source could be a film of silicone fluid that could form as a result of repeated contact of the closures with the underside of the shelves during the stoppering process [28]. This cleaning method should involve the condenser chamber to ensure the removal of any hydrocarbons deposited on the surface as a result of backstreaming from the vacuum system. The subject of backstreaming will be addressed later in this chapter.

In using a detergent or other water-based cleaning agent, one is always concerned with the possibility of inadequate rinsing that could leave residuals of the cleaning agent that could then become a source of product contamination. The latter can be overcome by rinsing the detergent from the dryer. The dryer will be considered free of the detergent when the *p*H of the final rinse with WFI is equal to the *p*H of the WFI.

It is important here to advise the reader that there are means, which will be considered later in this chapter, by which the presence of a hydrocarbon film on the surface of the dryer can be prevented so that the number of manual cleanings can be kept to a minimum. More stringent quality control (QC) inspections of the glassware may help to reduce the generation of glass-related particulate matter. This is applicable to the next two dryer cleaning methods.

Clean-in-Place

Murgatroyd [28] states that *clean-in-place* (CIP) has not been successful in cleaning freeze-dryers. The method uses a cleaning fluid(s) to remove particulate matter and soluble product. The shelves and other hardware in the drying chamber impede the effectiveness of the CIP system and raises the issue that CIP merely moves the *insoluble material into a corner where it cannot be swept away* . . . [28]. In addition, the CIP method requires the use of large volumes of cleaning and rinsing fluids [28]. In preparing this text, no publication in the literature was found that equated the volume of water (often WFI) to complete CIP to the size (shelf-surface area) of the dryer. Such a relationship would be very useful to those purchasing a dryer in estimating not only the cost of the dryer but also the additional expense associated with producing WFI.

I have little confidence at this time, except in the case of removing soluble material, that CIP is an adequate means for producing a clean freeze-dryer. This concern is based on the use of only water as the cleaning medium. One only has to attempt to clean their automobile by merely spraying it with water to realize that, after the car has dried, there is still a visible film of dirt on the surface. In order to remove such a film using CIP, one would have to first treat the surfaces with a solution that would emulsify and remove any particulate matter and hydrocarbon film that may be on the dryer surface. Another concern is the effective rinsing of connections, such as pressure gauges, once contaminated with a strong cleaning solution. The presence of such cleaning solution may not impact the performance of a capacitance manometer gauge, but it could affect the performance of a thermocouple or Pirani gauge. Large quantities of rinse water would then be required to ensure that the cleaning compound was completely removed from the system. In using the CIP method, one must also make certain that the design and construction of the dryer will provide for complete drainage of the dryer. One might also be faced with the issue of the impact that the repeated use of such a cleaning method would have on the environment.

Flooding

Flooding completely fills the dryer with the cleaning fluid. The cleaning fluid is then drained from the system. This method would be effective in removing soluble contaminants from the dryer; however, it may not be effective in removing hydrocarbon deposits from the dryer surface. A treatment with a cleaning solution may prove useful in removing the hydrocarbon deposits but may not be effective in removing particulate matter from the dryer. The volume of water required to remove the cleaning solution would be, in the case of a large production dryer, enormous. As with the CIP method, one would have to be sure that the design and construction of the dryer allows for complete drainage of the system. This method, while simple in design, could prove very expensive because of the cost of the rinse water. Other drawbacks would include the effective cleaning of the top of the chamber (possible air pockets), and the weight of the fluid places additional floor loading requirements on the equipment [28]. A prime source of such air pockets would be the pressure gauges on the top of the chamber or condenser. Gauges or ports located on the side of the chamber could also present a problem with regards to the formation of an air pocket and the effective rinsing of the connection.

Commentary

I regret taking such a negative position regarding the latter two cleaning methods, but the lack of objective publications concerning the cleaning of freeze-dryers by CIP and flooding necessitates revealing that potential pitfalls associated with such methods may outweigh the convenience of the methods.

THE CONDENSER

The main role of the condenser in the freeze-dryer is to remove water vapor from the system [2,6]. In essence, the condenser is a very efficient pump of condensible vapors, namely, water vapor. To understand just how efficient the condenser is as a condensable vapor pump, consider the sublimation of 1 L of ice over a period of 1 day at a pressure of 96 mTorr. At this pressure, the temperature of the water vapor would be -40°C. In order to simplify this example, consider that the chamber of the dryer is also at -40°C. Under these conditions, it is further assumed that the water vapor is acting as an ideal gas, and the sublimation of the ice will produce gas at a rate of 97.3 L/sec or 206 ft³/min. A mechanical pump having a pumping capacity of 200 ft³/min is a large vacuum pump that is used to evacuate production freeze-dryers. However, the same pumping speed could be obtained from a condenser surface of 1 m² maintained at -70°C and the power required for the sublimation of the ice (assuming the heat of condensation of water is equal to the heat of sublimation) would be about 32 W. Since the vapor pressure of ice at -70°C is about 2 mTorr, the efficiency of the condenser would be about 98%; if the ice were deposited uniformly over the entire condenser surface, it would attain a maximum thickness of just 1 mm, and a partial pressure of 2 mTorr of water vapor will enter the mechanical pump.

What is important for the reader to understand is that if the P_c were to be increased to 200 mTorr, by bleeding in a gas such a nitrogen, and the partial pressure of water vapor in the chamber were maintained at 100 mTorr, the capacity of the mechanical pump would have to be substantially increased to accommodate the additional nitrogen gas. In the case of the condenser, the increase in the P_c will result in the efficiency of the condenser approaching 100% because the water vapor must now leave the condenser surface by diffusion rather than by sublimation, as seen in the previous example.

The reader should also note that the water vapor throughput of a system, which was described in previous chapters, had little or no emphasis or importance placed on the condenser temperature. It is not that the condenser temperature is not important, but as long as P_c remains at a given value, the condenser temperature has no impact on the resulting T_p or drying rate. For the secondary drying process, the residual moisture was shown to be dependent on the partial pressure of the water vapor that is in equilibrium with the cake. In instances where it was difficult to reduce the residual moisture content in the cake, a series of nitrogen purges were shown to represent a means of acquiring the desired residual moisture content.

The remainder of this section will consider the design and construction of the condenser with respect to its impact on the lyophilization process or the performance of the dryer.

Condenser Configuration

In 1954, there was concern regarding the configuration of the condenser. There was a feeling at that time (which may persist to this day) that because ice is a poor conductor of heat, the effective temperature of the condenser will be much higher than that of the expanding refrigerant. Those who held this view felt that the area of the condensing surface should be large. At that time, and apparently even now, there were no data regarding the thermal conductivity of heat for ice at low temperatures. At temperatures $\geq -20°C$, the ice on the condenser tends to be transparent while, at lower temperatures, the ice becomes an opaque white in color [7].

As a general rule, the surface area of the condenser should be $\geq$ the shelf-surface area [2,21]. The distance between the condensing surfaces should be such that the buildup of ice does not impede (block) the flow of gases through the condenser [21]. The following will be a brief discussion of impact that the general configuration of the condenser surfaces can have on the lyophilization process or the performance of the dryer.

Plates

Condenser plates are made from two embossed metal sheets, generally stainless steel, that when joined together form a serpentine flow path for the refrigerant formed by direct expansion. A chief advantage of the plate design is that the total condensing surface area can be increased by increasing the number of plates. For flat plates, the plates are arranged in a parallel configuration such that there is a nonuniform thickness of ice across the condenser surface. This means that the vapor flow of the gases from the drying chamber must pass by the entire length of the condenser plate before entering the vacuum system. Thus, the ice thickness on the plates will be greater where the vapor flow first comes in contact with the condenser plates and be at a minimum near the inlet to the vacuum system. Should the condenser plates be configured such that a uniform ice thickness is formed across the surface area of the plate(s), then the probability of ice formation on the plate(s) nearest the vapor flow from the product will be the same as that near the inlet to the vacuum system. As a result, water vapor can now enter the vacuum system.

When there is a nonuniform distribution of ice across the condenser plates, one must be concerned with *bridging* (i.e., ice from one plate joins ice from an adjacent plate). The vapor flow is then diverted, and soon the condensed ice impedes the flow of gas from the chamber. The reduction in gas flow results in an increase in the P_c. The increased P_c will cause an increase in T_p, which can lead to collapse or meltback of the product. The increase in P_c can also result in a decrease in the drying rate and enhance the possibility of partial meltback of the product. Bridging of ice across the condenser plates will tend to occur during the primary drying process, which requires a low P_c. A low P_c would be a result of low ($< -40°C$) Tc. Merely increasing the distance between the plates would prevent bridging occurring at low pressures but permit water vapor to pass by the plates and enter the vacuum system when the chamber is at higher pressures. It will be shown later in this chapter that excess water in the vacuum pump can lead to a loss in pumping speed and an increase in P_c and T_p.

Another problem associated with the use of parallel plates as a condenser system is the potential for affecting the residual moisture of the final product. Unlike the door of the chamber that is a source of energy, especially during the freezing and primary drying processes, the plate condenser surfaces can, as a result of Stefan's law, cause vials having a direct line of sight with the plates of the condenser to be at a lower temperature than other vials in the chamber. At the completion of the secondary drying process, these vials would be at a lower temperature than other vials in the system. From the nature of adsorption isotherms that were illustrated in Chapter 9, cakes at a lower temperature, for a given partial pressure of water vapor, will have a higher residual moisture content.

In purchasing a dryer that employs a parallel plate design for the condenser, one must specify the range of water vapor flow for the condenser system in order to avoid bridging or contamination of the oil used to lubricate the vacuum pump. In addition, there should not be a line of sight between the condenser plates and any vials in the drying chamber.

Coils

Some condensers are configured as a coil(s) and fabricated from metal (usually stainless steel) tubing. The chief advantage of a coil is that it is generally fabricated from a single length of tubing and, for that reason, it is less likely to have a real leak and permit refrigerant to enter the dryer. The reason for less leaks stems from the connection to the tubing being made outside the dryer, whereas, for a plate condenser, the refrigeration connection must be made inside the dryer and, consequently, there is always the potential virtual leak of refrigerant gas into dryer. While there is no standardized condenser coil configuration, the ideal configuration, in my opinion, would be where the axis of the refrigeration coils would be normal to the path of the gases. With such a coil configuration, a baffle can be employed to direct the flow of the vapors over the coils such that there is a nonuniform distribution of ice thickness on the coil surface. In addition, such a baffle would also serve to prevent an optical path between the refrigerated coils and the vials on the shelves of the drying chamber.

Capacity

The capacity of a condenser system is sometimes expressed in terms of the maximum thickness of the ice uniformly distributed across the condenser surface [5]. I consider such a means for expressing the capacity of a condenser to be archaic because a nonuniform distribution of ice across the condenser is preferred. Condenser capacity should be given in terms of the maximum volume of water that can be deposited on the condenser at a specified P_c when water vapor is the major gas component.

The capacity of the condenser is sometimes also expressed in terms of the volume of the empty condenser chamber [5]. Such condenser capacities can be obtained only under dryer operating conditions that may not typify lyophilization operating parameters. As a result, under actual lyophilization parameters, only a fraction of the rated volume of water vapor can be accommodated by the condenser. If the actual capacity, rather than the rated capacity, is exceeded, bridging will generally occur, resulting in a loss of control of the P_c. An increase in P_c will also result in an increase in the T_p. As stated previously, an uncontrolled increase in the T_p can have a major impact on the quality of the final product.

Condenser Temperatures

Depending on the nature of the refrigerant and the types of compressors used, condenser temperatures can range from -50°C to -90°C [6]. For condenser temperatures limited to -50°C, the P_c is also limited to about 30 mTorr. If the pressures in the drying chamber are reduced to < 30 mTorr, then, as already pointed out, water vapor will start to sublimate from the condenser surface. This water vapor will enter the vacuum system and possibly contaminate the oil in the vacuum pump(s). However, if it is known that P_c will be ≥ 100 mTorr and the partial pressure of water vapor is ≥ 80 mTorr, then a condenser temperature of -60°C would be adequate. Since the actual P_c will depend on the thermal properties of the frozen matrix and the isothermal adsorption properties of the cake, the composition of the formulation will be the determining factor regarding the required performance specifications of the dryer.

Internal Condenser

The internal condenser means that the condensing surfaces share the same chamber as that of the shelves. The condensing surfaces, depending on the required capacity of the condenser, can be positioned on one or both sides of the shelves. In some designs, the condensing surface is under the bottom shelf. I have no knowledge of the condenser coils being located at the top of the chamber, although in the latter design condenser surfaces above the unusable shelf may have less impact on T_p than when under the bottom shelf. Both plates and coils have been successfully used as condenser surfaces for the internal condenser.

Most of the dryers using an internal condenser will have a metal (stainless steel) baffle that serves two functions. Because the difference between the product and condenser temperatures may be 50°C during primary drying and as much as 100°C during secondary drying, the first function of the baffle is that of an optical barrier. Such an optical baffle would prevent heat transfer between the condenser surfaces and the product and the shelves. The effect of an optical line-of-sight on the moisture content of the product has already been previously described. Without a baffle, there also would be an additional thermal load on the condenser and its refrigeration system; thereby, limiting the minimum operating temperature of the condenser. The higher the operating temperature of the condenser, the higher the P_c before water starts to sublimate from the condenser surfaces. A higher P_c during the secondary drying process may result in an extension of the secondary drying process by impeding the desorption of the water from the cake or result in a higher residual moisture in the final product.

The second principal function of the baffle plate is to direct the path of the gas to the condenser surface. In some designs, the baffle forces the flow of gases over the condenser surfaces such that there is a nonuniform distribution of ice across the condenser surfaces. Some baffles contain slits that will prevent a line of sight between the product and the condenser but will allow gas to pass through the baffle to form a uniform ice thickness on the condenser surface. I prefer the solid baffle design rather than one with slits because there is less chance of water vapor entering the vacuum system.

There are several major advantages to the internal condenser design. First, the dryer requires but one chamber, which significantly reduces the cost of the dryer. Since there is only one chamber, the unit can often be constructed as a single unit, which greatly simplifies shipping, reduces installation costs, and occupies less floor space. Finally, this type of condenser design allows for a high and uniform conductance of gases between the shelves and the condenser surfaces.

This design of the freeze-dryer condenser does have limitations and disadvantages. One of the major limitations is the size and capacity of the condenser. When the shelf area exceeds 120 ft^2 or the condenser capacity is greater than 100 L of water, then the increase in the overall dimensions of the drying chamber begins to have an impact on the cost of the construction of the dryer. In addition, it may no longer be practical to house both the chamber and the necessary refrigeration system as a single unit. The last limitation of this type of condenser system is that if the product is to be backfilled with nitrogen to a pressure < 1 atm (e.g., 600 Torr), in order to reduce the possibility of coring by the syringe, then the passing of ambient temperature gas over ice on the condenser surface could result in an increase in the partial pressure of water vapor in the chamber and the residual moisture content of the product.

External Condenser

An external condenser is one in which the condenser chamber is located in a separate chamber. The external condenser becomes economically feasible when the shelf-surface area of the dryer exceeds 120 ft^2 [21]. Since the condenser is located in a separate chamber, the chambers are connected by means of a port that contains an isolation valve. The relative gas conductance between the drying and condenser chambers in this system will tend to be lower than that for an internal condenser.

Careful consideration should be given to the nature of the lyophilized product and the lyophilization process. If the final product is hygroscopic in nature and stoppering will be performed at pressures just under 1 atm, then one should select an external dryer because the drying and condensing chambers can be isolated prior to the backfilling of the system. Isolating the drying chamber prior to backfilling eliminates the possibility of water vapor being removed from the condenser surfaces and transported back to the final product.

Isolation Valve

The isolation valve is located in the piping that connects the drying chamber to the condenser. While the valve is useful in isolating the drying chamber in order to perform a ROR test [6], there are other factors that can impact the lyophilization process and that must be considered when specifying such a valve.

Valve Configuration

When specifying the valve, one must also take into account the configuration of the condenser system. For example, if the condenser is configured as a coil, then the gases from the chamber must be deflected when they enter the condenser chamber so that they will pass over the refrigerated coils. Unless special provisions are made in the design of the condenser to direct the path of the vapors, a butterfly valve (a plate that turns 90° so that the only impedance to the gas flow will be the thickness of the valve plate) or gate valve (a valve that slides open, leaving the piping between the drying chamber and condenser an unobstructive path) will allow gases from the drying chamber to pass directly into the condenser chamber. If a popet valve (a type of valve that when closed will form a seal at the end of the piping from the condenser system) is used to isolate the condenser from the drying chamber, then when the valve is placed in an open position, it is moved back some distance from the valve seat. However, the valve will maintain its relative position with respect to the sealing surface. As a result, the gases that flow from the piping strike the surface of the valve and are deflected to the sides of the condenser chamber. The chief advantage of such a valve configuration is that it also serves as an optical baffle so that there is no line of sight between the product and the condenser plates to produce a product with a higher residual moisture.

For a parallel plate condenser, one is limited to using either a gate valve or a butterfly valve. In order to prevent a line of sight between the product and the condenser surfaces, one should specify the placement of a high conductance and optically dense chevron on the end of the piping that connects to the condenser system. The chevron will offer little impedance to the gas flow but will prevent any line of sight between the condenser plates and the product vials on the shelf of the dryer. It is hoped that chevron baffles will be used to prevent heat transfer between product vials and the cold condenser surface.

Gas Conductance

The gas conductance of a vacuum system was discussed in Chapter 11. In Table 11.4 it was shown that in specifying the freeze-dryer system, one must take into account the conductance of the piping and the capacity of the refrigeration system. Thus, there are two basic pitfalls to avoid. The first is that the dimensions of the piping are not adequate to handle the required throughput of the system, which can then result in an increase in P_c. The reader should by now realize that a loss in pressure control equates to a loss in T_p control during the primary drying process.

The second pitfall shown by Table 11.4 shows that even when the piping is adequate to handle the gas flow, one must be careful not to exceed the refrigeration capacity of the condenser. When the latter happens, and I can attest to it, the pressure increases and further overloads the capacity of the condenser, thus leading to further increases in P_c and T_p. Any attempt to refrigerate the shelves only takes further refrigeration from the condenser and increases the P_c. Needless to say, in a matter of minutes, the T_p has risen well beyond the Tc, making the final product unacceptable. The above experience taught me not to assume anything regarding the operation of a freeze-dryer—test before using.

VACUUM SYSTEM

The Need for a Vacuum System

It has been shown repeatedly that an important parameter in the control of the lyophilization process is the P_c. In order to commence with the sublimation of ice (primary drying process) and still maintain the T_p below the collapse or eutectic temperature, P_c must first be reduced to pressures that are lower than the vapor pressure of ice at the desired T_p (e.g., typically 100 and 200 mTorr [2]). In order to achieve such a pressure, it is necessary to reduce and maintain the partial pressures of the noncondensable gases (nitrogen, oxygen, argon, and carbon dioxide) in the chamber from 1 atm to values that are lower than the vapor pressure of the ice at the T_p. This is accomplished by means of a vacuum system. As previously stated, it is not within the scope of this text to consider the mechanisms involved in the vacuum system but to consider the impact that the vacuum system can have on the product during the course of the drying process.

For example, prior to the start of the primary drying process, it is general practice, at least with many production freeze-dryers, to use the main compressors of the system to chill the condenser to temperatures that are $\leq -20°C$ lower than that of the T_p. Only after the condenser has achieved a sufficiently low temperature will the vacuum system be used to evacuate the drying and condenser chambers to the desired P_c. It is during this time frame that the product must depend on a lower capacity auxiliary refrigeration system to maintain its temperature below Tc. Also during this time period, the shelves of the dryer and the product are being heated by radiant energy from the walls and door of the chamber. It is therefore imperative that the pumpdown time (i.e., that time period required by the vacuum system to evacuate the chamber to its operating pressure) not be excessive. With excessive pumpdown times (e.g., ≥ 1 hr), T_p may exceed Tc, and attempting to dry the product could lead to collapse or meltback of the cake [2]. For that reason, the key factor is not only that a vacuum system can evacuate a freeze-dryer to a given pressure; the rate of evacuation rate is also a key factor. The latter is dependent on the effective volume of the system, the real and virtual leak rates, and the pumping capacity of the vacuum system [13].

Foreline

The connection between the vacuum pump and the chamber is known as the foreline, and its design and the materials used in its construction can have a serious impact on the pumping capacity of the vacuum system. A significant reduction in the pumping capacity as a result of design of the foreline will prolong the pumpdown time and lowest attainable P_c. For example, the design of the foreline may call for the vacuum pumps to be some distance from the dryer and involve a number of $90°$ bends. It was stated in Chapter 11 that the approximate effect of each $90°$ bend is to increase the length of the pipe by 10 pipe diameters. An increase in pipe length can only reduce the conductance of the foreline and lead to a reduction in pumping capacity.

The forelines of some freeze-dryers still use an excessive amount of rubber hosing. In order to prevent the collapse of the walls of the tubing, as a result of the force exerted by the atmospheric pressure, a relatively small bore diameter is used which has poor gas conductance properties. While metal piping may be a better choice than rubber for a foreline, the mere fact that it is metal is no guarantee of its effectiveness. While threaded iron piping may be both inexpensive and easier to assemble, the use of threaded joints for a vacuum system strongly increases the potential for leaks in the foreline.

For the ideal vacuum foreline, the distance between the pumping system and the drying chamber should be kept to a minimum. The piping should be constructed out of stainless steel and equipped with bellows so that pump vibrations are not transferred to the foreline connections. Such vibrations may result, over a period of time, in either rupturing the vacuum connections or real leaks resulting from metal fatigue.

Backstreaming of Oil Vapors

At the beginning of the evacuation process, the nature of the gas flow to an oil-lubricated pump is viscous in nature, and oil vapors from the pump are prevented from leaving the pumping system. As the pressure is reduced and the gas flow becomes more molecular in nature, oil vapors can leave the pump and enter the condenser chamber. The latter process is known as *backstreaming* [29,30]. If precautions are not taken, oil vapors can find their way into the drying chamber. Backstreaming rates from a mechanical pump have been reported as high as 1×10^{-4} g/(cm^2 hr) [30].

The effects of backstreaming first became a major concern during the development of microcircuits and semiconductor integrated circuits (circa 1950), where the presence of surface contamination had a profound effect on the performance of the circuitry. It has only been recently that backstreaming has become a major concern in the pharmaceutical industry [29,31]. While backstreaming will not be a concern during the freezing or primary drying process because freezing is conducted at atmospheric pressure, while during primary drying there is sufficient flow from the vials to limit any hydrocarbon vapors from entering the containers. It is during the secondary drying process when gases enter and leave the container or vial by diffusion rather than by gas flow that one must be concern about the possible contamination by backstreaming. In addition, by its very nature, the lyophilized cake is a porous medium with a high surface area that would be conducive to the adsorption of hydrocarbon vapors. The effect of such vapors on the potency or stability of a lyophilized product will be product specific. It is recognized that it would be a difficult task to equate the lack of stability to hydrocarbon vapors backstreaming in a freeze-dryer. However, instead of attempting to show that the presence of the hydrocarbons is not contributing to the instability of the active constituent, I would prefer to resolve the issue by preventing backstreaming and then showing that hydrocarbon vapors are not present in the drying chamber. The following will consider the means for determining backstreaming and ways in which it can be greatly reduced or even eliminated.

Methods for Determining Backstreaming

The following are some methods for detecting the presence of oil vapors in a system as a result of backstreaming from the pumping system.

Qualitative Method. The measurement of the contact angle or diameter of a drop of water has already been shown to be a sensitive and qualitative means for determining whether a hydrocarbon film exists on a surface of the dryer. While this method indicates the presence of a surface film, one must exercise caution in assuming that the source stems entirely from the backstreaming of oil vapors.

Ponderal Method. This method determines the presence of backstreaming by measuring the mass of the gas deposited on a surface [29]. There are two basic methods for making this determination One method uses the increase in weight of a cold metal plate having a surface area of 10 cm^2 to ascertain the deposit of hydrocarbon film. Sensitivities of 1×10^{-4} g have been determined by this method.

A second method uses a cold quartz plate. With this method, the sensitivity of detecting the thickness of a film can be increased by a factor of 1,000, i.e., 1×10^{-7} g. The change in the resonant frequency of a quartz crystal can be related to the mass of the deposited hydrocarbon film [32]. It is unfortunate that the temperature of the quartz crystal was not specified. Since water vapor will always be present to some degree in a freeze-drying chamber, one would have to correct for the presence of multilayers of water vapor that may become adsorbed on the surface of the crystal, otherwise one would obtain a false high value for backstreaming of hydrocarbons into the system. Although using a quartz crystal in an oscillating electrical circuit certainly has its appeal, the uncertainty of the effect of the adsorption of water vapor or changes in temperature [32], depending on its location, leaves some doubt as to its application as a means of determining the backstreaming of oil vapors in a freeze-dryer.

Mass Spectrometry. The use of mass spectrometry (see Chapter 11) can provide both qualitative and quantitative determination of the oil vapors in a system [29,33]. Larrat and Sierakowski [29] connected the input of their quadrupole mass spectrometer to the foreline and the drying chamber. Their results indicated that hydrocarbon vapors were detected in the foreline but not in the drying chamber. It would appear logical to anticipate that backstreaming would favor lighter oil fractions (i.e., lower molecular weight) that can be generated by the thermal breakdown of the oil during the operation of the pump. However, one must be careful in reaching such an assumption based on mass spectrum data. Because these authors found m/e peaks at 41–43, 55–57, and 68–71, one should not infer that their source stemmed from the lighter oil fractions. The reason being is that I determined the cracking or fragmentation pattern for mechanical pump fluid [33], but the fragmentation pattern, similar to that of Larrat and Sierakowski [29], was obtained from fresh oil in a glass container in a dryer chamber. In spite of some evidence of backstreaming of oil vapors in the condenser, I have not observed, from a mass spectrum, the presence of hydrocarbon vapors in the drying chamber during a lyophilization process. Although a vapor film has been observed on the surface of the shelves of a dryer because of a high contact angle of water droplets on the surface, it has not been possible to ascertain the source of such vapors, e.g., backstreaming oil vapors, heat-transfer fluid, or outgassing from the closures.

Adsorption Method. López et al. [31] investigated the backstreaming of hydrocarbon vapors in a freeze-dryer and the impact that such vapors had on a lyophilized product. These authors first established that the absorption spectrum for rotary pump fluid showed a characteristic peak at 262 nm and minimum at 254 nm. They demonstrated that is was possible to determine quantitatively the oil deposited on the plate from the absorption spectrum. In order to eliminate possible contamination of the test plates from oil already in the drying chamber, the chamber was replaced with a clean experimental chamber that was fitted with a special connection between the chamber and the condenser to accommodate various absorbent materials (molecular sieve and activated charcoal or 0.22 μm filters). Their results are very interesting because they showed that, in spite of the presence of absorbent materials, test plates positioned in the test chamber still indicated the backstreaming of oil vapors from the mechanical pump oil.

Means for Eliminating or Preventing Backstreaming

While recognizing that backstreaming of oil vapors from a vacuum system can occur, one should be aware how to eliminate or prevent such a source of product contamination.

Drying Process. There is no doubt that the best and most economical means for the prevention of backstreaming is a properly designed lyophilization process. López et al. [31] established that the backstreaming of oil vapors was dependent on P_c. Their results showed that backstreaming is quite prevalent at pressures < 100 mTorr. There is a decrease in the rate of backstreaming at a pressure of 100 mTorr, but pressures ≥ 200 mTorr are required in order to significantly reduce the rate of backstreaming of the oil vapors. López et al. [31] showed that increasing the chamber pressure to 400 mTorr greatly decreased the backstreaming of oil vapors into the drying chamber during a 40 hr drying process. Amoignon [30] also showed, by means of mass spectrometric data, that by increasing the flow of gases in the chamber, one can reduce the rate of backstreaming to as low as 1×10^{-11} g/(cm^2 hr).

The reader should be cautioned not to arbitrarily increase P_c, especially during the primary drying process, to ≥ 200 mTorr or higher: in order to lyophilize a product at 200 mTorr, one would require a formulation with a Tc > -30°C; a P_c of 400 mTorr requires a Tc on the order of -20°C. It has been my experience that such collapse temperatures for formulations are the exception rather than the rule. Increasing P_c to 400 mTorr during the primary drying of a formulation having a Tc < -40°C would surely lead to collapse or even meltback of the cake. Thus, for those formulations with a low Tc, one must seek other means for preventing or eliminating the backstreaming of oil vapors into the drying chamber. The fact that it is necessary to seek other means to prevent product contamination as a result of backstreaming is strong testimony of the impact that the formulation can have on both the drying process and the considerations regarding the freeze-drying equipment.

Mechanical Pump. The function of the mechanical pump is to remove gases from the freeze-dryer and compress them such that they can be discharged to the atmosphere. The selection of the vacuum pump should be taken into consideration with respect to its backstreaming properties. In selecting a vacuum pump, one should consider not only its pumping capacity (cfm) and the blank-off pressure (lowest pressure obtainable) but also its operating temperature. A typical pump operating temperature is about 65°C. The pump temperature can be increased by operating the pump with the ballast value fully opened. An increase in the operating temperature of the pump will not only increase the vapor pressure of the oil but also can increase the rate of thermal degradation of the oil. The combining of these two pump characteristics will only lead to an increase in the rate of backstreaming at a given pressure [29].

Roots Blower Type Pumps. A Roots blower type pump consists of 2 or more massive steel vanes rotating at speeds of 3,000 to 4,000 rpm [27,32]. The pump can, but is not recommended, discharge directly into the atmosphere and for that reason is typically positioned between the mechanical pump and the foreline. The Roots blower will compress the gas from the system, such that the inlet pressure to the mechanical pump is increased. In this manner, the overall pumping speed of the vacuum system (Roots blower and mechanical pump) is increased. An increase in the inlet pressure to the mechanical pump, for P_c < 100 mTorr, will not only increase the pumping speed of the mechanical pump but also serve to reduce backstreaming. However, depending on the nature of the blower pump, once the inlet pressure to the mechanical pump is reduced to low operating pressures, e.g., 100 mTorr, then, in spite of the presence of the rotating vanes, the potential for backstreaming of oil vapors into the foreline will increase. Thus, the effectiveness of the Roots blower is limited to a flow of gas from the chamber that will maintain an effective inlet pressure to the mechanical pump of ≥ 200 mTorr.

Dry Pumps. The dry vacuum pump does not require large quantities of oil for lubrication or for maintaining an effective vacuum seal [29]; for that reason, backstreaming is virtually eliminated. These pumps are commercially available; however, they tend to be more expensive and sometimes require the use of a Roots blower type pump to increase the pumping speed. The demand for such pumps was precipitated by the needs of the electronics industry and, it is hoped that it will not be long before design improvements of these pumps and market competition will make them an attractive alternative to the present oil seal vacuum systems. When these pumps become the standard vacuum system for freeze-dryers, then the concern of backstreaming of hydrocarbon vapors in the drying chamber will be removed.

Barriers. One of the mechanisms involved in the backstreaming process is the migration of an oil film along the walls of the foreline. Because oil will not wet Teflon®, the insertion of a Teflon® ring in the foreline pipe can impede the migration of the oil and reduce the rate of backstreaming [30]. However, it is my opinion that as the pressure in the foreline is reduced, oil vapor may well bypass the Teflon® barrier.

Traps. The backstreaming of hydrocarbons from the pumping system can be trapped in the foreline by using adsorption type materials such as molecular sieve, activated carbon, and alumina [27,29]. The efficiency of such traps depends on the nature of the absorbing media and can vary from 50% to 99%. When nickel- or platinum-activated alumina is heated, it can convert the backstreaming hydrocarbons to CO_2 and water vapor with a 99% efficiency. A possible drawback to the use of such a system is that there may be a 10%–20% reduction in the pumping speed of the system [29]. Other authors [27,31] have shown that absorption traps are not always an effective means for preventing the backstreaming of hydrocarbon vapors.

For many years, cold traps maintained at liquid nitrogen temperatures have proven an effective means of preventing the backstreaming of hydrocarbons or the transport of other vapors to the pumping system [27]. At one time, I was making measurements of a vacuum chamber with a highly accurate mercury McLeod gauge. The mercury vapors from the gauge were prevented from entering the test chamber by means of a glass liquid nitrogen cold trap. It was not long, perhaps only after a few measurements, that a black deposit was observed forming in the cold trap. At the end of the day, the deposit had become dense and opaque in nature. Care had to be exercised to valve off the cold trap to the system prior to the removal of the liquid nitrogen refrigerant. When the trap warmed up, the black opaque deposit disappeared, which meant that the condensed mercury was returned to the McLeod gauge. It was important that prior to using the McLeod gauge, the trap had to be cooled to liquid nitrogen temperatures in order to prevent any transport of mercury vapors when the valve was opened.

Because of the need to increase the solubility of some compounds, solvents such as ethanol and acetonitrile are sometimes used in the formulation. Liquid nitrogen cold traps are often used in an attempt to prevent such vapors from entering in the pump oil or being discharged into the environment. However, in spite of the cold trap, the vapors still manage to enter the pumping system. The reason these gases bypass the cold trap is a result of the design of the trap, not that the vapor pressure of the compounds at liquid nitrogen temperatures is excessive. It is during the initial pumpdown of the system when, even at –40°C, the vapor pressures of these solvents may be in excess of 1 Torr, and the gas flow through the trap will be viscous in nature. Because of the nature of the vapor flow, some of the solvent gases do not come into contact with a cold surface and, therefore, will enter the pumping system.

In using a cold trap, one should be aware of the vapor pressure of the trapped gas at ambient temperature. For solvents such as ethanol and acetonitrile, the vapor pressure of these compounds at ambient temperatures is less than 1 atm. While nitrogen gas appears to be the most common gas used to control P_c, the inert gas argon has been used to control P_c. If argon is used in conjunction with a liquid nitrogen cold trap, the argon will condense in the trap. When the trap is allowed to warm up, the argon pressure will increase, and one could be faced with a dangerous explosive system [34].

Using a liquid nitrogen cold trap in conjunction with a freeze-dryer could result in the trap being in operation for not just a few hours but perhaps one or more days. One little realized fact about the use of liquid nitrogen is that the boiling point of oxygen is -183°C [24]. Since oxygen has a higher boiling temperature than nitrogen and oxygen constitutes 20% of the atmosphere, liquid oxygen will begin to collect in the cold trap. The quantity of condensed oxygen will depend on the design of the trap and the time the liquid in the trap is exposed to the atmosphere. The danger here is that if large quantities of liquid oxygen are present in the liquid nitrogen, then upon warming of the trap there would be a sudden release of a large quantity of oxygen gas into the room with the potential of a fire in any operating electrical device or equipment. Such a hazard is further enhanced when individuals cannot resist the urge to pour the liquid from the trap onto the floor and watch it evaporate with an associated cloud of water vapor. By doing so, they can release large volumes of potentially dangerous oxygen gas into the room.

Pump Oil. The quality of the oil used in the pump is also an important consideration when it comes to backstreaming [29]. A good pump oil will consist of high molecular weight molecules. Lower quality and often cheaper pump oils will contain lower molecular weight oil fractions that can be a source of backstreaming. The reader is cautioned not to attempt to replace the hydrocarbon oil with a lower vapor pressure (higher quality) silicone oil. I know of one such attempt which resulted in the pumps (five) ceasing up after only a short operating time. To the best of my knowledge, the pumps were damaged beyond repair.

Gas Ballast

Most mechanical oil seal vacuum pumps are equipped with a *gas ballast* valve. When placed in the open position, the gas ballast valve will permit air to be injected into the gases being compressed so as to increase the pressure. By increasing the pressure in the pump, condensable gases, such as water, will not reach its vapor pressure and condense into the pump oil. A buildup of water vapor in the pump oil will not only reduce the pumping speed but also lead to corrosion of the pump. It is sometimes recommended by equipment manufacturers and others to open the gas ballast valve during the primary drying process [6]. It is my opinion that the necessity to open the gas ballast valve during primary drying is a result of an inadequate design of the condenser system or using pressures that are well below the vapor pressure of ice deposited on the condenser surfaces. As previously stated, the opening of the ballast valve will cause an increase in the temperature of the pump oil, which can increase the potential for backstreaming of hydrocarbon vapors.

If it is necessary to use the gas ballast to remove gases condensed in the pump oil, then it is suggested that the reader consider an *air scavenge* or *sparging* method developed by Nicole [35]. Nicole showed that it is possible to remove condensable gases from a mechanical pump by bubbling air through the oil reservoir, as illustrated by Figure 13.9. In order to be effective, the end of the tube should be provided with a fitted sparger to produce small gas bubbles and increase the contact between the liquid in the reservoir tank and the gas. Nicole used a section of a gas line filter, designed for the gas line of an automobile, and soldered it to the end of a copper tube. The airflow was adjusted so that oil did not bubble out of the vent [35].

The effectiveness of this method was demonstrated by the addition of 4 mL of acetone to the oil reservoir of a vacuum pump when the foreline pressure was 15 mTorr. After the acetone was added to the oil, the foreline pressure increased to 35 mTorr for 85 min and 23 mTorr for 23 hr. Sparging the reservoir with air for 2 hr resulted in a foreline pressure of 15 mTorr. Using this technique, one is able to remove and prevent the buildup of condensable compounds in the oil reservoir

Figure 13.9. The Method to "Air Scavenge" Condensable Gases from the Oil Reservoir of a Mechanical Vacuum Pump

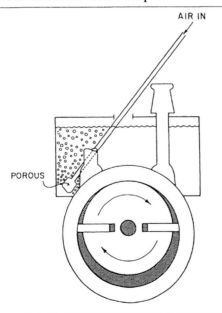

section of the pump without having to heat the oil of the pump and reducing the pumping speed of the vacuum system. In using this method, it is recommended that the sparged gases not be discharged directly into the room nor should this method be used to discharge large quantities of solvents into the atmosphere.

INSTRUMENTATION

The function of the instrumentation of a freeze-dryer is to measure and control three basic process parameters: i.e., P_c, T_s, and time. With the rapid advance and development of sophisticated computer hardware and software, I have mixed feelings regarding the use of such technology with respect to the lyophilization process. While I am no stranger to the use of computers in conjunction with the control of the freeze-dryer, I was among the first to publish the potential use of the computer on freeze-drying equipment [36]; nevertheless there is concern that the pendulum may have swung too far in favor of complete computer control, and perhaps the industry has lost sight of the need for both manual and automated controlled systems [23]. In addition, an effort to program the nonlinear output of pressure gauges has led to using ambiguous pressure units like percent vacuum.

Vacuum Gauges

Although the description of vacuum gauges used in conjunction with the lyophilization process was described in detail in Chapter 11, this section will be mainly concerned with their impact on the lyophilization process.

Mercury McLeod Gauge

At one time, the Mercury McLeod gauge was often used to determine P_c during a lyophilization process. This gauge was replaced by electronic gauges because the output of these latter gauges could be recorded and also used with a vacuum bleed system to control P_c. The use of the McLeod gauge or any gauge containing mercury with a freeze-dryer used in the manufacture of pharmaceutical products opens the possibility, because the gauge is constructed of glass, of large quantities of mercury entering the freeze-dryer as a result of accidental breakage of the gauge when the dryer is under vacuum. The steps one would have to take to remove mercury from the dryer and verify its complete removal are unthinkable. This should serve as a warning to the reader to avoid pressure measurements of a freeze-dryer, for whatever purpose, with a mercury manometer.

In addition, because mercury vapors are dangerous, such gauges should never be used with a freeze-dryer in the manufacture of pharmaceutical products [34]. The reported Occupational Safety and Health Administration (OSHA) limits for continuous exposure to mercury in air at 20°C in the United States is just 0.05 mg/m³ [34]. Translating this to a partial pressure, it comes to 0.0046 mTorr. The vapor pressure of mercury at 20°C is 1.2 mTorr [17], thus the partial pressure of mercury in the gas entering a freeze-dryer every time a McLeod gauge reading is taken (see Chapter 11) is 261 times greater than the limit that set by OSHA for continuous exposure. With typical condenser temperatures < –50°C, mercury will collect on the condenser surfaces. While some of the mercury will pass down the drain, there is a possibility that some mercury will remain. The latter statement raises the question as to whether long-term usage of such a gauge would result in a gradual buildup of mercury vapors in the dryer.

Pirani and Thermocouple Gauges

It is disheartening to learn that other authors [6,7] continue to encourage the use of a thermal conductivity gauge such as a Pirani or thermocouple gauge to measure P_c. These authors recognize that the accuracy of these gauges is dependent on composition of the gases in the drying chamber, and it has been shown in previous chapters that the composition of the gas in the drying chamber can vary from mainly nitrogen to primarily water vapor and finally primarily nitrogen. However, their rationale for the continued use of this class of gauge ranges from the gauges are stable, easy to calibrate, and have a fast response time to the *extreme*—accurate pressure measurements during a lyophilization process are seldom required. In the first instance, it makes little difference if a gauge is calibrated using one gas (nitrogen), and the gauge reads significantly false high (perhaps as much as 50%) in the presence of another gas (water vapor). While it is agreed that extreme accuracy is not necessary for the lyophilization process (±3% of the actual reading should be sufficient) such accuracy cannot be maintained by thermal conductivity gauges throughout the entire lyophilization process. Without confidence in the pressure measurements, how does one validate the lyophilization process or transfer the drying process from one dryer to another?

Because of the increased use of solvents in formulations, mainly those in the field of biotechnology, there is yet another aspect that should be taken into account when using a thermal conductivity gauge: The filament temperature should be below any ignition temperature of any organic solvent/oxygen mixture [34]. For example, the vapor pressure of ethanol, which is sometimes used in formulations to enhance solubility, is 390 mTorr at –40°C and increases to > 1,000 mTorr at –30°C, while the flash temperature is 38.9°C [24]. Since the temperature of the filament of a thermocouple or Pirani gauge is > 100°C, one must recognize the potential hazard of a fire or, worse yet, an explosion triggered by a high pressure of the solvent and the hot and perhaps autocatalytic (platinum wire) nature of the gauge filament. While there is no knowledge of a fire or explosion resulting from the presence of a solvent in a freeze-dryer and a thermal conductivity gauge, the reader should take heed and not depend on the odds.

Capacitance Manometer Gauge

The capacitance manometer gauge is commonly used for monitoring and controlling the P_c during the drying process. This gauge has sufficient accuracy, and its operation is independent of the nature of the gases in the chamber. It is recommended that during the first six months of operation, the zero of the capacitance manometer gauge be checked frequently (every one to two months). Once the stability of the gauge has been established, then a check of the zero of the gauge should be sufficient once every six months [22]. The need for zeroing this type of gauge was discussed in Chapter 11.

Process Control Systems

Process control systems include automatic resistivity and computer control systems. An important aspect of any control system is to reproduce a process. This means that the lyophilized product from a process that is conducted today with half a batch of formulation will be the same as that made tomorrow with a full batch of the same formulation. It is recommended that the reader avoid control systems that are designed to think for themselves. Use of such systems means that the process is in the hands of the programmer and is not following the temperature, pressure, and time intervals as determined by a development program. It is my opinion and Rowe's [5] that it is simply not possible, because of complexity of the lyophilization and wide variety of possible thermal properties of the formulations and necessary cake properties, for anyone to write a general computer program that will automatically meet the individual needs of each formulation. It would appear to be far easier to computer design a complete automobile than to ascertain an effective lyophilization process based only on the knowledge of the constituents in the formulation.

In addition to the above, this section will also consider individual control systems for pressure and temperature along with the means used for data collection.

Automatic Resistivity Control System

The automatic resistivity control system, a rather early version of a control system, was selected for review because it exemplifies how diligent one must be in evaluating a control system. The reader should understand that while I may be critical of the control system, I firmly believe that those who developed the system did so in good faith. The change in the resistivity (ρ) of a product was the underlying principle on which the control system was based. However, resistance (Ω) was used to control the heating of the product during the drying process. The actual control of the system was based on the deflection of a galvanometer, whose deflection was used to switch the power on or off for heating the shelves (see Figure 13.6). As a result, the entire control system was based on the resistance of just one vial. The heating to the shelf is turned off as soon as the resistance of the product dropped below a given value. The low resistance value would stem from mobile water in the interstitial region of the frozen matrix. By maintaining a given resistance value, it was felt that the instrument kept the product at the highest temperature compatible with the frozen state and maintained at its highest vapor pressure to reduce the sublimation time [37].

The reader should note that the control system specifies the resistivity of the frozen matrix; however, a plot of the data, as shown by Figure 13.10, shows the apparent T_p as a function of time. It is agreed that ρ is the correct process control parameter but it would not be possible to determine ρ during the drying process when the system being measured consists of two regions, i.e., a frozen region and a partially dried cake. It should be remembered that the units of ρ have a dimensional term (i.e., $\Omega \cdot cm$), so one would have to use the parameter resistance rather than resistivity. In order to show batch-to-batch consistency, resistance should also be plotted as a function of time. It should be noted in Figure 13.6 that there is no apparent upper cooling shelf, thus the product will be heated by radiation from the top of the dryer walls. The drying process in this system would differ from one with an unusable shelf.

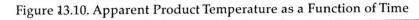

Figure 13.10. Apparent Product Temperature as a Function of Time

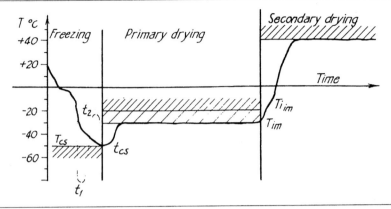

(Source: *Bulletin of the Parenteral Drug Association,* "Advances in Lyophilization Technology," Vol. 31, No. 5, Sept./Oct. 1977, Page 228, Figure 3. Reprinted with permission.)

The reader will note that instead of monitoring T_p, it was the resistance of the product that was being determined. A decrease in resistance was taken to signify a *softening* of the matrix. When a low resistance value is encountered, a feedback circuit would reduce the pressure in the chamber and the heat to the shelves. By making the above operating parameter changes, T_p is lowered, and rigidity is restored by an increase in the resistance of the matrix. When the resistance is increased to a given value, the pressure is increased and heat is again applied to the shelves. The reader should note that the above control system [38] is a system that determines the drying process rather than a system that defines a given drying process. As a result, there will be a range of drying processes for the product, and validation of the process could prove to be a rather difficult task.

Computer Control System

I made the common error of proposing a computer algorithm in which the drying of the product is not time dependent but a function of the drying parameters [36]. The process may be transferred from one dryer to another without altering the program. While the program may be transferred from one dryer to another, this does not mean that the same lyophilization process will be transferred because the effective drying time would be dependent on the quantity of the product present and the characteristics of the freeze-drying system. For example, using the above algorithm, one could have five freeze-dryers with five different lyophilization processes. If this should occur, then one would be faced with validating the lyophilization process for each dryer. It is my opinion that the main function of the computer system is to reproduce the same drying conditions from batch to batch, independent of the size or manufacturer of the freeze-dryer. In order to verify reproducibility, one must have a measure of the T_p because of its dependency on the two controlling factors, i.e., P_c and T_s.

We are often mesmerized by rapid advancements in the computer field, especially when we run into software problems that seem insurmountable. It is often our failure to understand the instructions that is the source of the problem. As a result, we tend to accept, almost without question, the output of a computer control system. I once witnessed a very impressive demonstration of a computer control system for a vacuum freeze-dryer. Freeze-dryer components changed color when they were put into operation. For example, the heat-transfer fluid passing through the shelves changed color when indicating cooling or warming of the shelves. However, there was one thing wrong with this dramatic demonstration: the computer control was not connected to the freeze-dryer. Such an observation immediately raises the question that if the computer could simulate the flow of the heat-transfer fluid, then what assurances does the operator have to know if the operation of the dryer on the monitor screen is real or just being simulated? While the topic of the installation qualification of the dryer is beyond the scope of this text, the reader should take heed and challenge the computer display to ensure that what is appearing on the screen is also taking place in the freeze-dryer.

Our dependence on the computer has reached such a level that one can no longer obtain full manual control over the dryer. It may be desirable, for example, to check for virtual leaks in the condenser system. However, for some dryers, the vacuum system will not operate unless the condenser temperature is below some given value. Such an automatic control feature could prevent the detection of a virtual leak in the condenser system. Thus, in the absence of manual control of the dryer, one may not be able to perform system testing that is necessary for determining a safe environment for the product or ensuring the necessary operating conditions for the lyophilization process.

The following represents opinions expressed by other authors but not necessarily shared by me. The reason for pointing out such views is to provide the reader with issues that they may be faced with in using or specifying a computer control system for a freeze-dryer.

User-Friendly. Murgatroyd [6] expressed a view that present-day control systems were *totally user-friendly* and that problems with the operation of the dryer can now be diagnosed and rectified from a remote site, e.g., a laptop computer. It was not clear just what kind of problems can be corrected via a remote computer system. Perhaps, one may be able to reset the control program after a brief power failure. But certainly there are going to be some manual tasks that simply cannot be corrected by a laptop computer, such as the replacement of a vacuum pump or compressor.

Computer Control Segments. Control programs provide the user with the ability to divide the control parameters (T_s and P_c) in a given part of the lyophilization process (freezing, primary drying, and secondary drying) into a number of segments, e.g., 6 to 12 depending on the design of the control system [6]. The fact that the computer permits a step in the lyophilization process to be segmented does not mean that more steps are equated with better control. In the three examples of a lyophilization process that were presented in Chapters 7, 8, and 9, each step in the segment was based on rationale obtained from the thermal properties of the formulation or the nature of the final product. The random selection of a series of steps to control the T_s and P_c that is obtained by trial and error reduces the lyophilization process to an *art* rather than a *science*. The basic problem one faces with a process based on trial and error is that there is no supporting rationale. When a lyophilization process based on *art* fails, the resulting product can get pretty ugly.

Temperature Control

I have already expressed my reasons for controlling the lyophilization process based on the T_s rather than the temperature of the shelf fluid. Unless there is a provision for monitoring the T_s of each shelf, one is often faced with controlling the entire process from the output of one temperature sensor. For example, Grieff [8] controlled the shelf temperature from a thermocouple attached to the middle shelf of the dryer. The controlling thermocouple was not positioned in the middle of this shelf but close to its edge. The placement of the thermocouple must be such that it is representative of the T_s encountered by the vials or the trays. The positioning of the sensor on the very edge or on the underside of the shelf will give temperatures more indicative of the shelf fluid rather than the shelf surface.

Using resistance thermal devices (RTDs) as sensors, Nail and Doorlag [39] advocated that the T_s and T_p should be controlled during the primary drying process. In addition, their control system provided for a *feedback signal from the product temperature sensors that help guard against eutectic melting*. By this means, T_p can be prevented from exceeding the Tc throughout the entire primary drying process. In such a system, the initial T_s will greatly exceed that of the T_p during the major portion of the primary drying process. As primary drying nears completion, and the thermal load on the shelves is reduced, T_p will increase when the gas-ice interface falls below the sensor. At this point in the process, T_s is reduced and becomes equal to T_p at the completion of the primary drying process. Although this will result in a lengthening of the time to complete the primary drying process, it will not eliminate any risk of collapse or partial meltback in some product cakes. Because the T_s will approach the T_p and the heat-transfer rate will approach zero, one would find it difficult to ascertain, even with knowledge of the frequency distribution of C_o, the duration of the primary drying process.

Data Collection System

Before the advent of the computer, data for the lyophilization process were generally recorded on a multipoint strip chart recorder. The recorder could be programmed to display as many as 24 individual inputs. While most of the inputs were from the output of temperature sensors, some recorders were equipped with dual function paper that would allow the recording of both temperature and pressure. The recorder operated as a *null* indicator (i.e., the recorder would move along a contact until the millivolt output of the contract was equal to the input of the temperature or pressure sensor). Needless to say, the time necessary to record all 24 sensors could require a time span of about 25 to 30 min.

Since the complete lyophilization process can require several days, one had to set the chart speed to a low value, e.g., at a chart speed of 2 in./hr, the length of the chart paper would about 12 ft. A reduction of the chart speed to 1 in./hr would reduce the length of the chart paper to 6 ft, but the numbers would be printed so close together that it would be difficult to distinguish individual sensor temperatures.

Although the individual numbers may be assigned different colors, the overprinting of the number often left a black band across the chart paper. Only near the end of the primary drying process was one able to identify individual sensors.

Although the data from the chart recorder were compared to other chart recordings, the data were not, as a general rule, replotted for a closer analysis. Thus, the chart recorder served as a record of the control parameters and T_p for the lyophilization process, but its use as a diagnostic tool was rather limited.

The chart recorder soon gave way to digital strip chart recorder. This instrument is still used widely in the industry and provides the output of the various temperature and pressure sensors in the form of rows of numbers. Once again, the operator must select the time period for recording data. Even at a printout time segment of 5 min, the data produced by the recorder amounted to reams of paper. One again is often overwhelmed by the sheer number of pages. Because of the large volume of data and, what may seem as endless sheets of paper, the data from this recorder were often treated like that of the strip chart recorder, stored but not examined. Later versions of this recorder did provide means for obtaining a plot of the data.

The data from the computer control system now have the added feature that the digital data are stored on either a hard drive or floppy disk. With the data in this form, one can now quickly analyze the results using data processing software packages such as Excel® or Lotus 1-2-3®. But once again, the operator must select the time segment, e.g., 15 sec to several minutes [6]. Although a hard drive can store gigabytes of data, one may find that the volume of data is more than the data processing software can handle or the data may exceed the memory of the printer.

It is recommended that the data storage be conducted on an *event* and time basis as used in the drying process monitor (DPM) system [9] and the D_2 and DTA instrument described in Chapter 6. With such a data collection system, an event would be recorded only when there has been a significant change in the output of the temperature or pressure sensors. The latter would require knowledge of the frequency distribution of the output signal during calibration of the sensor. By this method, the recorded value resulted from a change in the output of the sensor that would be considered well beyond the normal chance of the previous reading. By this means, data such as the degree of supercooling can be observed during the freezing process. If there is no significant change in the output of a sensor for a given time period, e.g., several minutes, then the data would be recorded to show that the data collection program was still in operation. Thus, with this approach, only the essential data will be recorded regarding the process.

THE FUTURE OF LYOPHILIZATION

The following will be a discussion regarding the future of lyophilization in the 21st century. It is hoped that during the years ahead that the practice of the *art* will fade into obscurity and the science and technology will shed new light onto this field of lyophilization. In order to make such a transformation, we must set forth a series of key objectives. These objectives must serve as the foundation on which the technical renaissance of lyophilization will rest.

More effort must be made to include the lyophilization aspects of a formulation as a part of the acceptance criteria. The practice of *just get it (the formulation) to work* or *that's someone else's problem* must give way to the double *p*—product and process.

It is time for this field to abandon the humorous notation that *three samples are science* [40] that was made in jest but often followed with faith and adopt a more rational approach based on statistics and supported by meaningful data.

Finally, freeze-dryer manufacturers must design and construct freeze-dryers based on the real needs of the process rather than on unsubstantiated and preconceived notions of needs of the industry. It is time for the user to understand and define their equipment requirements, and a need for manufacturers to standardize rather than to specialize.

New Objectives

The following are what I and Gershon [41] feel are key objectives that must be met if the transformation of lyophilization from an *art* to a *science* is not going to be just a dream but rather a reality.

It is time for all parties (societies, manufacturers of equipment and lyophilized products, and other interested individuals) associated with the lyophilization process to adopt statistical standards and standard methods that will be applicable to the lyophilization process. Such standards, if effective, will lead to improved quality of lyophilized products, a reduction in costs, and a general increase in productivity. It will also increase the sale of freeze-drying equipment because of the greater confidence in the process and its associated equipment.

Equipment manufacturers should be held accountable for the quality and performance of their equipment. Take the control of the lyophilization process from the hands of the software programmer and return it to the process development department.

Strive for excellence by finding problems and then take corrective actions [41]. Avoid the *if I don't know, I don't have to fix* syndrome. If it cannot be done, do not look for ways to justify not doing it but effective means for doing it. One example is determining the T_p in dryers equipped with automatic loading equipment. Why was it important to measure T_p in the past years but not now? The answer is that new technology has evolved that prevents the measurement of T_p rather than replace it with a means that is far more accurate and reliable.

Create a management environment that encourages its personnel to strive for better product quality [41]. Discard the notion that *if it's worth doing, it's not worth doing well* with the concept of *do it right*.

Insist on standards regarding the measurement of the operating parameters. There should be one universal pressure unit such as the Pascal rather than the ambiguous Torr and bar or the absurd percent vacuum as pressure units. Standardize on terms used in the industry such as leak rate and condenser capacity and eliminate such vague terms as *full vacuum*.

The ultimate goal is to develop standards of confidence in assessing the performance and quality of a product that will, in time, eliminate the need for a regulatory agency. The public is better served by an industry that can regulate itself rather than one that has to be closely regulated by a government agency.

Formulation

Because the formulation governs not only the process parameters but also the very design of the equipment, it is paramount that more effort be placed on developing formulations with acceptable thermal properties and cake properties.

Thermal Properties

The formulation should have sufficient supercooling properties that will ensure a uniform frozen matrix with little or no glaze on the cake surface. The frozen matrix of the formulation should be such that the cake produced during the primary drying process will have sufficient porosity and that there will be a minimal amount of impedance to gas flow from the sublimating surfaces. Tc should have values that are greater than $-35°C$. This would enable the minimum pressure during primary drying to be ≥ 100 mTorr. Although it was shown in the above that backstreaming of the oil vapors can occur at pressures near 100 mTorr, the use of a Roots blower pump in the foreline should provide a discharge pressure to the mechanical pump of ≥ 200 mTorr and prevent backstreaming from the vacuum system.

Cake Properties

The solid content of the formulation should be given special consideration. The reduction of the amount of water in the formulation, i.e., increasing the mg of constituents per mL, may prove counterproductive if a dense cake is formed. Such a cake may impede the flow of water vapor from the product during both the primary and secondary drying processes. By decreasing the concentration of the constituents, a more porous cake can be formed. With an increase in the porosity of the cake, there will be a decrease in the impedance to gas flow and a possible reduction in total process time. But there are limits with regard to dilution of the formulation. Reducing the concentration to lower than 2% wt/v could result in a cake with a poor self-supporting structure. Such a cake structure can lead to a loss of product from the container.

There needs to be more rationale given to determining the concentration of the *other* constituents, i.e., constituents other than the active constituent. It is a common practice to express the concentration of the other constituents in terms of % wt/v. If we are going to make an effort to treat lyophilization as a science, then we should begin with expressing concentrations in terms of molarity, not percentages. Perhaps by using molarity, those developing the formulation will become more cognizant of the role that a given constituent is playing in forming a stable lyophilized formulation. What is not being taken into account, which was introduced in this text, is the concentration of the constituents in the interstitial region of the matrix. The realization of the large change in concentration of a constituent that is taking place during the freezing process may further our understanding of the role that each constituent plays in forming the final product.

The industry needs better analytical instrumentation for determining the thermal characteristics of the formulation during both the freezing and warming processes. Such analytical instrumentation should provide us with an insight as to just how the interstitial region is formed: Is it homogeneous or heterogenous in nature?

If the region is heterogeneous, then we should know how the constituents are distributed.

More basic research is required for understanding the formation of the interstitial region, first using a simple system and then expanding it to more complex systems. We need to rid ourselves of reporting results of the thermal properties of formulations having constituents that are not readily available to all investigators unless such results contribute in someway to our general knowledge of the freezing process.

Statistical Approach

Throughout this text, I have stressed the need for putting the various aspects of the lyophilization process on a statistical basis. In order to accomplish this, the following are key changes that are needed in order that the lyophilization process can be based on a statistical approach [41]:

- Adopt a new approach to the process: No longer should commonly accepted delays, mistakes, poor work habits, and defective materials be accepted.

- Cease the dependence on mass inspection by demanding that statistical evidence of quality be built into the process. Two important areas regarding the process have been pointed out already in this text: the effect of the frequency distribution of the thermal properties of the formulation and the C_0 on the drying process.

The Scientific Forum

There is a need to change the basic content of the papers submitted for publication. Papers whose results cannot be reproduced should be discouraged in favor of papers that are written for the advancement of science and technology.

Remove the name(s) of the author(s) from the submitted paper so that the peer review is performed on an objective rather than a subjective basis. Editors should institute firm deadlines for returning manuscripts. Why should some papers be accepted in a matter of months when other papers require over a year, in spite of the fact that the author(s) responded promptly to the reviewer's comments? The peer review should not be conducted by telephone or private conversation.

There should be a general call for papers for all conferences and symposiums, and the scope of the topics to be covered should be clearly stated. Organizers of such public meetings should conduct such proceedings on an unbiased basis, where acceptance or rejection of a paper should be done solely on merit. The proceedings of the meeting should be published prior to the meeting so that attendees need not wait months to receive a written copy of the presentations.

Societies should encourage an open discussion and participation in their activities regarding the lyophilization process. Differences in opinions or concepts should be heard and not excluded for political or personal reasons because *dissidence does breed discovery* [42].

Batch Versus Continuous

Lyophilized products are currently being manufactured by the batch. While freeze-dried coffee is perhaps the only product that is manufactured on a continuous basis, it is really not a lyophilized product. Lyophilized coffee has a light yellow color rather than a dark brown. In addition, the methods used to manufacture coffee would hardly be acceptable for heat-sensitive injectable products (i.e., T_s is greater than 100°C). The remainder of this section will first consider a batch system and, in particular, the use of automatic loading and unloading systems. Based on a typical lyophilization process, the feasibility of a continuous freeze-drying system will be shown only to be possible if and only if there can be a major reduction in process time. A prototype continuous dryer will be described and its production output compared to that of a typical batch production dryer.

The Batch System

Automatic loading equipment capable of loading and unloading a 300 ft³ dryer in 2–3 min [43] has become increasing popular in recent years. These systems are capable of loading either vials or vials in trays directly on the shelves of the dryer. One of the chief advantages for such loading devices is to significantly reduce the number of people in the clean room, and thereby approach the conditions of isolation technology during the loading and unloading operation [28]. In this way, the probability of biological contamination of the product is greatly reduced. However, because such systems are not at present equipped with a means of loading product sensor vials into the system, one must now choose between biological validation and process validation. While there is no doubt that such automatic loading equipment will certainly reduce the potential for biological contamination of lyophilized products, I am not aware of widespread batch rejection because of biological contamination as a result of loading and unloading a lyophilized product in an aseptic Class 100 room. If automatic loading and unloading equipment is used as a matter of expediency, then one must question the need for expediency over validation of the lyophilization from a knowledge of T_p.

The Continuous System

In 1975, there was considerable interest in continuous freeze-drying systems. At that time, the system used trays to move the product from one stage to another. Since the trays did not come in direct contact with the shelves, the heat-transfer mechanism to the trays was mainly through radiation. The thermal radiation was produced from horizonal heating plates that were arranged to create isothermal zones through which the product trays would pass in a stack configuration. Each stack consisted of 15 trays [44].

It is of interest to note that, in his keynote address before the annual meeting of the Parenteral Drug Association on 3 November 1994, Polster [45] expressed concerns regarding the use of emerging technologies. He predicted that those manufacturers (of pharmaceutical products) who resisted technological advances would suffer at the hands of their competitors. He envisioned that in the coming 21st century, manufacturing will be performed using complete barrier technology. In the case of lyophilization, he predicted that the process would be continuous and require less than 8 hr [45].

Consider what would be involved to lyophilize a pharmaceutical or biotechnology product in 10 mL vials on a continuous basis [46], using the following lyophilization process:

Filling and loading 37,000 vials at 120 vials/min	6 hr
Freezing (ambient to –40°C)	12 hr
Primary drying process	48 hr
Secondary drying process	24 hr
Backfilling and stoppering	<u>1 hr</u>
Total process time	91 hr

The hypothetical continuous system will consist of three separate chambers. The first chamber will be maintained at –40°C and at 1 atm. The second chamber will be maintained at the T_s and P_c defined for the primary drying process. Secondary drying will be conducted in a third chamber maintained at the final T_s and P_c. The system will accommodate three trays of vials (each containing 168 vials) abreast.

With a system capable of filling 3 trays at a time, it would require approximately 6 hr to complete the filling process and another 6 hr to complete the freezing process. At a width capable of 3 trays across, the length of the freezing chamber would be 73 ft. Assuming that each vacuum lock can accommodate only 3 trays abreast, then the length of the primary drying chamber would be 48 ft, while the length of the secondary drying chamber will be 24 ft. The estimated total length of such a system would be 145 ft or about the height of a 7-story building.

Assuming that it would require 1 hr to load and unload a vacuum lock, the minimum time to transport all of the trays from the freezing chamber into the primary drying chamber would be 72 hr. An additional 48 hr would be necessary to complete primary drying and 24 hr to complete the secondary drying. The total process time would be 158 hr, or 1.7 times the process performed in a batch freeze-dryer.

The dimensions of the dryer can be reduced. For example, the length of the freezing chamber can be reduced by one-half; however, in so doing, the overall process time will be increased to 181 hr.

If the above continuous dryer required sterilization, then the entire system would have to be fabricated as a pressure-coded vessel. This would certainly add to the cost of the dryer. The cost of such a freeze-dryer would certainly exceed the cost of present dryers having the same shelf-surface capacity. When the additional processing time is factored in, the continuous dryer based on present lyophilization processes would hardly appear as an attractive alternative to present technology.

Only if the lyophilization process time could be reduced by a factor of 1/5th to 1/10th would a continuous freeze-drying system be deemed feasible. Reducing the time of the lyophilization process would also lead to reduced dimensions for the dryer. Because of the increase in the drying rate, the cake must have a sufficient self-supporting structure during primary drying to withstand the force exerted by the flow of gas from the vial. Therefore, it is unlikely that a formulation having a solid content less than 2% would be a candidate for lyophilization in a continuous dryer.

Tests conducted in my laboratory (unpublished) have shown that it is possible to significantly increase the heat-transfer rate (Q) to the vials as defined by

$$Q = C_o(T_r - T_p) \quad \frac{\text{cal}}{\text{sec}} \tag{3}$$

where C_o is the heat-transfer coefficient which, for typical glass vials, will range from 1×10^{-4} cal/(sec °C) to about 1×10^{-2} cal/(sec °C), while T_r is the temperature of the reference container and T_p is associated with the temperature of the product [9]. Because Q represents the energy that will be associated with the freezing and drying process, a significant increase in Q would result in a reduction in the time to complete the lyophilization process. However, Q must be increased without increasing T_p or one could exceed the Tc. The increase in Q at a constant T_s can be accomplished by increasing C_o by a factor of 5 to 10 times. Means for increasing the value of C_o has been obtained.

By increasing the magnitude of C_o for the vials, one would also establish a narrow frequency distribution of C_o values. By narrowing the frequency distribution of the C_o values, one does not have to contend with partial meltback resulting from vials having relatively low C_o values with respect to the mean $\overline{C_o}$ or complete meltback from vials having C_o that are significantly greater than $\overline{C_o}$. By narrowing the frequency distribution, one inspects the final product for defective containers rather that defective product resulting from the lyophilization process.

A Prototype of a Continuous Freeze-Dryer

Based on a 10-fold increase in C_o, the following is a general description of a proposed continuous freeze-drying system and a general comparison with a typical manufacturing dryer of equal capacity. The prototype continuous dryer is illustrated in Figure 13.11 [46]. The following describes the general steps illustrated by Figure 13.11:

- *Formulation.* The formulation is prepared. Rather than being limited to the capacity of the dryer, the batch size for the formulation will be determined by its stability at a storage temperature prior to the filling of the vials.

- *Preinspection.* The vials are visually inspected for defects as a result of handling.

- *Cleaning.* The vials are cleaned and then enter a Class 100 room that interfaces with the barrier technology system.

- *Tray Loading.* The vials are loaded onto trays that are bar coded. The barcoded tray will allow complete tracking of that tray of vials throughout the entire lyophilization process.

- *Depyrogenation.* The vials are placed in an oven having a particle count equivalent to that of a Class 100 room.

- *Isolation.* The depyrogenated vials are unloaded into a pressurized, sterile Class 100 or lower room. It is in this area that the vials are filled with

Figure 13.11. Prototype Design for a Robotics-Controlled Continuous Freeze-Dryer System

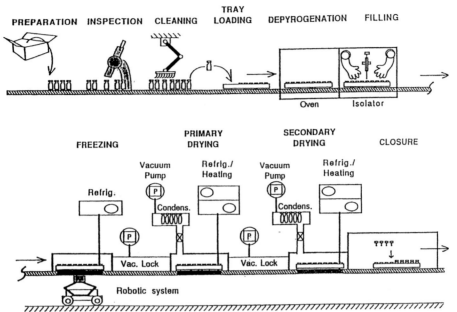

[Source: Reprinted with permission from *American Biotechnology Laboratory*, volume 14, number 11, page 55, 1996. Copyright 1996 by International Scientific Communications, Inc.]

the prescribed fill volume. Before leaving this area, a scanner reads the bar code on the side of the tray.

- *Freezing.* By means of a robot, the tray is placed on one of the tracks of the dryer. Selection of a given track will be discussed later. The tray is then transported into the freezing section. The freezing section may consist of one or more stages depending on the nature of the freezing function. The freezing section is pressurized, and means are used to maintain frost-free shelf surfaces.

- *Vacuum Lock #1.* Upon completion of the freezing process, the tray is placed in a vacuum lock. The lock is sealed, and the pressure is reduced to that of the primary drying process. When the primary drying pressure is attained, then the tray is transported into the primary drying chamber.

- *Primary Drying.* It should be noted that during the transport of the trays, the bottom of the trays does not come in contact with the shelves, thereby eliminating generation of particulate material as a result of tray movement. The completion of primary drying is verified by calorimetric monitoring of the process.

- *Vacuum Lock #2.* This lock separates the primary and secondary drying chambers. When the lock pressure is equal to that of the secondary drying chamber, the tray is moved into the secondary drying chamber.

- *Secondary Drying.* Secondary drying can be performed either in temperature segments or at one shelf temperature. The use of dry gas purges eliminates tray-to-tray moisture transport.

- *Vacuum Lock #3.* The pressure in this lock is first increased for placement of the stoppers on the vials. The pressure is further increased to allow the tray to be transferred back into the isolation barrier for crimp sealing. ·

- *Product Labeling.* The labels are applied with the same bar code that is on the tray so that the product can be tracked back to the very conditions that were used to lyophilize the formulation.

- *Tray Transport.* In Figure 13.11, the trays move through the continuous dryer from left to right. It should be noted that the system is programmed such that a tray cannot enter a vacuum lock unless there is an open position in the next segment of the dryer.

- *Multitrack System.* A multitrack system has parallel tracks that will accommodate a common tray size. In this system, each track contains a separate set of vacuum locks but share a common chamber for the freezing, primary drying, and secondary drying processes. The movement of the trays on each of the tracks, by robots, is controlled and monitored by a separate computer system that is tied to the master computer system.

   The multitrack system allows for increasing the throughput of a product by employing more than one track. More importantly, the multitrack system allows the lyophilization of more than one dosage form of a given formulation. In this example, one track may process the formulation in a 1 mL dosage form. A second track would be dedicated to a 5 mL fill volume, while a third track may have vials with a 50 mL fill volume. Each fill volume will pass through the dryer according to a validated lyophilization process. By this means a large batch of formulation could be prepared even when its storage time at the filling temperature is limited.

A Comparison of a Conventional and a Continuous Dryer

A comparison of the general features of a continuous dryer with respect to that of a conventional dryer having about the same capacity is shown in Table 13.1 [46]. An increase in C_0 will increase the drying rate, which is reflected in the overall dimensions of the dryer.

A second feature, while not shown in Figure 13.11, is that each of the drying segments will have two separate condenser systems. This feature overcomes an inherent difficulty that is associated with conventional freeze-dryers, i.e., operating the dryer at maximum efficiency. In the conventional dryer, one is often faced with either not fully utilizing the capacity of the condenser system or, if the condenser is used to its maximum capacity, not fully utilizing all of the shelf-surface area. The batch size for the continuous dryer is limited only by the stability of the formulation at the filling temperature.

Table 13.1. A Comparison of the General Characteristics of a Conventional
Freeze-Dryer and (Estimated) Continuous Freeze-Dryer

Feature	Conventional	Continuous (Estimated)
General material	Stainless steel (316L)	Stainless steel (316L)
Square footage	220	34 (three track)
Shelves (number)	9	3
Condenser capacity (H_2O)	400 kg	Unlimited
No. of 23-mm vials*	37,000	3888
No. of trays (1 × 1 ft)	220	27
No. of trays monitored per batch	9	All
Dimensions (ft):		
Length	25	12
Width	11	4
Height	11	3
Weight (empty)	>20 tons	<1 ton
Complete barrier technology	No	Yes
Sterilization between batches	Yes	No
Mixed fill volumes	No	Yes
Estimated dryer cost**	$1.5 million	<$1 million

*Based on total capacity of the shelf surface at any time.

**The estimated cost of the dryers does not take into account the additional cost for the purchase and operation of the required subsidiary equipment (e.g., clean steam generator for sterilization or WFI system for clean-in-place systems). Subsidiary equipment costs for the conventional dryer would greatly exceed those required for the continuous system.

[Source: Reprinted with permission from *American Biotechnology Laboratory*, volume 14, number 11, page 55, 1996. Copyright 1996 by International Scientific Communications, Inc.]

If a serious malfunction of the equipment should cause a complete loss of all the product in the dryer, the loss in the conventional dryer would be 100%, whereas the loss in the continuous dryer would be limited to about 10% of the total batch.

A Comparison of Required Utilities

A comparison of the estimated utilities for a conventional and continuous freeze-dryer is listed in Table 13.2 [46]. Although the power consumption on the conventional dryer is shown to be greater than that of the continuous dryer, Table 13.2 does not fully reveal the difference in the cost of utilities. In the conventional dryer, the

Table 13.2. A Comparison of the Required Services of a Conventional Freeze-Dryer and (Estimated) Continuous Freeze-Dryer

Feature	Conventional	Continuous (Estimated)
Electrical power:		
No. of phases, 60 Hz	3	1
Voltage460	230	
Wattage230 kW	4.5 kW	
Cooling water	200 gal/min	None
Steam sterilization @ 140°C for 1.5 hr	370 kg/batch	30 kg
Compressed air	50–100 psi	50–100 psi

[Source: Reprinted with permission from *American Biotechnology Laboratory*, volume 14, number 11, page 55, 1996. Copyright 1996 by International Scientific Communications, Inc.]

shelves are often required to reach a $T_s \leq -40°C$. Not only must the dryer reduce the total mass of the shelves to such a low temperature but also there will be heat transfer to the walls of the chamber and the chamber door. After completing the freezing process, energy must be supplied to the product to complete the drying process and to the shelves, door, and walls of the dryer in order to bring the entire dryer system back to near ambient temperature.

A Comparison of Production Parameters and Capacity

Since the continuous dryer will not be opened to the atmosphere in order to unload the vials, it is possible to show that the entire system will maintain its sterility. Perhaps the only time that one would have to clean and sterilize the dryer would be when there is a change in the composition of the product. Table 13.3 [46] shows that it is possible for the capacity of 1 continuous dryer to have an equivalent capacity of 2-1/2 conventional dryers. But what may be of even greater importance is that the product in each of the trays will be monitored throughout the entire lyophilization process, whereas conventional freeze-dryers using an automatic loading and unloading system do not have a means, at the time this text was published, to monitor the process for a single T_p on each of the shelves of the dryer.

Will the continuous freeze-dryer replace conventional dryers? The answer would have to be *no* because there will be products that may be only manufactured once a year, and it would not be economically feasible to develop a lyophilization process that will be run once a year in a continuous dryer. The continuous freeze-dryer would be best suited for high-volume products, such as vaccines for children. It would be a major step for mankind if every child could be spared the effects of enduring measles, mumps, and perhaps chicken pox.

Table 13.3. A Comparison of the Production Capacity of a Conventional Freeze-Dryer and (Estimated) Continuous Freeze-Dryer

Feature	Conventional	Continuous (Estimated)
Cleaning and sterilization	24 hr	24 hr
Filling*	6 hr (120 vials/min)	432 vials/hr
Freezing*	12 hr	2 hr
Primary drying*	48 hr	5 hr
Secondary drying*	24 hr	2 hr
Backfilling and stoppering	1 hr	1 hr
Unloading and crimp sealing	6 hr	1 hr
Condenser frost	12 hr	0 hr
No. of vials produced per five-day period	37,000	40,700 (67,800)**
No. of vials produced per month	166,500	295,000 (492,000)

*Denotes the same lyophilization process time for the conventional production dryer used in the above discussion on continuous freeze-dryers using current lyophilization processes.

**Values in parentheses are for a five-track continuous dryer (length: 12 ft, width: 6 ft, and height: 3 ft).

[Source: Reprinted with permission from *American Biotechnology Laboratory*, volume 14, number 11, page 55, 1996. Copyright 1996 by International Scientific Communications, Inc.]

SYMBOLS

CFU	colony forming unit
CFU_d	density of the microorganism in the surrounding environment
C_o	heat-transfer coefficient
D_m	microorganism density per unit time
m/e	mass to charge ratio
Q	heat-transfer rate
Q_v	viscous gas flow through the leak path expressed in terms of atm. cfm
P_c	chamber pressure
R_m	rate at which microorganisms can enter a drying chamber
ROR	rate of pressure rise
Tc	collapse temperature

T_p product temperature

T_s shelf-surface temperature

V_e effective volume of the chamber

REFERENCES

1.   D. Grieff, in International Symposium on freeze-drying of biological products, Washington D.C., 1976 (S. Karger, Basel, 1977), *Develop. Biol. Standard.*, 36: p. 105.

2.   H. R. Powell, loc. cit., p. 117.

3.   L. Le Floc'h, loc. cit., p. 131.

4.   J. H. Leary and E. A. Standford, loc. cit., p. 221.

5.   T. W. G. Rowe, loc. cit., p. 79.

6.   K. Murgatroyd, in *Good Pharmaceutical Freeze-Drying Practices* (P. Cameron, ed.), Interpharm Press, Inc., Buffalo, Ill., 1997, p. 59.

7.   R. I. N. Greaves, in *Biological Applications of Freezing and Drying* (R. J. C. Harris, ed.), Academic Press Inc., New York, 1954, p. 87.

8.   D. Grieff, *J. Parenter. Sci. and Technol.*, 44: 119 (1990).

9.   T. A. Jennings and H. Duan, *PDA J. Pharm Sci & Technol.*, 46: 272 (1995).

10.   T. A. Jennings, *J. Parenter. Sci. and Technol.*, 44: 22 (1990).

11.   T. A. Jennings, *Pharm. and Cosmetic Equip.*, April, 1998, p. 38.

12.   T. A. Jennings, *J. Parenter. Sci. and Technol.*, 36: 151 (1982).

13.   P. F. Waltrick, *J. Parenter. Sci. and Technol.*, 40: 66 (1986).

14.   D. Hoffman, *J. Vac. Sci. Technol.*, 16: 71 (1979).

15.   K. Kinnarney, in *Good Pharmaceutical Freeze-Drying Practices* (P. Cameron, ed.), Interpharm Press, Inc., Buffalo Grove, Ill., 1997, p. 205.

16.   R. Erlandsson, *J. Vac. Sci. Technol.*, 19: 748 (1981).

17.   *CRC Handbook of Chemistry and Physics* (R. C. Weast, ed.), CRC Press, Inc., 1984, Boca Raton, Fla.

18.   J. P. Connelly and J. V. Welch, *J. Parenter. Sci. & Technol.*, 47: 70 (1994).

19.   G. L. Fowler, *J. Vac. Sci. Technol.*, A5: 390 (1987).

20.   B. R. F. Kendall, *J. Vac. Sci. Technol.*, 20: 248 (1982).

21.   S. L. Morgan and M. R. Spotts, *Pharmaceutical Technology*, Nov. 1979, p. 95.

22.   T. A. Jennings, A. Scheer, A. Emodi, L. Puderbach, S. King, and T. Norton, *PDA J. Parenter. Sci. & Technol.* 50: 205 (1996).

23.   H. Seager, C. B. Taskis, M. Syrop and T. J. Lee, *J. Parenter. Sci. and Technol.*, 39: 161 (1985).

24.   *Handbook of Chemistry and Physics*, 31th ed. (C. D. Hodgman, ed.), 1949, Chemical Rubber Publishing Co., Cleveland, Ohio.

25.   T. A. Jennings, A. Scheer, A. Emodi, L. Puderbach, S. King, and T. Norton, *PDA J. Parenter. Sci. & Technol.* 50: 180 (1996).

26.   F. DeVecchi, in *Validation of the Aseptic Pharmaceutical Process* (F. J. Carleton and J. P. Agalloco, eds.), Marcel Dekker, Inc., New York, 1986, p. 125.

27. S. Dushman, *Scientific Foundations of Vacuum Technique*, 2nd ed. (J. M. Lafferty, ed.), John Wiley & Sons, Inc., New York, 1962.

28. K. Murgatroyd, in *Good Pharmaceutical Freeze-Drying Practices* (P. Cameron, ed.), Interpharm Press, Inc., Buffalo, Ill., 1997, p. 1.

29. P. Larrat and D. Sierakowski, *Pharm. Engineering*, Nov./Dec., p. 71 (1993).

30. J. Amoignon, in *Freeze-Drying and Advanced Food Technology* (S. A. Goldblith, L. Rey, and W. W. Rothmayr, eds.), Academic Press, New York, 1975, p. 445.

31. F. V. López, I. P. Solís, and F. A. Castro, *J. Parenter. Sci. & Technol.*, 36: 259 (1982).

32. R. W. Berry, P. M. Hall, and M. T. Harris, *Thin Film Technology*, Van Nostrand Reinhold Company, New York (1968).

33. T. A. Jennings, *Parenter. Drug Assoc., Inc.*, 34: 62 (1980).

34. R. N. Peacock, *J. Vac. Sci. Technol.*, A11: 1627 (1993).

35. P. P. Nicole, *J. Vac. Sci. Technol.*, 17: 1384 (1980).

36. T. A. Jennings, *Parenter. Drug Assoc., Inc.*, 32: 273 (1978).

37. B. Couriel, *Parenter. Drug Assoc., Inc.*, 31: 227 (1977).

38. L. R. Rey, in International Symposium on freeze-drying of biological products, Washington, D.C., 1976 (S. Karger, Basel, 1977), *Develop. Biol. Standard.*, 36: p. 19.

39. S. L. Nail and E. Doorlag, in *Automation of Pharmaceutical Operations* (D. J. Fracde, ed.), Pharmaceutical Technology, Springfeld, Ore., 1984, p. 283.

40. R. E. Madsen, *PDA Letter*, September, 36: 19 (1998).

41. M. Gershon, *J. Parenter. Sci. and Technol.*, 45: 41 (1991).

42. T. A. Jennings, *J. Parenter. Sci. and Technol.*, 42: 118 (1988).

43. B. Couriel, *Parenter. Drug Assoc., Inc.*, 34: 358 (1980).

44. J. Lorentzen, in *Freeze-Drying and Advanced Food Technology* (S. A. Goldblith, L. Rey, and W. W. Rothmayr, eds.), Academic Press, New York, 1975, p. 429.

45. R. K. Polster, *J. Parenter. Sci. & Technol.*, 48: 52 (1994).

46. T. A. Jennings, *Amer. Biotech. Lab.*, Oct. 1996, p. 53.

Index